Food and Industrial Bioproducts and Bioprocessing

Food and Industrial Bioproducts and Bioprocessing

Edited by

Nurhan Turgut Dunford
Department of Biosystems and Agricultural Engineering and Robert M. Kerr Food & Agricultural Products Center
Oklahoma State University
Stillwater
Oklahoma
USA

WILEY-BLACKWELL
A John Wiley & Sons, Ltd., Publication

Wiley-Blackwell is an imprint of John Wiley & Sons, formed by the merger of Wiley's global Scientific, Technical and Medical business with Blackwell Publishing.

Editorial Offices
2121 State Avenue, Ames, Iowa 50014-8300, USA
The Atrium, Southern Gate, Chichester, West Sussex, PO19 8SQ, UK
9600 Garsington Road, Oxford, OX4 2DQ, UK

For details of our global editorial offices, for customer services and for information about how to apply for permission to reuse the copyright material in this book please see our website at www.wiley.com/wiley-blackwell.

Library of Congress Cataloging-in-Publication Data

Food and industrial bioproducts and bioprocessing / edited by Nurhan Turgut Dunford.
p. cm.
Includes bibliographical references and index.
ISBN 978-0-8138-2105-4 (hard cover : alk. paper)
1. Biological products. 2. Biotechnology. 3. Biomass. I. Dunford, Nurhan Turgut, 1953–
TP248.2.F725 2012
660.6–dc23

2011035806

A catalogue record for this book is available from the British Library.

Wiley also publishes its books in a variety of electronic formats. Some content that appears in print may not be available in electronic books.

Set in 10/12pt Times by SPi Publisher Services, Pondicherry, India
Printed and bound in Malaysia by Vivar Printing Sdn Bhd

1 2012

Contents

Preface

Petroleum-derived products have dominated the markets for decades because of the ease of production and economies of scale. In recent years diminishing petroleum resources, volatile political environments in some of the major petroleum producing countries and environmental concerns inspired a paradigm shift. Today significant resources have been dedicated to the development of bioproducts from renewable sources. Research and development efforts to harness the unique chemical and physical properties of plants and microorganisms to produce ecologically benign products that outperform their non-renewable counterparts have accelerated. Ever increasing consumer demand for "chemical free", "healthy" and "natural" foods incited the food industry to reevaluate the conventional food ingredients and processing techniques and adapt new and advanced production systems.

This book, which contains 14 chapters, provides a comprehensive review of the latest developments in food and industrial bioproducts and bioprocessing techniques. Although it is an important topic, biofuels are not covered in the book. This book is designed as a reference source for scientists, students, and government and industry personnel who are interested in the recent developments and future opportunities in food and industrial bioproducts and relevant bioprocessing techniques. The contributing authors of the book from Australia, Canada, Denmark, Germany and the USA are internationally renowned experts in their fields and their contributions to the book are invaluable. I would like to express my sincere gratitude to the authors for accepting my invitation to contribute and completing their chapters in a timely manner. The comments received from the external reviewers, James T.C. Yuan, Ibrahim Banat, Sue, Nokes, David Cowan, Randy Berka, Mark R. Marten, B. Dave Oomah, J.L. Willett, Laurent Bazinet, Dan Farkas, Richard Ashby, Thrandur Helgason, Donghai Wang, Cristina Sabliov, Wenqiao (Wayne) Yuan, Michael J. Haas, Sang-Hyun Pyo, Krister Holmberg, Aaron L. Brody, Amos Richmond and Mike Packer, were extremely helpful. I would like to thank all the reviewers for generously spending time to review the chapters. Certainly their contributions enhanced the quality of the book.

I would also like to thank the staff at Wiley-Blackwell for their help and guidance which made the successful completion of this project possible. I am grateful to my son, Michael John, for his patience, understanding and encouragement during the preparation of this book.

Nurhan Turgut Dunford

Contributors

Hasan Atiyeh
Department of Biosystems
and Agricultural Engineering
Oklahoma State University
Stillwater, Oklahoma, USA

Scott E. Baker
Chemical and Biological Process
Development Group
Pacific Northwest National Laboratory
Richland, Washington, USA

Chockry Barbana
Food Research and Development Centre
Agriculture and Agri-Food Canada
Saint-Hyacinthe, Quebec, Canada

Natasha Berry
Department of Chemistry and Biology
Ryerson University
Toronto, Ontario, Canada

Susan I. Blackburn
CSIRO Marine and Atmospheric Research
and Energy Transformed Flagship
Hobart, Tasmania, Australia

Joyce Irene Boye
Food Research and Development Centre
Agriculture and Agri-Food Canada
Saint-Hyacinthe, Quebec, Canada

Jonathan Y. Chen
School of Human Ecology
Texas Materials Institute Material
Science & Engineering Program
The University of Texas at Austin
Austin, Texas, USA

Ling-Zhi Cheong
Department of Engineering
Aarhus University
Aarhus, Denmark

John N. Coupland
Department of Food Science
The Pennsylvania State University
University Park, Pennsylvania, USA

Nurhan Turgut Dunford
Department of Biosystems and
Agricultural Engineering and
Robert M. Kerr Food & Agricultural
Products Center
Oklahoma State University
Stillwater, Oklahoma, USA

Ryan J. Elias
Department of Food Science
The Pennsylvania State University
University Park, Pennsylvania, USA

George F. Fanta
US Department of Agriculture
Agricultural Research Service
National Center for Agricultural
Utilization Research
Peoria, Illinois, USA

Sergey N. Fedosov
Department of Engineering
Aarhus University
Aarhus, Denmark

Frederick C. Felker
US Department of Agriculture
Agricultural Research Service
National Center for Agricultural
Utilization Research
Peoria, Illinois, USA

Zheng Guo
Department of Engineering
Aarhus University
Aarhus, Denmark

Gündüz Güzel
Department of Engineering
Aarhus University
Aarhus, Denmark

Douglas G. Hayes
Department of Biosystems Engineering and Soil Science
University of Tennessee
Knoxville, Tennessee, USA

Ram C.R. Jala
Department of Engineering
Aarhus University
Aarhus, Denmark

Sue A. Karagiosis
Chemical and Biological Process Development Group
Pacific Northwest National Laboratory
Richland, Washington, USA

Bena-Marie Lue
Department of Engineering
Aarhus University
Aarhus, Denmark

Michael A. R. Meier
Karlsruhe Institute of Technology
Institute of Organic Chemistry
Karlsruhe, Germany

Dérick Rousseau
Department of Chemistry and Biology
Ryerson University
Toronto, Ontario, Canada

Fernando Sampedro
USDA ARS Eastern Regional Research Center
Wyndmoor, Pennsylvania, USA

Randal L. Shogren
US Department of Agriculture
Agricultural Research Service
National Center for Agricultural Utilization Research
Peoria, Illinois, USA

Oğuz Türünç
Karlsruhe Institute of Technology
Institute of Organic Chemistry
Karlsruhe, Germany

John K. Volkman
CSIRO Marine and Atmospheric Research and Energy Transformed
Hobart, Tasmania, Australia

Mark R. Wilkins
Department of Biosystems and Agricultural Engineering
Oklahoma State University
Stillwater, Oklahoma, USA

Xuebing Xu
Department of Engineering
Aarhus University
Aarhus, Denmark

Rickey Yada
Department of Food Science
University of Guelph
Guelph, Ontario, Canada

Fei Yu
Department of Agricultural and Biological Engineering
Mississippi State University
Mississippi State, Mississippi, USA

Umut Yucel
Department of Food Science
The Pennsylvania State University
University Park, Pennsylvania, USA

Howard Q. Zhang
USDA ARS Western Regional
Research Center
Albany, California, USA

Abbreviations

ADMET	Acyclic diene metathesis polymerization
AOT	Sodium dioctyl sulfosuccinate
CALA	Candida antarctica lipase A
CALB	Candida antarctica lipase B
DAG	Diacylglycerol
DVB	Divinyl benzene
E	Enzyme
EP	Enzyme-product complex
ES	Enzyme-substrate complex
ESBO	Epoxidized Soybean Oil
FA	Fatty acid
FAME	Fatty acid methyl ester
G (or GLY)	Glycerol
HAP	Hazardous air pollutant
hPL	Human pancreatic lipase
IPN	Interpenetrating network
KmS	Michaelis constant
LOI	Limiting oxygen index
LPL	Lysophospholipids
MAG	Monoacylglycerol
METU	Methyl undec-10-enoate
NMMO	N-methylmorpholine-N-oxide
P	Product
PC	Phosphatidylcholine
PG	Partial Acylglycerol
PGA	Poly(glycolic acid)
PHA	Polyhydroxyalkanoate
PHB	Poly(3-hydroxybutyrate)
PHBV	Poly(3-hydroxybutyrate-co-3-hydroxyvalerate)
PL	Phospholipid
PLA	Poly(lactic acid)
PLA_1	Phospholipase A_1
PLA_2	Phospholipase A_2
PLC	Phospholipase C
PLCD	Phospholipase D
PLLA	Poly(l-lactic acid)
PU	Polyurethane

PVA	Poly(vinyl alcohol)
RM	Rhizomucor meihei
S	Substrate
S*	Micellar substrate
TAG	Triacylglycerol
TLL	Thermomyces lanuginosum
UA	Undec-10-enoic acid
V	Velocity
VOC	Volatile organic compounds

1 Traditional and Emerging Feedstocks for Food and Industrial Bioproduct Manufacturing

Nurhan Turgut Dunford

1.1 INTRODUCTION

Many industrial products, such as dyes, inks, paints and plastics, were made from biomass generated by trees, vegetables or other crops during the early 1900s. By 1970, petroleum-based products had largely replaced bio-based products. The utilization of plant-based materials decreased from about 35% to less than 16% between 1925 and 1989 (Forward, 1994). Waning interest in bio-based products was due to the relative ease and lower cost of manufacturing similar products from petrochemicals.

The petrochemical industry has been very successful in developing new products (more than 100 000 commercial products) (Metzger and Eissen, 2004). About 2.6 million barrels per day of petroleum equivalent are used for production of chemicals and industrial building blocks. More than 95% of the world's petrochemical production is derived from oil or natural gas (Weissermel and Arpe, 1997). Excessive reliance on non-renewable energy and resources is the major problem facing petrochemical industry today. In 2001 it was projected that the global oil reserves would last for about 40 years (Metzger and Eissen, 2004). Oil production is expected to reach its maximum in this decade, at the latest by 2015–2020, and then slowly decrease. According to Gavrilescua and Chisti, the issues that make the petrochemical industry unsustainable in the long run are: (1) utilization of manufacturing techniques that are not environmentally benign or safe, (2) production of toxic by-products and waste, (3) products are not readily recyclable and biodegradable after their useful life, and (4) social benefits of the production are not broadly accessible due to excessive regional concentration of production (Gavrilescua and Chisti, 2005).

Nearly one billion of the current world population (the total is about six billion) live in the industrialized countries. The world population is expected to reach to about nine billion by 2050. It is anticipated that the population growth will mainly soar in the developing countries (Metzger and Eissen, 2004). As the population and the standard of living increase, demand for food and other goods will substantially grow, consequently exerting tremendous pressure on resources. Today it is true that "hunger is a problem of poverty rather than absolute food scarcity" (Koning *et al.*, 2008). Yet, the global demand for food production will more than double by 2050, competing for resources needed to grow biomass for other purposes, including biofuels and bio-based non-food industrial products (Koning *et al.*, 2008). A combination of further increases in crop yields (about 2% per year) and doubling

Food and Industrial Bioproducts and Bioprocessing, First Edition. Edited by Nurhan Turgut Dunford.

or tripling of resource use efficiencies (especially of nitrogen and water productivity in biomass production systems) will be necessary to meet the rapidly growing demand for food, feed and industrial bioproducts over the next 20–30 years (Spiertz and Ewert, 2009).

About 224×10^9 tonnes of dry biomass is generated globally as a result of photosynthesis (Champagne, 2008). Today, forestry products and agricultural crops are the major feedstocks for bioproduct manufacturing. Utilization of agricultural residues, forestry, animal and municipal solid wastes and marine vegetation as feedstock could ease the pressure on agricultural land needed to grow food. However, the effect of excessive biomass removal on ecosystems has to be examined very carefully.

In this chapter, current and potential feedstocks for food and bioproduct manufacturing will be reviewed under three categories: grain, oilseed and lignocellulosic biomass, which will include grasses and trees. Microalgae, emerging as a biomass source, will be covered in another chapter of this book.

1.2 GRAIN CROPS

Grain crops, specifically cereal crops, are major feedstocks for the food and fermentation industry because of their high starch and protein content. Cereal crops are by far the most important crops cultivated globally. In 2009 about 2.5 billion tonnes of cereals were produced worldwide (FAO, 2010). Wheat, corn, barley and sorghum are the common starch sources that have been traditionally used in food and industrial bioproduct manufacturing. Straw and stocks from cereal crops are also important as lignocellulosic feedstock for bioproduct manufacturing.

1.2.1 Wheat

1.2.1.1 *Production*

Wheat is one of the major grain crops produced, consumed and traded worldwide. About 683 million tonnes of wheat is produced globally each year (Table 1.1). China, India, the USA and the Russian Federation are among the largest wheat growers (FAO, 2010).

It is believed that einkorn, which was developed from a wild grass native to western Asia, was the first type of wheat cultivated (Atwell, 2001; Orth and Shellenberger, 1988). Four species, *Tiriticum*: *T. monococcum, T. turgidum, T. timopheevi and T. aestivum*, are the commercially important wheat cultivars today. Among these, *T. turgidum* and *T. aestivum*, which are mainly used for bread and pasta making, respectively, are the most widely grown wheat species (Pomeranz, 1988). Enhancement of nutritional composition and value of wheat through biotechnology is an area that is gaining ever increasing scientific attention. It has been shown that a gene, GPC-B1, found in wild wheat but lost its functionality during domestication has the potential of increasing protein and micronutrient content of cultivated wheat by 10–15% (Uauy *et al.*, 2006). Novel wheat varieties with high amylose content have been developed by using the RNAi gene silencing technique that suppresses the expression of two wheat genes, SBEIIa and SBEIIb (Regina *et al.*, 2006). These genes produce starch branching enzymes and play important role in the starch synthesis pathway. The suppression of these two genes produced a wheat variety with high resistant starch (amylose) content and low glycemic index (GI). This new wheat variety could potentially provide health benefits to people with bowel, diabetes and obesity problems.

Table 1.1 Wheat, corn, sorghum and barley production in 2008 (million metric tons).

Wheat		Corn		Sorghum		Barley	
World	683.4	World	826.3	World	66.8	World	155.1
China	112.5	USA	307.1	USA	12.0	Russian Federation	23.2
India	78.6	China	166.0	Nigeria	9.3	Ukraine	12.6
USA	68.0	Brazil	58.9	India	7.9	France	12.2
Russian Federation	63.8	Mexico	24.3	Sudan	3.9	Germany	12.0

Classification of wheat for commercial purposes is based not on variety but on grain properties such as softness/hardness, winter/spring growth habit, red/white bran and protein content. Hard wheat has a hard kernel and produces high protein content flour suitable for making bread and noodles. Soft wheat has lower protein content than hard wheat and is mainly used for making cakes, biscuits and pastries. Winter wheat is planted in late summer or fall and takes the advantage of fall moisture for germination. It is grown in regions where soil does not completely freeze and kill the crop. As the name implies, spring wheat is sown in spring and harvested in late summer. Yields for winter wheat tend to be higher than that for spring wheat due to the risks associated with summer harvest (Pomeranz, 1988). Color, white or red, refers to the color of the outer layers of grain. Depending on milling extraction rate (bran removal rate), the color of the wheat flour can be quite dark, affecting the appearance of the final product. Major wheat exporting countries have their own wheat grading standards that are based on test weight, protein content, moisture, foreign material content and so on (Bushuk and Rasper, 1994). The US Grain Standards Act is enforced by inspectors under the supervision of the US Department of Agriculture (USDA, 2006).

Although wheat straw production yield depends on variety and agronomic and climatic factors, in general 1.3 kg of straw per kg of grain is produced for the most common varieties (Montane *et al.*, 1998). It is estimated that over 90 million metric tons (tonnes) of wheat straw is produced annually in the United States. Considering that world wheat production was about 683 tonnes in 2008 (Table 1.1), global wheat straw production would be 888 tonnes in the same year. About 500 kg of straw per acre needs to be left on the soil surface during wheat harvest for erosion control of steeply sloped ground (Mckean and Jacobs, 1997). It is apparent from these numbers that a significant amount of wheat straw is available for value-added product development.

1.2.1.2 *Chemical composition*

Chemical composition of wheat fractions varies significantly by variety, agronomic and climatic conditions and separation techniques used. Grain consists of 40–50% of the biomass produced by the crop. Wheat grain structure is very complex and composed of many layers (Godon, 1994). Broadly speaking wheat grain consists of endosperm, bran and germ, which account for 81–84%, 14–16% and 2–3% of the grain, respectively (Atwell, 2001). Whole wheat grain is generally processed by dry milling to obtain flour. Various forms of bran, germ and the "clean-out" of the screen room represent almost 25% of the original wheat grain.

Carbohydrates make up about 60–80% of the dry grain weight. Starch is the main carbohydrate found in endosperm or flour. Other carbohydrates present in wheat include free sugars, glucofructans, cellulose and hemicellulose. Arabinose and xylose are the major free sugars.

Proteins are the second largest group of compounds in endosperm (12–15%). Cereal proteins are classified based on their solubility characteristics; water soluble albumins (5–10%), dilute salt-solution soluble globulins (5–10%), aqueous alcohol soluble prolamins (40–50%) and dilute acid or alkali soluble glutelins (30–40%) (Godon, 1994). Cereal proteins, similar to other plant proteins, are low in some of the essential amino acids, for example lysine. Glutamic acid is the major amino acid in wheat.

Lipids are the minor components of wheat grain (2–3%) and consist of polar and nonpolar components. Triacylglycerides (TAG) make up majority of the nonpolar lipids that are rich in unsaturated fatty acids. Huge variations in linoleic acid content of wheat, 45–75% of total fatty acids, were reported among five market classes of wheat (Davis *et al.*, 1980). Polar lipids include glycolipids and phospholipids (Godon, 1994).

The mineral content of wheat grain varies between 1 and 3%. Even though the mineral content is not very high, wheat could provide significant amount of minerals as it is readily found in most daily diets. Magnesium, phosphorous and potassium are the most abundant minerals. Phosphorous is mostly present in the organic form phytic acid. It has been reported that agronomic condition does not have a significant effect on the mineral composition of wheat grain (Godon, 1994). Wheat grain is rich in vitamins, niacin (about 6 mg/100 mg) and tocopherols (about 20 mg/100 g) (vitamin E).

Wheat bran and germ fractions are rich sources of a number of phytonutrients, including policosanol (PC), phytosterols (PS), α-tocopherol and phenolic acids. A number of studies have shown that PS and PC reduce serum low density lipoprotein (LDL) cholesterol levels (Ostlund, 2002; Hirai *et al.*, 1984; Quilez *et al.*, 2003; Aleman *et al.*, 1994; Castano *et al.*, 2000; Gouni-Berthold and Berthold, 2002). Antioxidant properties of wheat bran and germ extracts are well known (Zhou *et al.*, 2004). It has been reported that dietary wheat bran provides protection against colorectal cancer (Qu *et al.*, 2005). This property is due to the presence of phenolic acids, lignans and flavonoids in wheat bran. Wheat germ contains approximately 11% oil (Dunford and Zhang, 2003; Eisenmenger and Dunford, 2008; Dunford, 2005). The oil contains a number of bioactive compounds, such as tocopherols (Eisenmenger and Dunford, 2008; Dunford and Zhang, 2003), polyunsaturated fatty acids, PS (Chen *et al.*, 2009a), and PC (Chen *et al.*, 2009b; Irmak and Dunford, 2005; Irmak *et al.*, 2005). Wheat germ oil (WGO) is the richest natural source of vitamin E (Kahlon, 1989). It has been reported that wheat germ oil improves human physical fitness; this effect is attributed to its high PC content (Cureton, 1972). There are numerous research studies indicating that 5–20 mg/day of PC consumption is effective in lowering total cholesterol (17–21% reduction) and low density lipoprotein (21–29% reduction) levels and increasing high density lipoprotein (HDL) (8–15% increase) by inhibiting cholesterol synthesis and increasing LDL processing (Menendez *et al.*, 1994, 1999). There is scientific evidence that PC has additional beneficial effects on smooth muscle cell proliferation, platelet aggregation and LDL peroxidation (Taylor *et al.*, 2003). PC formulations are being used as "antifatigue drugs" (Cureton, 1972). Currently, beeswax and sugar cane are the main sources of PC. It is important to note that there have been reports indicating that the cholesterol lowering properties of PC are not reproducible in studies performed in United States (Varady *et al.*, 2003). A better understanding of the effects of PC on disease prevention and treatment requires large scale independent animal and clinical studies involving various ethnic groups and subjects with different health histories.

Wheat straw is a lignocellulosic material containing about 35–40% cellulose, 30–35% hemicellulose and 10–15% lignin (Harper and Lynch, 1981). Wheat straw also contains both lipophilic and hydrophilic extractives, which may be released and/or interfere during pulping and pretreatment of feedstock prior to hydrolysis of carbohydrate polymers to their monomeric sugars prior to microbial fermentation (Sun *et al.*, 2003; Sun and Tomkinson, 2003; Sun and Sun, 2001). Wheat straw contains many bioactive compounds, including PC and PS (Sun and Sun, 2001; Irmak *et al.*, 2005; Dunford and Edwards, 2010). Recovery of these high value bioactive compounds during or prior to bioconversion of wheat straw to bioproducts could improve the feasibility of the conversion process and may improve the efficiency of the hydrolysis of cellulose and hemicellulose and the subsequent fermentation process.

1.2.1.3 *Utilization*

Wheat is a vital component of human diet in many countries. In the United States about 17–18% of total daily calories are acquired from foods derived from wheat. In some other countries wheat-based foods may provide two thirds, even more of the daily caloric intakes (Bushuk and Rasper, 1994). In the United States about 70% of wheat is utilized for human consumption, 24% as animal feed and the remainder for seed or industrial products. The main reason for using a small fraction of wheat in industrial products is due to high demand for wheat for food and feed applications.

Annetts and Audsley (Annetts and Audsley, 2003) defined the term "biorefinery" as "a factory consisting of a collection of processes which takes agricultural inputs from the surrounding area, and produces a wide range of products which are specifically targeted at different market uses and are refined to their specification". Wheat is an excellent crop to demonstrate the "biorefinery" concept. Current and emerging processing techniques present tremendous opportunities for converting wheat fractions – flour, milling industry by-products (bran, shorts and germ) and straw – in to a broad range of bioproducts. In fact, the economic viability of multiproduct wheat biorefinery systems has been examined by various groups (Dorado *et al.*, 2009; Annetts and Audsley, 2003; Clark, 2007; Dunford, 2007; Koutinas *et al.*, 2004). For example, a process for producing a generic wheat-based fermentation feedstock with two liquid streams, one rich in glucose and one rich in nitrogen, was developed and successfully used for the production of ethanol, lactic acid, pigment and glycerol (Webb and Wang, 1997; Koutinas *et al.*, 2004).

Currently excess straw is baled for use as livestock bedding or low-grade animal feed providing minimal economic return. Straw represents a significant opportunity for fiber substitution. Pulp from straw is already being partially substituted for wood fiber in some paper and paperboard products (Mckean and Jacobs, 1997). Advanced cellulose hydrolysis technologies are being developed for converting straw to biofuels and other industrial bioproducts (Szczodrak, 1988; Ahring *et al.*, 1996; Saha *et al.*, 2005). Recovery and conversion of surface waxes and other bioactive compounds present in wheat straw into high value products, such as insecticides, nutracueticals and cosmeceutical ingredients, may improve the economic feasibility of other commodity or lower value products generated in a biorefinery (Dunford and Edwards, 2010; Irmak *et al.*, 2005; Clark, 2007). Waxes derived from sustainable plant products have a growing market for the replacement of both synthetic (petroleum derived) and animal product derived waxes including lanolin obtained from sheep's wool (Clark, 2007).

1.2.2 Corn

1.2.2.1 *Production*

Maize or corn (Zea mays) has a long history as a domesticated crop. Today corn is the most widely grown cereal crop in the world. In 2008, world corn production was 826.3 million tonnes (USDA-FAS, 2008). It is estimated the grain yield will reach 1.2 tonnes/ha

(190 bu/acre) by 2030 from the current 940 kg/ha (150 bu/acre), assuming that current 1% annual genetic gain in corn grain yield can be sustained over time (Duvick and Cassman, 1999). In an average year corn is grown on 30–32 million hectares in the United States. Yet, in 2007, 36.5 million hectares (90 million acres) was dedicated to corn because of the high demand for ethanol production. Presuming that production area is maintained at the 2007 level, United States' corn production could reach 430 million tonnes (approximately 17 billion bu/year) accounting for the future yield increases (Dhugga, 2007). Even under these conditions United States would be far short of producing enough corn needed to produce enough ethanol to replace gasoline and meet the increasing demand of grain corn for feed and food (Dhugga, 2007).

1.2.2.2 *Chemical Composition*

Corn is categorized as dent, flint, flour, sweet, pop or pod based on its kernel characteristics.

The majority of corn grown in the United States is yellow dent, yellow referring to the color of endosperm. Chemical composition of corn kernel varies significantly depending on type, variety, environmental conditions and agronomic practices used during crop production (Reynolds *et al.*, 2005). Starch is the most economically important component of corn. Typical starch content of dent type corn is about 60% (NCGA, 2010). Mature kernels of some corn hybrids may contain as high as 75% starch (w/w, dry basis). Morphological, rheological, functional and thermal properties of starches from different corn types vary significantly (Li *et al.*, 1994). Amylopectin and amylose are two structural components of starch. Regular corn starch consists of 75% branched amylopectin and 25% linear amylose while waxy corn hybrids may contain 100% amylopectin. A clear understanding of starch properties in different corn types and lines is vital for selecting corn for breeding purposes and producing starch with specific properties that are required by manufacturing industry.

Sweet corn is primarily grown for fresh consumption. Young sweet corn kernels are succulent because of a mutant recessive sugary-1 gene that retards the conversion of sucrose into starch during endosperm development (Dickerson, 2003). Immature sweet corn contains about 10% sucrose, which is rapidly converted to starch after harvest. Kernels can lose up to 50% of their sucrose at ambient temperature within 24 hours after harvest.

The protein content of corn kernel is about 10%. Zein is the major storage protein and comprises 45–50% of the total protein in corn. Zein cannot be digested efficiently by humans and other non-ruminants. Since corn is an important staple in many countries, its protein quality is important. Corn kernels tend to be low in lysine and tryptophan, two of the eight essential amino acids, requiring corn-based diets to be supplemented with other proteins such as bean proteins. Corn kernels containing the Opaque-2 gene are shown to have lower amounts of zein and higher amounts of lysine and tryptophan in their endosperm than standard dent corn (Paulis *et al.*, 1991). High lysine corn containing increased levels of lysine and tryptophan has been developed. Although high lysine corn has demonstrated nutritional advantage over field corn it still has several disadvantages. It has soft texture, dull appearance and very little hard endosperm, which make high lysine corn difficult to harvest and prone to attack by pests. The value of high lysine corn in poultry rations is limited because of its insufficient methionine content. Furthermore, high lysine corn has lower crop yield and must be segregated from other corn varieties to preserve protein quality (Dickerson, 2003).

Field corn contains 4–6% oil. Corn oil is a healthy energy source for both humans and livestock because of its high polyunsaturated fatty acid, specifically linoleic acid content. High oil corn, which contains 7–8% oil, is preferred by cattle feeders for its high calorie content, which promotes greater animal weight gain per unit of feed. Nevertheless, corn varieties with oil content higher than 6% tend to have lower crop yields.

1.2.2.3 *Utilization*

In 2007, about 98% of the world corn production was consumed as food or feed (USDA-FAS, 2008). The majority of the corn grown in the United States is used as feed. Nearly 43% of the 2009 US corn production was used as feed or left as residue on the field (NCGA, 2010). The same year biofuel, particularly the ethanol production industry, utilized almost 32% of the corn grown in United States. The shares of food/cereal and high fructose corn syrup production in corn consumption were relatively low, 5 and 12 million tonnes, respectively. The high fructose corn syrup production industry takes up about 3–4% of total United States corn production.

Besides its food and feed use, corn has numerous other industrial applications. According to the US National Corn Growers Association, there are more than 4200 different uses for corn products (NCGA, 2010). Although zein, a major storage protein in corn, is not directly used for human consumption it has many potential industrial applications in fiber, adhesive, coating, ceramic, ink, cosmetic, textile, chewing gum and biodegradable plastics production (Shukla and Cheryan, 2001). Biofuels, starch and high value products such as recombinant pharmaceutical proteins and specialty chemicals are some of the economically important corn-based products (Naqvi *et al.*, 2011).

A significant fraction of United States corn is used to produce ethanol. Corn is converted into ethanol primarily by two processes: wet milling and dry grinding. In wet milling, the corn kernel is fractionated into germ, fiber, and starch resulting in several co-products. The starch portion of the corn kernel is converted to ethanol, while the protein, fiber and oil are passed through to the by-products. Only one co-product, distillers' dried grains with solubles (DDGS), is produced when dry milled corn is used for ethanol production. Every bushel of corn (approximately 25 kg) processed for sweeteners, oil, or ethanol generates nearly 7 kg of protein and fiber-rich residues (Leathers, 2003). Currently these by-products are used in low value applications. It has been suggested that ethanol production industry by-products have the potential to be used as inexpensive fermentation media for production of polysaccharides and carotenoids by yeasts or yeast-like fungi (Leathers, 2003). The potential of DDGS for value-added product development has also been explored. Utilization of DDGS as a substrate for biobutanol production by various clostridia species is an area that is drawing attention (Ezeji and Blaschek, 2008).

Today corn is the leading platform to synthesize high value molecules, including pharmaceuticals, through biotechnology. Corn has a number of advantages over other plants for molecular pharming in crops. These advantages include its GRAS (Generally Regarded As Safe) status, well characterized genetic properties, responsiveness to *in vitro* manipulation and gene transfer, well established agricultural infrastructure and efficient biomass production (Ramessar *et al.*, 2008). Initially new traits for herbicide tolerance and pest or disease resistance were introduced into corn. Recent research on metabolic pathways that produce primary and secondary metabolites, specifically the compounds that are beneficial for human health and nutrition, has led to the introduction of novel traits in to corn. In a recent article utilization of advanced biotechnology tools to produce a broad range of high value molecules in corn was reviewed (Naqvi *et al.*, 2011). Some of the compounds successfully produced in transgenic corn include amino acids (Galili and Höfgen, 2002), very long chain polyunsaturated fatty acids (Napier *et al.*, 2006), vitamins (Giuliano *et al.*,

2008; Zhu *et al.*, 2008; Chen *et al.*, 2006; Ishikawa *et al.*, 2006) and minerals (Drakakaki *et al.*, 2005). The emerging regulations that govern how transgenic corn plants and their products are grown, used and traded were also discussed by Naqvi and co-workers (2011).

Utilization of corn stover for bioproducts has advantages over alternative crops that are being examined as feedstock, because corn stover is already produced with grain and does not require dedicated land. The challenge is to determine the sustainable levels of stover that can be collected without adversely affecting ecosystems. Corn fiber is a by-product of the wet-milling process that separates fiber, gluten and germ from starch. Stillage residues from starch fermentation are folded into the corn fiber to produce corn gluten feed and sold as low-value cattle feed. New co-products from corn fiber and gluten feed could also add value to the corn processing industry. For example, it has been shown that corn fiber is a rich source of oryzanol (ferulic acid esters of phytosterols), which has cholesterol lowering properties (Jain *et al.*, 2008). Both corn fiber oil and gum have potential for functional food applications (Yadav *et al.*, 2007).

1.2.3 Barley

1.2.3.1 *Production*

Barley is among the four largest cereal crops (with wheat, maize and rice) grown in the world (Newman and Newman, 2008). In 2008 global barley production was about 155.1 million tonnes (FAO, 2010). The top barley growing countries are the Russian Federation, Ukraine, France and Germany. The United States was ranked as the ninth highest barley grower in the world with 5.3 million tonnes of production in 2008.

Barley is a grass that belongs to the Poaceae family, Triticeae tribe and genus Hordeum (Newman and Newman, 2008). One of the most important mutations associated with the domestication of wild barley to cultivated barley was non-brittle rachis formation, which resulted in efficient harvest without loss of grains (Pourkheirandish and Komatsuda, 2007). Genetic variation in starch structure and composition determine the end uses of barley. Traditionally, barley breeders focused on the improvements that benefit malting and the brewing industries rather than the feed market because malting barley receives premium price (Ullrich and Eslick, 1978). Recently, feed quality of barley has been receiving attention from the breeders, partly because barley is becoming a very important feed crop due to the increasing industrial uses and uncertainty in the availability of corn as feed (Rudi *et al.*, 2006).

The first transgenic barley plants were developed in 1994 (Jähne *et al.*, 1994). Early barley transformation studies aimed at improved grain quality for feed and beverages (Jensen *et al.*, 1996; Horvath *et al.*, 2000) or disease resistance (McGarth *et al.*, 1997; Leckband and Lörz, 1998). These studies have demonstrated that functional recombinant proteins such as enzymes can also be produced in transgenic barley, specifically in grain.

1.2.3.2 *Chemical Composition*

Barley is classified in several ways depending on agronomic properties, chemical composition and end use. The most common barley types are: spring and winter, two- and six-row, hulled and hulless, malting and feed, normal, waxy and high amylose starch, high lysine, high β-glucan, and proanthocyanidin-free barley (Baik and Ullrich, 2008). The chemical composition of barley grain varies significantly with genotype, agronomic practices and environmental conditions (Griffey *et al.*, 2009; Aman *et al.*, 1985; Oscarsson *et al.*, 1996). Barley hull comprises about 13% of the kernel and consists mostly of cellulose, hemicelluloses (xylans), lignin and a small amount of protein (Andersson *et al.*, 1999). Hulls adhere to the caryopsis of the hulled barley while they are not attached or loosely attached to the grain surface of hulless barley. According to Bhatty (1999), hulless barley would ideally have less than 5% adhering hulls. The thickness of the hulls varies. Thick hulls adhere to the caryopsis less firmly than thin hulls. Presence or absence of hulls significantly affects grain composition. Hulless barley generally has lower ash and dietary fiber but higher starch, protein and oil content due to the absence of the hull.

Carbohydrates comprise about 80% of the barley grain. Starch is the major grain component and barley may contain up to 65% starch (Song and Jane, 2000). Starch and protein contents of hulless barley can be as high as those of field corn. A study on the chemical composition of 92 Swedish barley varieties showed that two-rowed barley varieties had slightly higher starch content while six-rowed barleys had higher protein and fiber (Aman *et al.*, 1985). Waxy barley varieties contain 5–8% less starch than that of non-waxy/regular barley varieties (Bhatty, 1999). Starch type and properties have a significant effect on barley end use. Amylopectin comprises 72–78% of the total starch in barley (Bhatty, 1999). Waxy barley varieties contain very high levels of amylopectin. The existence of barley cultivars with 100% amylopectin has been reported (Bhatty, 1997). Zero or waxy, normal and high amylose barley contain 0–5%, 20–30% and up to 45% amylose (based on grain weight), respectively (Baik and Ullrich, 2008). Starch granule size in hulless barley ranges from 2 to 30 μm (Bhatty, 1999; You and Izydorczyk, 2002). Among four types of hulless barley (normal, high, waxy and zero amylose), normal amylose type has the greatest amount of large granules (74.7%). Waxy, zero and high amylose starches consist of 66.4, 43.9 and 19.4% large granules, respectively (You and Izydorczyk, 2002).

The major non-starch carbohydrates in barley comprise (1,3)(1,4)-β-D-glucans and arabinoxylans. β-Glucans, which are mainly present in the endosperm cell walls (Oscarsson *et al.*, 1996), consist of high molecular weight linear chains of β-glucosyl residues polymerized through β-(1-3) and β-(1-4) linkages (Newman and Newman, 2008). The high β-glucan content of barley (2.5–11.3%) is notable (Izydorczyk and Dexter, 2008). High amylose and waxy barley have been reported to contain higher β-glucan than normal amylose type (You and Izydorczyk, 2002). β-Glucan is partially soluble in aqueous solutions due to the molecular, structural and solubility differences of polysaccharides present in its chemical structure (Newman and Newman 2008). The water soluble part of β-glucan produces high viscosity starch slurries that can cause problems during industrial processing. High viscosity mash increases pumping costs and complicates production. Low β-glucan content in the grain leads to low viscosity and little need of expensive enzymes to break it down for efficient processing and fermentation. Cellulose (1,4-β-D-glucan), fructans, arabinoxylans, glucommannan, galactomannan, arabinogalactan and a number of simple sugars and oligosaccharides are also present in barley grain in relatively small quantities.

Although phenolic compounds are minor constituents in barley, they play an important role in the nutritional quality of the grain. Phenolic compounds bind proteins, carbohydrates and minerals, thereby affecting the nutritional and functional value of the bound constituents. The major nutritional concern is the ability of condensed and hydrolysable tannins to bind strongly to large proteins, specifically to proteins high in proline, thereby reducing protein digestibility. Significant differences were found in flavanol and total phenol content among diverse barley genotypes (Griffiths and Welch, 1982). The variations in phenolic content were not correlated with grain size, malting quality and oil and protein contents. Barley grain also contains numerous polyphenolic compounds, including catechin and proanthocyanidin (Aastrup *et al.*, 1984). Proanthocyanidins cause undesirable haze in beer. Fortunately, about 600 proanthocyanidin-free barley mutants have been isolated and used in breeding programs (Jende-Strid, 1993). Oxidation of phenolic compounds to *o*-quinones results in discoloration of the products made with barley after cooking (Sapers, 1993).

1.2.3.3 *Utilization*

It is estimated that about 85% of the world barley production is utilized as feed (Mäkinen and Nuutila, 2004). Malting is the second largest application for barley grain. Only 2% of barley is used for food production in the United States. However, in regions with extreme climates, such as Himalayan nations, Ethiopia and Morocco, barley remains an important food source (Baik and Ullrich, 2008).

Low protein content barley grain is preferred for malting. In general, protein and starch contents are negatively correlated in crops. Higher starch content in barley grain leads to higher extract content in malt. Low protein is ideal for starch production while higher protein content is desirable in feed barley. The health benefits of barley include reduction of blood LDL cholesterol level, glycemic index and body mass, which lead to control of heart disease and type-2 diabetes. Interest in incorporating low tannin barley in human diet is increasing because of its high nutritional value (Newman and Newman, 2008). The beneficial effects of barley are due to the presence of several bioactive compounds, such as β-glucans, tocopherols and tocotrienols in the grain (Baik and Ullrich, 2008). Indeed the US Food and Drug Administration (FDA) approved a health claim verifying that barley contains high levels of β-glucans which help to prevent coronary heart disease when consumed by humans (FDA, 2005). Although today the main source of barley β-glucan in functional foods is oats, barley β-glucan has a potential for new formulations. Co-products from barley milling can be used to recover phenolic compounds and other antioxidants that can be incorporated into functional foods and nutraceuticals, such as high performance sports drinks and dietary supplements, anti-aging cosmetics and sun-screen lotions (Griffey *et al.*, 2009; Newman and Newman, 2008). Barley has the potential as a platform to produce various enzymes, therapeutic proteins and novel high value chemicals through genetic modifications (Jensen *et al.*, 1996; Jende-Strid, 1993; Horvath *et al.*, 2000).

Interest in using barley, particularly winter hulless winter barley, for bioproduct and biofuel production is growing specifically outside the Corn Belt area in the United States (Septiano *et al.*, 2010; Griffey *et al.*, 2009). Ground whole barley grain and flour have been successfully converted to bioethanol (Sohn *et al.*, 2007; Septiano *et al.*, 2010; Gibreel *et al.*, 2009). Production of valuable high protein content DDGS as a by-product improves the feasibility of barley as feedstock for alcohol production for fuel and solvent uses (Ingledew *et al.*, 1995).

1.2.4 Sorghum

1.2.4.1 *Production*

Sorghum is a member of the grass family Poaceae. It is a drought-tolerant crop that grows well with minimal input. In regions with low rainfall, less than 900 mm annual rainfall, sorghum out-performs corn, making it an appealing crop in semiarid regions of the world (Zhan *et al.*, 2003). Total world production of barley was 66.8 million tonnes in 2008 (FAO,

2010). The United States, Nigeria, India and Sudan are the largest sorghum growers. Annual sorghum production in the United States was about 12 million tonnes in 2008.

In the 1960s, there were concerns about the low level of genetic diversity in the existing grain sorghum cultivars due to breeding practices used at the time. The Sorghum Conversion Program, which was created as a US Department of Agriculture (USDA) and Texas A&M University partnership, introduced genetic variation from tropical species into cultivated sorghum lines grown in the United States to alleviate the problem (Vandenbrink *et al.*, 2010). As a result, 701 sorghum varieties were released by the Sorghum Conservation Program in 2004. The S. *bicolor* genome was sequenced in 2009 (Paterson *et al.*, 2009). This achievement could lead to further improvements in sorghum chemical and physical properties, including its nutritional composition, starch and cellulose hydrolysis yields and fermentation efficiency.

In terms of their use sorghum is classified into three groups: grain, grass/forage and sweet sorghum (Monk *et al.*, 1984). Grain sorghums are grown for their seeds which are rich in starch. Grassy sorghums grow fast and can reach ten feet (about 3.1 m) in height, generating significant amount of biomass. Sweet sorghums have sweet juicy stems and produce the highest amount of biomass among the sorghum varieties. Over three decades ago it was proposed that development of high energy sorghum (HES) hybrids that would combine characteristics of both sweet and grain sorghums could increase sorghum production potential (Monk *et al.*, 1984). Eventually in 2008 two sorghum varieties, BTx623, a grain sorghum, and RIO, a sweet sorghum, were crossed to create recombinant inbred lines (RILS) (Murray *et al.*, 2008a, 2008b). RILS have been used to genetically map quantitative traits, such as wet and dry yields, and the amounts of structural carbohydrates, such as cellulose, lignin and hemicelluloses (Vandenbrink *et al.*, 2010). It is expected that new sorghum lines will lead to the discovery and introduction of specific genes for traits such as lower lignin and higher starch content, resulting in higher biomass hydrolysis yields during pretreatment of sorghum for bioproduct manufacturing.

1.2.4.2 *Chemical composition*

The structure of sorghum grain is similar to that of other cereal grains. The endosperm, germ and pericarp account about 85%, 9.55% and 6.5% of the whole grain, respectively. Like corn, sorghum has a large germ relative to its endosperm size. Oil content in sorghum (3.4%) is higher than that in wheat kernel (2.2%) (Kent and Evers, 1994). The endosperm is rich in starch (higher than 80%) and relatively poor in protein (about 10%) and lipid (less than 1% lipid) (Serna-Saldivar and Rooney, 1995). The prolamins, storage proteins in sorghum, also referred to as kafirins, account for about 80% of the total grain proteins (Taylor and Dewar, 2001). Kafirins are low in essential amino acids, specifically in lysine which comprise only 0.2% of the total amino acids in kafirin. The World Health Organization (WHO) recommendation for lysine content in sorghum protein is 5.5 mg lysine/g protein (FAO, 1973).

Cell wall components of sorghum, lignin, cellulose, protein and hemicellulose, vary significantly by variety (Corredor *et al.*, 2009). Endosperm cell wall and non-starch polysaccharides are rich in water-insoluble glucuronoarabinoxylans that affect the end use of this crop (Verbruggen *et al.*, 1995). Sweet sorghum contains about equal quantities of soluble (glucose and sucrose) and insoluble carbohydrates (cellulose and hemicellulose) (Monk *et al.*, 1984). Sweet sorghum produces both grain and large amounts of sugar-rich stems. The sugar in stems is mainly comprised of sucrose (up to 55%, dry basis), fructose and glucose (Billa *et al.*, 1997).

Grain sorghum contains a number of bioactive compounds, including phenolic compounds, phytosterols and policosanols. Benzoic and cinnamic acid derivatives, tannins and anthocynanins are the major phenolic acids and flavonoids in sorghum (Hahn *et al.*, 1983; Waniska *et al.*, 1989; Krueger *et al.*, 2003; Awika and Rooney, 2004). These compounds are secondary metabolites which play an important role in plant defense mechanism. Some sorghum cultivars that are commonly known as "bitter", "bird proof", "bird resistant" or "brown", are rich in condensed tannins (proanthocyanins) which protect the plants against pests, insects and diseases. Based on its tannin content sorghum is categorized as type I (no significant amount of tannins that are extractable with acidified methanol) and type II (contains tannins that are extractable with acidified methanol but not with pure methanol) and type III (contains tannins that are extracted with either methanol or acidified methanol) (Dykes and Rooney, 2006). The difference between type II and III sorghums is the location of tannins in the grain structure. Tannins in type II sorghums are deposited in the vesicles within the testa layer, whereas in type III sorghum tannins are deposited along the cell walls of the testa and some are present in the pericarp. In type II sorghums acid is needed to break up the structure of vesicles to release tannins (Earp *et al.*, 2004). Examples of sorghum cultivars within this classification are: Type I, TX2911 (red perocarp), Type II, Early Hegari and type III, Early Sumac (Price *et al.*, 1978; Awika and Rooney, 2004). Sorghum is also classified based on the color of the grain and total extractable phenol content. White sorghum, also referred to as food-type, contains no detectable amount of tannins or anthocyanins and has a very low level of total extractable phenol. Red sorghum does not contain tannins but has a red pericarp with significant amount of extractable phenols. As the name implies, black sorghum has a black pericarp and contains very high levels of anthocyanins. Brown sorghum contains considerable amount of tannins and has a pigmented testa (seed coat) and pericarp with varying degrees of pigmentation (Awika and Rooney, 2004). Currently 99% of the sorghum grown in United States is tannin free. A small amount of tannin sorghum grown in the United States is identity preserved seed stock.

The United Nations (UN) World Health Organization (WHO) identified iron (Fe), zinc (Zn) and pro-vitamin A (β-carotene) as limiting micronutrients in human diet. Deficiency of these micronutrients is highest in South and Southeast Asia and sub-Saharan Africa (SSA), where sorghum is consumed as a staple food by millions of people (Reddy *et al.*, 2005). In 2004, the HarvestPlus Challenge Program was launched with the objective of developing biofortified crops (HarvestPlus, 2011). Within this program 84 sorghum lines were analyzed for their micronutrient contents. Significant differences in iron (20.1–37.0 ppm), zinc (13.4–31.0 ppm) and phytate (3.8–13.5 mg/g) contents were found among the sorghum samples examined. There were also variations in β-carotene content among the samples. Only a trace amount of β-carotene was present in non-yellow lines while lines with yellow endosperm contained 0.56–1.13 ppm β-carotene. The information about the variations in chemical composition among sorghum lines is valuable to plant breeders because this information can be used to increase the micronutrient content of sorghum. Furthermore, data on the chemical composition of sorghum lines is also important for choosing the right feedstock for specific sorghum applications.

1.2.4.3 *Utilization*

Food applications of sorghum have been reviewed extensively (Taylor and Dewar, 2001; Taylor *et al.*, 2006; O'Kennedy *et al.*, 2006). White food sorghum grain is milled into flour and incorporated into snacks, cookies (biscuits), bread and ethnic foods. In the United States food products made with white sorghum are marketed as substitutes for wheat-based products, targeting people with celiac disease who are allergic to wheat gluten (Ciacci *et al.*, 2007). Traditional sorghum varieties with moderate tannin content are commonly grown and used for food and alcoholic beverages in Africa. The agronomic advantages of traditional varieties outweigh the reduced nutrient availability and astringency of tannin sorghum. In some African cultures tannin sorghums are preferred because of their satiating properties. It is believed that pigmented sorghum varieties promote the health of unborn babies and have therapeutic affects against digestive system related diseases (Taylor *et al.*, 2006).

Similar to other cereal grains, sorghum is rich in starch; as such, it is suitable for use in fermentation. A large portion of grain sorghum is used for malting in Africa. There has been extensive research and development work on enzymes used for sorghum malting and malting and brewing technologies (Owuama, 1999; Taylor *et al.*, 2006). It is estimated that about 200 000 tonnes of sorghum are malted and 3000 million liters of sorghum beer is produced annually in Southern Africa alone (Taylor and Dewar, 2001).

Sweet sorghum is classified as sugar and syrup-types. Syrup-type sweet sorghum is rich in glucose and used for syrup, wine, lactic acid and alcohol production (Billa *et al.*, 1997). Sorghum syrup is made by pressing the juice off the stems with rollers and boiling it down to the desired consistency. Residual sweet sorghum stems after juice extraction are used as animal feed and silage. Sugar-type sweet sorghum, which mainly contains sucrose, can be used for crystal sugar production. Grain sorghum is used to produce industrial bioproducts, such as starch, biopolymer films and coatings. Isolation of starch from sorghum is done by wet milling (Munck, 1995). However, pigments in tannin sorghum stain the starch, making the final product undesirable. Bleaching and low cost abbreviated wet milling have been used to improve sorghum starch color (Yang and Seib, 1995; Beta *et al.*, 2000). It is anticipated that new developments in wet-milling techniques and sorghum breeding will increase the industrial uses of sorghum starch. Utilization of grain fractions from sorghum wet milling for production of activated carbon (Diao *et al.*, 2002) and health beneficial food ingredients enriched in PS and PC have been explored (Singh *et al.*, 2003).

Good film-forming properties make sorghum kafirin a potential feedstock for biopolymer production. It has been demonstrated that kafirin films plasticized with glycerol and polyethylene glycol had similar tensile and water vapor barrier properties to the films made from commercial maize zein plasticized in the same way (Da Silva and Taylor, 2005).

Efficacy of coating fruits with kafirin to delay ripening, reduce stem-end withering and increase shelf-life has also been investigated (Taylor *et al.*, 2006). Kafirin coating tests with pears were successful but similar studies carried out with litchis resulted in excessive darkening and formation of white deposits on the fruit. Recovery of sorghum wax from DDGS and its utilization as edible coating have also been examined (Weller *et al.*, 1998). However, edible coating applications with sorghum wax and kafirin require further research prior to commercialization.

1.3 OIL AND OILSEEDS

Oilseed crops are major agricultural commodities grown globally. Total world oilseed production was about 450 million tonnes during the 2010–2011 crop year (USDA, 2011). It is expected that oilseed production will increase to 460 million tonnes in 2011–2012. About 376 million tonnes of oilseeds were crushed for oil in 2010. Currently rapeseed, canola, soybean, cottonseed, peanuts and sunflower seed are the most important commercial oilseeds.

1.3.1 Rapeseed/Canola

1.3.1.1 *Production*

Rapeseed is the major oilseed crop produced worldwide. Total world rapeseed production was 58.1 million tonnes in 2008 (FAO, 2010). Canada, China, India, Germany are the top rapeseed growing counties (Table 1.1).

Rapeseed belongs to *Cruciferae* family and *Brassica* genus. The original rapeseed cultivars had high levels of erucic acid in the oil and glucosinolates in the meal. Rapeseed oil feeding trials carried out with rodents indicated that high levels of erucic acid in oil led to fatty deposits in heart and skeletal muscles and impaired growth of the animals (Przybylski *et al.*, 2005). The presence of glucosinolate in rapeseed meal was damaging to poultry, swine and ruminants. Isothiocyanates and other sulfur-containing compounds formed during hydrolysis of glucosinolates interfere with iodine uptake by the thyroid gland, contribute to liver disease and reduce animal growth and weight gain. Utilization of oilseed meal (seed residue after oil extraction) as animal feed is critical for the economic viability of oilseed processing operations. The erucic acid and glucosinolate content of the seeds had to be reduced for the rapeseed meal to be acceptable as animal feed. In the 1970s the world's first low erucic acid and low glucosinolate cultivar of *Brasicca napus*, often called double-zero rapeseed, was developed through breeding. The term "canola" was registered by the Western Canadian Oilseed Crushers in 1978 and then transferred to the Canola Council of Canada in 1980. Canola referred to those cultivars containing less than 5% erucic acid in the oil and 3 mg/g aliphatic glucosinolates in the meal. In 1986, the definition of canola was amended to *Brasicca napus* and *Brasicca rapa* lines with less than 2% erucic acid in the oil and less than 30 μmol/g glucosinolates in the oil-free meal. Today the name "Canola" is mainly used in the American continent and Australia. In Europe "rapeseed" is the term commonly used for both original high and low erucic acid rapeseed cultivars.

There have been very important developments in breeding and genetic modifications of oilseeds, resulting in increased crop yields and oil content, incorporation of fatty acids of nutritional importance that are usually obtained from other sources and expression of industrial fatty acids that are currently sourced from petrochemicals or from low-yielding plants. Recently, plant breeders have been paying more attention to increasing contents of minor seed components, such as carotenoids, vitamin A (Shewmaker *et al.*, 1999), tocopherols, vitamin E (Marwede *et al.*, 2004), sinapate esters (Hüsken *et al.*, 2005; Zum Felde *et al.*, 2006) and phytosterols (Amar *et al.*, 2008) in rapeseed. Traditional rape (*Brassica napus* L.) has a black seed coat. Yellow seeded rape types, which have thinner seed coat and higher oil and protein content, are available today (Rahman, 2001).

1.3.1.2 *Chemical composition*

Chemical compositions of oilseeds vary extensively with variety, climate, soil, agricultural inputs (i.e. seed quality, fertilizer and other agrochemicals) and agronomic practices. Hence, the seed compositional data given in this chapter should be used as a point of reference. Oil content of rapeseed, which is a heritable characteristic, may vary between 36 and 50% (dry matter basis). Protein content of rapeseed is negatively correlated with its oil content (Zhao *et al.*, 2006). Canola varieties contain 17–26% protein (Aider and Barbana, 2011). Ash and fiber contents of rapeseed are around 4 and 23% based on dry matter, respectively (Yoshie-Stark and Wäsche, 2004).

Similar to the other edible oils rapeseed oil is primarily composed of TAG (90–99%). The free fatty acid (FFA) content of oils is an indication of oil quality. High FFA content in crude oil results in high refining losses. Canola and rapeseed oils may contain between 0.4 and 1.8% FFA (Przybylski *et al.*, 2005). Fatty acid composition of vegetable oils determines their end use and health effects. Erucic acid (C22:1) reduction in rapeseed oil resulted in a significant increase in oleic acid (C18:1) content. Presence of high levels of linolenic acid (C18:3) reduces oxidative stability of oils. Canola varieties with low linolenic acid content (2%) have been developed through breeding to increase oil stability. Other canola varieties with modified fatty acid content include high oleic (60–85%), lauric (39%), stearic (C18:0) (up to 40%), palmitic (C16:0) (10%) and gamma linolenic acids (up to 40%) (Przybylski *et al.*, 2005). Canola oil contains significant amount of tocopherols (770 mg/kg), which mainly consist of α- and β-isomers. Total phytosterol content of canola oil is about 8.8 mg/kg. Brassicasterol and campesterol are the major phytosterols present in canola and rapeseed oil (Schwartz *et al.*, 2008).

1.3.1.3 *Utilization*

Liquid canola oil is widely used in salad dressings and for cooking. In the United States liquid canola is preferred in salad dressings because of its low saturated fatty content. Indeed canola oil has the lowest saturated fatty acid content among the common vegetable oils. For frying applications canola oil is lightly hydrogenated to improve its oxidative stability. Soft (tub) and hard (stick) margarines, shortenings and baking and pastry margarines are also produced with partially hydrogenated and liquid canola oil. In Canada high erucic acid rapeseed oil is used only in special food applications. Fully hydrogenated oil is added to peanut butter at 1–2% levels to prevent oiling (Przybylski *et al.*, 2005).

In Europe rapeseed growing countries, particularly Germany, Austria and France, produce substantial amount of fatty acid methyl esters and use it as fuel (biodiesel). Use of rapeseed-based methyl esters as solvent in paints and coatings is also increasing. Methyl esters have a number of advantages as to petroleum-based solvents. They are biodegradable, have very low volatility and viscosity and high flash point and good water wetting and penetration properties. Because of its relatively low polyunsaturated fatty acid content rapeseed oil is a better feedstock for base for engine oils, heat transfer and hydraulic fluids, inks and demolding agents than soybean oil (Przybylski *et al.*, 2005). Erucamide, which is derived from erucic acid and used as antiblock and slip-promoting agents in plastic film production, is one of the most important utilization areas for HEAR oil (Murphy, 1996). Brassylic and pelargonic acids derived from HEAR oil are used for polymer production (nylon, polyester and melamine resin coatings).

1.3.2 Soybeans

1.3.2.1 *Production*

Cultivated soybean (Glycine max) originated in China. Today soybeans are widely grown and utilized around the world. In 2008, about 231 million tonnes of soybeans were produced globally (FAO, 2010). The United States leads the world soybean production, followed by Brazil, Argentina and China (Table 1.2).

Variety selection is one of the most important considerations for crop production. Maturity rating, yield, resistance or tolerance to disease, aphids, stand ability and soybean

Table 1.2 Rapeseed/canola, soybean, sunflower and cottonseed production in 2008 (million metric tons).

Rapeseed/canola		**Soybean**		**Sunflower**		**Cottonseed**	
World	58.1	World	230.6	World	35.7	World	65.4
Canada	12.6	USA	80.7	Russian Federation	7.3	China	15.0
China	12.1	Brazil	59.2	Ukraine	6.5	India	7.6
India	5.8	Argentina	46.2	Argentina	4.7	Pakistan	3.8
Germany	5.2	China	15.6	China	1.8	Brazil	2.5

cyst nematode resistance are some of the criteria used to select suitable soybean varieties for production. Soybeans are classified based on their agronomic properties and hilum color. Soybean development is affected by temperature, day length and hours of sunshine. Varieties adapted to a particular region are assigned a maturity group number. During the summer, day length increases from south to north in United States. Varieties in maturity group 00 are adapted to the northernmost regions and those in group VIII to the southernmost region. As the maturity group number increases, the lengths of the vegetative and reproductive stages of development are extended.

The hilum is the point at which the soybean seed attaches to the pod. Soybean hilum color can be yellow, imperfect yellow, grey, buff, brown, black or imperfect black. Yellow hilum soybeans are generally preferred for the export market. Soybeans are also classified as food and oil beans based on utilization.

Seed coat and hilum color, bean size and grade, organoleptic properties and chemical composition are important characteristics for selection of soybeans for food applications. For example, Japanese food manufacturers prefer soybeans with clear to light colored hilum, large seed size (18 g/100 seeds), creamy white cotyledons and 40% or higher protein content beans for making miso and tofu. Many soybean varieties have been developed for specific food applications, such as tofu, natto, miso and edamame beans (Liu, 1997).

Human health and environmental concerns about genetically modified organism (GMO) lead to consumer demand for identity-preserved and non-GMO soybeans. Identity preservation is the segregation of a crop variety from planting through to delivery to the end user. This concept is used in food-grade soybeans production. Soybeans contain at least three lipoxygenase isozymes, which are responsible for the production of unpleasant grassy and beany flavors. In the early 1990s, soybeans lacking lipoxygenase enzyme and having less bean-like flavor were developed (Moreira *et al.*, 1993). Both traditional breeding and genetic engineering tools have also been used to improve other soybean properties. Soybeans with higher protein, sucrose and fermentable sugar content and yield and lower oligosaccharide content and altered oil fatty acid compositions have been developed. Recent studies focus on increasing stearidonic acid (Eckert *et al.*, 2006) and α-tocopherol contents (Van Eenennaam *et al.*, 2003), removal of the Gly m Bd 30 K protein, which is a soybean allergen (Herman *et al.*, 2003), and reduction of phytic acid (Shi *et al.*, 2007) in soybeans.

1.3.2.2 *Chemical composition*

Similar to the other crops, soybean chemical composition varies significantly with environmental and agronomic conditions and variety. It has been reported that the oil content of soybeans varied more than its protein content from year to year (Ren *et al.*, 2009). Draught stress caused a significant increase in soybean protein content while oil content decreased.

Soybeans contain about 40% protein (dry basis) and are rich in amino acids lysine and threonine. However, low content of the essential amino acid L-methionine limits nutritive value of soy protein. Globulins, referred to as glycinin and conglycinin, are the most abundant protein types in soybeans. Health benefits of soybean proteins have been studied extensively. Positive effects of soybean proteins on hypertension and heart disease, osteoporosis, bone health and certain cancers have been reported (Friedman and Brandon, 2001). The US Food and Drug Administration (FDA) has authorized a health claim on the role of soy protein in reducing risk of coronary heart disease. The FDA rule permits a statement on the soybean containing product labels indicating that "Diets low in saturated fat and cholesterol that include 25 g of soy protein a day may reduce the risk of heart disease" (FDA, 1999).

Soybeans are believed to be responsible for 90% of the food-related allergenic reactions (L'Hocine and Boye, 2007). Allergens are proteins. Although significant advances have been made in the identification and characterization of soybean allergens, it is still not entirely clear which soy proteins cause allergic reactions. At least 16 allergenic soy proteins have been identified (L'Hocine and Boye, 2007). It might be possible to remove some of the known soybean allergens through traditional breeding and genetic engineering. However, considering that most proteins have metabolic, storage or protective functions in seeds, removal or modification of some of the allergenic proteins may impair the nutritional and functional properties of soybeans.

Soybeans contain compounds that impair activity of enzymes trypsin, chymotrypsin, carboxypeptidase, elastase, and α-amylase. The presence of enzyme inhibitors, Bowman-Birk inhibitor (BBI) (an inhibitor of chymotrypsin and trypsin) and Kunitz trypsin inhibitor (KTI), in unheated soybeans hinders the activity of protease enzymes trypsin and chymotrypsin and, consequently, limits protein hydrolysis during digestion. Rat feeding studies with some of the soybean derived inhibitors led to development of pancreatic lesions and caused neoplasia or tumor formation (McGuiness *et al.*, 1987). Interestingly, BBI has the opposite effect, it has been shown to have cancer prevention properties. It has been postulated that BBI may block the formation of oxygen free radicals by stimulated neutrophils, inhibit tumor promotion and prevent the digestion of proteins to amino acids, thus depriving rapidly growing cancer cells of essential amino acids (Troll *et al.*, 1986).

Soybeans contain other bioactive proteins, cytochrome-c, lectin, lipoxygenase and urease. Lectins are glycoproteins and known as hemagglutinins because of their red blood cell agglutinating properties. Although growth inhibiting properties of lectins have been established in rat studies, removal of lectins from a soy extract did not result in significant improvement in the weight gain of rats (Friedman and Brandon, 2001).

Even though the secondary plant metabolites, isoflavones, saponins, phytic acid, flatus-producing oligosaccharides and goitrogen, are minor components of soybeans, they have an impact on the nutritive value of soybeans. For example, high phytic acid content in soybeans could impair mineral absorption (Friedman and Brandon, 2001). Significant variations in total isoflavone content among soybean cultivars have been reported, 1161–2743 μg/g, a 2.5-fold variation from the lowest to highest isoflavone content (Wang *et al.*, 2000a). There has been great interest in the role of isoflavones in prevention and treatment of chronic diseases. (Messina and Messina, 2000). However, there are many inconsistent research studies on the health benefits of isoflavones (Tikkanen and Adlecreutz, 2000; Anderson *et al.*, 1999; Washburn *et al.*, 1999). The disparities in the research results were attributed to the different amounts and types as well as different ratios of the 12 isoflavones (3 aglycones, 3 glucosides, 3 acetyl glucosides, and 3 malonyl glucosides) used in the diets (Friedman and

Brandon, 2001). The individual isoflavones may have different biological activity and certain combination of isoflavones may act synergistically or antagonistically (Rayburn *et al.*, 1995). The presence of other phytoestrogens (biochanin A, coumesterol, and formononetin) in the soybean diet could also influence the results (Kudou *et al.*, 1991).

Soybeans contain about 20% oil on dry basis. Refined soybean oil contains more than 99% TAG. There are large variations in the fatty acid composition of soybean oil obtained from different genotyped and beans grown under different agronomic conditions. The reported range of main fatty acids are as follows: C16:0, 8–17%; C18:0, 3–30%; C18:1, 25–60%; C18:2, 25–60%; and C18:3, 2–15% (Hammond and Glatz, 1989). Soybean oil contains health beneficial lipid soluble vitamins and phytosterols. Crude soybean oil contains about 2 mg/kg vitamin K_1 or phylloquinone, which have blood coagulation properties (Piironen *et al.*, 1997). Tocopherols and phytosterols comprise 0.12 and 0.24% of the crude oil (Mounts *et al.*, 1996).

1.3.2.3 *Utilization*

Soybean oil is an important edible oil extensively used in many food formulations and as cooking and frying oil by the food industry. Salad dressings, margarine shortening and confectionary coatings are commonly made with soybean. Food applications of soybean oil have been extensively reviewed by other authors (Liu, 1997; Hammond *et al.*, 2005).

Oriental or traditional soy foods are classified into two groups, fermented and non-fermented foods. Popular non-fermented foods include tofu, soy milk and sprouts. Miso, tempeh, natto and soy sauce are the most common fermented soy products. Traditionally, soy protein has not been a major part of the Western diet. Yet, in recent years soy protein has been getting a lot of attention as an ingredient in infant formulas, extended meat products, baked goods and dairy foods. Soy flour is used in high protein breakfast cereals to improve amino acid balance in the product. Soy protein is the main ingredient in many specialty products, such as nutritional or weight loss beverages, confectioneries, simulated nutmeats, spreads and puddings. Protein concentrates and isolates derived from soybean are often used as emulsifiers in soups and sauces. Edible films are typically made with proteins, lipids or carbohydrates or a combination of the three. They provide protection to foods and drugs against desiccation and oxidative and microbial deterioration. Yuba is a soy-milk-based film and a traditional food known for its unique flavor and texture (Liu, 1997).

Soy-based biopolymers for packaging applications and drug delivery systems have been extensively studied (Swain *et al.*, 2004; Guerrero *et al.*, 2011; Schmidt *et al.*, 2005; Sue *et al.*, 1997; Chen *et al.*, 2008; Ogale *et al.*, 2000). Poor mechanical properties and the hygroscopic nature of the films derived from soy protein limit their applications (Gennadios *et al.*, 1994). Physical, chemical and enzymatic modifications and compositing with hydrophobic materials are being explored to improve the properties of soy films (Stuchell and Krochta, 1994; Rhim *et al.*, 2007).

Utilization of soy-based plastics (Paetau *et al.*, 1994; Wang *et al.*, 2007), adhesives (Hamarneh *et al.*, 2010; Prasittisopin and Li, 2010), composites (Huang and Netravali, 2007; Husic *et al.*, 2005), elastomers (Wang *et al.*, 2010; Jong, 2006) and coatings (Madrigal-Carballo *et al.*, 2010) are being considered for manufacturing agricultural equipment and automobile parts and infrastructure and building materials (Wool *et al.*, 2000). Textile fibers, foams for fire extinguishers, resins as binders and lubricants are other products that can be made from soy straw, protein and oil (Altuna *et al.*, 2010; Reddy and Yang, 2009; Situ *et al.*, 2007; Lathi and Mattiasson, 2007; Honary, 1996).

Lecithin is comprised of phosphatides that are recovered during crude oil degumming processes. Soybean lecithin has many uses because of its emulsifying, wetting, colloidal and antioxidant properties. Lecithin finds applications in foods, beverages, animal feed, health and nutrition products, cosmetics and industrial coatings (Liu, 1997). In animal feed, lecithin helps to stabilize the product, promotes fat absorption in the digestive system and increases the energy efficiency of feed. Industrial applications of lecithin include releasing agents in plastics industry, dispersing agent in inks and paints, anti-sludge additive in motor lubricants, anti-gumming agent in gasoline and emulsifier, spreading agent and antioxidant in textile and rubber industries.

The US-based United Soybean Board (USB) funds the research, development and commercialization of new industrial uses for soybeans through soybean check-off programs. Numerous soybean-based products at commercial or development stage can be found on the USB web page (http://www.soynewuses.org/Default.aspx).

1.3.3 Other Oilseeds

1.3.3.1 *Camelina*

Camelina sativa (L.), also known as false flax or gold of pleasure, is an annual plant that belongs to the Brassicaceae family. Camelina was an important crop in Europe until it was replaced with rapeseed. There was no commercial production of camelina in the United States until 2004. A rapid increase in production has been observed since 2004 reaching approximately 30 000 ha in 2009 (Berti *et al.*, 2011).

Oil and protein contents of camelina seeds vary between 32 and 46% and 25 and 29%, respectively (Vollmann *et al.*, 2007). Camelina oil is composed of 90% unsaturated fatty acid of which 25–42% is α-linolenic acid (18:3), 13–21% linoleic acid (18:2), 14–20% oleic acid (18:1), 12–18% eicosenoic acid (20:1), and 2–4% erucic acid (22:1) (Berti *et al.*, 2011).

The increased interest in applications of ω-3 fatty acids in food and pharmaceutical industries attracted a lot of attention to camelina seed oil because of its high α-linolenic acid (ω-3) and natural antioxidants contents (Pilgeram *et al.*, 2007). Because of its high γ-tocopherol content, camelina oil is very stable has longer shelf-life than many of the other commodity oils. Camelina oil and meal can be used in animal rations, food formulations, cosmetics, soaps, and in other industrial applications, such as fatty acid methyl esters and jet fuel production. Camelina oil is registered as a commercial food oil in many European countries and Canada (Ghamkhar *et al.*, 2010).

1.3.3.2 *Cotton*

Cotton (*Gossypium, H.L.)* is mainly produced for its fiber (cotton lint). The cotton plant also produces seed that contains about 20–22% oil and 20–23% protein. The traditional cotton varieties contain gossypols, which are comprised of sesquiterpenoid phenolic aldehydes and related compounds. Gossypols are toxic to humans and monogastric animals and present in the seed, foliage and roots of cotton, where they protect the plant against insects and pathogens. The presence of gossypol in cottonseed limits its use as food and feed. Glandless cotton lacking gossypol has been developed and available commercially (Sunilkumar *et al.*, 2006).

Cottonseed oil has always been a premium oil for food applications, specifically for frying because of its neutral flavor and extended fry life. A few years ago snack manufacturers moved away from cottonseed oil because of its relatively high saturated fatty acid content and consumer concerns over the adverse effects of saturated fats. Yet, since the announcement of the FDA trans-fat labeling rule (FDA, 2003a), a reversal has been observed and cottonseed oil is now back in high demand as trans-free oil. Cottonseed oil is widely used in food formulations that require solid fat. Cottonseed oil forms β-prime crystals that provide a smooth consistency, fine texture and plasticity to the products, leading to a pleasant mouth feel. Zero calorie fat substitutes are also made from cottonseed oil (Akoh, 1998).

Non-food applications of cottonseed oil include soap, lubricant, sulfonated oils, pharmaceutical, coating and rubber production. A relatively small amount of cottonseed oil is used during leather, textile, ink, polishes, plastics and resin manufacturing (O'Brien *et al.*, 2005).

Extraction of cottonseed oil produces meal, hulls and linters as co-products in addition to oil. Meal is primarily used as a protein source for ruminant animals because it increases the butterfat content of milk. Hulls are used as source of roughage for cattle and a substrate for mushroom production. Linters are rich in cellulose and used for battings, felt liners and insulation material (O'Brien *et al.*, 2005).

1.3.3.3 *Sunflower*

Sunflower, *Helianthus annuus* L., belongs to the Compositae family. Traditionally, calcium rich seeds were roasted and ground into flour and incorporated in foods, florets were extracted to obtain a yellow dye, oil was utilized in ceremonial body paintings and residual plant parts were dried and used as building material (Putt, 1997). Sunflowers are classified in three groups: oilseed, confectionary and ornamentals, which are planted in gardens to attract birds. Some sunflower cultivars with large seeds are roasted for snack foods or blended with other grains to make bird seed.

Whole sunflower seeds contain about 40% oil and 25% protein. Small black sunflower seeds may contain up to 55% oil. In the 1980s sunflower became a major oilseed crop after the introduction of high yielding varieties. Linoleic (66–72%) and oleic acids (16–20%) are the major fatty acids in oil extracted from regular sunflower oil seeds. Saturated fatty acids, mainly palmitic and stearic acids, comprise less than 15% of the total fatty acids in the oil.

Today sunflower varieties with modified fatty acid composition are available; high oleic sunflower (80–87% oleic acid and 7–8% saturated fatty acids) and mid-oleic sunflower (NuSun, 50–75% oleic acid and less than 10% saturated fatty acids). Sunola and Sunwheat are two hybrids that were developed to overcome harvesting problems faced with traditional varieties. These hybrids are 25–35% shorter than regular sunflower, allowing the use of the same machinery used to harvest other grains and canola (Grompone, 2005). With the growing emphasis on beneficial effects of polyunsaturated fatty acids in human diet, sunflower oil became a premium oil for many food applications (Carvalho *et al.*, 2006). Remaining meal after oil extraction is used as livestock feed for ruminants, pigs and poultry.

1.3.3.4 *Safflower*

Safflower (*Carthamus tinctorius* L.) is also an important oil source. Safflower seed is covered with a tough fibrous hull that protects the kernel. Whole seed and kernel contain 37–45% and 53–63% oil, respectively. Regular safflower contains mainly linoleic (77%) and oleic acids (15%). Oleic safflower (77% oleic and 15% linoleic acid) was originally marketed as an ingredient for mayonnaise and a replacement for peanut oil used by the snack food industry (Goodman, 1964). Later, oleic safflower became an ingredient in infant formulations because of its high oxidative stability. In recent years oleic safflower oil has been preferred over linoleic safflower oil in many products or used in blends.

Historically safflower was used as a source of dye and food coloring, in cosmetics and for medicinal purposes. Safflower contains both yellow and red (carthamin) pigments that are suitable for coloring foods and dying cotton and silk (Saito *et al.*, 1983). Safflower seeds, pollens, florets and oil have been used in traditional medicine (Hanania *et al.*, 2004). For example safflower oil, called oleum cnicium, is a milder substitute for castor oil, which is used as an over-the-counter laxative. It has been reported that charred safflower oil may treat sores and rheumatism and has applications as an antidote for poisons and an agent to induce sweating to cure fever (Smith, 2005). Safflower seeds are widely used as bird feed because of their high oil content. Ornaments, wreaths and jewelry made from safflower seeds and flowers are very popular in Egypt. The spineless safflower is a source of food coloring and is planted as garden flower in Spain. Dried safflower florets are used as a saffron substitute.

1.3.3.5 *Castor*

Castor (*Ricinus communis*) beans contain ricin, ricinine and other allergens that are toxic to humans and animals. Hence, they are not suitable for edible applications and animal feed unless the toxins are inactivated by steam or other chemical means. Seeds contain about 40–55% oil. Ricinoleic acid (12-hydroxy-9-octadecenoic acid) is the major fatty acid in castor oil (89%). Linoleic (4.2%), oleic (3.0%), stearic (1%), palmitic (1%), dihydroxystearic (0.7%), linolenic (0.3%) and eicosanoic (0.3%) acids are also present in minor amounts. The high viscosity of the oil, which is mainly due to the hydrogen bonding of its hydroxyl groups, makes it a good feedstock for paints, coatings, inks and lubricants. Castor oil is suitable for chemical modifications because of the presence of double bonds and hydroxyl groups in the ricinoleic acid structure (Ogunniyi, 2006). Hydrogenated castor oil (HCO), which is a wax-like material, is used in cosmetics, hair products, ointments, preparation of hydrostearic acid and its derivatives, and as wax substitutes and polishes. HCO is also used as a paint additive, solid lubricant and pressure mold release agent. Sulfated castor oil, also known as Turkey red oil, is an effective wetting agent that is used in cotton and linen dyeing and finishing. Dehydrated castor oil (DCO) is desirable in the preparation of alkyd resins that are, in turn, used in paints, enamels, lacquers and varnishes with high gloss and good adhesion and wetting properties (Ogunniyi, 2006). Undecylenic acid and heptaldehyde, which are important intermediates in the preparation of perfume formulations, have been produced via pyrolysis of castor oil (Das *et al.*, 1989). E-2-Nonenal, which is used as an ingredient in natural flavors, fragrances, polyurethane elastomers and foam production, has been synthesized from castor oil (Kula *et al.*, 1994; Lyon and Garret, 1973).

1.3.3.6 *Palm*

Both palm and palm kernel oil are derived from the fruits of oil palm (*Elaeis guineessis* Jack.), which grows well in tropical climates. Oil palm is the highest oil yielding vegetable crop (5 tonnes/ha) with a commercial life-span of about 25 years. Palm oil is extracted from the outer flesh (mesocarp) of the ripe fruit and palm kernel oil is produced from the nuts. About 20 tonnes of palm oil and 5 tonnes of kernel oil can be produced from 100 tonnes of oil palm fruit.

Traditional uses of oil palm were very versatile, and every plant part was utilized and nothing was wasted. Oil palm fruits were used to prepare special dishes and boiled to obtain cooking oil. Oil was burned to provide light and used as soothing ointment and for soap making. The hard shells of the kernel provided fuel for fire and leaves were utilized for making roof thatching, fencing, mats or brooms. Ropes made from the leaf stalk fiber were woven into baskets and fish traps. Sap from the immature flower stalks was used to prepare soft drinks or fermented to make palm wine, vinegar and other alcoholic drinks (Henderson and Osborne, 2000).

Today palm oil is widely used in food applications and preferred for frying and baking applications because of its good oxidative stability and high solid fat content. Palm oil contains about 50% saturated (42–48% palmitic and 4–5% stearic acids) and 50% unsaturated fatty acids (37–41% linoleic and 9–11% linolenic acids). The fatty acid composition of palm kernel oil resembles that of the coconut oil rather than that of palm oil. Palm kernel oil is rich in lauric (about 48%), myristic (16%) and oleic (15%) acids. Both palm oil and palm kernel oil are commercially separated into stearin (solid) and olein (liquid) fractions for special applications. The stearin fraction obtained from palm kernel can be used as a cocoa butter substitute. The olein fraction is used in baked goods and soap manufacturing. Imitation palm-oil-based cheese, hand and body lotion, fatty acid methyl esters for use as fuel or solvent, and epoxidized palm oil to produce plasticizers and stabilizers for conventional polyvinyl chloride plastics are some of the other products that are produced from palm oil (Basiron, 2005).

Kernel cake residue after palm kernel oil extraction is a good source of livestock feed. The other palm fruit bunch residues fire the boilers to generate steam and electricity for the processing plant. Trunks, fronds and kernel shells have even been used for the production of furniture, building materials (panel board), pulp and paper. Oil palm fiber extracted from the empty fruit bunches (43–65% cellulose and 13–25% hemicellulose) is being tested for production of biocomposites (Shinoj *et al.*, 2011).

1.3.3.7 *Ground nut*

Ground nut (*Arachis hypogaea*), also known as peanut, earth nut, and monkey nut, is an important food crop. The oil contents of four major United States market types of peanuts, runner, Virginia, Valencia and Spanish, vary from 44 to 56%. Major fatty acids of peanut oil are oleic (36–67%) and linoleic (15–46%) acids. In 1969 a peanut breeding line containing about 80% oleic acid and 2% linoleic acid was identified (Norden *et al.*, 1969). The incorporation of high oleic genes into new peanut breeding lines resulted in the SunOleic cultivar, a high oleic variety released by the Florida Agricultural Experiment Station in 1995 (Gorbet and Knauft, 1997). The SunOleic peanut variety has a favorable high oleic acid content and, consequently, extended shelf life (Jonnala *et al.*, 2005b). Although they have not been commercialized yet genetically modified peanut varieties have been developed (Chenault *et al.*, 2005). No substantial difference was found when chemical composition of genetically modified peanut varieties were compared to their parent cultivars (Jonnala *et al.*, 2005a, 2006b).

In 2003 the US FDA approved the first qualified health claim for a food for immediate use on package labels. The claim states that "Scientific evidence suggests but does not prove that eating 1.5 ounces per day of most nuts, as part of a diet low in saturated fat and cholesterol, may reduce the risk of heart disease" (FDA, 2003a). Numerous studies have shown that including nuts in the diet can reduce the risk of heart disease (Haumann, 1998; Kris-Etherton *et al.*, 2001; Higgs, 2002). Peanuts are a good source of a wide range of nutrients and bioactive compounds with health benefits. Many of the health beneficial bioactive compounds present in nuts are associated with the oil fraction (Jonnala, 2004).

Nuts contain tocopherols, tocotrienols, phytosterols, and many different flavonoids, including isoflavones and quercetin.

About two-thirds of total world peanut production is crushed for oil. Peanuts are found in a wide variety of products and are an integral part of many cuisines all over the world. Unlike other countries where most of the peanuts produced are used for oil, cake and meal, the prime market in the United States is peanut butter. About half of the United States peanut crop is made into peanut butter, which is one of America's favorite foods. Other forms of peanut consumption in the United States are confectionary (25%) and snack/salted peanuts (10%). Peanuts are even considered as a potential food for space travel (Haumann, 1998).

1.3.3.8 *Crambe*

Crambe (*Crambe abyssinica Hochst.* ex Fries) is an annual oilseed crop that has great potential as a non-edible industrial feedstock for bioproduct development. Whole seeds contain about 35% oil that is rich in erucic acid (55–60%). Crambe oil can be used in a wide range of products including pharmaceuticals, detergents, and cosmetics, as well as lubricants, rubber additives, a new type of nylon, bases for paints and coatings, high temperature hydraulic fluids, dielectric fluid and waxes (Leonard, 1992; Lazzeri *et al.*, 1994; Wang *et al.*, 2000; Muuse *et al.*, 1992). Similar to the other cruciferous oil-bearing seeds, crambe seeds contain glucosinolates, which limit their use in edible applications and as feed for non-ruminant animals. Yet, the US FDA has approved the use of defatted crambe meal as a beef cattle protein supplement at levels up to 4.2% of the total weight of rations (Wang *et al.*, 2000b). Crambe seed has relatively high fiber content (about 18%) compared to other cruciferous seeds. The seed coat is rich in lignin (20%) and α-cellulose (60%). Crambe seed coat can be utilized as an efficient absorbent material due to the absence of a strong fibrous structure.

1.3.3.9 *Other oilseeds*

Lesqueralla and vernonia are non-food oilseed crops that have been evaluated for their potential as feedstocks for bioproduct and biofuel production. Furthermore, these oilseeds contain unusual fatty acids that are not suited for human consumption but have unique chemical properties that are valuable for specific industrial applications. Ricinoleic acid (an hydroxy acid) in lesquerella and vernolic acid (an epoxy fatty acid) in vernonia are excellent feedstocks for production of bioplastics and plasticizers (Battey *et al.*, 1989). Oils from non-food oilseed also have various functionalities as they are. For example, when lesquerella oil was mixed with engine oil at low concentrations (0.25%) it had superior performance in reducing wear and damage in diesel engines compared to castor, soybean, and rapeseed methyl esters (Goodrum and Geller, 2004). Lesquerella oil can also be used as an ingredient for a number of other bioproducts, such as lubricants, motor oils, plastics, inks, and adhesives. Lesquerella seed coat and meal contain a gum that is useful in coatings and food thickeners.

Eruca sativa Mill, *Limnanthes alba* Benth., *Euphorbia iagascae* Sprengel, *Dimorphotheca pluviafis* (L.) Much., and *Calendula oficinalis* L. are other oil bearing crops that are getting attention as potential industrial feedstocks for bioproduct development. *Eruca sativa* seed oil is rich in erucic acid and is being evaluated as an alternative to rapeseed. Oil from *Limnanthes alba* is extremely resistant to oxidation and contains very long chain fatty acids. *Euphorbia lagascae* and *Dimorphotheca pluviafis* can provide unique oils that contain highly reactive and functionalized fatty acids, such as epoxy and hydroxydiene fatty acids, to the oleochemical industry.

Calendula oficinalis seed oil properties are similar to that of tung oil with high viscosity and refractive index and an excellent air-drying tendency, which make these oils excellent feedstock for binders in paints and coatings (Muuse *et al.*, 1992). Although these crops have been around for a long time further research is needed to exploit their full potential for bioproduct development.

1.4 LIGNOCELLULOSIC BIOMASS

The plant cell walls of lignocellulosic biomass are a complex mixture of polysaccharides, pectin and lignin. Lignocellulosic materials are mainly composed of cellulose (insoluble fibers of β-1,4-glucan), hemicellulose (non-cellulosic polysaccharides, including xylans, mannans, and glucans), and lignin (a mixture of complex polyphenolic compounds). In some cases where fermentation is used to convert biomass into bioproducts, lignocellulosic biomass components, specifically cellulose and hemicellulose, need to be converted to fermentable sugars that can be readily fermented to commodity or specialty chemicals by suitable microorganisms. However, chemical bonding between lignin and cellulose and hemicellulose makes depolymerization of polysaccharides into their respective monomers very difficult. To mitigate this problem, lignin content, composition, hydrophobicity and cross-linking may be modified by selective plant breeding and transgenic techniques. Reducing the lignin content of the plant through biotechnology could be a viable option as well. However, considering that lignin plays an important role in plant structure, reduction of lignin in only specific tissues or cell types rather than in the whole plant could be a better approach (Shadle *et al.*, 2007). For example, retaining lignin in xylem cell walls but lowering lignin content in storage parenchyma cell walls may improve digestibility without compromising xylem function (Byrt *et al.*, 2011; Simmons *et al.*, 2010).

There are two major sources of lignocellulosic biomass: agricultural crops and forest resources. Plants with C3 photosynthesis generate high yields of biomass, but take longer to grow and have a higher content of lignin than C4 plants. On the other hand, C4 plants have high light, water and nitrogen use efficiency compared with C3 species. C3 plants, such as poplar (*Populus*), eucalyptus, loblolly pine (*Pinus taeda*), willow (*Salix*) and silver maple (*Acer saccharum*), are some of the forest resources that are getting attention as feedstock for bioproduct manufacturing. Poplar contains 40% cellulose, 14% hemicellulose and 20% lignin and can be grown as a short-rotation woody crop because it grows relatively rapidly at high density. Poplar can produce 9–16 dry tonnes per hectare (4–7 dry tonnes per acre) biomass annually over a 6–10-year rotation (Alig *et al.*, 2000). Willow yields 10–18 tonnes/ha/year of biomass depending on the growth conditions and has higher cellulose content, about 60%, than poplar. But hemicellulose and lignin content of poplar and willow are similar (Labrecque and Teodorescu, 2005). Eucalyptus, native to Australia but also grown throughout the world including California and Florida, has been evaluated as a lignocellulosic feedstock and shows great potential (Rockwood *et al.*, 2008). It is important that residues generated during forest product harvesting and processing, as well as biomass that could become available through initiatives to reduce fire hazards and improve forest health, are exploited to their full capacity to meet the lignocellulosic feedstock needs of a bio-based industry (Wright *et al.*, 2005).

C4 plants are ideal as feedstock for bioproduct manufacturing. The grass (Poaceae) species mostly use the C4 photosynthetic pathway. A multi-institutional research project funded by the US Department of Energy assessed thirty-four herbaceous species on a wide range of soil types at thirty-one different sites spread over seven states in crop producing regions of the United States during the late 1980s and early 1990s (Wright *et al.*, 2005). Several species, including sorghums, reed canarygrass, wheatgrasses and switchgrass were identified as having merit for further development. It was also recommended that perennial grasses be given high research priority.

Switchgrass (*Panicum virgatum*), a C4 perennial grass, produces a significant amount of biomass (5–11 tonnes/ha); it has relatively low lignin (about 6%) and high cellulose and hemicellulose contents, 32% and 36%, respectively (Schmer *et al.*, 2008; Karp and Shield, 2008). The US Department of Energy has selected switchgrass as one of the dedicated high energy crops for biofuel production. Advantages of switchgrass over other potential biofuel feedstocks include relatively high and reliable productivity across a wide geographical range, suitability for growth on marginal quality land, perennial growth habit and low water and nutrient requirements (Sanderson *et al.*, 1996). Yet, it may take up to two years to establish switchgrass for production.

Miscanthus (*Miscanthus sacchariflorus* and *Miscanthus sinensis*), which used to be an ornamental plant in the United States, is receiving increased attention as a dedicated energy crop in the United States and Europe (Karp and Shield, 2008; Villamil *et al.*, 2008). Similar to switchgrass, miscanthus is an herbaceous perennial crop that is capable of producing a significant amount of lignocellulosic biomass. Yet, miscunthus may take two to three years to establish, delaying its full production. It has been reported that miscanthus could produce significantly higher biomass than switchgrass (about 18 vs. 14 tonnes/ha) (Boehmel *et al.*, 2008). Furthermore, miscanthus has a higher cellulose content (about 58%) than switchgrass (about 32%) (Sanderson *et al.*, 2007). Reed canarygrass (*Phalaris arundinacea* L.), napiergrass (*Pennisetum purpureum* Schumach.), bermudagrass (*Cynodon* spp.), aleman grass (*Echinochloa polystachya*), elephant grass (*Pennisetum purpureum*), fox tail millet (*Setaria italica*), straw and stalks of grain crops and sweet sorghum, and sugarcane bagasse are also being examined for their viability to produce lignocellulosic biomass for bioethanol production. Extensive reviews on chemical composition and yield potential of these crops have been published (Sanderson *et al.*, 2007; Simmons *et al.*, 2008; Karp and Shield, 2008; Byrt *et al.*, 2011; Wright *et al.*, 2005).

1.5 CONCLUSIONS

According to the 2006 Vision Goals established by the Biomass Research & Development Technical Advisory Committee, 25.1 million tonnes of products would be produced from biomass by 2030 (BTAC, 2006). Expansion of the bio-based product manufacturing industry will exert tremendous pressure on feedstock resources and environment. Advancements in biomass production systems and efficient and environmentally benign manufacturing technologies are essential for the global success of sustainable bio-based economies. Increasing feedstock needs will require significant improvements in traditional crop production yields, development of alternative and dedicated crops and efficient use of biomass resources. Biotechnology will play a crucial role in the development of new or modified traditional crops with desirable physical, chemical and agronomic properties, increased production yields and even in the expression of health beneficial and high value compounds at elevated levels in crops for biofarming. Demand for food and feed sources will continue to grow with already increasing global population. Hence, development of non-food alternative crops that can grow on marginal quality soil with minimal inputs (water and nutrient) is a high priority.

REFERENCES

Aastrup S, Outtrup H, Erdal K. 1984. Location of the Proanthocyanidins in Patterns of Barley Grain. Carlsberg Res. Commun. 49:105–109.

Ahring BK, Jensen K, Nielsen P, Bijerre AB, Schmidt AS. 1996. Pretreatment of Wheat Straw and Conversion of Xylose and Xylan to Ethanol by Thermophilic Anaerobic Bacteria. Bioresour Technol. 58:107–113.

Aider M, Barbana C. 2011. Canola Proteins: Composition, Extraction, Functional Properties, Bioactivity, Applications as a Food Ingredient and Allergenicity – a Practical and Critical Review. Trends in Food Science & Technology 22(1):21–39.

Akoh CC. 1998. Fat Replacers. Food Technol. 52(3):47–53.

Aleman CL, Mas R, Hernandez C, *et al.* 1994. A 12-Month Study of Policosanol Oral Toxicity in Sprauge Dawley Rats. Toxicol. Lett. 70:77–87.

Alig RJ, Adams DM, McCarl BA, Ince PJ. 2000. Economic Potential of Shortrotation Woody Crops on Agricultural Land for Pulp Fiber Production in the United States. Forest Prod. J. 50:67–74.

Altuna FI, Espósito L, Ruseckaite RA, Stefani PM. 2010. Syntactic Foams from Copolymers Based on Epoxidized Soybean Oil. Composites Part A: Applied Science and Manufacturing 41(9):1238–1244.

Aman P, Hesselman K, Tilly AC. 1985 The Variation in Chemical Composition of Swedish Barleys. J Cereal Sci. 3:73–77.

Amar S, Becker HC, Möllers C. 2008. Genetic Variation and Genotype X Environment Interactions of Phytosterol Content in Three Doubled Haploid Populations of Winter Rapeseed. Crop Sci. 48:1000–1006.

Anderson JJ, Anthony MS, Cline JM, Washburn SA, Garner SC. 1999. Health Potential of Soy Isoflavones for Menopausal Women. Publ. Health Nutr. 2:489–504.

Andersson AAM, Elfverson C, Andersson R, Regner S, Aman P. 1999. Chemical and Physical Characteristics of Different Barley Samples. Journal of the Science of Food and Agriculture 79(7):979–986.

Annetts JE, Audsley E. 2003. Modeling the Value of a Rural Biorefinery-Part II: Analysis and Implications. Agricultural Systems 76:61–76.

Atwell WA. 2001. Wheat Flour. St Paul, MN: Eagan Press.

Awika JM, Rooney LW. 2004. Sorghum Phytochemicals and Their Potential Impact on Human Health. Phytochemistry 65:1199–1221.

Baik B-K, Ullrich SE. 2008. Barley for Food: Characteristics, Improvement, and Renewed Interest. Journal of Cereal Science 48(2):233–242.

Basiron Y. 2005. Palm Oil. In: Shahidi, F. (editor). Bailey's Industrial Oil & Fat Products. Hoboken, NJ: John Wiley & Sons, Inc. pp. 333–429.

Battey JF, Schmid KM, Ohlrogge JB. 1989. Genetic Engineering for Plant Oils: Potential and Limitations. Trends in Biotechnology 7(5):122–126.

Berti M, Wilckens R, Fischer S, Solis A, Johnson B. 2011. Seeding Date Influence on Camelina Seed Yield, Yield Components, and Oil Content in Chile. Industrial Crops and Products 24(2):1358–1365.

Beta T, Rooney LW, Marovatsanga LT, Taylor JRN. 2000. Effect of Chemical Treatments on Polyphenols and Malt Quality in Sorghum. Journal of Cereal Science 31:295–302.

Bhatty RS. 1997. Milling of Regular and Waxy Starch Hull-Less Barleys for the Production of Bran and Flour1. Cereal Chem. 74(6):693–699.

Bhatty RS. 1999. B-Glucan and Flour Yield of Hull-Less Barley. Cereal Chemistry 76(2):314–315.

Billa E, Koullas DP, Monties B, Koukios EG. 1997. Structure and Composition of Sweet Sorghum Stalk Components. Ind Crops Prod. 6:297–302.

Boehmel C, Lewandowski I, Claupein W. 2008. Comparing Annual and Perennial Energy Cropping Systems with Different Management Intensities. Agricultural Systems 96:224–236.

BTAC. 2006. Vision for Bioenergy & Biobased Products in the United States. Biomass Technical Advisory Committee. http://www1.eere.energy.gov/biomass/pdfs/final_2006_vision.pdf

Bushuk W, Rasper VF. 1994. Wheat Production, Properties and Quality. New York, NY: Chapman & Hall.

Byrt CS, Grof CPL, Furbank RT. 2011. C4 Plants as Biofuel Feedstocks: Optimising Biomass Production and Feedstock Quality from a Lignocellulosic Perspective. Journal of Integrative Plant Biology 53(2):120–135.

Carvalho IS, Miranda I, Pereira H. 2006. Evaluation of Oil Composition of Some Crops Suitable for Human Nutrition. Industrial Crops and Products 24(1):75–78.

Castano G, Mas R, Fernandez JC, Fernandez L, Alvarez E, Lezcay M. 2000. Efficacy and Tolerability of Policosanol Compared with Lovastatin in Patients with Type II Hypercholesterolemia and Concomitant Coronary Risk Factors. Current Therapeutic Research 61(3):137–146.

Champagne P. 2008. Biomass. In: Champagne, P. (editor). Future Energy:Improved , Sustainable and Clean. Oxford: Elsevier Ltd. pp. 151–170.

Chen L, Remondetto G, Rouabhia M, Subirade M. 2008. Kinetics of the Breakdown of Cross-Linked Soy Protein Films for Drug Delivery. Biomaterials 29(27):3750–3756.

Chen S, Li H, Liu G. 2006. Progress of Vitamin E Metabolic Engineering in Plants. Transgenic Res. 15: 655–665.

Chen Y, Dunford NT, Edwards J, Carver B, Goad C. 2009a. Genotype and Environment Affect Phytosterol Content and Composition of Wheat Cereal Chem. 86(1):96–99.

Chen Y, Dunford NT, Edwards J, Carver B, Goad C. 2009b. Policosanol Content and Composition of Wheat Varieties as Affected by Environment. J. Sci. Food Agric. 89:310–314.

Chenault KD, Burns JA, Melouk HA, Payton ME. 2005. Field Reaction to Sclerotinia Blight among Transgenic Peanut Lines Containing Antifungal Genes. Crop Sci. 45(2):511–515.

Ciacci C, Maiuri L, Caporaso N, *et al.* 2007. Celiac Disease: In Vitro and in Vivo Safety and Palatability of Wheat-Free Sorghum Food Products. Clinical Nutrition 26(6):799–805.

Clark JH. 2007. Perspective Green Chemistry for the Second Generation Biorefinery – Sustainable Chemical Manufacturing Based on Biomass. J Chem Technol Biotechnol. 82:603–609.

Corredor DY, Salazar JM, Hohn KL, Bean S, Bean B, Wang D. 2009. Evaluation and Characterization of Forage Sorghum as Feedstock for Fermentable Sugar Production. Appl. Biochem. Biotechnol. 158:164–179.

Cureton TK. 1972. Influence of Wheat Germ Oil as a Dietary Supplement in a Program of Conditioning Exercises with Middle-Aged Subjects. In: Cureton, TK & Thomas, CC (editors). The Physiological Effects of Wheat Germ Oil on Human Exercise. NJ: Springfield.

Da Silva LS, Taylor JRN. 2005. Physical, Mechanical and Barrier Properties of Kafirin Films from Red and White Sorghum Milling Fractions. Cereal Chemistry 82:9–14.

Das G, Trivedi RK, Vasishtha AK. 1989. Heptaldehyde and Undecylenic Acid from Castor Oil. J. Am. Oil Chem. Soc. 66:938–941.

Davis KR, Litteneker N, LeTourneau D, Cain RF, Peters LJ, McGinnis J. 1980. Evaluation of the Nutrient Compisition of Wheat. I. Lipid Constituents. Cereal Chem. 57:178–184.

Dhugga KS. 2007. Maize Biomass Yield and Composition for Biofuels. Crop Sci. 47:2211–2227.

Diao Y, Walawender W, Fan L. 2002. Activated Carbons Prepared from Phosphoric Acid Activation of Grain Sorghum. Biosource Technology 81:45–52.

Dickerson GW. 2003. Specialty Corns. New Mexico State University, College of Agriculture and Home Economics, Cooperative Extension Service.

Dorado MP, Lina SKC, Koutinasa A, Dua C, Wang R, Webb C. 2009. Cereal-Based Biorefinery Development: Utilisation of Wheat Milling By-Products for the Production of Succinic Acid. J. Biotechnol. 143:51–59.

Drakakaki G, Marcel S, Glahn RP, Lund L, Periagh S, Fischer R. 2005. Endosperm Specific Co-Expression of Recombinant Soybean Ferritin and Aspergillus Phytase in Maize Results in Significant Increases in the Levels of Bioavailable Iron. Plant Mol Biol. 59:869–880.

Dunford NT. 2005. Germ Oils from Various Sources. In: Shahidi, F. (editor). Bailey's Industrial Oil and Fat Products. 6th edn. Hoboken, NJ: John Wiley & Sons, Inc. pp. 195–231.

Dunford NT. 2007 Power Beyond the Flour in a Wheat Biorefinery System. Wheat Quality Council Annual Meeting. Kansas City.

Dunford NT, Edwards J. 2010. Health Beneficial Bioactive Components of Wheat Straw as Affected by Genotype and Environment. Bioresource Technology 101:422–425.

Dunford NT, Zhang M. 2003. Pressurized Solvent Extraction of Wheat Germ Oil. Food Research International 36:905–909.

Duvick DN, Cassman KG. 1999. Post-Green Revolution Trends in Yield Potential of Temperate Maize in the North-Central United States. Crop Sci. 39:1622–1630.

Dykes L, Rooney LW. 2006. Sorghum and Millet Phenols and Antioxidants. Journal of Cereal Science 44(3):236–251.

Earp CF, McDonough CM, Awika J, Rooney LW. 2004. Testa Development in the Caryopsis of *Sorghum Bicolor* (L.) Moench. J. Cereal Sci. 39:303–311.

Eckert H, LaVallee B, Schweiger BJ, Kinney AJ, Cahoon EB, Clemente T. 2006. Co-Expression of the Borage Δ6 Desaturase and the Arabidopsis Δ15 Desaturase Results in High Accumulation of Stearidonic Acid in the Seeds of Transgenic Soybean Planta 224(5):1050–1057.

Eisenmenger M, Dunford NT. 2008. Bioactive Components of Commercial and Supercritical Carbon Dioxide Processed Wheat Germ Oil. J. Am. Oil Chem. Soc. 85:55–61.

Ezeji T, Blaschek HP. 2008. Fermentation of Dried Distillers' Grains and Solubles (DDGS) Hydrolysates to Solvents and Value-Added Products by Solventogenic Clostridia. Bioresource Technology 99(12): 5232–5242.

FAO. 1973. Energy and Protein Requirements FAO Nutritional Meeting Report Series No 52, WHO Technical Report Series No. 522. Rome, Italy.

FAO. 2010. Food and Agricultural Organization (FAO) Faostat Data Base. FAO.

FDA. 1999. Food and Drug Administration Final Rule. Food Labeling Health Claims: Soy Protein and Coronary Heart Disease. Fed. Regist. 64(206):57700–57733.

FDA. 2003a. Federal Register – 68 Fr 41433 July 11, 2003: Food Labeling; Trans Fatty Acids in Nutrition Labeling; Consumer Research to Consider Nutrient Content and Health Claims and Possible Footnote or Disclosure Statements; Final Rule and Proposed Rule. US Food and Drug Administration, Department of Health and Human Services. pp. 41433–41506.

FDA. 2003b. Qualified Health Claims: Letter of Enforcement Discretion – Nuts and Coronory Heart Disease. US Food and Drug Administration. http://www.cfsan.fda.gov/~dms/qhcnuts2.html (accessed 6 November 2010.

FDA. 2005. FDA Allows Barley Products to Claim Reduction in Risk of Coronary Heart Disease. http://www.fda.gov/NewsEvents/Newsroom/PressAnnouncements/2005/ucm108543.htm (accessed 3 October 2010).

Forward P. 1994. Beyond Ethanol: Industrial Uses of Agricultural Materials. A Background and Opportunities Paper. Ottawa, Canada: Food Bureau, Market and Industry Services Branch, Dept of Agriculture and Agri-Food.

Friedman M, Brandon DL. 2001. Nutritional and Health Benefits of Soy Proteins. J. Agric. Food Chem. 49(3):1069–1086.

Galili G, Höfgen R. 2002. Metabolic Engineering of Amino Acids and Storage Proteins in Plants. Metab Eng. 4:3–11.

Gavrilescua M, Chisti Y. 2005. Biotechnology – a Sustainable Alternative for Chemical Industry. Biotechnology Advances 23:471–499.

Gennadios A, McHugh TH, Weller CL, Krochta JM. 1994. Edible Films and Coatings Based on Proteins. In: Krochta, JM, Baldwin, E & Nisperos-Carriedo, MO (editors). Edible Films and Coatings to Improve Food Quality. Lancaster, PA: Technomic Publishing Co.

Ghamkhar K, Croser J, Aryamanesh N, Campbell M, Kon'kova N, Francis C. 2010. Camelina (*Camelina Sativa* (L.) Crantz) as an Alternative Oilseed: Molecular and Ecogeographic Analyses. Genome 53(7):558–567.

Gibreel A, Sandercock JR, Lan J, *et al*. 2009. Fermentation of Barley by Using Saccharomyces Cerevisiae: Examination of Barley as a Feedstock for Bioethanol Production and Value-Added Products. Appl. Environ. Microbiol. 75(5):1363–1372.

Giuliano G, Tavazza R, Diretto G, Beyer P, Taylor MA. 2008. Metabolic Engineering of Carotenoid Biosynthesis in Plants. Trends Biotechnol. 26:139–145.

Godon B. 1994. The Biochemical Composition of Cereals. In: Godon, B & Willm, C (editors). Primary Cereal Processing. New York, NY: VCH Publishers Inc. pp. 61–84.

Goodman DH. 1964. Safflower: Utilization and Significance in Nutrition and Allergy. J. Allergy 35(1):38–42.

Goodrum JW, Geller DP. 2004. Influence of Fatty Acid Methyl Esters from Hydroxylated Vegetable Oils on Diesel Fuel Lubricity. Bioresource Technology 96:851–855.

Gorbet DW, Knauft DA. 1997. Registration of "Sunoleic 95r" Peanut. Crop Sci. 37(July-August):1392.

Gouni-Berthold I, Berthold HK. 2002. Policosanol: Clinical Pharmacology and Therapeutic Significance of a New Lipid-Lowering Agent. American Heart Journal 143(2):356–365.

Griffey C, Brooks W, Kurantz M, *et al*. 2009. Grain Composition of Virginia Winter Barley and Implications for Use in Feed, Food, and Biofuels Production. Journal of Cereal Science 51(1):41–49.

Griffiths DW, Welch RW. 1982. Genotypic and Environmental Variation in the Total Phenol and Flavanol Contents of Barley Grain. J. Sci. Food Agric. 33:521–527.

Grompone MA. 2005. Sunflower Oil. In: Shahidi, F (editor). Bailey's Industrial Oil and Fat Products. Hoboken, NJ: John Wiley & Sons, Inc. pp. 655–730.

Guerrero P, Stefani PM, Ruseckaite RA, de la Caba K. 2011. Functional Properties of Films Based on Soy Protein Isolate and Gelatin Processed by Compression Molding. Journal of Food Engineering 105(1):65–72.

Hahn DH, Rooney LW, Faubion JM. 1983. Sorghum Phenolic Acids, Their Hplc Separation and Their Relation to Fungal Resistance. Cereal Chemistry 60:255–259.

Hamarneh AI, Heeres HJ, Broekhuis AA, Sjollema KA, Zhang Y, Picchioni F. 2010. Use of Soy Proteins in Polyketone-Based Wood Adhesives. International Journal of Adhesion and Adhesives 30(7):626–635.

Hammond EG, Glatz BA. 1989. Biotechnology Applied to Fats and Oil. In: King, R & Cheethan, PSJ (editors). Developments in Food Technology. New York, NY: John Wiley & Sons, Inc. pp. 173–217.

Hammond EG, Johnson LA, Su C, Wang T, White PJ. 2005. Soybean Oil. In: Shahidi, F (editor). Bailey's Industrial Oil and Fat Products. Hoboken, NJ: John Wiley & Sons, Inc.

Hanania M, Duarte R, Livingstone D, Droluk T, Lewis T, Kitain E. 2004. Folk Medicine/Herbal Remedies: Topical Safflower Plant Extract for Chronic Pain: An Open-Label Prospective Study. The Journal of Pain 5(3, Supplement 1):S52–S52.

Harper SH, Lynch JM. 1981. The Chemical Components and Decomposition of Wheat Straw Leaves, Internodes and Nodes. J. Sci. Food Agric. 32:1057–1062.

HarvestPlus. http://www.harvestplus.org/ (accessed 1 March 2011).

Haumann BF. 1998. Peanuts Finds Niche in Healthy Diet. Inform. 9(8):746–752.

Henderson J, Osborne DJ. 2000. The Oil Palm in All Our Lives: How This Came About. Endeavour 24(2):63–68.

Herman EM, Helm RM, Jung R, Kinney AJ. 2003. Genetic Modification Removes an Immunodominant Allergen from Soybean. Plant Physiology and Biochemistry 132(1):36–43.

Higgs J. 2002. The Beneficial Role of Peanuts in the Diet – an Update and Rethink! Peanuts and Their Role in Chd. Nutr. Food Sci. 32:214–218.

Hirai K, Ohno Y, Nakano T, Izutant K. 1984. Effects of Dietary Fats and Phytosterol on Serum Fatty Acid Composition and Lipoprotein Cholesterol in Rats. J. Nutr. Sci. Vitaminol. 30:101–112.

Honary LAT. 1996. An Investigation of the Use of Soybean Oil in Hydraulic Systems. Bioresource Technology 56(1):41–47.

Horvath H, Huang J, Wong O, *et al.* 2000. The Production of Recombinant Proteins in Transgenic Barley Grains. Proceedings of the National Academy of Sciences USA. 97:1914–1919.

Hüsken A, Baumert A, Strack D, Becker HC, Möllers C, Milkowski C. 2005. Reduction of Sinapate Ester Content in Transgenic Oilseed Rape (Brassica Napus L.) by Dsrnai-Based Suppression of Bnsgt1 Gene Expression. Mol. Breed. 16:127–138.

Huang X, Netravali A. 2007. Characterization of Flax Fiber Reinforced Soy Protein Resin Based Green Composites Modified with Nano-Clay Particles. Composites Science and Technology 67(10):2005–2014.

Husic S, Javni I, Petrovic ZS. 2005. Thermal and Mechanical Properties of Glass Reinforced Soy-Based Polyurethane Composites. Composites Science and Technology 65(1):19–25.

Ingledew WM, Jones AM, Bhatty RS, Rossnagel BG. 1995. Fuel Alcohol Production from Hull-Less Barley. Cereal Chemistry 72(2):147–150.

Irmak S, Dunford NT. 2005. Policosanol Contents and Compositions of Wheat Varieties. J. Agric. Food Chem. 53(14):5583–5586.

Irmak S, Dunford NT, Milligan J. 2005. Policosanol Contents of Beeswax, Sugar Cane and Wheat Extracts. Food Chemistry 95:312–318.

Ishikawa T, Dowdle J, Smirnoff N. 2006. Progress in Manipulating Ascorbic Acid Biosynthesis and Accumulation in Plants. Physiol Plant. 126:343–355.

Izydorczyk MS, Dexter JE. 2008. Barley [Beta]-Glucans and Arabinoxylans: Molecular Structure, Physicochemical Properties, and Uses in Food Products – a Review. Food Research International 41(9):850–868.

Jähne A, Becker D, Brettschneider R, Lörz H. 1994. Regeneration of Transgenic, Microspore-Derived, Fertile Barley. Theoretical and Applied Genetics 89:525–533.

Jain D, Ebine N, Jia X, *et al.* 2008. Corn Fiber Oil and Sitostanol Decrease Cholesterol Absorption Independently of Intestinal Sterol Transporters in Hamsters. The Journal of Nutritional Biochemistry 19(4):229–236.

Jende-Strid B. 1993. Genetic Control of Flavonoid Biosynthesis in Barley. Hereditas 119:187–204.

Jensen LG, Olsen O, Kops O, Wolf N, Thomsen K, von Wettstein D. 1996. Transgenic Barley Expressing a Protein-Engineered, Thermostable (1,3–1,4)-B-Glucanase During Germination. Proceedings of the National Academy of Sciences USA. 93:3487–3491.

Jong L. 2006. Effect of Soy Protein Concentrate in Elastomer Composites. Composites Part A: Applied Science and Manufacturing 37(3):438–446.

Jonnala RS. 2004. Nutritional Composition of Peanuts (*Arachis Hypogaea* L.) as Affected by Conventional Breeding and Genetic Modifications. Food Science. Stillwater, OK: Oklahoma State University.

Jonnala RS, Dunford NT, Chenault K. 2005a. Nutritional Composition of Genetically Modified Peanut Varieties. J. Food Sci. 70(4):S254–256.

Jonnala RS, Dunford NT, Dashiell KE. 2005b. New High Oleic Peanut Cultivars Grown in the Southwestern U.S. J. Am. Oil. Chem. Soc. 85:125–128.

Jonnala RS, Dunford NT, Chenault K. 2006. Tocopherol, Phytosterol and Phospholipid Compositions of Genetically Modified Peanut Varieties. J. Sci. Food Agric. 86:473–476.

Kahlon TS. 1989. Nutritional Implications and Uses of Wheat and Oat Kernel Oil. Cereal Foods World 34:872–875.

Karp A, Shield I. 2008. Bioenergy from Plants and the Sustainable Yield Challenge. New Phytologist 179:15–32.

Kent NL, Evers AD. 1994. Technology of Cereals. Kidlington, UK: Elsevier Science.

Koning NBJ, Van Ittersum MK, Becx GA, *et al.* 2008. Long-Term Global Availability of Food: Continued Abundance or New Scarcity? Wageningen Journal of Life Sciences, Journal of the Royal Netherlands Society for Agricultural Sciences (KLV) 55(3):229–292.

Koutinas AA, Wang R, Webb C. 2004. Evaluation of Wheat as Generic Feedstock for Chemical Production. Industrial Crops and Products 20:75–88.

Kris-Etherton PM, Zhao G, Binkoski AE, Coval SM, Etherton TD. 2001. The Effects of Nuts on Coronary Heart Disease Risk. . Nutr. Review 59:103–111.

Krueger CG, Vestling MA, Reed JD. 2003. Matrix-Assisted Laser Desorption/Ionization Time-of-Flight Mass Spectrometry of Heteropolyflavan-3-Ols and Glucosylated Heteropolyflavans in Sorghum (Sorghum Bicolor (L.) Moench). J. Agric. Food Chem. 51:538–543.

Kudou S, Fleury Y, Welti D, *et al.* 1991. Malonyl Isoflavone Glycosides in Soybean Seeds (Glycine Max Merrill). Agric. Biol. Chem. Eng. Commun. 55:2227–2233.

Kula J, Sikora M, Dabrowski R. 1994. Further Study on the Onepot Synthesis of (E)-2-Nonenal from Castor Oil. J. Am. Oil Chem. Soc. 71:545–546.

L'Hocine L, Boye JI. 2007. Allergenicity of Soybean: New Developments in Identification of Allergenic Proteins, Cross-Reactivities and Hypoallergenization Technologies. Critical Reviews in Food Science and Nutrition 47:127–143.

Labrecque M, Teodorescu TI. 2005. Field Performance and Biomass Production of 12 Willow and Poplar Clones in Short-Rotation Coppice in Southern Quebec (Canada). Biomass Bioener. 29:1–9.

Lathi PS, Mattiasson B. 2007. Green Approach for the Preparation of Biodegradable Lubricant Base Stock from Epoxidized Vegetable Oil. Applied Catalysis B: Environmental 69(3–4):207–212.

Lazzeri L, Leoni O, Conte LS, Palmieri S. 1994. Some Technological Characteristics and Potential Uses of Crambe Abyssinica Products. Industrial Crops and Products 3(1–2):103–112.

Leathers TD. 2003. Bioconversions of Maize Residues to Value-Added Coproducts Using Yeast-Like Fungi. FEMS Yeast Research 3:133–140.

Leckband G, Lörz H. 1998. Transformation and Expression of a Stilbene Synthetase Gene of Vitis Vinifera L. In Barley and Wheat for Increased Fungal Resistance. Theoretical and Applied Genetics 96:1004–1012.

Leonard EC. 1992. High-Erucic Vegetable Oils. Industrial Crops and Products 1(2–4):119–123.

Li J, Berke TG, Glover DV. 1994. Variation for Thermal Starch Properties of Starch in Tropical Maize Germplasm. Cereal Chem 71:87–90.

Liu K. 1997. Soybeans: Chemisty, Technology, and Utilization. New York, NY: Chapman & Hall.

Lyon CK, Garret VH. 1973. New Castor Oil-Based Urethane Elastomers. J. Am. Oil Chem. Soc. 50:112–114.

Madrigal-Carballo S, Lim S, Rodriguez G, *et al.* 2010. Biopolymer Coating of Soybean Lecithin Liposomes Via Layer-by-Layer Self-Assembly as Novel Delivery System for Ellagic Acid. Journal of Functional Foods 2(2):99–106.

Mäkinen K, Nuutila AM. 2004. Barley Seed as a Production Host for Industrially Important Proteins. AgBiotechNet 6(119):1N–8N.

Marwede V, Schierholt A, Möllers C, Becker HC. 2004. Genotype X Environment Interactions and Heritability of Tocopherols Content in Canola. Crop Sci. 44:728–731.

McGarth PF, Vincent JR, Lei C-H, *et al.* 1997. Coat Protein-Mediated Resistance to Isolates of Barley Yellow Dwarf in Oats and Barley. European Journal of Plant Pathology 103:695–710.

McGuiness EE, Morgan RG, Wormsley KG. 1987. Fate of pancreatic nodules induced by raw soya flour in rats. Gut 28:207–212.

Mckean WT, Jacobs RS. 1997. Wheat Straw as a Paper Fiber Source. Prepared for Recycling Technology Assistance Partnership (ReTAP) by The Clean Washington Center and Domtar Inc. http://www.cwc.org/paper/pa971rpt.pdf (accessed 1 August 2010).

Menendez R, Fernandez SI, Del Rio A, *et al.* 1994. Policosanol Inhibits Cholesterol Biosynthesis and Enhances Ldl Processing in Cultured Human Fibroblasts. Biol Res. 27:199–203.

Menendez R, Fraga V, Amor AM, Gonzales RM, Mas R. 1999. Oral Administration of Policosanol Inhibits in Vitro Copper Ion-Induced Rat Lipoprotein Peroxidation. Physiol. Behav. 67:1–7.

Messina M, Messina V. 2000. Soyfoods, Soybean Isoflavones, and Bone Health: A Brief Overview. J. Ren. Nutr. 10:63–68.

Metzger JO, Eissen M. 2004. Concepts on the Contribution of Chemistry to a Sustainable Development. Renewable Raw Materials. C. R. Chimie 7:569–581.

Monk RL, Miller FR, McBee GG. 1984. Sorghum Improvement for Energy Production. Biomass 6(1–2):145–153.

Montane D, Farriol X, Salvadó J, Jollez P, Chornet E. 1998. Application of Steam Explosion to the Fractionation and Rapid Vapor-Phase Alkaline Pulping of Wheat Straw. Biomass and Bioenergy 14(3):261–276.

Moreira MA, Tavares SR, Ramos V, de Barros EG. 1993. Hexanal Production and Tba Number Are Reduced in Soybean [*Glycine Max* (L.) Merr.] Seeds Lacking Lipoxygenase Isozymes 2 and 3. J. Agric. Food Chem. 47:103–106, 109.

Mounts TL, Abidi SL, Rennick KA. 1996. Effect of Genetic Modification on the Content and Composition of Bioactive Constituents in Soybean Oil. J. Am. Oil Chem. Soc. 73(5):581–586.

Munck L. 1995. New Milling Technologies and Products: Whole Plant Utilization by Milling and Separation of the Botanical and Chemical Components. In: Dendy, DAV (editor). Sorghum and Millets: Chemistry and Technology. St Paul, MN: American Association of Cereal Chemists. pp. 223–281.

Murphy DJ. 1996. Engineering Oil Production in Rapeseed and Other Oil Crops. Trends in Biotechnology 14(6):206–213.

Murray SC, Rooney WL, Mitchell SE, *et al.* 2008a. Genetic Improvement of Sorghum as a Biofuel Feedstock: II. Qtl for Stem and Leaf Structural Carbohydrates. Crop Sci. 48:2180–2193.

Murray SC, Sharma A, Rooney WL, *et al.* 2008b. Genetic Improvement of Sorghum as a Biofuel Feedstock: I. Qtl for Stem Sugar and Grain Nonstructural Carbohydrates. Crop Sci. 48:2165–2179.

Muuse BG, Petrus Cuperus F, Derksen JTP. 1992. Composition and Physical Properties of Oils from New Oilseed Crops. Industrial Crops and Products 1(1):57–65.

Napier JA, Haslam R, Caleron MV, Michaelson LV, Beaudoin F, Sayanova O. 2006. Progress Towards the Production of Very Long-Chain Polyunsaturated Fatty Acid in Transgenic Plants: Plant Metabolic Engineering Comes of Aeg. Physiol Plant. 126:398–406.

Naqvi S, Ramessar K, Farré G, *et al.* 2011. High-Value Products from Transgenic Maize. Biotechnology Advances 29:40–53.

NCGA. 2010. World of Corn. National Corn Growers Association. http://www.ncga.com (accessed 6 January 2011).

Newman RK, Newman CW. 2008. Barley for Food and Health: Science, Technology and Products. Hoboken, NJ: John Wiley & Sons, Inc.

Norden AJ, Lipscomb RW, Carver WA. 1969. Registration of Florunner Peanuts (Reg. No. 2). Crop Sci. 9:850.

O'Brien RD, Jones LA, King CC, Wakelyn PJ, Wan JP. 2005. Cottonseed Oil. In: Shahidi, F (editor). Bailey's Industrial Oil & Fat Products. Hoboken, NJ: John Wiley & Sons, Inc. pp. 173–279.

O'Kennedy MM, Grootboom A, Shewry PR. 2006. Harnessing Sorghum and Millet Biotechnology for Food and Health. Journal of Cereal Science 44(3):224–235.

Ogale AA, Cunningham P, Dawson PL, Acton JC. 2000. Viscoelastic, Thermal, and Microstructural Characterization of Soy Protein Isolate Films. J. Food Sci. 65(4):672–679.

Ogunniyi DS. 2006. Castor Oil: A Vital Industrial Raw Material. Bioresource Technology 97(9): 1086–1091.

Orth RA, Shellenberger JA. 1988. Origin, Production, and Utilization of Wheat. In: Pomeranz, Y (editor). Wheat: Chemistry and Technology. St Paul, MN: American Association of Cereal Chemists. pp. 1–45.

Oscarsson M, Andersson R, Salomonsson AC, Åman P. 1996. Chemical Composition of Barley Samples Focusing on Dietary Fibre Components. J Cereal Sci. 24(2):161–170.

Ostlund RE. 2002. Phytosterols in Human Nutrition. Annual Review of Nutrition 22(1):533–549.

Owuama CI. 1999. Brewing Beer with Sorghum. Journal of the Institute of Brewing 105:23–34.

Paetau I, Chen CZ, Jane J. 1994. Biodegradable Plastics Made from Soybean Products. I. Effects of Crosslinking and Cellulose Incorporation on Mechanical Properties and Water Absorption. J. Environ. Polym. Degrad. 2:211–217.

Paterson AH, Bowers JE, Bruggmann R, *et al.* 2009. The Sorghum Bicolor Genome and the Diversification of Grasses. Nature 457:551–556.

Paulis JW, Bietz JA, Lambert RJ, Villegas EM. 1991. Identification of Modified High-Lysine Maize Genotypes by Reversed-Phase High-Performance Liquid Chromatography. Cereal Chem. 68:361–365.

Piironen V, Koivu T, Tammisalo O, Mattila P. 1997. Determination of Phylloquinone in Oils, Margarines and Butter by High-Performance Liquid Chromatography with Electrochemical Detection. Food Chemistry 59(3):473–480.

Pilgeram AL, Sand DC, Boss D, *et al.* 2007. Camelina Sativa, a Montana Omega-3 Fatty Acid and Fuel Crop. In: Janick, J & Whipkey A (editors). Issues in New Crops and New Uses. Alexandria, VA: ASHS Press. pp. 129–131.

Pomeranz Y. 1988. Wheat: Chemistry and Technology. St Paul, MN: American Association of Cereal Chemists Inc.

Pourkheirandish M, Komatsuda T. 2007. The Importance of Barley Genetics and Domestication in a Global Perspective. Annals of Botany 100:999–1008.

Prasittisopin L, Li K. 2010. A New Method of Making Particleboard with a Formaldehyde-Free Soy-Based Adhesive. Composites Part A: Applied Science and Manufacturing 41(10):1447–1453.

Price ML, Van Scoyoc S, Butler LG. 1978. A Critical Evaluation of Vanillin Reaction as an Assay for Tannin in Sorghum. J. Agric. Food Chem. 26:1214–1218.

Przybylski R, Mag TK, Eskin NAM, McDonald BE. 2005. Canola Oil. In: Shahidi, F (editor). Bailey's Industrial Oil and Fat Products. Hoboken, NJ: John Wiley & Sons, Inc. pp. 61–121.

Putt ED. 1997. Early History of Sunflower. In: Scheiter, AA (editor). Sunflower Technology and Production. Madison, WI: ASA, CSSA and ASSA. pp. 1–19.

Qu H, Madl RL, Takemoto DJ, Baybutt RC, Wang W. 2005. Lignans Are Involved in the Antitumor Activity of Wheat Bran in Colon Cancer Sw480 Cells1,2. Journal of Nutrition 135(3):598.

Quilez J, Garcia-Lorda P, Salas-Salvado J. 2003. Potential Uses and Benefits of Phytosterols in Diet: Present Situation and Future Directions. Clinical Nutrition 22(4):343–351.

Rahman MH. 2001. Production of Yellow-Seeded Brassica Napus through Interspecific Crosses. Plant Breed. 120(463–472).

Ramessar K, Capell T,Christou P. 2008. Molecular Pharming in Cereal Crops. Phytochem. Rev. 7:579–592.

Rayburn JR, Friedman M, Bantle JA. 1995. Synergistic Interaction of Glycoalkaloids R-Chaconine and R-Solanine on Developmental Toxicity in Xenopus Frog Embryos. Food Chem. Toxicol. 33:1013–1019.

Reddy BVS, Ramesh S, Longvah T. 2005. Prospects of Breeding for Micronutrients and B-Carotene-Dense Sorghums. SAT eJournal 1(1):1–4.

Reddy N, Yang Y. 2009. Natural Cellulose Fibers from Soybean Straw. Bioresource Technology 100(14):3593–3598.

Regina A, Bird A, D. T, Bowden S, *et al.* 2006. High-Amylose Wheat Generated by Rna Interference Improves Indices of Large-Bowel Health in Rats. Proceedings of the National Academy of Sciences USA 103(10):3546–3551.

Ren C, Bilyeu KD, Beuselinck PR. 2009. Composition, Vigor, and Proteome of Mature Soybean Seeds Developed under High Temperature. Crop Sci. 49:1010–1022.

Reynolds TL, Nemeth MA, Glenn KC, Ridley WP, Astwood JD. 2005. Natural Variability of Metabolites in Maize Grain: Differences Due to Genetic Background. J. Agric. Food Chem. 53:10061–10067.

Rhim J-W, Lee JH, Ng PKW. 2007. Mechanical and Barrier Properties of Biodegradable Soy Protein Isolate-Based Films Coated with Polylactic Acid. Food Science and Technology 40(2):232–238.

Rockwood DL, Rudie AW, Ralph SA, Zhu JY, Winandy JE. 2008. Energy Product Options for Eucalyptus Species Grown as Short Rotation Woody Crops. Int. J. Mol. Sci. 9:1361–1378.

Rudi H, Uhlen AK, Harstad OM, Munck L. 2006. Genetic Variability in Cereal Carbohydrate Compositions and Potentials for Improving Nutritional Value. Animal Feed Science and Technology 130(1–2):55–65.

Saha BC, Iten LB, Cotta MA, Wu YV. 2005. Dilute Acid Pretreatment, Enzymatic Saccharification and Fermentation of Wheat Straw to Ethanol. Process Biochemistry 40:3693–3700.

Saito K, Takahashi Y, Wada M. 1983. Enzymic Synthesis of Carthamin in Safflower. Biochimica et Biophysica Acta – General Subjects 756(2):217–222.

Sanderson MA, Reed RL, McLaughlin SB, *et al.* 1996. Switchgrass as a Sustainable Bioenergy Crop. Bioresource Technology 56(1):83–93.

Sanderson MA, Adler PR, Martin NP. 2007. Biomass, Energy, and Industrial Uses of Forages. In: Barnes, RF, Nelson, CJ, Moore, KJ & Collins, M (editors). Forages. 6th ed. Ames, IA: Iowa State University Press. pp. 635–647.

Sapers GM. 1993. Browning of Foods: Control by Sulfites, Antioxidants, and Other Means. Food Technol. 47:75–84.

Schmer MR, Vogel KP, Mitchell RB, Perrin RK. 2008. Net Energy of Cellulosic Ethanol from Switchgrass. Proceeding National Academy of Sciences USA 105:464–469.

Schmidt V, Giacomelli C, Soldi V. 2005. Thermal Stability of Films Formed by Soy Protein Isolate-Sodium Dodecyl Sulfate. Polymer Degradation and Stability 87(1):25–31.

Schwartz H, Ollilainen V, Piironen V, Lampi A-M. 2008. Tocopherol, Tocotrienol and Plant Sterol Contents of Vegetable Oils and Industrial Fats. Journal of Food Composition and Analysis 21(2):152–161.

Septiano W, Dunford NT, Wilkins W, Edwards J. 2010. Ethanol Production from Winter Hulless Barley. Biological Engineering 2(4):211–219.

Serna-Saldivar S, Rooney LW. 1995. Structure and Chemistry of Sorghum and Millets. In: Dendy, DAV (editor). Structure and Chemistry of Sorghum and Millets. St Paul, MN: American Association of Cereal Chemists. pp. 69–124.

Shadle G, Chen F, Reddy MSS, Jackson L, Nakashima J, Dixon RA. 2007. Down-Regulation of Hycroxycinnamoyl Coa: Shikimate Hycroxycinnamoyl Transferase in Transgenic Alfalfa Affects Lignification, Development and Forage Quality. Phytochemistry 68:1521–1529.

Shewmaker CK, Sheehy JA, Daley M, Colburn S, Ke DY. 1999. Seed-Specific Overexpression of Phytoene Synthase: Increase in Carotenoids and Other Metabolic Effects. Plant J. 20:401–412.

Shi J, Wang H, Schellin K, Li B, *et al.* 2007. Embryo-Specific Silencing of a Transporter Reduces Phytic Acid Content of Maize and Soybean Seeds. Nature Biotechnology 25(8):930–937.

Shinoj S, Visvanathan R, Panigrahi S, Kochubabu M. 2011. Oil Palm Fiber (Opf) and Its Composites: A Review. Industrial Crops and Products 33(1):7–22.

Shukla R, Cheryan M. 2001. Zein: The Industrial Protein from Corn. Industrial Crops and Products 13:171–192.

Simmons BA, Loque D, Blanch HW. 2008. Next-Generation Biomass Feedstocks for Biofuel Production. Genome Biology 9:242.

Simmons BA, Loqu´e D, Ralph J. 2010. Advances in Modifying Lignin for Enhanced Biofuel Production. Curr. Opin. Plant Bio. 13:1–8.

Singh V, Moreau RA, Hicks KB. 2003. Yield and Phytosterol Composition of Oil Extracted from Grain Sorghum and Its Wet-Milled Fractions. Cereal Chemistry 80:126–129.

Situ Y, Hu J, Huang H, Fu H, Zeng H, Chen H. 2007. Synthesis, Properties and Application of a Novel Epoxidized Soybean Oil-Toughened Phenolic Resin. Chinese Journal of Chemical Engineering 15(3):418–423.

Smith J. 2005. Safflower Oil. In: Shahidi, F (editor). Bailey's Industrial Oil & Fat Products. Hoboken, NJ: John Wiley & Sons, Inc. pp. 491–536.

Sohn M, Himmelsbach DS, Barton FE, Griffey CA, Brooks W, Hicks KB. 2007. Near-Infrared Analysis of Ground Barley for Use as a Feedstock for Fuel Ethanol Production. Applied Spectroscopy 61:1178–1183.

Song Y, Jane J. 2000. Characterization of Barley Starches of Waxy, Normal, and High Amylose Varieties. Carbohydrate Polymers 41(4):365–377.

Spiertz JHJ, Ewert F. 2009. Crop Production and Resource Use to Meet the Growing Demand for Food, Feed and Fuel:Opportunities and Constraints. NJSA 56(4):281–300.

Stuchell YM, Krochta JM. 1994. Enzymatic Treatments and Thermal Effects on Edible Soy Protein Films. J. Food Science 59(6):1332–1337.

Sue HJ, Wang S, Jane JL. 1997. Morphology and Mechanical Behaviour of Engineering Soy Plastics. Polymer 38(20):5035–5040.

Sun RC, Salisbury D, Tomkinson J. 2003. Chemical Composition of Lipophilic Extractives Released During the Hot Water Treatment of Wheat Straw. Bioresource Technology 88(2):95–101.

Sun RC, Sun XF. 2001. Identification and Quantitation of Lipophilic Extractives from Wheat Straw. Industrial Crops and Products 14:51–64.

Sun RC, Tomkinson J. 2003. Comparative Study of Organic Solvent and Water-Soluble Lipohilic Extractives from Wheat Straw I: Yield and Chemical Composition. J. Wood Sci. 49:47–52.

Sunilkumar G, Campbell LM, Puckhaber L, Stipanovic RD, Rathore KS. 2006. Engineering Cottonseed for Use in Human Nutrition by Tissue-Specific Reduction of Toxic Gossypol. Proceedings of the National Academy of Sciences USA 103 (48):18054–18059.

Swain SN, Biswal SM, Nanda PK, Nayak PL. 2004. Biodegradable Soy-Based Plastics: Opportunities and Challenges. Journal of Polymers and the Environment 12(1):35–42.

Szczodrak J. 1988. The Enzymatic Hydrolysis and Fermentation of Pretreated Wheat Straw to Ethanol. Biotechnol Bioeng. 32:771–776.

Taylor JC, PRapport L, Lockwood GB. 2003. Octacosanol in Human Health. Nutrition 19:192–195.

Taylor JRN, Dewar J. 2001. Developments in Sorghum Food Technologies. Advances in Food and Nutrition Research. Academic Press. pp. 217–264.

Taylor JRN, Schober TJ, Bean SR. 2006. Novel Food and Non-Food Uses for Sorghum and Millets. Journal of Cereal Science 44(3):252–271.
Tikkanen MJ, Adlecreutz H. 2000. Dietary Soy-Derived Isoflavone Phytoestrogens. Could They Have a Role in Dietary Heart Disease Prevention? Biochem. Pharmacol. 60:1–5.
Troll W, Frankel K, Wiesner R. 1986. Protease Inhibitors: Their Role as Modifiers of Carcinogenic Processes. Adv. Exp. Med. Biol. 199:153–165.
Uauy C, Distelfeld A, Fahima T, Blechl A, Dubcovsky J. 2006. A Nac Gene Regulating Senescence Improves Grain Protein, Zinc, and Iron Content in Wheat. Science 314:1298–1301.
Ullrich SE, Eslick RF. 1978. Lysine and Protein Characterization of Spontaneous Shrunken Endosperm Mutants of Barley. Crop Sci. 18:809–812.
USDA. 2006. United States Standards for Wheat. http://archive.gipsa.usda.gov/reference-library/standards/810wheat.pdf (accessed 9 August 2010).
USDA. 2011. Oilseeds: World Markets and Trade. United States Department of Agriculture, Foreign Agricultural Service.
USDA-FAS. 2008. World Corn Production, Consumption and Stocks Report. Washington, DC: USDA.
Van Eenennaam AL, Lincoln K, Durrett TP, *et al.* 2003. Engineering Vitamin E Content: From Arabidopsis Mutant to Soy Oil. The Plant Cell 15:3007–3019.
Vandenbrink JP, Delgado MP, Frederick JR, Feltus FA. 2010. A Sorghum Diversity Panel Biofuel Feedstock Screen for Genotypes with High Hydrolysis Yield Potential. Industrial Crops and Products 31(3):444–448.
Varady KA, Wang Y, Jones PJH. 2003. Role of Policosanols in the Prevention and Treatment of Cardovascular Disease. Nutr. Rev. 61(11):376–383.
Verbruggen MA, Beldman G, Voragen AGJ. 1995. The Selective Extraction of Glucuronoarbinoxylans from Sorghum Endosperm Cell Walls Using Barium and Potassium Hydroxide Solutions. J. Cereal Sci. 21:271–282.
Villamil MB, Silvis AH, Bollero GA. 2008. Potential Miscanthus' Adoption in Illinois: Information Needs and Preferred Information Channels. Biomass and Bioenergy 32(12):1338–1348.
Vollmann J, Moritz T, Kargl C, Baumgartner S, Wagentristl H. 2007. Agronomic Evaluation of Camelina Genotypes Selected for Seed Quality Characteristics. Industrial Crops and Products 26(3):270–277.
Wang C, Sherrard M, Pagadala S, Wixon R, Scott RA. 2000a. Isoflavone Content among Maturity Group 0 to II Soybeans. J. Am. Oil Chem. Soc. 77:483–487.
Wang YP, Tang JS, Chu CQ, Tian J. 2000b. A Preliminary Study on the Introduction and Cultivation of Crambe Abyssinica in China, an Oil Plant for Industrial Uses. Industrial Crops and Products 12(1):47–52.
Wang H, Jiang J, Fu L. 2007. Properties of Molded Soy Protein Isolate Plastics. Journal of Applied Polymer Science 106(6):3716–3720.
Wang W, Guo Y, Otaigbe JU. 2010. Synthesis, Characterization and Degradation of Biodegradable Thermoplastic Elastomers from Poly(Ester Urethane)S and Renewable Soy Protein Isolate Biopolymer. Polymer 51(23):5448–5455.
Waniska RD, Poe JH, Bandyopadhyay R. 1989. Effects of Growth Conditions on Grain Molding and Phenols in Sorghum Caryopsis. Journal of Cereal Science 10:217–225.
Washburn S, Burke GL, Morgan T, Anthony M. 1999. Effect of Soy Protein Supplementation on Serum Lipoproteins, Blood Pressure, and Menaupausal Symptoms in Perimenopausal Women. Menopause 6:7–13.
Webb C, Wang R. 1997. Development of a Generic Fermentation Feedstock from Whole Wheat Flour. In: Campbell, GM, Webb, C & McKee, SL (editors). Cereals: Novel Uses and Processes. New York, NY: Plenum Press. pp. 205.
Weissermel K, Arpe HJ. 1997. Industrial Organic Chemistry, 3rd edn. Weinheim: VCH.
Weller CL, Gennadios A, Saraiva RL, Cuppett SL. 1998. Grain Sorghum Wax as an Edible Coating for Gelatine-Based Candies. Journal of Food Quality 21:117–128.
Wool RP, Khot SN, LaScala JJ, *et al.* 2000. All Natural Composites for the Transportation Industry. Polym. Mater. Sci. Eng. 83:12.
Wright L, Perlack R, Graham R, Stokes B, Turhollow A, Erbach D. 2005. Biomass as Feedstock for a Bioenergy and Bioproducts Industry: The Technical Feasibility of a Billion-Ton Annual Supply. Oak Ridge National Laboratory, USDOE & USDA.
Yadav MP, Johnston DB, Hotchkiss JAT, Hicks KB. 2007. Corn Fiber Gum: A Potential Gum Arabic Replacer for Beverage Flavor Emulsification. Food Hydrocolloids 21(7):1022–1030.
Yang P, Seib PA. 1995. Low-Input Wet-Milling of Grain Sorghum for Readily Accessible Starch and Animal Feed. Cereal Chem. 72(5):498–503.
Yoshie-Stark Y, Wäsche A. 2004. In Vitro Binding of Bile Acids by Lupin Protein Isolates and Their Hydrolysates. Food Chemistry 88(2):179–184.

You S, Izydorczyk MS. 2002. Molecular Characteristics of Barley Starches with Variable Amylose Content. Carbohydrate Polymers 49(1):33–42.

Zhan X, Wang D, Tuinstra MR, Bean S, Seib PA, Sun XS. 2003. Ethanol and Lactic Acid Production as Affected by Sorghum Genotype and Location. Industrial Crops and Products 18(3):245–255.

Zhao J, Becker HC, Zhang D, Zhang Y, Ecke W. 2006. Conditional Qtl Mapping of Oil Content in Rapeseed with Respect to Protein Content and Traits Related to Plant Development and Grain Yield. Theor Appl Genet. 113(33–38).

Zhou K, Laux JJ, Yu L. 2004. Comparison of Swiss Red Wheat Grain and Fractions for Their Antioxidant Properties. J. Ag. Food Chem. 52:1118–1123.

Zhu C, Naqvi S, Breitenbach J, Sandmann G, Christou P, Capell T. 2008. Combinatorial Genetic Transformation Generates a Library of Metabolicpphenotypes for the Carotenoid Pathway in Maize. Proceedings of the National Academy of Sciences USA 105:18232–18237.

Zum Felde T, Becker HC, Möllers C. 2006. Genotype X Environment Interactions, Heritability, and Trait Correlations of Sinapate Ester Content in Winter Rapeseed (*Brassica Napus* L). Crop Sci. 46:2195–2199.

2 Recent Processing Methods for Preparing Starch-based Bioproducts

George F. Fanta, Frederick C. Felker
and Randal L. Shogren

2.1 INTRODUCTION

The physical and chemical properties of starch and the commercial applications of starch and starch derivatives have been extensively reviewed (Jarowenko, 1970; Whistler and Daniel, 1997). Starch is a polysaccharide composed of repeating 1,4-α-D-glucopyranosyl units, often referred to as anhydroglucose units or simply AGU (Figure 2.1); it is isolated from the seeds and tubers of living plants as discrete granules with diameters ranging from about 1 to 100 μm, depending on the plant source. Corn starch is the most commonly used starch of commerce and granules of corn starch have diameters in the order of 10–20 μm. Starch granules typically contain both linear polysaccharides (amylose) and highly-branched polysaccharides with 1,6-α-D linkages at the branch points (amylopectin). The amylose:amylopectin ratio of normal dent corn starch is about 25:75; however, high amylose starch varieties with apparent amylose contents of about 70% are also commercially available, as well as the waxy variety that contains essentially 100% amylopectin. The molecular weight of amylose is in the order of several hundred thousand, whereas amylopectin is over an order of magnitude higher in molecular weight.

Although starch granules are hydrophilic and absorb water readily, they are insoluble in water at room temperature. Granule swelling is minimal due to hydrogen bonding and crystallinity within the granule. Native starch granules are birefringent and show a Maltese cross pattern when viewed microscopically with polarized light. Starch granules are partially crystalline; the areas of crystallinity consist of double-helical clusters of amylopectin side chains. Starch granules exhibit different X-ray diffraction patterns, depending upon the starch variety. Cereal starches such as corn, wheat and rice exhibit the A-type pattern, while the B-type is characteristic of tuber starches, such as potato. B-type diffraction patterns are also observed for high amylose starch varieties. Some starches, such as tapioca, exhibit the C-type pattern, which is a combination of the A- and B-type patterns. V-type patterns are observed for helical inclusion complexes formed from amylose and lipophilic materials, such as the native lipids normally present in concentrations of less than 1% in cereal starch granules.

Gelatinization takes place when starch granules are heated in water; during this process the granules absorb water, swell irreversibly, and lose their crystallinity and birefringence. Techniques such as differential scanning calorimetry (DSC), X-ray diffraction and light microscopy have been used to study this process. Gelatinization typically occurs between

Food and Industrial Bioproducts and Bioprocessing, First Edition. Edited by Nurhan Turgut Dunford.

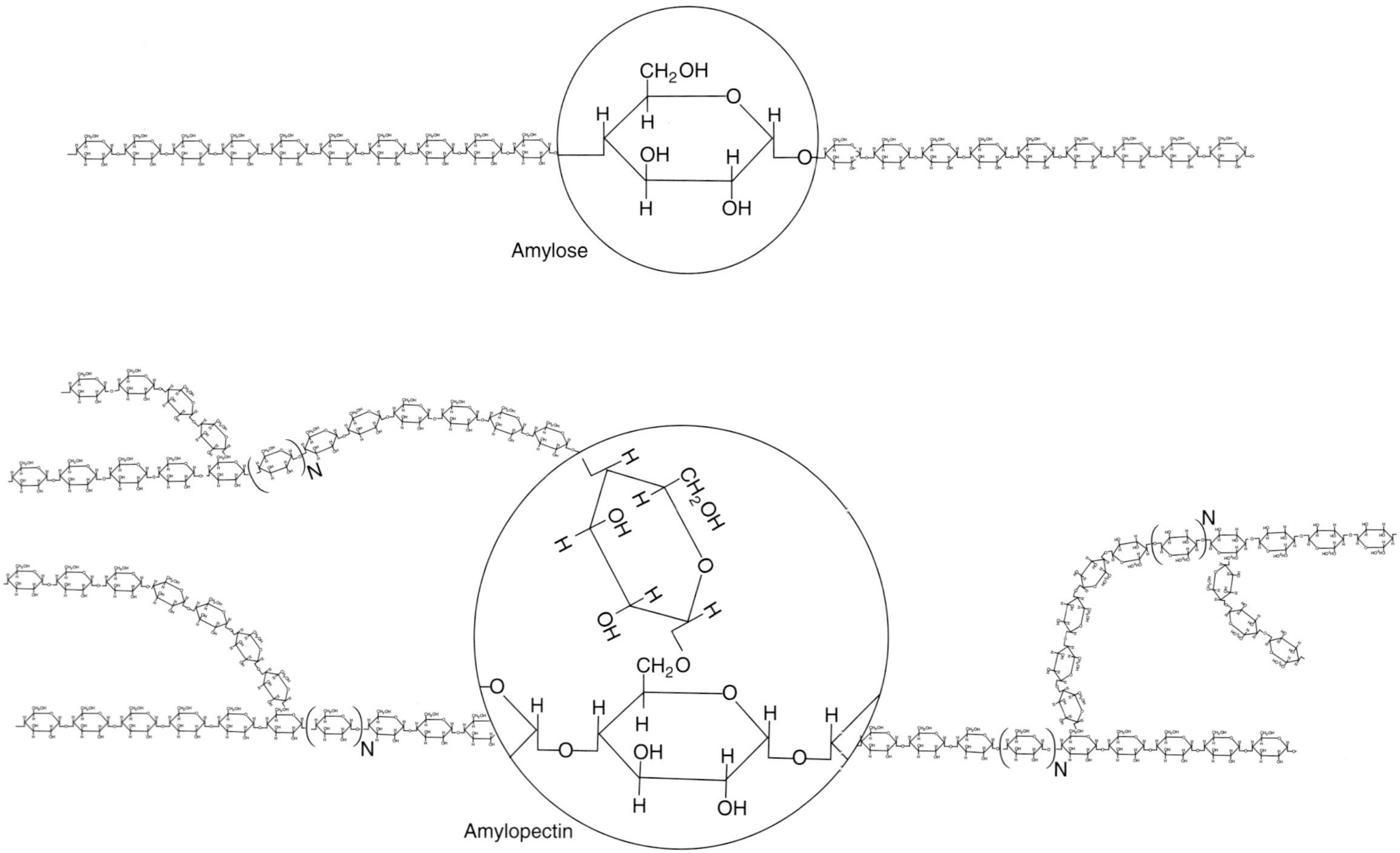

Figure 2.1 Structure of amylose and amylopectin.

60 and 70 °C, depending on the starch variety. Loss of birefringence and the endothermic changes in enthalpy (ΔH) observed with DSC are two methods typically used to characterize the gelatinization process. Mechanisms proposed to explain the gelatinization of starch have been summarized by Jenkens and Donald (1998).

Heating aqueous starch dispersions above the gelatinization temperature causes the starch granules to further swell and disintegrate, and increased amounts of starch (particularly the amylose component) dissolve as the temperature is increased. At temperatures of about 80–90 °C, smooth aqueous dispersions or pastes are obtained. Although amylose dissolves in hot water, complete solubility of amylopectin is not usually achieved unless an autoclave or a high-pressure steam jet cooker is used to obtain temperatures greater than 100 °C. Thick pastes or gels are formed when hot, aqueous starch dispersions are cooled due to the retrogradation of dissolved amylose. Retrogradation also occurs with amylopectin, although the process is much slower due to its highly-branched structure. Some additives, such as sodium sulfate, sucrose and glucose, inhibit the gelatinization of starch; whereas other additives, such as alkali metal hydroxides and urea, lower the temperature at which gelatinization and solubility take place.

Treatment of starch with acids, enzymes or oxidizing agents (such as sodium hypochlorite) are commonly used to reduce the molecular weight of starch and thus decrease the viscosity and modify the gelling properties of aqueous starch dispersions. Starches that have been partially depolymerized by acids or enzymes are commonly referred to as dextrins. Conversely, the swelling and solubility of starch granules can be reduced by cross-linking.

The hydroxyl groups on starch provide reactive sites for the formation of ester and ether derivatives, many of which are produced commercially. The number of hydroxyl groups derivatized is defined by the degree of substitution (DS). Since the glucopyranose repeating unit has three hydroxyl groups available for reaction, a DS of three indicates that derivatization of all available hydroxyl groups has taken place. Partial derivatization is indicated by DS values lower than three. Starch has also been derivatized by graft polymerizing vinyl and acrylic monomers onto the starch backbone, and reviews on this subject have been published elsewhere (Fanta and Bagley, 1977; Fanta, 1999).

There are numerous applications for starch in food products due to its thickening and gelling properties (Jarowenko, 1970; Whistler and Daniel, 1997). Waxy varieties of starch, which contain no amylose, are used in applications where gel formation due to retrogradation would be undesirable. Conversely, high amylose starches that retrograde readily have been added to food products for their enzyme-resistant properties after retrogradation. Pregelatinized starches are used in products where thickening is required but cooking must be avoided, for example, in instant puddings and pie fillings.

Starches and starch derivatives are also used for non-food applications (Jarowenko, 1970; Whistler and Daniel, 1997). For example, pregelatinized starches are used as components in oil well drilling muds and in cold water-dispersible adhesives. Acid modified and oxidized starches are used in the textile industry as sizes and fabric finishes to increase yarn strength and abrasion resistance; they are also used as adhesives and in the manufacture of paper. A number of commercially-important products, such as glucose, ethanol and cyclodextrins, are derived from starch by enzymatic and fermentation processing. There are also numerous applications for starches that have been derivatized through reaction of the hydroxyl groups. For example, hydroxyethyl and hydroxypropyl starch ethers have been used as paper coatings, sizes and adhesives. Cationic starch ethers, where the ether substituents contain tertiary amine or quaternary ammonium groups, are used in paper manufacture, principally for fiber and pigment retention. These starch derivatives have also been used as flocculating agents for water treatment and purification. Phosphate and acetate esters of starch have food and non-food applications. The solubility of the triacetate ester of starch in some organic solvents extends its range of commercial applications.

Although many processes have been used for the conversion of starch into commercially-viable products, newer processing methods are being developed that could supplement the well-established methods now used. Examples of these methods are (1) annealing and heat–moisture treatment of starch granules, (2) pressure treatment of starch granules, (3) use of microwave heating and ultrasound, (4) use of supercritical fluids, (5) reactive extrusion, and (6) steam jet cooking at high temperatures and pressures. This review will focus on the current literature in these areas.

2.2 ANNEALING AND HEAT–MOISTURE TREATMENT

Annealing and heat–moisture treatments are physical, hydrothermal methods that have been used to modify the properties of starch granules without destroying the granule structure (Jacobs and Delcour, 1998; Stute, 1992; Zavareze and Dias, 2011). Both treatments involve heating starch granules for various time periods at specific temperatures and moisture levels, and both treatments are carried out at temperatures above the glass transition temperature of the amorphous regions of the starch granule, but below the gelatinization temperature at the moisture levels used.

The process of annealing has been reviewed by Tester and Debon (2000). Annealing is a process that occurs when starch granules are heated in excess water (>60% w/w) or at an intermediate water content of about 40% w/w. Temperatures used for annealing are typically in the 50–60 °C range (Jacobs and Delcour, 1998). Annealing does not alter the morphology of the starch granules and no soluble carbohydrates are leached from the granules during the annealing process. Also, annealing does not result in significant changes in crystal type and degree of crystallinity, as measured by X-ray diffraction. The effect of annealing on the properties of starch granules is an increase in the gelatinization temperature and a decrease in the gelatinization temperature range. In maize starches, the largest effects observed by DSC were with starches having the highest amylose contents. Annealing takes place most rapidly at temperatures just below the gelatinization temperature, which depends upon the water content of the starch during the annealing process. The effects of annealing on the pasting properties of starch granules and their susceptibility to acid and enzymatic hydrolysis depend upon the botanical source of the starch. For example, with wheat, potato, oat and lentil starches, annealing decreases the granule swelling power and also decreases the amount of amylose leached from the granules when the treated samples are heated in excess water.

Heat–moisture treatment is a process that occurs when starch granules are heated at moisture contents lower than those used for annealing (i.e., below 35%). As in the annealing process, temperatures are above the glass transition temperature but below the gelatinization temperature of the starch at the moisture level used. Temperatures ranging from 84 to 120 °C have been used (Jacobs and Delcour, 1998). As observed for the annealing process, starch granule morphologies are not significantly changed. A reported effect of the heat–moisture treatment of potato starch is a change in the X-ray diffraction pattern from the B-type to A- or C-type. The A-type pattern for cereal starches was not changed by heat–moisture treatment, although new V-type peaks were observed in the X-ray diffraction patterns, presumably due to the formation of amylose inclusion complexes with the native lipids normally present in small amounts in these granules. Gelatinization temperatures were increased by heat–moisture treatment, and the endotherms observed by DSC were broadened. The pasting properties of starches were also affected by heat–moisture treatment. Viscosity increases during pasting were observed at higher temperatures and final viscosities of the pasted starches were lower, compared with untreated starch samples. As with annealing, decreased granule swelling was observed. The effect of heat–moisture treatment on the hydrolysis of starch with acids and enzymes varied with the botanical source of the starch.

In summary, explanations suggested for the effects of annealing and heat–moisture treatment on the properties of starch granules include changes in crystallinity and crystallite growth, increased order and interactions between amylose and amylopectin in the amorphous fraction, and increased interactions between the amorphous and crystalline areas of the granule. Since annealing and heat–moisture treatments are non-chemical in nature and significantly affect the gelatinization and pasting properties of starches, most applications for these processing methods will probably be in the food area. For example, annealing and heat–moisture treatments have been examined as methods for increasing the percentages of slowly-digestible and resistant starches in food products (Chung *et al.*, 2009).

2.3 HIGH-PRESSURE TREATMENT

The literature in this area has been reviewed by Fernandez-Martin and co-workers (2005) and Pei-Ling co-workers (2010). Pressures used in the high-pressure treatment of starch have typically ranged from about 100 to 1000 MPa (987–9869 atm). If sufficient water is present and the applied pressure is sufficiently high, some starches can be gelatinized at room temperature or below, as evidenced by the loss of birefringence or the reduction or absence of gelatinization endotherms in DSC thermograms. Compared to heat-induced gelatinization, pressure-gelatinized starch granules are less swollen and retain more of their original granular appearance.

Numerous articles have been published that describe the effects of high pressure on starch properties. Gelatinization of wheat starch and potato starch as a function of pressure and water content was studied (Muhr and Blanshard, 1982); it was observed that high pressure did not affect the microscopic appearance of the granules or their staining behavior. Reduction in gelatinization temperature was greatest at higher pressures and water contents, and wheat starch was more easily gelatinized than potato starch. Later research also showed that starches with A-type X-ray diffraction patterns (such as corn, wheat and rice) were affected the most by high-pressure treatment (Ezaki and Hayashi, 1992). Starches with B-type X-ray diffraction patterns (such as potato) were more resistant to pressure, and C-type starches (such as sweet potato and tapioca) showed intermediate properties. The X-ray diffraction patterns of normal corn starch, defatted normal corn starch, waxy corn starch and rice starch showed few changes up to 200 MPa; however, after treatment at 500 MPa, considerable destruction of the A-type pattern for corn starch was observed (Hibi *et al.*, 1993). The B-type pattern for potato starch was not significantly changed. To demonstrate the role of water in the changes observed in X-ray diffraction patterns, corn starch and potato starch were dried to moisture contents of 5.6–7.6% and pressure treatments were carried out in hexane at 500 MPa for 60 min. Both the A-type and B-type patterns remained the same, showing that in the absence of water the crystalline structure was not changed.

Volume increases in wheat starch and potato starch granules during the compression and release phases were measured microscopically (Douzals *et al.*, 1996a). Gelatinization of wheat starch began at pressures above 300 MPa and the largest increase in granule volume was observed between 300 and 420 MPa. Gelatinization was 100% at 600 MPa. Pressure-induced changes were less for potato starch; there was no change in gelatinization enthalpy at pressures up to 600 MPa. Below 300 MPa, the compressibility of 16% starch–water suspensions was close to that of pure water; whereas over 300 MPa the volume reduction due to compression was higher for the starch-containing suspensions at the same pressure. It was concluded from these experiments that the water molecules that enter the starch granules during pressure-induced gelatinization occupy a reduced volume, compared to pure liquid water. It was also shown that this volume reduction was maintained after the wheat starch gels were returned to atmospheric pressure (Douzals *et al.*, 1996b).

The swelling index and the release of amylose from the starch granules depended on whether starch was gelatinized by heat or by high pressure (Douzals *et al.*, 1998). When iodine titrations were carried out on the soluble fractions obtained from gelatinizations at 5% starch solids, more iodine binding (and therefore more amylose release) was observed with thermal gelatinization than with high-pressure gelatinization. This agrees with microscopic observations that showed that the granule structure was better preserved during high-pressure gelatinization. Young's modulus measurements carried out on 30% solids starch gels showed that the gels produced by high-pressure gelatinization were weaker than those produced by thermal gelatinization, since less amylose (that can retrograde at low temperature) was released from the granules.

Effects of pressure–temperature combinations on the gelatinization of wheat starch granules were determined, and a pressure–temperature gelatinization diagram was constructed (Douzals *et al.*, 2001). Similarities between starch gelatinization and protein unfolding under pressure were observed. Baks co-workers (2008) also measured the degree of gelatinization as a function of pressure, temperature and treatment time for wheat starch–water mixtures and obtained phase diagrams relating the degree of gelatinization to both pressure and temperature. Stute co-workers (1996) reported the effects of high-pressure treatment on the gelatinization and properties of a number of different starch varieties. Above a certain pressure, an additional DSC peak was observed at about 50 °C; this was identified as a retrogradation peak, in agreement with the fact that no water-soluble amylose was observed in pressure-treated starches. When wheat starch was treated at 600 MPa for 15 min at 20 °C, a minimum of about 50% moisture was necessary for pressure-induced gelatinization to occur. Higher gelatinization temperatures were needed in the presence of sugar or ethanol. Starches that exhibited restricted swelling during pressure-induced gelatinization also had lower viscosities when they were pasted by heating in water; and due to their restricted swelling, smooth-textured pastes were obtained at high starch concentrations (15–30%). The high-pressure treatments of pea starch (Fernandez-Martin *et al.*, 2008), buckwheat starch (Vallons and Arendt, 2009a), sorghum starch (Vallons and Arendt, 2009b), and Japonica rice starch (Tan *et al.*, 2009) were also studied.

Katopo co-workers (2002) pressure-treated normal, waxy, and high amylose maize starches, as well as tapioca, potato and rice starches at 690 MPa (100 000 psi) at room temperature under the following treatment conditions: (1) starch in powder form with original moisture contents of 10–16%; (2) starch suspended in ethanol (1:1 ethanol:starch); and (3) starch suspended in water (1:1 and 2:1 water:starch). Crystalline structures, thermal and pasting properties, molecular weight distributions, and morphologies of the treated starch granules were investigated. X-ray diffraction showed that high-pressure treatment converted the starches that originally displayed the A-type pattern into starches having X-ray patterns that resembled the B-type. With the exception of high amylose maize starch, the degree of gelatinization was higher with 2:1 water:starch than with 1:1. No gelatinization peaks were detected in A-type starches pressure-treated at the 2:1 ratio, indicating total loss of crystalline structure and molecular order. DSC thermograms after high-pressure treatment showed a newly-developed peak that resembled that of retrograded B-type starch. Starches having the B-type pattern were not changed by high-pressure treatment, and the high-pressure treatment did not change the molecular weight distribution of the starch. Treatment of starches in powder form and as suspensions in ethanol resulted in a reduction in the peak intensity of the X-ray diffraction pattern; however no conversion from the A-type to the B-type pattern was observed. Changes in pasting properties due to pressure treatment varied with starch type; and when pressure-treated in water, increased pasting temperatures and decreased paste viscosities were observed. The effect of high-pressure treatment on normal corn, waxy corn, wheat, and potato starches at their ambient moisture contents of about 14–20% was investigated by Liu co-workers (2008) at pressures up to 1500 MPa. DSC showed that high-pressure treatment reduced the gelatinization temperature as well as the corresponding gelatinization enthalpy. These changes in properties were irreversible and were maintained during storage for six months at 25 °C. The birefringence observed in the untreated- and pressure-treated starches was not visually different, and the X-ray diffraction patterns were not significantly changed by high-pressure treatment. However, scanning electron microscopy (SEM) showed that pressure treatments caused some changes in the morphology and surface appearance of the granules.

A high-pressure cell was developed that enabled optical examination of starch granules *in situ* at pressures up to 300 MPa (Bauer *et al.*, 2004). Potato (B-type), wheat (A-type) and tapioca (C-type) starches were dispersed in water at concentrations of 0.8%. Although the maximum pressure of 300 MPa was not high enough to cause swelling of potato starch and tapioca starch, most wheat starch granules swelled during pressure treatment. Pressure treatment of iodine-stained wheat and tapioca starches resulted in discoloration of the granules, whereas potato starch discolored only slightly. Discoloration during pressure treatment was attributed to the loss of amylose-iodine complex from the starch granules and the accumulation of complexed amylose in the aqueous phase surrounding the granules, which resulted in an observed darkening of this phase. Resistant starch yields of about 12% were obtained by high-pressure treatment processes and were in the same range as yields obtained by conventional heat gelatinization (Bauer *et al.*, 2005).

When aqueous 10% dispersions of potato starch were subjected to high-pressure treatment at 20 °C and 600 MPa, NMR spectra of the treated samples showed signal resonances corresponding to amorphous sites within the starch structure (Błaszczak *et al.*, 2005a). FTIR spectra also showed that high pressure treatment affected the intensity of bands corresponding to the amorphous and more ordered areas of the starch granules. DSC thermograms showed a decrease in gelatinization temperature after high-pressure treatment and a decrease in total enthalpy with increased treatment time. SEM images suggested that the starch granule surface was more resistant to high-pressure treatment than the interior of the granule. When 30% aqueous dispersions of waxy corn starch and high amylose corn starch were pressure-treated at 20 °C and 650 MPa, changes in the starch structure were observed in NMR spectra, and X-ray diffraction patterns of pressure-treated high amylose starch showed a decrease in crystallinity (Blaszczak *et al.*, 2005b). An amorphous structure was observed for waxy corn starch after only three minutes at high pressure. These results confirmed the NMR data and indicated that changes in structure were mainly related to the crystalline areas of the starch granule. After pressure treatment for nine minutes at 650 MPa, the birefringence remaining in the high amylose starch granules tended to be located in the outer portions of the granules. While most of the high amylose starch granules retained their granule shape after pressure treatment, SEM images of pressure-treated waxy corn starch showed complete breakdown of the granule structure. High-pressure treatment for nine minutes did not alter the molecular weight distribution of high amylose corn starch. However, a polydisperse product was obtained when waxy corn starch was pressure-treated in the same manner.

Since waxy starches are more sensitive to high-pressure treatment than amylose-containing starches, the properties of waxy rice and waxy corn starches, pressure-treated at 500 MPa, were studied to determine their texturizing properties for potential applications in food products (Simonin *et al.* 2009). Whereas normal corn starch granules swelled gradually with increasing time under pressure and retained their granule structure, the waxy corn starch granules lost their granular structure during pressure treatment. Waxy rice starch was the most sensitive to high-pressure treatment; after ten minutes at 500 MPa the granular structure was lost and the gel reached its maximum viscosity.

The retrogradation kinetics and water dynamics of wheat starch, pressure-treated in water at 25 °C and 621 MPA, were investigated and compared with thermal gelatinization in a boiling water bath for 30 min (Doona *et al.*, 2006). The pressure-treated and heat-treated wheat starch granules showed similar degrees of swelling, and partial disintegration was observed for the heat-treated granules. Similarities between the pressure-treated and heat-treated starches observed by microscopy were attributed to the higher starch concentrations (40%) used in this study. DSC was used to determine enthalpy changes in the starches during storage for 14 days at 4 °C. The pressure-treated starch showed a lower increase in enthalpy during storage than the heat-treated starch, suggesting less retrogradation. NMR analyses of proton relaxation times suggested different water dynamics in the pressure-treated and heat-treated wheat starches.

The effect of starch concentration on the pressure-induced gelatinization of potato starch was determined by pressure treating aqueous dispersions containing 10–70% potato starch at 40 °C and 400–1200 MPa (Kawai *et al.*, 2007a). DSC was used to determine the extent of gelatinization and showed that gelatinization increased with increased pressure and water content. Retrogradation was observed in mixtures with starch solids of 30–60%, and was observed at lower treatment pressures when the water content was increased. DSC was also used to determine whether longer treatment times would affect the gelatinization and retrogradation of potato starch (Kawai *et al.*, 2007b). Treatment pressures used in this study were 600, 800, and 1000 MPa, starch concentrations were 10–70%, the treatment temperature was 40 °C, and the treatment times were 18 and 66 h. Although gelatinization and retrogradation were significantly affected by the treatment pressure and water content, increasing the pressure-treatment time from 1 to 66 h had little effect on the observed enthalpy changes.

Błaszczak co-workers (2007) examined the high-pressure treatment of high amylose starch and waxy maize starch, as well as their 1:1, 1:3, and 3:1 mixtures. Starches were dispersed in water to give 30% starch solids, and the dispersions were pressure-treated at 20 °C for nine minutes at 650 MPa. Treatment of high amylose starch resulted in only 10.8% gelatinization, whereas 85.9% gelatinization was observed for waxy maize starch. Granule disintegration was also observed with waxy maize starch and a gel-like dispersion was obtained. When mixtures of the two starches were pressure treated, amylopectin formed a continuous gel phase, while the intact high amylose starch granules acted as fillers in these mixed systems. In a subsequent study, native and high-pressure treated waxy maize and high amylose starches were studied by electron paramagnetic resonance spectroscopy (EPR) to determine the nature, number, and stability of free radicals generated when the starches were heated to 180–230 °C (a temperature range commonly used for food preparation) (Blaszczak *et al.*, 2008). Waxy maize starch exhibited a greater ability to generate free radicals than high amylose starch; however, the presence of water and the high-pressure pretreatment of the starches reduced the amount of free radicals observed.

Aqueous 10% dispersions of normal rice, waxy rice, normal corn, waxy corn, tapioca and potato starches were pressure-treated at 400 and 600 MPa for 30 min at 20 °C and the dispersions were characterized by their pasting properties, degree of swelling, and changes in birefringence (Oh *et al.*, 2008a). As observed by earlier workers, potato starch was least affected by pressure treatment and retained its birefringence after treatment for 30 min at 600 MPa. Under the same conditions, tapioca starch and the waxy starches showed complete gelatinization; normal corn starch and rice starch were partially gelatinized. In a subsequent publication, the effects of treatment pressure (100–700 MPa), treatment time (up to 30 min) and treatment temperature (10–60 °C) on the gelatinization of normal and waxy rice starches were described (Oh *et al.*, 2008b). Pasting behavior, viscosity, degree of swelling, changes in birefringence and leaching of amylose and amylopectin were studied. Degree of gelatinization during pressure treatment depended upon the type of starch and the pressure, temperature, and duration of the treatment. Although normal and waxy rice starches both lost all birefringence after treatment at 500 MPa, they exhibited different gelatinization characteristics. Normal rice starch maintained more of its granular structure and the particulate solid obtained consisted of swollen starch granules as opposed to granule ghosts (Obanni and BeMiller, 1995). Also, the leaching of soluble starch was minimal. In contrast, waxy rice starch appeared to contain granule ghosts and no additional changes related to gelatinization were observed on further heating.

The effects of dissolved salts and sugars on the pressure-induced gelatinization of wheat, tapioca and corn starches were generally comparable to those observed when the starches were gelatinized thermally (Rumpold and Knorr, 2005). For example, addition of sucrose led to a reduction in the degree of gelatinization of all three starches; at a sucrose concentration of about 20%, gelatinization was almost completely suppressed. The effects of different solutes on gelatinization were attributed to their effect on the availability of free water. When normal corn starch was suspended in 2 M sodium chloride or in 50% w/w sucrose, as opposed to water, these two solvents showed a protective effect against starch gelatinization during high-pressure treatment at 600 MPa (Kweon *et al.*, 2008a), and the protective effects of sucrose were more dramatic than those of sodium chloride. Kweon co-workers (2008b) also studied the pressure-induced gelatinization of wheat starch in sodium chloride solutions of varying concentrations. Although high-pressure treatment at all sodium chloride concentrations from 0.1 to 5 M resulted in reduced gelatinization compared to water alone, maximum stabilization with respect to gelatinization was observed with sodium chloride concentrations near 2 M. When aqueous sodium dodecyl sulfate solution was added to potato starch, and the sample was pressure-treated at 1.5 GPa for one hour at 40 °C, an amylose-sodium dodecyl sulfate helical inclusion complex was formed, as indicated by DSC and the appearance of a V-type X-ray diffraction pattern (Yamamoto *et al.*, 2002). Use of this technique to prepare amylose complexes of commercial interest, such as flavorings and emulsifiers, was suggested.

Selmi co-workers (2000) compared the amyloglucosidase-hydrolyses of corn and wheat starches that were pressure-treated for 15 min at 600 MPa and 25 °C, or 400 MPa and –20 °C, with starches that were thermally gelatinized by heating in a water bath at 80 °C for 15 min. Although a higher conversion to glucose after five minutes was observed for the thermally-gelatinized corn starch, a conversion of about 90% was observed after 30 min for all starch samples. With wheat starch the two high-pressure treatments gave higher equilibrium yields of glucose than the thermally-gelatinized sample.

Since batch-wise, high-pressure treatment of starch for finite time periods has limited practicality with respect to commercial processing, Wang co-workers (2008) studied the pressure treatment of maize starch using a commercial, high-pressure homogenizer with a pressure range up to 150 MPa. Aqueous starch suspensions were prepared at 1.0% solids and were subjected to single-pass, high-pressure treatment at 60, 100 and 140 MPa. DSC showed a decrease in gelatinization temperature and gelatinization enthalpy with increasing homogenization pressure, and X-ray diffraction indicated a loss of crystallinity after treatment at 140 MPa.

In summary, high-pressure treatment reduces the temperature at which the gelatinization of starch takes place, and the effects of high pressure vary with the type of starch used. Cereal starches such as corn, wheat and rice, that exhibit A-type X-ray diffraction patterns, are affected the most by high-pressure treatment, whereas starches with B-type diffraction patterns, such as potato, are more resistant to the effects of high pressure. Also, waxy starches are more sensitive to high-pressure treatment than amylose-containing starches. A major application of this processing technique will be the tailoring of starch properties for food applications, for example, food products processed with minimal heating. Compared to thermal gelatinization, the starch granule structure is better preserved when high-pressure gelatinization is used, and high-pressure treatment therefore has the potential for creating new food products with unique textures. The shelf-life of foods can also be improved by high-pressure treatment without the detrimental side effects of heat on food texture, flavors, and heat-sensitive nutrients. Since covalent bonds are not affected by high-pressure treatment, high pressures have only a minimal effect on low molecular weight materials such as flavors and nutrients, whereas protein-based sources of food spoilage, such as enzymes and microorganisms, can be denatured and inactivated by high-pressure treatment without the application of heat.

2.4 MICROWAVE PROCESSING

The use of microwave heating to carry out chemical reactions between organic compounds has been studied since the mid-1980s and several reviews of this technology have been published (Caddick, 1995; Corsaro *et al.*, 2004; Gabriel *et al.*, 1998; Galema, 1997; Lidström *et al.*, 2002). The modification of starch by microwave radiation was reviewed by Tomasik and Zaranyika (1995).

Lewandowicz co-workers (1997) studied the microwave heating of potato and tapioca starches in the absence of chemical reactants. Time–temperature profiles of starch–water systems during microwave heating were established, and the effects of microwave radiation on the properties and structure of the heated starch granules were determined. The increase in temperature with irradiation time depended upon the moisture content of the starch. At low moisture contents (about 1–5%), a rapid rise in temperature was observed whereas a slower temperature rise was observed with moisture contents of about 7–15%. When the moisture content exceeded about 20%, the temperature leveled off at about 80–90 °C and remained fairly constant, even with prolonged heating times. A correlation was found between the moisture content during microwave heating and the pasting properties of the heated starch granules. With moisture contents over 20%, there was a rise in the gelatinization temperature of the microwave-heated starches; the pasting properties changed from those typical of tuber starches to those resembling cereal starches. There was also a reduction in starch granule solubility and a change in the X-ray diffraction pattern of potato starch was observed. Although SEM showed some distortion of starch granules after microwave treatment, the granule structure was maintained. Similar studies were carried out with wheat starch, corn starch and waxy corn starch, in which the moisture contents were adjusted to 30% prior to microwave irradiation (Lewandowicz *et al.*, 2000). Investigations of pasting properties showed that microwave heating of wheat starch and corn starch for 60 min caused a reduction in viscosity during pasting whereas there was little effect on the viscosity of waxy corn starch. The degree of crystallinity of wheat starch and normal corn starch was decreased by microwave heating whereas the crystallinity of waxy corn starch remained almost unchanged. Dyrek co-workers (2007) used electron spin resonance (ESR) to investigate the free radicals formed in corn starch, potato starch and oxidized potato starch when the starches were heated by microwave irradiation and by conventional thermal heating. Both methods of heating generated free radicals with ESR signals.

A number of publications describe the use of microwave heating to form ester and ether derivatives of starch. Koroskenyi and McCarthy (2002) reacted potato starch with acetic anhydride and sodium acetate (an acetylation catalyst) for five minutes to yield starch acetate with a DS of 0.37. Evaporation of acetic anhydride during microwave heating prevented the use of longer radiation times. Succinate esters of starch were also prepared and a succinate ester with a DS 0.59 was obtained in the presence of glacial acetic acid solvent. Ether derivatives of starch that contained tertiary amine and quaternary amine substituents were prepared by microwave-heating starch with aqueous solutions of 2-(diethylamino)ethyl chloride and (3-chloro-2-hydroxypropyl)trimethylammonium chloride under alkaline conditions for 1–10 min. DS values varied from 0.21 to 1.54, depending on experimental conditions and were comparable to the DS values obtained by conventional heating for two hours at 70 °C. Cationic amine derivatives of starch are used extensively as retention aids in papermaking. Singh and Tiwari (2008) also used microwave heating to prepare methyl ethers of starch; complete methylation of soluble starch was observed in less than five minutes by heating starch with iodomethane and potassium hydroxide.

Shogren and Biswas (2006) used microwave heating to prepare starch acetates and determined the effects of reaction conditions and starch type (waxy, normal, or high amylose) on product properties such as water solubility, swelling, and viscosity. Reactions were carried out with mixtures of dry starch, acetic acid and acetic anhydride in a sealed Teflon vessel. Maximum temperatures varied from 120 to 175 °C, and DS values for the resulting starch acetates varied from 0.19 to 0.77. Molecular weights of the starch acetates were lower than those of the native starches. Heating was also carried out in a conventional block heater that was adjusted to give a temperature profile similar to that used with microwave heating. With a heating time of six minutes, the starch acetate prepared by microwave heating had a DS of 0.72 whereas a DS of 0.56 was obtained with conventional block heating. The iodine-catalyzed acetylation of starch with acetic anhydride in the absence of acetic acid was also studied (Biswas *et al.*, 2008). Since iodine is a good absorber of microwave radiation, it was assumed that addition of iodine would accelerate the microwave-heated reaction. Acetic anhydride was used with three different concentrations of starch, and varying amounts of iodine were added. Mixtures were then microwave-heated to 100 °C and held at this temperature for two minutes. In the absence of iodine, the DS was only 0.03. However, the DS increased with increased amounts of iodine and values approaching three could be obtained if sufficient iodine was added. Acetylation under these conditions reduced the molecular weight of starch, probably because of the action of iodine as an oxidizing agent. Although esterification with acetic anhydride occurred readily, reaction with propionic anhydride was more difficult, even with iodine catalysis. A similar method of acetylation was used by Sánchez-Rivera and co-workers (2010) to prepare acetate esters of banana starch.

Microwave heating was also used by Xing and co-workers (2006) to prepare maleate esters from dry corn starch. Starch samples with different moisture contents were mixed with maleic anhydride and heated, either in a microwave oven for different periods or in a conventional oven at 100 °C for 1–6 h. Shorter reaction times and higher DS values were obtained with microwave heating. Some moisture in the starch sample was necessary to obtain good reaction efficiency; maximum efficiency was observed with a water content of 12.1%. SEM showed that esterification did not greatly change the morphology of the starch granules, since not enough water was present to cause gelatinization. Biswas and co-workers (2006a) observed a rapid reaction of corn starch with maleic anhydride when the reaction mixtures were microwave-heated in DMSO in the presence of pyridine. The effects of reaction time, microwave power, and the amounts of maleic anhydride and pyridine used were determined; addition of pyridine yielded products with DS values up to 0.30. DMSO is a good solvent for microwave reactions, since it readily absorbs microwave radiation and heats up rapidly. When the reactivities of maleic anhydride, succinic anhydride, and octenyl succinic anhydride were compared in DMSO without addition of pyridine, succinic anhydride and octenyl succinic anhydride gave starch esters with a DS of 0.3, compared to a DS of only 0.1 with maleic anhydride. Compared to conventional block heating, microwave heating produced starch esters with higher DS values, and in slightly higher yields. Jyothi and co-workers (2005) prepared succinate esters from cassava starch; the highest DS of 0.051 was obtained when the starch sample had a moisture content of 20% and the heating time was seven minutes at 120 °C. SEM showed that esterification did not change the outward appearance of the starch granules. Citric acid esters of cassava starch were also prepared (Jyothi *et al.*, 2007). DS values for the resulting esters ranged from 0.005 to 0.063, with the highest obtained by heating starch with 0.45 mole of citric acid per mole of anhydroglucose unit (AGU) for seven minutes at 160 °C. Examination of the pasting properties showed that paste viscosities increased with increasing DS and the highest peak viscosity was observed for the sample with DS 0.063.

Microwave heating was also used to prepare starch derivatives by reaction with urea, thiourea, biuret and thiosemicarbazide. Siemion and co-workers (2004) blended air-dried potato starch, containing residual adsorbed water, with either urea or biuret. The mixtures were then microwave-heated to obtain starch derivatives in which hydrogen atoms of the hydroxyl groups were substituted with $-CONH_2$ or $-CO-NH-CO-NH_2$, respectively. Reaction products of starch and urea have been suggested as additives in the diets of ruminant animals as a source of metabolized nitrogen. Reactions of thiourea with potato starch were also carried out with both microwave and conventional heating (Siemion *et al.*, 2005a). The reaction with thiourea was similar to that observed with urea; however, the reaction products underwent further decomposition and dextrins contaminated with isothiocyanates were obtained. Potato starch was also microwave-heated with thiosemicarbazide; TGA, DSC and FTIR were used to characterize the resulting products (Siemion *et al.*, 2006). Cationic starches were similarly prepared by reacting starch with semicarbazide hydrochloride (Siemion *et al.*, 2005b).

Microwave heating was also used to prepare inorganic derivatives of starch. Potato starch was sulfated with pyridine. SO_3 complex in a microwave-assisted solid state process, and a maximum DS of 1.05 was obtained (Staroszczyk *et al.*, 2007a). Although conventional oven heating gave about the same DS as microwave heating, longer reaction times were required. Granular potato starch was also esterified by reaction with the sodium salts of selenous and selenic acids (Na_2SeO_3 and Na_2SeO_4) (Staroszczyk *et al.*, 2007b). Molar ratios of AGU:selenium salt were 1:1 or 1:0.1, and mixtures were heated in a microwave oven at either 450 W or 800 W for 20 min. A maximum DS of about 0.03 was observed and the granular appearance of the starch was retained. Microwave heating in the solid state was used by Gui-Jie and co-workers (2006) to cross-link corn starch with sodium trimetaphosphate (STMP). Varying amounts of STMP and sodium carbonate, in water solution, were added to corn starch and the samples were dried at 40 °C for 24 h to give a water content of about 8%. About three minutes at 270 W was required to obtain a DS of 1.89×10^{-2}, whereas conventional oven heating required 24 h to obtain a DS of 0.70×10^{-2}. Cross-linking by microwave heating caused no significant changes in starch granule morphology. Since STMP has low toxicity, the cross-linked starches can be used as thickeners in food products. Zinc derivatives of starch were prepared by solid state reactions of potato starch with sodium tetrahydroxozincate [$Na_2Zn(OH)_4$], which was prepared by the reaction of zinc oxide with sodium hydroxide (Staroszczyk and Janas, 2010).

Microwave heating was employed in procedures used to prepare starch graft copolymers. For example, with microwave heating, free radical graft polymerization of acrylamide onto potato starch could be achieved with low concentrations of potassium persulfate initiator (Singh *et al.*, 2006). Maximum grafting was observed with 0.1 g of potato starch in 25 ml of water, and with acrylamide and potassium persulfate concentrations of 0.10 M and 0.0025 M, respectively. Heating for 60 s at 720 W produced a final temperature of 98 °C. Graft polymerization was not observed with these reactant concentrations under conventional polymerization conditions. Starch graft copolymers were also prepared by microwave heating aqueous mixtures of potato starch and caprolactone for three minutes at 390 W (Koroskenyi and McCarthy, 2002). Guan and co-workers (2006) used a different method of grafting (based on Maillard-type reactions between amino groups of a protein and the reducing carbonyl groups of a polysaccharide or sugar) to prepare co-polymers of soluble starch and soy protein isolate. Reactions were carried out by mixing soy protein isolate with sugars, dextran or soluble starch and then rapidly heating the mixtures to about 90 °C in a microwave oven. The resulting protein-polysaccharide co-polymers were shown to have emulsifying and antimicrobial properties.

Finally, microwave heating was used to prepare silver nanoparticles, using starch as both a stabilizer (or capping agent) and a reducing agent for the conversion of silver ions to nanoparticulate metallic silver (Sreeram *et al.*, 2008). A solution of starch was first prepared at a concentration of 1% and silver nitrate solution was then added. The following methods of heating were then used for the reduction of silver ion: (1) uncontrolled heating to about 80 °C; (2) controlled heating at 2 °C/min; and (3) microwave heating for 30–120 s. Microwave heating produced the smallest silver nanoparticles (average particle size: 12 nm) and the particle size was more monodisperse, compared to the other two methods. Also, with microwave heating, silver ion was reduced to metallic silver within 120 seconds, compared to 60 min for the other methods. The fact that the nanoparticles of silver did not aggregate showed that starch was acting as both a reducing agent and stabilizer. End-use applications for silver nanoparticles include their use as catalysts and antimicrobials, and also their use in medical and biological applications.

In summary, microwave heating allows reactions to be carried out with shorter reaction times, more uniform heating, and less chance for overheating, since microwave radiation heats only the reactants and solvent and not the reaction vessel itself. In contrast, heating with conventional heat transfer equipment is slow and uneven, and temperature gradients can easily result in overheating and product decomposition. Pressurized microwave reactors have been designed for laboratory use and numerous publications describe the use of microwave heating to prepare starch derivatives, such as esters and ethers, reaction products of starch with urea, thiourea and thiosemicarbazide, starch phosphates, and starch graft copolymers. It is predicted that future years will see an increase in the use of microwave heating to prepare new starch derivatives.

2.5 PROCESSES USING ULTRASOUND

The effect of ultrasonic irradiation on liquids, and on liquid dispersions of suspended solids, has been reviewed by Suslick and co-workers (Suslick, 1990; Suslick and Flanagan, 2009). Review articles have also been published on the effects of ultrasound on polymers in solution (Basedow and Ebert, 1977) and on carbohydrates (Kardos and Luche, 2001). The effects of passing ultrasound through a liquid are the result of acoustic cavitation, which is the formation of micrometer-sized cavities or bubbles in the liquid followed by the growth of these cavities, and their subsequent collapse by implosion. When the cavities collapse, small areas of intense local heating and high pressure are generated, followed by very high rates of cooling. Cavities are formed in the liquid because the passage of ultrasound through the liquid medium exerts areas of negative pressure in the liquid; when this negative pressure exceeds the tensile strength of the liquid, molecular separation occurs, creating cavities or bubbles. Cavity formation often results from nucleation in areas of the liquid having reduced tensile strength, for example, in gas-filled crevices of suspended particles, or in micro-bubbles of gas already present in the liquid medium. When the cavities reach a certain size, implosion of the cavities occurs. It is the implosion process that generates the small areas of intense thermal energy and pressure in the otherwise cold liquid.

Cavitation occurs in a different manner near a solid–liquid interface. Whereas the cavities formed in pure liquids remain spherical during collapse, cavity collapse occurs asymmetrically near a solid surface and high-speed jets of liquid are generated during the implosion process that can damage and degrade the surface of the solid. The solid must be several times larger than the collapsing cavity for this process to occur; this phenomenon is therefore not observed in the presence of very small particles. When small particles are present, however, the shock waves generated by cavity collapse can cause interparticle collisions, resulting in damage of the suspended particles. High-intensity probes (50–500 W/cm^2), as well as lower-intensity units, such as ultrasonic cleaning baths, are commercially available for laboratory use. Earlier literature on the modification of starch by ultrasonic irradiation has been reviewed by Tomasik and Zaranyika (1995).

The effect of ultrasound at frequencies of 280 and 960 kHz on potato starch granules was examined under a wide range of treatment conditions by Gallant and co-workers (1972) and Degrois and co-workers (1974). Potato starch was dispersed in water, and treatments were carried out under vacuum and in atmospheres of air, oxygen, hydrogen and carbon dioxide. Light microscopy and SEM were used to determine the effects of these ultrasonic treatments on starch granule structure. Under vacuum or under an atmosphere of carbon dioxide, ultrasonic irradiation produced little or no effect on the morphology of the potato starch granules. However, under an atmosphere of hydrogen, deep pits in the granules were observed. In oxygen or air, the pits were not as deep and more surface damage was observed.

The extent of granule erosion thus depended on the nature of the atmospheric gas under which the irradiations were carried out, and it was observed that the sizes of the pits were inversely proportional to the solubility of the atmospheric gas in water. The extent of granule damage increased with increasing irradiation time but decreased with increased starch concentration in the aqueous starch suspension. Granule damage was more extensive at the lower frequency of 280 kHz.

Luo and co-workers (2008) examined the effect of ultrasonic treatment on the properties of maize starch with different amylose contents. Waxy maize starch, normal maize starch and amylomaize starch containing about 50% amylose were used. Starches were adjusted to 70% moisture content and ultrasonic treatment was carried out for 30 min at 60 °C. Although treatment with ultrasound did not alter the size or shape of the starch granules, pores were observed on the surfaces of normal and waxy maize starch granules, and fissures could be seen on the granule surfaces of amylomaize starch. Treatment with ultrasound increased the swelling power and solubility of all three starches. X-ray diffraction patterns of the ultrasound-treated starches were similar to those of the corresponding native starches, indicating that the crystalline structures were essentially unchanged by the ultrasonic treatment. Syneresis of water from frozen and thawed starch gels decreased after ultrasonic treatment of normal and waxy maize starches, but increased after ultrasonic treatment of amylomaize starch. The paste–viscosity curve for sonicated waxy maize starch was nearly the same as that observed for native starch; however, normal maize starch showed a small lowering of viscosity due to ultrasonic treatment, and a larger decrease in viscosity was observed for amylomaize starch. It was concluded that ultrasonic treatment preferentially degrades the amorphous regions of the starch granule and that amylose is more easily degraded than amylopectin. The changes resulting from ultrasonic treatment thus depend not only on the crystal structure of starch, but also on its amylose content. Jambrak and co-workers (2010) examined the effects of ultrasound on the physical properties of corn starch. Aqueous starch suspensions were treated at 24 kHz using either an ultrasonic probe with intensities ranging from 34 to 73 $W\,cm^{-2}$ or an ultrasonic bath with an intensity of 2 $W\,cm^{-2}$. With the ultrasonic probe, temperatures after sonication varied from about 18 to 56 °C, depending on the intensity. Rheological parameters, turbidities, and swelling powers of the treated samples were compared with those observed for untreated starch.

The ultrasonic degradation of waxy rice starch was studied by Isono and co-workers (1994). An ultrasonic homogenizer was used at 29 kHz and samples were continuously pumped through the cell and returned at a rate of 0.7 l/min. The pasting properties of samples sonicated at 40, 50, and 60 °C indicated that ultrasonic degradation was accelerated at the temperature where gelatinization begins. A sonication temperature of 60 °C was, therefore, used for the remainder of the study. Molecular weight decreased with increasing sonication time and the rate of ultrasonic degradation was independent of polymer concentration at concentrations below 7%. With long sonication times, the molecular weight approached a constant value; the molecular weight distribution tended to be fairly narrow.

A number of articles have been published on the effect ultrasound on aqueous solutions of starch and on aqueous dispersions of gelatinized starch granules. Azhar and Hamdy (1979) heated potato starch in sodium acetate buffer at concentrations of 1.00, 0.75, and 0.50%, and the dispersions were sonicated at 1–2 °C at 20 ± 0.4 kHz. A rapid decrease in viscosity was observed during the first five minutes of sonication; the viscosity then leveled off at longer sonication times. Even though the sonicated swollen granules were disintegrated by the ultrasonic treatment, and the starch molecular weight was reduced, no reducing sugars were detected; the rate of enzymatic hydrolysis by β-amylase was not affected.

Jackson and co-workers (1988) used mild sonication conditions (20 s at 55 °C) to disrupt swollen starch granules. This treatment increased the solubility of starch in water without extensive depolymerization and allowed starch samples to be characterized by HPLC-SEC without the use of DMSO or alkaline aqueous solvent systems. Sonication for longer than 80 s substantially altered the HPLC-SEC profile.

A patent was issued (Nitsch 1995) on the use of ultrasound to produce high yields of starch degradation products with narrow molecular weight distributions. Molecular weights of the sonicated starches could be adjusted by varying the duration and intensity of the sonication treatment, and contamination with undesired low molecular weight components could be largely avoided. Proposed applications for these sonicated starch products included their use as blood plasma substitutes and as components in the preparation of pharmaceutical products. Schittenhelm and Kulicke (2000) studied the ultrasonic degradation of hydroxyethyl starch, starch acetate and carboxymethyl amylose in aqueous solution. Since the high-shear cavitation resulting from ultrasonic treatment causes bond cleavage at the center of gravity in polymeric materials, this treatment process was an effective method for preparing degraded polymers of varying molecular weight for use in establishing structure–property relationships. Comparison of cellulose and starch derivatives showed that the coiled structure of hydroxyethyl starch was degraded more slowly and yielded a higher limiting molar mass than the linear derivatives of cellulose.

Chung and co-workers (2002) prepared starch pastes by heating 5% aqueous slurries of mung bean, potato and rice starches for five minutes at 95 °C; and the hot starch pastes were then sonicated for 0.25–5 min. Treatment with ultrasound reduced the viscosities of the starch dispersions, reduced the amount of insoluble starch remaining after centrifugation, and increased the clarity of the starch dispersions. Lipatova and co-workers (2002) prepared aqueous corn starch pastes at 90 °C and the resulting starch pastes (1–8% starch concentration) were then treated with ultrasound at a frequency of 22 kHz for 5–60 s. Degradation of the starch pastes was accompanied by a decrease in viscosity, an increase in transparency, and an increase in the amount of water-soluble material. For comparison, a portion of the starch paste was treated with ultrasound and another portion was thermally treated at 105 °C. With thermal treatment, an 80% disruption of starch granules in a 5% starch paste was observed after 2 h, whereas with ultrasonic treatment the same amount of disruption was observed in only 40 s. Compared to thermal treatment, lower viscosities and lower optical densities were observed with ultrasonic treatment at 40 °C. The sharp decrease in viscosity was attributed to disintegration of swollen starch granules in the first few seconds of ultrasonic treatment.

Czechowska-Biskup and co-workers (2005) studied the effects of 360 kHz ultrasound on aqueous solutions of chitosan and corn starch, and observed that ultrasonic treatment reduced the molecular weights of both polysaccharides. Comparison of results obtained in the presence and absence of *tert*-butanol (a scavenger for hydroxyl radicals) suggested that degradation was caused by both mechanochemical effects and reactions of the polysaccharides with hydroxyl radicals generated during the cavitation process. Chain scission depended upon polymer concentration, ultrasonic power, and the gas used to saturate the aqueous solution. Ultrasound-induced chain scission of starch proceeded with lower yield than that observed with chitosan, probably due to the different chain conformation of starch, compared to the rod-like conformation of chitosan. Ida and co-workers (2008) studied the sonication of starch pastes prepared by heating 5% and 10% aqueous dispersions of waxy maize, potato, tapioca and sweet potato starches at 90 °C for 60 min. When 10% pastes prepared from waxy maize starch were sonicated at 80, 60 and 40 °C, the lowest final viscosities were observed at the lowest sonication temperatures, indicating that cavitation was more effective at reduced temperatures. Examination of the effect of sonication frequency on the viscosity of a 5% waxy corn starch paste, using a bath-type sonicator at frequencies of 183, 143, 99, and 44 kHz, showed that the final viscosities decreased as the frequency was lowered. Gel permeation chromatography and light scattering showed that the molecular weight of waxy maize starch decreased sharply with sonication times up to 30 min, and then decreased slowly. NMR spectra showed that although viscosities decreased during sonication, the chemical structure of the polymer chains was not altered. Cheng and co-workers (2010) studied the effects of ultrasonic treatment at 20 kHz on the properties of aqueous dispersions of gelatinized maize starch, as well as the properties of cast films prepared from the sonicated dispersions. Treatment with ultrasound greatly reduced the viscosities of the gelatinized starch dispersions, and fragments of granule ghosts present in the gelatinized dispersions were disintegrated by sonication. Cast films with good properties were obtained from the sonicated dispersions.

As an alternative to dispersing, gelatinizing or dissolving starch in water prior to treatment with ultrasound, Lima and Andrade (2010) pretreated high amylose maize starch (70% amylose) by first blending with 25, 30, 35, and 40% glycerol and then melt-processing the mixtures at 100 °C. After melt-processing, the samples were extracted with ethanol to remove glycerol and dried at 50 °C. The samples were then dispersed in water at 5 g/l at 50 °C and sonicated for 30 min at 10 °C at a frequency of 20 kHz. The largest reduction in intrinsic viscosity was observed for the sample melt-processed with 40% glycerol, and the higher degradation was attributed to a greater solubility of this sample in water prior to sonication. NMR spectra showed that melt processing and ultrasonic treatment did not change the chemical structure of the treated starch. Jackson (1991) examined the sonication of corn starch dispersions in 90:10 DMSO:water and observed complete solubility, as well as depolymerization of the starch, when the dispersions were sonicated for 15 or 20 s. Suguchi and co-workers (1994) prepared a 1% dispersion of wheat starch in 90% DMSO; after cooling to room temperature, the sample was sonicated at 40 °C for 300–2700 s. Gel filtration and reducing power were used to calculate molecular weights before and after sonication. The molecular weight was reduced by sonication, and the large amylopectin peak with $MW > 10^8$ changed to a broad peak with lower average molecular weight after sonication for 600 s. A large reduction in molecular weight was observed after 1800 sec, and a single sharp peak was observed.

Treatment with ultrasound has not only been used to disperse and dissolve starch in both aqueous and non-aqueous solvents, but has also been used to reduce the viscosities of aqueous polymer dispersions prepared by graft polymerizing water-insoluble polymers onto gelatinized starch. Gugliemelli and co-workers (1974) used ceric ammonium nitrate to initiate the graft polymerization of acrylonitrile onto gelatinized cationic starch (prepared by reacting wheat starch with 2-chlorodiethylaminoethyl hydrochloride); graft copolymers containing about 50% grafted polyacrylonitrile (PAN) were obtained. Ultrasonic treatment of the viscous, aqueous graft copolymer dispersions at 20 kHz for 1–3 min at 5–10 °C yielded low-viscosity, latex-like dispersions. The decrease in viscosity was attributed to the disruption of associations between individual particles of graft copolymer. Treatment with ultrasound had little effect on the molecular weight of grafted PAN. Low viscosity latexes were also obtained by ultrasonic treatment of graft copolymers prepared using chloroprene as the monomer (Gugliemelli and co-workers, 1976b). The scope of this research was further extended to include graft polymerizations of acrylonitrile onto a gelatinized cationic starch with quaternary amine substituents, as well as graft polymerizations using a mixed monomer system composed of acrylonitrile plus *t*-butylaminoethyl methacrylate (Gugliemelli *et al.*,

1976a). Latex-like dispersions were similarly obtained by sonicating graft copolymers obtained by polymerizing isoprene and acrylonitrile onto gelatinized wheat starch and onto a cationic starch derivatized with a quaternary amine (Gugliemelli *et al.*, 1979). In a series of later publications, starch graft copolymer latexes were prepared from methyl acrylate, butyl acrylate, methyl methylate and acrylonitrile, and the use of these latex-like dispersions as stabilizers to inhibit the water-erosion of soil was examined (Weaver and Fanta, 1987; Weaver, 1988, 1989, 1989). Laboratory tests showed that starch graft copolymers prepared from methyl acrylate and butyl acrylate showed good performance for this area of application.

Ultrasound has been used in the preparation of starch derivatives and also to modify starch for specific end-use applications. Mongenot and co-workers (2000) investigated the encapsulation of liquid cheese aroma in different carbohydrate matrices by spray drying, and used ultrasound to obtain stable emulsions with maltodextrin, which normally has poor emulsification properties. Spray-dried maltodextrin microcapsules were more effective for retaining cheese aroma when ultrasound was used for emulsification, as compared to high-speed homogenization. Huang and co-workers (2007) studied the effect of ultrasound on the properties and chemical reactivity of corn starch granules. Starch granules were suspended in water at a concentration of 30% and ultrasonic treatment times varied from 3 to15 min. Degree of hydrolysis (determined from the percentage of soluble saccharides formed in the aqueous starch suspensions) varied from 3.1% after three minutes sonication to 18.4% after 15 min. Little additional effect was observed with further treatment times. Although the pore sizes in sonicated granules increased as the degree of hydrolysis increased from 3.1 to 18.4%, there was little change in the birefringence patterns of the sonicated granules. The sonicated granules showed the native A-type X-ray diffraction pattern and there was little change as the degree of hydrolysis increased from 0 to 18.4%. The effect of ultrasonic treatment on the reactivity of the starch granules with propylene oxide was then determined. Although an increase in degree of hydrolysis from 0 to 3.1% did not significantly change the DS and reaction efficiency, a further increase in the degree of hydrolysis up to 15.7% resulted in a significant increase in DS. When the degree of hydrolysis was greater than 15.7%, decreases in DS and reaction efficiency were observed.

The effect of ultrasound on the preparation of acetate derivatives of yam starch was investigated by Zhang and co-workers (2008). Comparison of reactions carried out in the presence and absence of ultrasound showed that the use of ultrasound enabled the esterification reactions to be carried out with shorter reaction times and decreased amounts of catalyst. Čižová and co-workers (2008) used ultrasound and microwave irradiation to shorten the reaction time needed to synthesize water-soluble octenylsuccinates of carboxymethyl starch using DMSO as a solvent. Kim and co-workers (2008) prepared small starch particles by hydrolyzing the amorphous regions of waxy rice starch with *alpha*-amylase and then treating the enzyme-hydrolyzed starch with ultrasound in ethanol to fragment the starch granules. Practical applications for small starch particles in areas such as flavor delivery, fat substitutes, and reinforcing fillers for polymers are currently of interest. Grinberg and Gedanken (2010) also developed a process degrading starch into small fragments with ultrasound and then organizing these fragments into microsphere structures containing encapsulated insulin.

Treatment with ultrasound has also been used to enhance the reactivity of starch with enzymes. Barton and co-workers (1996) studied the effects of ultrasound on the activities of alpha-amylase and glucoamylase toward soluble potato starch. With α-amylase, increased reaction rates were observed with sonication at low concentrations of starch (8 g per liter); although at starch concentrations of 15–50 g per liter, both sonicated and un-sonicated samples gave similar results. With glucoamylase, ultrasound enhanced the reaction rate at starch concentrations up to 50 g per liter. Increases in reaction rate were attributed to an increase in the efficiency of mixing and diffusion of reaction components in the sonicated mixtures. Khanal and co-workers (2007) used ultrasound to treat slurries of corn to enhance the formation of glucose from corn starch by enzymatic hydrolysis with α-amylase and glucoamylase. Treatment with ultrasound decreased the size of corn particles and thus increased the formation of glucose, although prolonged use of high power settings reduced glucose formation due to denaturing of enzymes. This technology could find use in processes used for ethanol production.

Shewale and Pandit (2009) explored the use of ultrasound to enhance the hydrolysis of starch to glucose in sorghum flour using α-amylase and amyloglucosidase. Treatment at a frequency of 20 kHz for one minute decreased particle size and increased saccharification; the results were largely attributed to the increased availability of starch for hydrolysis due to ultrasound-assisted disruption of the protein matrix surrounding starch granules. Chan and co-workers (2010) examined the effect of sonication on the removal of proteins from the surfaces of corn starch and mung bean starch granules using 2% aqueous solutions of sodium dodecyl sulfate (SDS). Although the amount of surface protein was reduced by SDS treatment, no significant difference was observed between the results obtained with and without ultrasound. The X-ray diffraction patterns of SDS-treated starch samples were identical to those of the native starches, indicating that helical inclusion complexes with SDS were not formed during the treatment. Nikolić and co-workers (2010) investigated the possibility of improving glucose yields and ethanol productivity by using an ultrasound pretreatment in the production of bioethanol by saccharification and fermentation of corn meal in a batch system. Pretreatment of 1:3 corn meal:water mixtures at 40 kHz under the optimum conditions of five minutes at 60 °C increased the concentration of glucose and improved ethanol production.

Treatment with ultrasound was also investigated as a method for isolating starch granules from cereal grains without the use of some of the chemicals commonly used conventional wet-milling processes. The use of ultrasound in the isolation of rice starch was investigated by Wang and Wang (2004a). The surfaces of isolated starch granules were not damaged by sonication and, by combining a surfactant (particularly SDS) with the ultrasonic treatment, the yield of starch was increased and the amount of residual protein was decreased. The same authors investigated the effect of ultrasonic treatment on the efficiency of starch isolation when ultrasound was employed before, during, and after treatment with protease enzyme (Wang and Wang, 2004b). Starch isolation was improved by combining neutral protease with ultrasound, the preferred method of treatment was digestion with protease for 2 h followed by sonication for 15 or 30 min. Size exclusion chromatography and SEM showed no depolymerization of starch and no damage to starch granule surfaces. Zhang and co-workers (2005b) examined laboratory-scale wet-milling procedures using ultrasonic treatments at different points in the milling of corn starch; the results were compared with corn that was processed by conventional wet milling in the presence of sulfur dioxide. Starch yields obtained with ultrasound were comparable to those obtained by conventional wet milling and the starches isolated using ultrasonic treatment exhibited somewhat higher paste viscosities. Zhang and co-workers (2005a) also examined the use of ultrasound to enhance the yield of starch obtained from degermed corn flour and from hominy feed. Higher starch yields were obtained from the ultrasound-treated samples; grinding alone did not significantly enhance the yield of starch. Cameron and Wang (2006) observed that the combined action of neutral protease and sonication was effective in isolating starch from degermed corn flour. The preferred method of isolation was neutral protease digestion for 2h followed by sonication for 30min at 50°C. It was suggested that this method could replace sulfur dioxide and shorten the steeping time in the wet-milling process used for the isolation of corn starch.

In summary, irradiation of a liquid with ultrasound results in acoustic cavitation. This high-shear process results in intense local heating and high pressure followed by very high rates of cooling. The effect of ultrasound on granular starch, aqueous starch solutions, and aqueous dispersions of gelatinized starch has been studied. Since high-shear cavitation cleaves polymeric materials at the center of gravity, ultrasound has been used to prepare starch degradation products with narrow molecular weight distributions. Proposed applications for these degraded starch products include their use as blood plasma substitutes and as components in the preparation of pharmaceuticals. Treatment with ultrasound has also been used to prepare low-viscosity, latex-like dispersions from starch graft copolymers. When applied to soils, these aqueous dispersions functioned as stabilizers to inhibit water erosion. Ultrasound has been used in the preparation of starch derivatives, to enhance the reactivity of starch with enzymes used for ethanol production, and to facilitate the separation and isolation of starch granules from cereal grains during wet milling.

2.6 PROCESSING USING SUPERCRITICAL FLUIDS

When a fluid substance such as carbon dioxide or water is heated under pressure, the density of the liquid phase will decrease as the temperature is increased, whereas the density of the gas phase will increase with increased pressure. If a temperature–pressure phase diagram is constructed, and if we move upward along the curve separating the liquid phase from the gas phase, a temperature and pressure will be reached where the densities of the liquid and gas phases become identical and there will be no distinction between the gas and liquid phases. This point on this phase diagram is known as the critical point, and a supercritical fluid is any fluid that is at a temperature and pressure greater than those at the critical point. The critical temperature and pressure will vary according to the chemical structure of the fluid. For example, carbon dioxide (CO_2) has a critical temperature of 304 K and a critical pressure of 74 bar, whereas the critical temperature and pressure of water are 647 K and 221 bar, respectively (Williams *et al.*, 2002). Above the critical temperature, a gas cannot be liquefied by pressure.

Supercritical fluids have properties intermediate between those of gases and liquids. Densities and viscosities of supercritical fluids are less than those of liquids, and they are able to diffuse readily through solid substances. Their solvation power is dependent upon temperature and pressure and can be changed by altering these two variables. Although a number of substances have been used as supercritical fluids, carbon dioxide has been the most widely used because of its convenient critical temperature and pressure, as well as its low price, stability, non-flammability and low toxicity. Its functionality as a solvent lies between non-polar hydrocarbons and weakly-polar solvents. To improve the ability of carbon dioxide to dissolve polar molecules, a co-solvent such as methanol or ethanol may be added. One of the first practical uses for supercritical fluids was for the extraction of caffeine from coffee, and extraction of non-polar materials is a major application for these fluids. Supercritical fluids have been used as the mobile phase in chromatography and as solvents in chemical reactions and processing. They have also been used for the preparation of micometer-size particles and for the drying of biological specimens for microscopy, since solvent penetration is improved and the absence of surface tension in supercritical fluids allows samples to be dried without distortion. Supercritical fluids and their applications in biotechnology and the processing of food products have been reviewed by King (2000), Williams and co-workers (2002), and Darani and Mozafari (2009).

The gelatinization of cassava, wheat and potato starches in supercritical CO_2 was studied by da Cruz Francisco and Sivik (2002). Aqueous starch dispersions were prepared at 28% solids and the dispersions were treated with supercritical CO_2 for 20 min at pressures of 8 and 30 MPa. The percentage of gelatinized starch granules was determined using light microscopy to observe the loss in birefringence. Compared to gelatinization at atmospheric pressure, the percentage of gelatinized cassava and potato starch granules was greater when supercritical CO_2 was used at a pressure of 8 MPa, whereas less gelatinization was observed at 30 MPa. With wheat starch, higher percentages of gelatinized granules were observed at both CO_2 pressures (compared with the control gelatinization at atmospheric pressure), and a higher percentage of gelatinized granules was observed at the lower CO_2 pressure of 8 MPa. Pressurizing the starch at 8 MPa with nitrogen rather than CO_2 reduced the percentages of gelatinized granules for all three starches. The gelatinization behavior in supercritical CO_2 was attributed to the plasticizing effect of CO_2 on starch at low to moderate pressures.

The influence of supercritical fluid extraction on the starches obtained from ginger and turmeric was studied in order to evaluate the potential use of these extracted starches by the food industry after the ginger and turmeric tubers were extracted to obtain oleoresin and essential oil (Braga *et al.*, 2006). Extraction of ginger was carried out with supercritical CO_2 plus isopropyl alcohol as a co-solvent at 250 bar and 35 °C. Extraction of turmeric was carried out with CO_2 plus a co-solvent comprised of a 1:1 mixture of ethanol and isopropyl alcohol at 300 bar and 30 °C. SEM showed that supercritical fluid extraction did not modify the surface and overall morphology of the starch granules, and did not cause the granules to swell. Small effects of the extraction processes on the X-ray diffraction patterns, viscosities, and gelatinization temperatures of the starches were observed. Muljana and co-workers (2009) studied the gelatinization of potato starch (containing 16.2% water) in supercritical CO_2 at pressures up to 25 MPa and temperatures ranging from 50 to 90 °C. The intensities of infrared absorption bands associated with amorphous starch increased with both temperature and CO_2 pressure, indicating a higher degree of gelatinization. These observations were confirmed by DSC measurements. A maximum degree of gelatinization of 14% was observed at 90 °C and a CO_2 pressure of 25 MPa.

When experiments were carried out using nitrogen instead of supercritical CO_2, a maximum degree of gelatinization of only 3.9% was observed under the most extreme conditions, indicating that the gelatinization observed in supercritical CO_2 was not solely due the effect of hydrostatic pressure. The plasticizing effect of supercritical CO_2 was attributed to the dissolving of supercritical CO_2 in the starch granule matrix, which would induce swelling and plasticize the amorphous regions, thus accelerating the diffusion of water into the granule. X-ray diffraction was also used to evaluate changes in the degree of gelatinization of the treated samples. Although the diffraction pattern for samples treated with supercritical CO_2 at 90 °C was similar to that of the native starch, the area of the amorphous region increased due to partial gelatinization, and higher CO_2 pressures produced lower orders of crystallinity. SEM showed no changes in the appearance and morphology of the starch granules due to the supercritical CO_2 treatment.

The solubility and the diffusion coefficients of CO_2 in starch–water mixtures were determined by Chen and Rizvi (2006a) using native and pregelatinized corn starches for their experiments. Starch mixtures containing 50:50, 75:25, and 100:0 ratios of pregelatinized starch to native corn starch were prepared, and sufficient water at 50 °C was then added to obtain a water content of 40%. The starch–water mixtures were then placed in a diffusion cell and the solubility and diffusion coefficients of CO_2 were determined at pressures up to16 MPa at 50 °C. The experimentally-measured solubility accounted for the estimated swollen volume; the solubility increased with pressure. The solubility of CO_2 showed no dependence on the degree of gelatinization of the starch. The diffusion coefficient of CO_2 increased with the concentration of CO_2 (which was pressure dependent) and decreased with increasing degrees of starch gelatinization.

Supercritical CO_2 has been used to extract various materials from the starch matrix. Artz and Myers (1995) prepared a model corn starch system to evaluate extraction with supercritical CO_2 as a method for analyzing emulsifiers in food products. Twin-screw extrusion was used to mix acetylated monoglycerides with corn starch at a concentration of 3%, and extractions of the mixtures were then carried out. Supercritical fluid extractions with CO_2 and with CO_2 containing 5% methanol as a co-solvent resulted in higher extraction percentages than those observed when extractions were carried out with chloroform/methanol. A European patent application describes a process for removing off-flavors, off-odors and off-colors from starches by extraction of the starches with supercritical fluids (Koxholt *et al.*, 2003). Due to environmental concerns and the safety hazards associated with conventional methods of extraction using organic solvents, Hubbard and co-workers (2004) investigated alternative methods for the extraction of lipids from wheat flour using supercritical CO_2. Different pressures, temperatures and concentrations of co-solvent were used for the extractions. The best results were obtained with 12% ethanol as a co-solvent, a pressure of 51.7 MPa, and a temperature of 80 °C. Under these conditions, the amount of extracted lipid was similar to that obtained by conventional extraction using American Association of Cereal Chemists (AACC) and American Oil Chemists' Society (AOCS) methods. Ashraf-Khorassani and co-workers (2005) examined supercricital fluid extraction with CO_2 for its ability to remove impurities from several common pharmaceutical excipient powders, including corn starch. Solutions of formic acid, formaldehyde and hydrogen peroxide were added to the excipients (or sand control), and the samples were extracted at 450 atm and 40 °C. Surface-bound impurities were removed more effectively than imbedded impurities, and extractions from starch were less effective than extractions from some of the other excipients tested. Extraction of hydrogen peroxide was not effective, since supercritical CO_2 is a poor solvent for this material.

Peterson and co-workers (2007) studied the critical fluid extraction of native lipids from corn starch using 80:20 CO_2:ethanol and 100% ethanol as solvents. The effects of lipid extraction on the pasting properties of the extracted starch were determined, and the extraction results and pasting properties were compared with those obtained with corn starch that was defatted using refluxing 75:25 (v/v) *n*-propanol:water. Since the temperatures and pressures used in this study were below the critical point for ethanol, this procedure was designated as a critical fluid-extraction rather than an extraction with a supercritical fluid. No extractable lipid was obtained when supercritical fluid extractions were carried out with pure CO_2. More native lipid was extracted with pure ethanol than with 80:20 CO_2:ethanol, and improved extractions of native lipid were observed when the moisture content of the starch was increased from 10 to 19%. Compared to refluxing 75:25 *n*-propanol:water, critical fluid extraction resulted in incomplete removal of native lipid, and was not selective with respect to the chemical composition of the lipid removed. However, despite the incomplete removal of native lipids, critical fluid extractions of corn starch reduced the swelling and deformability of the starch granules when they were heated in water, and also altered the paste viscosities observed at different concentrations. In a subsequent investigation, critical fluid extractions with 75:25 ethanol:water were shown to be more effective in removing native lipids from corn starch than previous extractions that did not include water as a component in the solvent system (Peterson *et al.* 2008). The efficiency of lipid extraction increased with the solvent/starch ratio, and over 99% of the native lipid could be removed from normal dent corn starch using a solvent/starch ratio of 11.9 ml/g. Comparison of critical fluid-extracted starches with starch that was conventionally defatted by extraction with refluxing 75:25 *n*-propanol:water showed that neither of these extraction procedures caused any changes in the X-ray diffraction pattern, the birefringence of the starch granules, and the morphology of the granules, as observed by SEM. Pasting curves of the extracted starches exhibited plateau-like peak viscosities, as opposed to the sharper peak viscosities observed with native corn starch. Maximum viscosities and shapes of the pasting curves depended on the concentration of starch, the amount of residual native lipid in the starch sample, and the method used for lipid extraction (i.e., critical fluid vs. refluxing 75:25 *n*-propanol:water).

A process was developed for preparing extruded foams or puffed products from starch-containing materials using supercritical fluids in the extrusion process (Rizvi and Mulvaney, 1992; Mulvaney and Rizvi, 1993). In this process, aqueous starch dispersions are first cooked by extrusion at high temperature and pressure. The cooked starch is then cooled in the extruder barrel to less than 100 °C (to prevent foaming and expansion of the extrudate by steam). In the final step, supercritical CO_2 is injected into the barrel of the extruder just before the cooked starch is forced through the extruder die (or nozzle). Under these conditions, the added CO_2 dissolves in the cooked, aqueous starch dispersion and serves as a blowing agent for the foaming process. Nucleation for the formation of bubbles is caused by the sudden drop in pressure at the die, and bubble growth then occurs by diffusion of CO_2 into the nucleated bubbles. Further expansion of the extrudate then takes place, until the foamed structure is set by post-extrusion oven drying. Extruded foams with regularly-shaped pores can thus be produced at temperatures lower than 100 °C, which would not be possible if conventional processing methods using steam as the blowing agent were used. Since lower temperatures are used in the supercritical fluid process, starch degradation is reduced, and heat-sensitive materials such as flavorings and colorants can be dissolved in the supercritical CO_2 and easily mixed into the cooked starch dispersion before it passes through the die. Extruder wear is also reduced at the lower temperatures used. The conditions used in the supercritical fluid process produce denser, harder foamed products with a more uniform cell distribution than those obtained by conventional steam puffing.

To aid in the design and control of the supercritical CO_2 extrusion process, Singh and co-workers (1996) studied the diffusivity of carbon dioxide in extruded gelatinized starch at 343 K and pressures up to 117 bar (1700 psi). The diffusivity decreased with pressure but was not affected by moisture content in the range 34.5–39%. Lee and co-workers (1999) investigated the effects of various processing parameters on the physical properties of extruded corn starch extrudates that were expanded using supercritical CO_2 injection. The effects of barrel temperature, screw speed and water addition on the specific mechanical energy input for the process were examined, as well as the expansion ratio, water absorption, water solubility, breaking stress and elastic modulus of the extruded products. Alavi and co-workers (1999) determined the effect of adding different levels of egg white protein and whey protein concentrate on the expansion and cellular structure of starch-based supercritical fluid extrudates. Since post-extrusion drying is used in the process, the effect of drying temperature on the expansion of extrudates and their cellular structure were also studied.

Cell size and uniformity as well as cell density were determined. Addition of proteins reduced the shrinkage of extrudates before oven drying and reduced the bulk density of the extruded products. The cell size distribution was more uniform than that observed with steam-expanded products. Supercritical fluid extrudates also had a relatively non-porous skin surrounding the internal cells, which reduced water penetration (a desirable property for a breakfast cereal). Compared to whey protein concentrate, it was observed by Gogoi and co-workers (2000) that addition of egg white gave a softer skin and a fragile but well-formed cellular structure. Also, when drying was carried out between 70 and 100 °C, the structure was more homogeneous when egg white was added.

To better understand the supercritical extrusion process, Alavi and co-workers (2003a) developed a mathematical model for bubble growth, as well as models for the flow of starch melt through the extruder die, the bulk diffusion of CO_2 and water to the atmosphere, and the heat transfer in the extruded product. The same authors also published simulation results and comparisons with experimental data (Alavi *et al.*, 2003b). Data obtained with extrudates of pregelatinized corn starch and potato starch with 4–7% added whey protein concentrate were used for validation. Since the high diffusivity of CO_2 favors escape of the gas to the environment (and thus reduces the amount of CO_2 available for diffusion into the bubbles) Alavi and Rizvi (2005) examined two approaches for minimizing this problem. The first approach was to increase the rate of nucleation of CO_2 bubbles, which was achieved by reducing the diameter of the extruder nozzle to obtain a higher rate of pressure drop when the starch–CO_2 melt was forced through the nozzle. This approach increased the expansion ratio by as much as 160%. The second approach was to reduce the temperature of the melt in order to decrease the diffusion coefficient of CO_2; this was achieved by introducing a cooling zone in the extruder barrel prior to entry of the melt into the nozzle. A 34% increase in expansion ratio was observed when the melt temperature was decreased from 60 to 40 °C.

Chen and Rizvi (2006b) used a slit-die rheometer fitted to a twin extruder to measure the viscosity and expansion characteristics of starch-based melts plasticized with water and supercritical CO_2. Supercritical CO_2 was shown to be an effective plasticizer for starch–water mixtures, and the increase in free volume due to addition of supercritical CO_2 was shown to be an effective method for viscosity reduction. Mariam and co-workers (2008) prepared extruded foams having thermal properties (such as heat capacity, thermal conductivity and thermal diffusivity) comparable to commercial products. Pregelatinized corn starch was extruded with different concentrations of whey protein and supercritical CO_2, and the expansion and density of the extruded foams were varied by changing the amount of whey protein added and the injection rate of supercritical CO_2. Cho and Rizvi (2009a) studied foams prepared from pregelatinized corn starch containing five different concentrations of whey protein isolate and extruded at four different levels of supercritical CO_2. The melt rheology of the extrudates was determined with an on-line slit die rheometer; X-ray microtomography was used to determine the microstructural features of the foams, such as average cell diameter, cell wall thickness, and void fraction. Rheological properties were then correlated with the expansion characteristics and microstructures of the foams.

In a second article by Cho and Rizvi (2009b), microstructural data were correlated with mechanical properties, as determined using compression testing and three-point bending tests. Cell sizes, observed in cross-sections of the extrudates, decreased with radial distance from the center; while in the longitudinal direction, the cell shapes were more elliptical than spherical, and were aligned along the extrusion direction. Both the density of the extrudate and the ratio of cell wall thickness to cell diameter were found to be good predictors of compressive and flexural mechanical properties. Compressive modulus data suggested that the anisotropy of cell shape, due to cell elongation in the longitudinal direction, affected the mechanical properties of the extruded products.

Since starch microcellular foams are readily dispersible in water, they cannot be used for end-use applications in aqueous environments. To reduce the water sensitivity of these foams, Ayoub and Rizvi (2008) studied the cross-linking of foams with epichlorohydrin during their preparation by supercritical CO_2 extrusion. A mixture of native and pregelatinized wheat starch was used; the in-barrel moisture content of the starch melt was maintained at about 45%. Three levels of epichlorohydrin were used and sodium hydroxide was added to the formulation to provide the alkaline pH needed for the cross-linking reaction to occur. The relationship between cross-link density and cell size was studied and a decrease in average cell size was observed with increasing concentrations of epichlorohydrin. The cross-linked starch foams prepared by this process were insoluble in water at room temperature. Patel and co-workers (2009) prepared cross-linked and uncross-linked starch foams by first using supercritical CO_2 extrusion to prepare the foams (with and without epichlorohydrin) and then immersing the extrudates in ethanol/water solutions of varying concentration to solvent exchange the water remaining in the foams with ethanol. For comparison, molded aqueous starch gels (aquagels) were also solvent exchanged in a similar manner. The aquagel-based products showed a finer micropore structure than the products obtained by extrusion, they also had lower density and higher brightness than the extruded samples.

The effect of supercritical fluid extrusion of ground sorghum grain on the enzymatic production of ethanol was investigated by Zhan and co-workers (2006). Analysis of supercritical CO_2-extruded sorghum showed increases in measureable starch content and free sugar content; high levels of gelatinized starch were also observed. The supercritical-extruded and non-extruded sorghum were further liquefied, saccharified and fermented to ethanol. The yield of ethanol from supercritical-extruded sorghum was greater than that obtained from non-extruded material, indicating that supercritical extrusion was an effective pretreatment method that could improve the bioconversion of sorghum starch into ethanol. Improvements in the bioconversion process were attributed to the enhanced release of fermentable starch from the protein matrix, a decrease in fiber content, and the fine, porous structure produced by the extrusion process.

Supercritical fluids have been used as reaction media for the preparation of starch derivatives. Yalpani (1993) prepared a phosphate ester derivative of amylose (phosphorus content, 3.7%) by heating an aqueous mixture of amylose and sodium tripolyphosphate in supercritical CO_2 for 1.5 h at 60 °C. Oxidized starches were obtained when corn starch and maltodextrins were treated with 19:1 mixtures of carbon dioxide and oxygen for two hours. Carbonyl absorption bands in infrared spectra provided evidence for oxidation. A European patent application (Harris *et al.*, 1999) describes a process for chemically modifying starch to produce starch derivatives, such as esters and ethers, in supercritical CO_2. The process yields starch derivatives with a high degree of substitution without the use of organic solvents and by-products of the reaction are easily separated. For example, high amylose starch was esterified with acetic anhydride in supercritical CO_2 using sodium acetate catalyst at 115 °C and 4000 psi for 2.5 h to yield starch acetate with a DS of 2.47. The acetylation of starch in supercritical CO_2 was also studied by Muljana and co-workers (2008). When supercritical CO_2 was compared with water as a solvent for the acetylation of potato starch, better results were obtained with supercritical CO_2 with respect to degree of substitution and acetic anhydride selectivity. Green and co-workers (2000) prepared silicon-containing composites from starch by adding $SiCl_4$ using supercritical CO_2 as the carrier medium. Characterization of the resulting composites showed that 12–15% silicon, in the form of SiO_2, was homogeneously dispersed throughout the starch matrix.

Duarte and co-workers (2009a) used supercritical fluid technology to prepare starch-based, porous matrices for use as scaffolds for tissue engineering. This technique (referred to as immersion precipitation) involves extracting polymer solutions in organic solvents with supercritical CO_2. Since the solvent chosen to dissolve the polymer is soluble in supercritical CO_2, whereas supercritical CO_2 is a non-solvent for the polymer itself, a dry, porous polymer structure is formed in the process. The polymers used in this study were commercial blends of starch and poly-L-lactic acid (50:50 and 30:70 by weight); the solvents used were dichloromethane and chloroform. Process variables such as polymer concentration, temperature, and CO_2 pressure were examined. This technique was also used to prepare porous starch–poly(L-lactic acid) composite scaffolds loaded with inorganic particles (Bioglass) for use in bone tissue engineering (Duarte *et al.*, 2009b). A similar method was used to obtain a dry, porous structure from a commercial 30:70 wt-% blend of corn starch and poly(ε-caprolactone) that was dissolved in chloroform at a concentration of 15% (Duarte *et al.*, 2010). The effect of pressure (80–150 bar) and temperature (45 and 55 °C) on the morphologies of the scaffolds was studied by SEM and micro-computed tomography. Mechanical properties of the scaffolds were also studied, as well as the adhesion, morphology, viability and proliferation of cells.

A biomass conversion system using supercritical water was developed by Saka and Ueno (1999) for the hydrolysis of cellulose and starch to glucose and levoglucosan. The treatment temperature was 500 °C and the pressure 35 MPa. Water in this supercritical state can function as an acid and thus hydrolyze the polysaccharides to produce low molecular weight sugars. Although the major subject of this study was cellulose, corn starch, which is easier to hydrolyze, was examined for comparison. Cellulose was liquefied by the supercritical water treatment in about ten seconds, whereas starch was liquefied in about five seconds. Glucose yields from cellulose II, cellulose I and starch were 48%, 32% and 33%, respectively. Water under supercritical conditions has also been used to degrade biomass into gaseous products such as hydrogen, CO_2, and hydrocarbons such as methane. This process has been commonly referred to as gasification and is receiving increased attention due to the current interest in the use of hydrogen as a fuel. Antal and co-workers (2000) used this technique to treat biomass feedstocks including corn starch and potato starch gels, wood sawdust suspended in a corn starch gel, and potato wastes. The organic material vaporized when the feedstock was rapidly heated in a tubular flow reactor to temperatures above 650 °C at pressures above the critical pressure of water; a packed bed of carbon in the reactor catalyzed the gasification reaction. The gaseous mixture formed in the process was composed of hydrogen, CO_2, methane, carbon monoxide, and traces of ethane; the composition of the mixture was influenced by the peak temperature of the reactor and the condition of the reactor wall. High yields of gas (>2 l/g) with a high content of hydrogen (57 mol-%) were observed at the highest temperatures. A thermodynamic model was developed by Tang and Kitagawa (2005) to predict the equilibrium composition of product mixtures. D'Jesús and co-workers (2005) studied the effect of $KHCO_3$ addition, particle diameter, and the dry matter content of aqueous dispersions on the gasification of corn starch and corn silage with supercritical water. Treatments producing hydrogen, CO_2, and hydrocarbons ranging from CH_4 to C_4H_{10} were carried out at a maximum temperature of 700 °C. In a subsequent study

(D'Jesús *et al.*, 2006) corn starch was used to evaluate the influence of potassium on the gasification efficiency. The gasification yield increased with increasing potassium concentration up to 500 ppm and did not increase further. Temperature was the most important process variable and the gasification yield increased when the temperature was increased from 823 to 973 K.

To summarize the significant research in this area, supercritical CO_2 has been used to extract impurities from starch, and the gelatinization of starch in supercritical CO_2 has been studied. Ethanol-containing critical fluids have been used to extract native lipids from granular corn starch. Starch-based extruded foams were prepared using supercritical CO_2 as the blowing agent in the foaming process. Under these processing conditions, foams with good properties were prepared at temperatures less than 100 °C. At these lower temperatures, starch degradation is reduced, extruder wear is minimized, and heat sensitive materials such as flavorings can be dissolved in supercritical CO_2 and easily mixed into the cooked starch prior to foam formation. Extrusion with supercritical CO_2 also improved the enzymatic conversion of starch to ethanol. Supercritical CO_2 has been used as reaction medium for preparing starch derivatives, such as esters and ethers, and as a solvent for preparing starch-based porous matrices for tissue engineering. When supercritical water was used to treat starch at 500 °C, hydrolysis of starch to low molecular sugars was observed due to the acidic properties of supercritical water at this temperature. Supercritical water has also been used at temperatures exceeding 650 °C to degrade polysaccharide-containing biomass into gaseous products such as hydrogen, CO_2, and low molecular weight hydrocarbons. This process has been referred to as "gasification" and is receiving increased attention due to the current interest in hydrogen as a fuel.

2.7 EXTRUSION PROCESSING

Extrusion equipment and its application to polymer and food processing in general have been described in a number of books and reviews (Guy, 2001; Kohlgruber, 2007; Rauwendaal, 2001; Riaz, 2000). Extruders generally consist of one or more steel screws powered by an electric motor contained inside a closed barrel. Solids (such as starch) and liquids (such as water) can be added at nearly any point in the extruder using appropriate metering feeders or pumps. Extruder barrels consist of sections which are normally heated separately to vary the temperature along the length of the extruder. Heat is also contributed by work supplied by the motor and is termed specific mechanical energy (SME). SME is quantified by the equation: $SME = \tau\omega/m$, where τ is torque, ω is rotation rate and m is mass flow rate. Extruder screws can be designed in a wide variety of ways to generate forwarding motion, compression, reverse flow, dispersive mixing and distributive mixing. Single-screw extruders are generally used when high-pressure, stable flows are needed for sheet or foam extrusion, while twin-screw extruders have excellent conveying and mixing capabilities.

Extrusion is a useful method for gelatinizing starch, blending starch with other components or conducting reactions in highly-viscous, lower-moisture systems. Additional functions of extrusion may include specific shape generation, sterilization, reduction in moisture or other volatile components, encapsulation and flavor generation. The food industry has long used extrusion to prepare ready-to-eat cereal snacks as well as animal feeds (Guy, 2001; Harper, 1978; Mercier *et al.*, 1989; Riaz, 2000). Much of the research on starch extrusion has been conducted to understand the structural changes and properties of starch in such systems (Akdogan, 1999; Barsby *et al.*, 2002; Colonna *et al.*,

1987; Lai and Kokini, 1991; Liu *et al.*, 2009; Wolf, 2010; Xie *et al.*, 2006a). More recently, interest has focused on rates of digestibility due to relationships to nutrition and disease (Björck and Asp, 1983; Sajilata *et al.*, 2006; Singh *et al.*, 2010; Svihus *et al.*, 2005). Since chemically-modified starches are used in foods, paper coatings and a myriad of other areas, many studies have been conducted on the chemical modification of starch in by extrusion (Carr, 1994b; Narkrugsa *et al.*, 1992; Wilpiszewska and Spychaj, 2008; Xie *et al.*, 2006b). More recently, there has been considerable interest in the extrusion of starch alone and with other polymers for use as biodegradable plastics (Averous, 2004; Dubois and Narayan, 2003; Janssen and Moscicki, 2009; Kalambur and Rizvi, 2006; Shogren *et al.*, 1993; Yu *et al.*, 2006). This section briefly describes some fundamentals related to starch extrusion, including the authors' work as well as some more recent studies conducted in the past 5–10 years.

Gelatinization or melting of starch during extrusion is dependent on several factors, including moisture content, temperature, pressure and shear or SME. These areas have been the subject of reviews (Lai and Kokini, 1991, Liu *et al.*, 2009; Xie *et al.*, 2006a). Amylopectin, amylose and amylose–lipid complexes form separate phases during heating, so these will have different melting temperature ranges (Shogren, 1992, 1994). Amylopectins from different sources have different branch lengths, and hence different types of native crystalline structures (A, B, C), and these can have different melting temperatures. In the absence of shear, thermal transitions are conveniently characterized by DSC. Such data on the effect of water or other diluent on the melting temperature of starch are often modeled by the lattice theory of Flory (1953):

$$1/T_m = 1/T_m^{\circ} + RV_2/\Delta H V_1(v_1 - \chi v_1^2)$$

where T_m° is the melting temperature for dry starch, V_2 and V_1 are the molar volumes of starch and water, v_1 is the volume fraction of water and χ is the starch–water interaction parameter. Typical values of T_m°, ΔH, and χ for the melting of the amylopectin phase of corn starch in water are 286 °C, 18 kJ/mol and 0.64, respectively (Shogren, 2000). Thus, melting temperatures for starch decrease as water content increases; this allows preparation of starches having different kinds and amounts of crystalline phases depending on extrusion temperature and moisture. At high extrusion temperatures, starch becomes completely amorphous, although amylose–lipid V-type complexes often form on cooling, since these complexes crystallize rather rapidly. Increasing screw speed, and hence shear (or SME), has also been found to increase the extent of starch gelatinization or melting (Xie *et al.*, 2006a). Increasing pressure has been found to decrease the melting temperature of starch in water at a rate of 0.075 °C/MPa (Douzals *et al.*, 2001). Thus, the effect of pressure is expected to be rather small at typical extrusion pressures of ~10 MPa. Efforts have been made to computationally simulate the melting, mixing, shearing and heating processes occurring during the extrusion of starch (Edi-Soetaredjo, 2010).

Other solutes or solvents have been used to partially or completely replace water to process and gelatinize starch by extrusion (Shogren, 1993). Some of the most common of these are glycerol, sorbitol, and urea (Shogren *et al.*, 1992). Since starch is a weak acid (pKa ~ 12.8), it is expected that mild bases, such as amines and amides, would tend to interact with starch and promote melting. Tomka (1990) showed that solvents having a solubility parameter of 15–25 cal$^{1/2}$/cm$^{3/2}$ were effective additives for the thermoplastic processing of starch. Recently, several ionic liquids have been shown to be good solvents for starch, but no extrusion work has yet been published in this area (Biswas *et al.*, 2006b).

Properties of starch extrudates, such as strength, crispiness, and brittleness, depend to a large extent on molecular weight, which is influenced to a great extent by extrusion conditions (Liu *et al.*, 2010; Van Den Einde *et al.*, 2003; Willett *et al.*, 1997). Increasing SME leads to decreased molecular weights and viscosities of the amylopectin component, while amylose is rather insensitive to shear. The high molecular weight fractions of amylopectin, in particular, tend to be reduced the most in molecular weight during extrusion.

Several applications of extruded starch in the food and pharmaceutical areas have been described in the recent literature. The preparation of starch-encapsulated drugs by extrusion for controlled-release applications has been described (Repka *et al.*, 2007; Trivedi *et al.*, 2007). Starch capsules which can replace those comprised of gelatin were prepared by extrusion and injection molding (Stepto and Tomka, 1987). Microencapsulated flavors, omega fatty acids and other food ingredients requiring protection and stabilization have been prepared by extrusion (Gray *et al.*, 2007; Qi and Xu, 1999; Shahidi and Han, 1993). Instant food powders have been prepared by extrusion with the benefit of reduced water and energy consumption over traditional drum drying processes, which operate at higher water contents (Schuchmann and Danner, 2000). Starches that are resistant to digestion with amylases are termed resistant starches, and these starches have been prepared by extrusion. The resistance or decrease in rate of digestion by amylases is due to the presence of residual, ungelatinized native crystalline amylopectin or to the formation of retrograded, crystalline amylose during subsequent storage of the starches in moist food (Sajilata *et al.*, 2006; Singh *et al.*, 2010). Thus, high amylose starches are often used to prepare resistant starches (Shrestha *et al.*, 2010). Extruded, acid-modified starches tend to yield more resistant starch after retrogradation, since lower molecular weight starch can more easily reorganize into crystalline structures (Hasjim and Jane, 2009). Health benefits of resistant starches include slow glucose absorption for improved metabolic control in diabetes, and the functionality of these starches as probiotics for colon cancer prevention and for reduction in cholesterol for the prevention of heart disease (Sajilata *et al.*, 2006).

Chemically modified starches are used in a myriad of applications, including paper coating, cardboard adhesives, foods, textile sizing, and various binders (BeMiller, 1997; Tharanathan, 2005; Tomasik and Schilling, 2004; Wurzburg, 1986). Chemical modifications serve to slow recrystallization, prevent syneresis, confer viscosity stability via cross-linking or add anionic or cationic groups. Typical commercially-available modified starches are low DS (<0.05) and include hydroxyethyl or hydroxpropyl starches, oxidized starches, cationic starches, starch acetates, starch phosphates, starch alkenylsuccinates and starch graft copolymers. These are usually prepared in aqueous, granular suspension with $>60\%$ water in batch reactors, although some reactions such as phosphorylation can be conducted in a dry state. The potential advantages of using reactive extrusion to conduct starch reactions is the ability of extruders to handle viscous, low-water states and thus enhance reaction efficiency and decrease by-products. Thermoplastic starch, formed during extrusion, tends to react more quickly than granular starch; and the distribution of the modifying groups is generally different. Extruders also offer continuous, rapid reactions in minutes at high temperatures and under pressurized conditions, when it is necessary to prevent the evaporation of volatiles. A disadvantage of extrusion processing is that it is generally more difficult to extract by-products from gelatinized starch than from granular starch suspensions.

Preparation of starch derivatives by reactive extrusion has been reviewed (Wilpiszewska and Spychaj, 2008; Xie *et al.*, 2006b). Reaction efficiencies up to 92% for the cationization of starch with Quat 188 in aqueous sodium hydroxide at 90 °C were achieved in a twin-screw extruder (Carr, 1994a), compared to typical reaction efficiencies of 85% for batch reactions. A similar process has been scaled-up and compared to a computational model (Berzin *et al.*, 2007a, 2007b). Oxidized starches having high water solubility were prepared by reactive extrusion of starch, hydrogen peroxide and ferrous–cupric sulfate catalyst (Wing and Willett, 1997). Graft copolymers of starch with acrylonitrile using ceric ammonium nitrate initiator were prepared by twin-screw extrusion; and the conversion of 74% was similar to that of a batch reaction (Yoon *et al.* 1994). Graft copolymers of starch with acrylamide with conversions of 95% and graft efficiencies of 75% were prepared by twin-screw extrusion using ammonium persulfate as initiator (Finkenstadt and Willett, 2005). There are numerous examples of starch esters prepared by reactive extrusion, and there have been many studies on the preparation of starch phosphates by reactive extrusion of dry or semi-dry mixtures of starch with orthophosphates or metaphosphates (Landerito and Wang, 2005; O'Brien *et al.*, 2009; San Martin-Martinez *et al.*, 2004; Seker *et al.*, 2003). Fatty acid esters of starch, including starch acetates, starch succinates, starch alkenylsuccinates, starch maleates and starch palmitate, have been prepared by extrusion of starch with anhydrides, sometimes in the presence of sodium hydroxide or sodium bicarbonate ($NaHCO_3$) as catalyst (Hannah and Fang, 2002; Murua-Pagola *et al.*, 2009; Raquez *et al.*, 2008; Rudnik and Zukowska, 2004; Tomasik *et al.*, 1995; Wang *et al.*, 1997).

Extrusion of starch and starch–polymer blends into biodegradable plastics has been the subject of extensive research and development over the past twenty years. Worldwide, starch-based plastics are produced commercially at the rate of about 300 million pounds/year, and these products represent about 0.1% of the total plastics market (Shen *et al.*, 2009). Reasons for the current interest in these starch-based materials have been the low cost of starch, the replacement of dwindling petroleum-based resources with renewable feedstocks, the biodegradability of these polymers (to alleviate problems associated with the disposal of plastic waste), and the creation of new markets for farm commodities (Gross and Kalra, 2002). Interest in starch as a plastic has continued despite some problems with the use of starch as a starting material, for example, poor water resistance, brittleness and low strength. Examples of some commercially successful starch-based products include starch foam packing peanuts and protective foam sheet, starch foam plates, and starch–polymer blends for compost bags and molded articles. Starch-based plastic materials have been the subject of several reviews (Averous, 2004; Averous and Halley, 2009; Bastioli, 2001; Chivrac *et al.*, 2009; Dubois and Narayan, 2003; Janssen and Moscicki, 2009; Kalambur and Rizvi, 2006; Shogren *et al.*, 1993; Yu *et al.*, 2006).

Starch foams for protective packaging are typically prepared by extrusion of starch with 15–20% moisture, a nucleating agent such as talc, and 5–20% of an added polymer (Nabar *et al.*, 2006; Pushpadass *et al.*, 2008; Sjoqvist *et al.*, 2009; Willett and Shogren, 2002). Extrusion temperatures are typically 120–180 °C. Water serves as both plasticizer and blowing agent. The low moisture content ensures that the glass transition temperature of the starch foam is sufficiently high to prevent collapse after expansion and cooling. The added polymer serves to enhance expansion and also, in the case of certain hydrophobic polymers such as PLA and PHBV, to increase water resistance. The latter occurs as a result of migration of the hydrophobic polymer to the surface of the foam (Willett and Shogren, 2002). Foams made from extruded high amylose starches generally have better expansion and resilience, due to increased entanglement of the long linear polymer chains. Starch foams have competed successfully with expanded polystyrene foam cushioning materials and have captured a significant share of that market.

Thermoplastic starch has been blended with a variety of other polymers via extrusion to improve moisture resistance and flexibility (Averous, 2004; Dubois and Narayan, 2003; Janssen and Moscicki, 2009)); Kalambur and Rizvi, 2006; Yu *et al*., 2006). Polyesters, such as polylactic acid (PLA), polyhydroxy butyrate-co-valerate (PHBV) and polycaprolactone (PCL), have been often used as additives in these blends, since they are also biodegradable. Intense shear forces within the extruder reduce the size of the starch domains to several micrometers. The synthetic polymers chosen for these starch–polymer blends will usually have compositions and rheological properties that will enable them to form the continuous phase in these blends, so that the starch phase will be encapsulated and the resistance to moisture will be improved. Compatibilizing agents are usually added to ensure good adhesion between the starch and polymer phases. Nanofillers, such as clays, have been found to enhance stiffness and moisture resistance (Chivrac *et al*., 2009). Examples of applications for starch–plastic blends include packaging films, mulch films, planting pots, food containers and hygiene articles (Bastioli, 2001).

In summary, extruders are currently used on a large scale to process starch and starch-rich grains into ready to eat snack foods, pastas, and so on. With the current interest in improving nutrition and decreasing chronic diseases, it seems likely that extruders will be used in the future to create healthy, convenient foods incorporating resistant starch along with fruits, vegetables and whole grains. Extruders are also currently used on a small commercial scale to make starch-based biodegradable foam peanuts, foam sheet cushioning, and compost bags. Interest in these areas will likely continue, due to depletion of petroleum as a resource for plastic materials. Since extruders can heat and process starch with low moisture contents, use of extruders to chemically or physically modify starch may expand as interest in water and energy conservation grows.

2.8 PROCESSING BY STEAM JET COOKING

Many applications of starch require heating (i.e., cooking) the starch granules in water to form aqueous dispersions or solutions, which can be used as liquids or gels, or dried by various techniques. Steam is widely used to provide the heat for cooking and the technology for cooking starch on an industrial scale has changed very little for much of the twentieth century. A continuous steam cooking apparatus was described in 1940 (Coppock, 1940) and a significant engineering improvement, described later by Winfrey and Black (1964), introduced the possibility of applying excess steam flow to improve the properties of the cooked starch dispersion. Modern devices, commonly known as steam jet cookers, can be used in the thermal mode to apply just enough steam to gelatinize or paste the starch granules (Kasica and Eden, 1992), or in the excess steam mode, in which the additional mechanical shear resulting from the passage of excess steam through the cooker reduces the viscosity and molecular weight of the starch (Klem and Brogly, 1981). It is widely known that the molecular and bulk properties of starch processed by excess steam jet cooking vary significantly from those of starch heated with less mechanical shear, and this fact has enabled the exploration of new applications and post-processing methods based on the technology of steam jet cooking. It has been found that factors such as initial slurry pH and steam pressure significantly affect the intrinsic viscosity of jet-cooked starch dispersions (Dintzis and Fanta, 1996). Other properties of excess steam jet cooked starch have also been investigated, including the mechanical and thermal degradation of starch (Byars, 2003), the effect of cooling rate and stirring conditions on

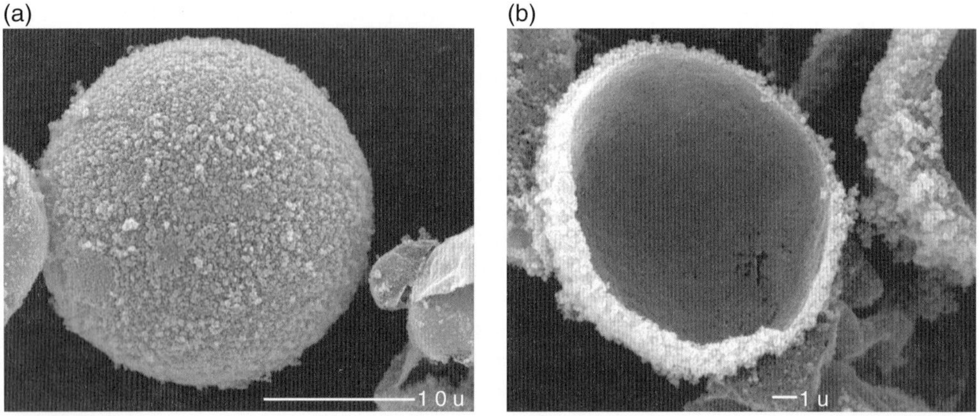

Figure 2.2 SEM images of entire (a) and fractured (b) starch coatings isolated from a jet-cooked starch–lipid composite, dehydrated with ethanol, and critical-point dried.

rheological properties (Byars *et al.*, 2003), and the effect of added oil on dispersion properties (Byars *et al.*, 2006).

In addition to widespread use of steam jet cooking for starch utilization in the papermaking and textile industries, jet cooking has also been used for producing starch gels for food use (Edwards *et al.*, 1998), for cooking cereal flours (Rankin *et al.*, 1976), for preparing complexes of starch with poly(ethylene-co-acrylic acid) (Christianson *et al.*, 1992), and for forming composites of starch with natural gums (Fanta and Christianson, 1996). More versatile and sophisticated jet-cooking devices continue to be developed (Shroyer, 1997; Cummins and Perry, 1998).

A significant discovery was made in the 1990s when a mixture of soy oil and corn starch was passed through an excess steam jet cooker. The resulting starch dispersion contained soy oil droplets about 1 to 10 μm in diameter; and the droplets did not coalesce with time, even after prolonged storage (Fanta and Eskins, 1995). The stability and lack of coalescence of the oil droplets in both liquid dispersions and dried solids was demonstrated, suggesting a specific starch–oil interaction (Knutson *et al.*, 1996). Dickinson (1988) proposed that starch separates from water solution and adsorbs onto the hydrophobic surfaces of oil droplets because water is a thermodynamically poor solvent for starch, and the adsorption of starch at the lipid–water interface leads to a reduction in the interfacial tension of the two-phase aqueous oil system.

The development of this technology was significantly enhanced by the issuance of two patents covering a wide range of starches, lipophilic materials, and processing conditions (Eskins and Fanta, 1997, 1999). Light microscopy of the composites revealed a boundary layer around the oil droplets (Eskins *et al.*, 1996); this absorbed layer was later shown to consist of starch and starch–lipid complexes (Fanta *et al.*, 1999a). Efficient complex formation was demonstrated in jet-cooked mixtures of high amylose starch and fatty acids (Fanta *et al.*, 1999b). The nature and composition of the adsorbed starch layers (or shells) was described (Fanta *et al.*, 2001a), and a patent on the starch shells and starch-coated droplets was issued (Fanta *et al.*, 2001b). SEM images of the adsorbed starch shells (Figure 2.2) were obtained by isolating the starch-coated oil droplets by centrifugation, extracting the oil with ethanol, and critical-point drying the remaining adsorbed starch. Further characterization of starch–lipid composite properties included oil content and pH effects (Garzon *et al.*, 2003a), oxidation of encapsulated oil under different storage conditions (Garzon *et al.*, 2004), methods of preparing composites with higher oil content (Fanta *et al.*, 2009a), and factors affecting the particle size of drum-dried composites (Singh, 2008).

Research on the chemical and physical properties of jet-cooked starch–oil composites supported the development of many applications for the technology, including fat replacers in ground beef patties (Warner *et al.*, 2001; Garzon *et al.*, 2003c), cookies (Garzon *et al.*, 2003b), soft-serve ice cream (Byars, 2002), and yogurt (Singh and Kim, 2009; Singh and Byars, 2009). Non-food applications demonstrating the efficacy of oil delivery in aqueous starch-based media included biodegradable polyurethane foams (Cunningham *et al.*, 1997), lubricants for water-based oil drilling muds (Fanta *et al.*, 2002a; Sifferman *et al.*, 2003) and lubricants for metal working (Biresaw, 2004; Biresaw and Erhan, 2002; Biresaw *et al.*, 2007; Kenar *et al.*, 2009). The jet-cooked starch–oil technology was also used to deliver a soybean oil-based UV absorbing agent to provide UV protection in cosmetic and agricultural applications, and was shown to increase the efficiency of UV absorption (Compton *et al.*, 2007).

In addition to adsorption onto the surfaces of oil droplets, starch was also shown to adsorb onto the hydrophobic surface of a polyethylene film, presumably by a similar mechanism (Fanta *et al.*, 2002b); the adsorption of starch rendered the film surface hydrophilic. Polyacrylonitrile could be graft polymerized onto the surface coating of starch, imparting the interesting property of reversible curling in response to humidity changes (Fanta *et al.*, 2003). The reverse process of depositing polyethylene dissolved in various organic solvents onto dried starch films was also observed and provided a means of imparting water resistance to the starch films (Fanta *et al.*, 2009b).

Another technology resulting from the excess steam jet cooking of starch was based on the observation that slow cooling of dilute solutions of jet-cooked starch resulted in the formation of crystalline spherulites composed of amylose–lipid complexes. Davies and co-workers (1980) also observed the formation of crystalline particles in cooked starch dispersions, but did not study their morphology and properties. Similar particles were noted, but not described, in other reports (Kitamura *et al.*, 1984; Zobel, 1988; Jane *et al.*, 1996; Heinemann *et al.*, 2003). The crystallization of amylose into various structures, including spherulites, has also been studied extensively in the context of investigations of *in vivo* starch granule biosynthesis (Buleon *et al.*, 2007; Nordmark and Ziegler, 2002, 2002; Ziegler *et al.*, 2003, 2005).

A detailed description of the different morphological types of spherulites obtained from dispersions of jet-cooked starch was made in 2002 (Fanta *et al.*, 2002c). Micrographs of the spherical, toroidal, and small-particle spherulites isolated under different processing and cooling conditions are shown in Figures 2.3 and 2.4. Conditions for spherulite formation were further investigated by using defatted corn starch and supplementing the starch with specific fatty acids (Fanta *et al.*, 2006). Spherulite yields of about 60%, based on total starch, were obtained with high amylose corn starch; different types of spherulites could be obtained by selecting specific cooling rates and stirring conditions (Fanta *et al.*, 2008). The identity of the complexed fatty acid ligands in these spherulites was determined by extraction with different solvent systems (Peterson *et al.*, 2005) and research has been carried out to understand the factors that determine the formation of a particular type of spherulite particle. It was also observed that crystal transformations can take place in these spherulites after their formation, due to changes in hydration and extraction conditions (Shogren *et al.*, 2006). Although the formation of crystalline spherulites from jet-cooked starch has been

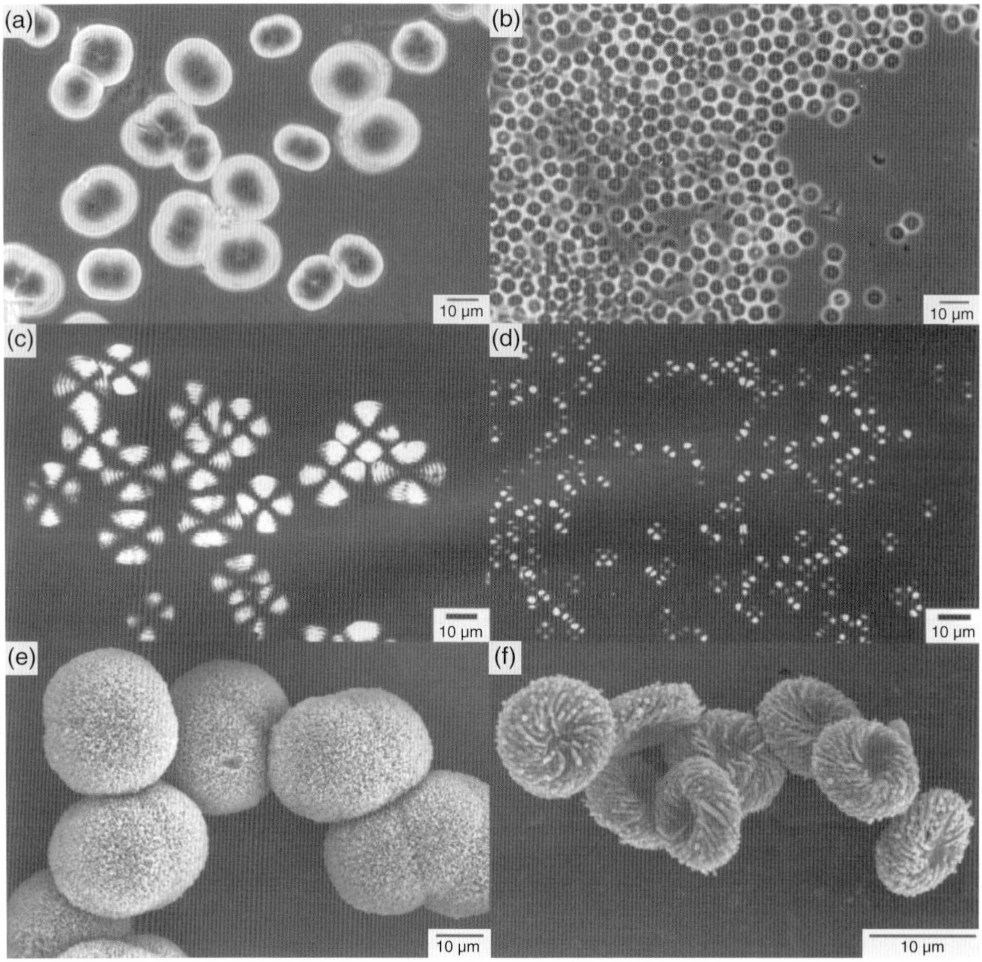

Figure 2.3 Photomicrographs of spherical (a,c,e) and toroid (b,d,f) spherulites obtained from slowly cooled, jet-cooked, high amylose corn starch. (a) and (b), phase contrast light micrographs. (c) and (d), polarized light micrographs showing birefringence. (e) and (f), scanning electron micrographs.

considered to be a problem in the papermaking industry, current research has shown that potentially valuable new forms of bio-based, biodegradable, particulate materials can be obtained if their formation can be controlled and increased by adjusting the methods used for processing.

In summary, the effect of excess steam jet cooking on starch properties has been investigated and this method of starch processing has been used to prepare starch–natural gum composites, spherulites from amylose–fatty acid complexes, and stable starch–oil composites that contain micrometer-sized oil droplets in an aqueous starch matrix. Commercial applications of the starch–oil composites in foods, lubricants and lotions are under investigation. A major advantage of steam jet cooking as a processing method is that high temperatures and intense mechanical shear can be continuously applied to aqueous starch slurries, as well as to aqueous mixtures of starch and non-starch components. Also, the equipment used for steam jet cooking

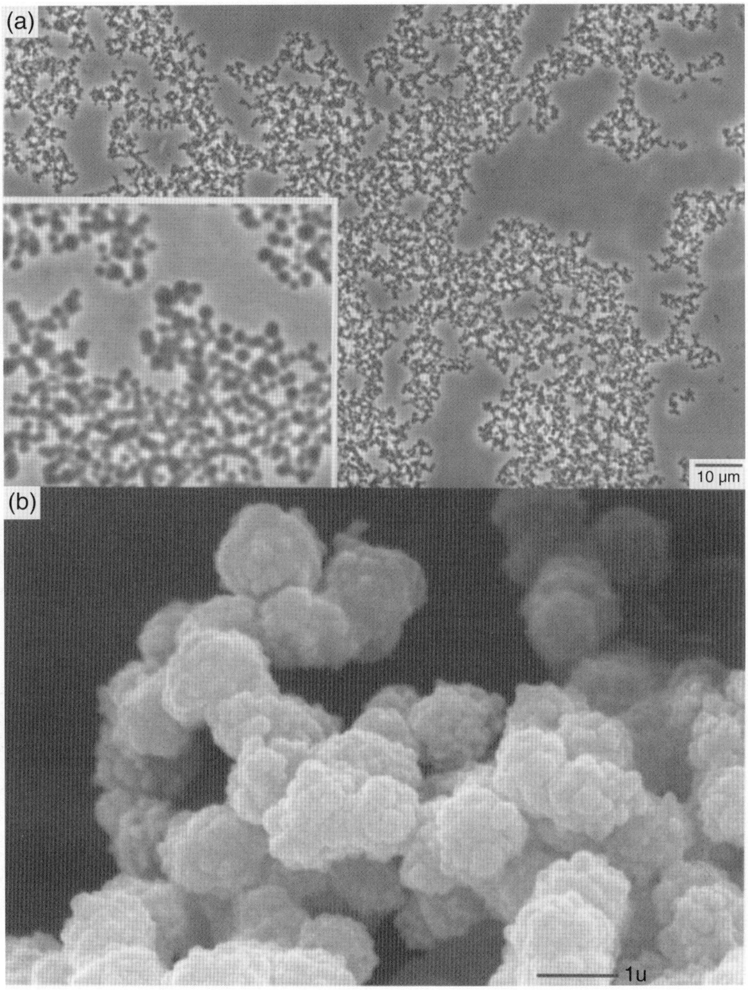

Figure 2.4 Small spherical particles formed from slowly cooled, jet-cooked wheat starch (identical in appearance with those formed with corn starch). (a), phase contrast light micrograph with higher magnification in inset. (b), Scanning electron micrograph.

can be easily scaled up to produce large quantities of material for new product development. The ability to use excess steam conditions to maximize mechanical shear enhances the versatility of this processing method and allows this method to be used with numerous feedstock combinations.

2.9 CONCLUSIONS

In this chapter, a number of non-traditional processing methods that have been developed to modify starch properties for a wide variety of end-use applications have been reviewed. For example, annealing and heat–moisture treatment have been shown to modify the swelling and pasting properties of starch granules; these techniques could be used to increase the percentage of slowly-digestible, resistant starch in food products. The application of high pressure has been shown to reduce the gelatinization temperature of starch; this processing method will enable starches in food products to be processed without the detrimental effects of high temperature on heat-sensitive food components. Also, protein-based sources of food spoilage, such as enzymes and microorganisms, can be denatured and inactivated by high-pressure processing.

Microwave heating for starch processing has been extensively investigated and is more rapid and uniform than conventional thermal heating. Reactions of starch have been shown to take place with shorter reaction times when microwave heating is used. Investigations of ultrasound in starch processing have shown that the high-shear cavitation resulting from ultrasonic treatment reduces the molecular weight of starch, and yields starch solutions with less insoluble material, lower viscosity, and increased clarity. Bond cleavage occurs at the center of gravity and produces degraded starch products with narrow molecular weight distributions and minimal contamination with low molecular weight material. Shorter reaction times have also been observed when ultrasound was used during the preparation of starch derivatives. Treatment with ultrasound has been shown to enhance glucose production from flour and corn meal, and this technology could lead to increased ethanol production during saccharification and fermentation. Incorporating ultrasound into the wet-milling process facilitated the isolation of starch granules from cereal grains.

Extraction with supercritical CO_2 was shown to be a viable process for removing undesirable contaminants from starch and for extracting lipids from wheat flour. This method of extraction eliminates the hazards and environment concerns associated with traditional extraction methods using organic solvents. A method was also developed for preparing expanded foams or puffed products from starch by adding supercritical CO_2 during extrusion. Supercritical CO_2 thus replaces steam as the blowing agent, resulting in foams with improved properties. Supercritical CO_2 has also been used as a reaction medium for preparing starch derivatives and starch-based matrices with potential applications as scaffolds for tissue engineering. A biomass conversion system using supercritical water has also been developed to convert starch and other biomass into gaseous products.

Extrusion is a rapid and continuous method for preparing new starch products and derivatives. There is interest in these materials because of the low cost of starch, the desirability of replacing petroleum-based materials with renewable feedstocks, the improved biodegradability of starch-based materials, and the creation of new markets for agricultural commodities. Steam jet cooking has been used to prepare stable starch–oil composites useful in food products and as water-based lubricants. Spherulites that could have practical end-use applications have also been prepared.

Future years should see an increase in the use of these new processing techniques and other non-traditional methods of processing will also be developed. The current interest in biodegradability and the replacement of petrochemical-based materials with agricultural products will result in increased research on starch as a starting material for new products.

REFERENCES

Akdogan H. 1999. High moisture food extrusion. Int J Food Sci Technol 34:195–207.

Alavi SH, Gogoi BK, Khan M, Bowman BJ, Rizvi SSH. 1999. Structural properties of protein-stabilized starch-based supercritical fluid extrudates. Food Res Internat 32:107–118.

Alavi SH, Rizvi SSH, Harriott P. 2003a. Process dynamics of starch-based microcellular foams produced by supercritical extrusion. I: Model development. Food Res Internat 36:309–319.

Alavi SH, Rizvi SSH, Harriott P. 2003b. Process dynamics of starch-based microcellular foams produced by supercritical extrusion. II: Numerical simulation and experimental evaluation. Food Res Internat 36:321–330.

Alavi SH, Rizvi SSH. 2005. Strategies for enhancing expansion in starch-based microcellular foams produced by supercritical expansion. Int J Food Prop 8:23–34.

Antal Jr MJ, Allen SG, Schulman D, Xu X, Divilio RJ. 2000. Biomass gasification in supercritical water. Ind Eng Chem Res 39:4040–4053.

Artz WE, Myers MR. 1995. Supercritical fluid extraction and chromatography of emulsifiers. J Amer Oil Chem Soc 72:219–224.

Ashraf-Khorassani M, Taylor LT, Waterman KC, Narayan P, Brannegan DR, Reid GL. 2005. Purification of pharmaceutical excipients with supercritical fluid extraction. Pharmaceut Devel Tech 10:507–516.

Averous L. 2004. Biodegradable multiphase systems based on plasticized starch. J Macromol Sci – Polym Rev 44:231–274.

Averous L, Halley PJ. 2009. Biocomposites based on plasticized starch. Biofuels Bioprod Biorefin 3:329–343.

Ayoub A, Rizvi SSH. 2008. Properties of supercritical fluid extrusion-based crosslinked starch extrudates. J Appl Polym Sci 107:3663–3671.

Azhar A, Hamdy MK. 1979. Sonication effect on potato starch and sweet potato powder. J Food Sci 44: 801–804.

Baks T, Bruins ME, Janssen AEM, Boom RM. 2008. Effect of pressure and temperature on the gelatinization of starch at various starch concentrations. Biomacromol 9:296–304.

Barsby TL, Frazier PJ, Donald AM. 2002. Starch: Advances in Structure and Function. New York, NY: Springer.

Barton S, Bullock C, Weir D. 1996. The effects of ultrasound on the activities of some glycosidase enzymes of industrial importance. Enz Microb Technol 18:190–194.

Basedow AM, Ebert KH. 1977. Ultrasonic degradation of polymers in solution. Adv Polym Sci Phys Chem 22:83–148.

Bastioli C. 2001. Global status of biobased packaging materials. Starch/Stärke 53:351–355.

Bauer BA, Hartmann M, Sommer K, Knorr D. 2004. Optical in situ analysis of starch granules under high pressure with a high pressure cell. Innov Food Sci Emerg Technol 5:293–298.

Bauer BA, Wiehle T, Knorr D. 2005. Impact of high hydrostatic pressure treatment on the resistant starch content of wheat starch. Starch/Stärke 57:124–133.

BeMiller JN. 1997. Starch modification: challenges and prospects. Starch/Stärke 49:127–131.

Berzin F, Tara A, Tighzert L, Vergnes B. 2007a. Computation of starch cationization performances by twin-screw extrusion. Polym Engin Sci 47:112–119.

Berzin F, Tara A, Vergnes B. 2007b. Optimization and scale-up of starch cationization in a twin screw extruder. Polym Engin Sci 47:814–823.

Biresaw G. 2004. Biobased dry-film metalworking lubricants. J Synth Lubr 21:43–57.

Biresaw G, Erhan SM. 2002. Solid lubricant formulations containing starch-soybean oil composites. J Amer Oil Chem Soc 79:291–296.

Biresaw G, Kenar JA, Kurth TL, Felker FC, Erhan SM. 2007. Investigation of the mechanism of lubrication in starch-oil composite dry film lubricants. Lubr Sci 19:41–550.

Biswas A, Shogren RL, Kim S, Willett JL. 2006a. Rapid preparation of starch maleate half-esters. Carbohydr Polym 64:484–487.

Biswas A, Shogren RL, Stevenson DG, Willett JL, Bhowmik PK. 2006b. Ionic liquids as solvents for biopolymers: acylation of starch and zein protein. Carbohydr Polym 66:546–550.

Biswas A, Shogren RL, Selling G, Salch J, Willett JL, Buchanan CM. 2008. Rapid and environmentally friendly preparation of starch esters. Carbohydr Polym 74:137–141.

Björck I, Asp N-G. 1983. The effects of extrusion cooking on nutritional value – A literature review. J Food Engin 2:281–308.

Błaszczak W, Fornal J, Valverde S Garrido L. 2005a. Pressure-induced changes in the structure of corn starches with different amylose content. Carbohydr Polym 61:132–140.

Błaszczak W, Valverde S, Fornal J. 2005b. Effect of high pressure on the structure of potato starch. Carbohydr Polym 59:377–383.

Błaszczak W, Fornal J, Kiseleva VI, Yuryev VP, Sergeev AI, Sadowska J. 2007. Effect of high pressure on thermal, structural and osmotic properties of waxy maize and Hylon VII starch blends. Carbohydr Polym 68:387–396.

Błaszczak W, Bidzińska E, Dyrek K, Fornal J, Wenda E. 2008. Effect of high hydrostatic pressure on the formation of radicals in maize starches with different amylose content. Carbohydr Polym 74:914–921.

Braga MEM, Moreschi SRM,.Meireles MAA. 2006. Effects of supercritical extraction on *Curcuma longa L.* and *Zingiber officinale R.* starches. Carbohydr Polym 63:340–346.

Buléon A, Véronèse G, Putaux J-L. 2007. Self-association and crystallization of amylose. Aust J Chem 60:706–718.

Byars J. 2002. Effect of starch-lipid fat replacer on the rheology of soft-serve ice cream. J Food Sci 67: 2177–2182.

Byars JA. 2003. Jet cooking of waxy maize starch: solution rheology and molecular weight degradation of amylopectin. Cereal Chem 80:87–90.

Byars JA, Fanta GF, Felker FC. 2003. The effect of cooling conditions on jet-cooked normal corn starch dispersions. Carbohydr Polym 54:321–326.

Byars JA, Fanta GF, Felker FC. 2006. The influence of oil on the properties of slowly-cooled jet-cooked normal corn starch dispersions. Carbohydr Polym 63:316–322.

Caddick S. 1995. Microwave Assisted Organic Reactions. Tetrahedron 51(38):10403–10432.

Cameron DK, Wang Y-J. 2006. Application of protease and high-intensity ultrasound in corn starch isolation from degermed corn flour. Cereal Chem 83:505–509.

Carr ME. 1994a. Preparation of cationic starch containing quaternary ammonium substituents by reactive twin-screw extrusion processing. J Appl Polym Sci 54:1855–1861.

Carr ME. 1994b. Preparation of starch derivatives by reactive extrusion. In: ANTEC 94: Plastics Gateway to the Future. Brookfield, CT: Society of Plastics Engineers. pp 1444–1448.

Chan H-T, Bhat R, Karim AA. 2010. Effects of sodium dodecyl sulfate and sonication treatment on physicochemical properties of starch. Food Chem 120:703–709

Chen K-H, Rizvi SSH. 2006a. Measurement and prediction of solubilities and diffusion coefficients of carbon dioxide in starch-water mixtures at elevated pressures. J Polym Sci, Part B, Polym Phys 44:607–621.

Chen K-H, Rizvi SSH. 2006b. Rheology and expansion of starch-water-CO_2 mixtures with controlled gelatinization by supercritical fluid extrusion. Internat J Food Prop 9:863–876.

Cheng W, Chen J, Liu D, Ye X, Ke F. 2010. Impact of ultrasonic treatment on properties of starch film-forming dispersion and the resulting films. Carbohydr Polym 81:707–711.

Chivrac F, Pollet E, Averous L. 2009. Progress in nano-biocomposites based on polysaccharides and nanoclays. Mater Sci Engin Res 67:1–17.

Cho KY, Rizvi SSH. 2009a. 3D microstructure of supercritical fluid extrudates. I: Melt rheology and microstructure formation. Food Res Internat 42:595–602.

Cho KY, Rizvi SSH. 2009b. 3D microstructure of supercritical fluid extrudates. II: Cell anisotropy and the mechanical properties. Food Res Internat 42:603–611.

Christianson DD, Fanta GF, Bagley EB. 1992. Complexes between starch and poly(ethylene-co-acrylic acid). Viscosity and gel rheology of jet cooked dispersions. Carbohydr Polym 17:221–226.

Chung H-J, Liu Q, Hoover R. 2009. Impact of annealing and heat–moisture treatment on rapidly digestible, slowly digestible and resistant starch levels in native and gelatinized corn, pea and lentil starches. Carbohyd Polym 75:436–447.

Chung KM, Moon TW, Kim H, Chun JK. 2002. Physicochemical properties of sonicated mung bean, potato, and rice starches. Cereal Chem 79:631–633.

Čížová A, Sroková I, Sasinková V, Malovíková A, Ebringerová A. 2008. Carboxymethyl starch octenylsuccinate: microwave- and ultrasound-assisted synthesis and properties. Starch/Stärke 60:389–397.

Colonna P, Buleon A, Mercier C. 1987. Physically modified starches. In: Galliard, T (editor). Starch: Properties and Potential. New York, NY: John Wiley & Sons, Inc. pp 79–114.

Compton DL, Kenar JA, Laszlo JA, Felker FC. 2007. Starch-encapsulated, soy-based, ultraviolet-absorbing composites with feruloylated monoacyl- and diacylglycerol lipids. Indust Crop Prod 25:17–23.

Coppock PD. 1940. Manufacture of starch paste. U.S. patent 2,202,573.

Corsaro A, Chiacchio U, Pistarà V, Romeo G. 2004. Microwave-assisted chemistry of carbohydrates. Curr Organ Chem 8:511–538.

da Cruz Francisco J, Sivik B. 2002. Gelatinization of cassava, potato and wheat starches in supercritical carbon dioxide. J Supercrit Fluids 22:247–254.

Cummins RD, Perry JA. 1998. Dual control mixing jet cooker. U.S. patent 5,743,638.

Cunningham RL, Gordon SH, Felker FC, Eskins K. 1997. Jet-cooked starch-oil composite in polyurethane foams. J Appl Polym Sci 64:1355–1361.

Czechowska-Biskup R, Rokita B, Lotfy S, Ulanski P, Rosiak JM. 2005. Degradation of chitosan and starch by 360-kHz ultrasound. Carbohydr Polym 60:175–184.

D'Jesús P, Artiel C, Boukis N, Kraushaar-Czarnetzki B, Dinjus E. 2005. Influence of educt preparation on gasification of corn silage in supercritical water. Indust Engin Chem Res 44:9071–9077.

D'Jesús P, Boukis N, Kraushaar-Czarnetzki B, Dinjus E. 2006. Gasification of corn and clover grass in supercritical water. Fuel 85:1032–1038.

Darani KK, Mozafari MR. 2009. Supercritical fluids technology in bioprocess industries: A review. J Biochem Technol 2:144–152.

Davies T, Miller DC, Procter AA. 1980. Inclusion complexes of free fatty acids with amylose. Starch/Stärke 32:149–158.

Degrois M, Gallant D, Baldo P, Guilbot A. 1974.Effects of ultrasound on starch grains. Ultrasonics 12:129–131.

Dickinson E. 1988. The role of hydrocolloids in stabilizing particulate dispersions and emulsions. In: Phillips GO, Williams PA, Wedlock DJ (editors). Gums and Stabilizers for the Food Industry 4; Proceedings of the 4th International Conference. Oxford: IRL Press. pp. 249–263.

Dintzis FR, Fanta GF. 1996. Effects of jet-cooking conditions upon intrinsic viscosity and flow properties of starches. J Appl Polym Sci 62:749–753.

Doona CJ, Feeherry FE, Baik M-Y. 2006. Water dynamics and retrogradation of ultrahigh pressurized wheat starch. J Agric Food Chem 54:6719–6724.

Douzals JP, Marechal PA, Coquille JC, Gervais P. 1996a. Microscopic study of starch gelatinization under high hydrostatic pressure. J Agric Food Chem 44:1403–1408.

Douzals JP, Marechal PA, Coquille JC, Gervais P. 1996b. Comparative study of thermal and high pressure treatments upon wheat starch suspensions. Prog Biotechnol 13:433–438.

Douzals JP, Perrier-Cornet JM, Gervais P, Coquille JC. 1998. High-pressure gelatinization of wheat starch and properties of pressure-induced gels. J Agric Food Chem 46:4824–4829.

Douzals JP, Perrier-Cornet JM, Coquille JC, Gervais P. 2001. Pressure-temperature phase transition diagram for wheat starch. J Agric Food Chem 49:873–876.

Duarte ARC, Mano JF, Reis RL. 2009a. Preparation of starch-based scaffolds for tissue engineering by supercritical immersion precipitation. J Supercrit Fluids 49:279–285.

Duarte ARC, Caridade SG, Mano JF, Reis RL. 2009b. Processing of novel bioactive polymeric matrixes for tissue engineering using supercritical fluid technology. Mater Sci Eng, C: Mater Biol Applic 29:2110–2115.

Duarte ARC, Mano JF, Reis RL. 2010. Supercritical phase inversion of starch-poly(ε-caprolactone) for tissue engineering applications. J Mater Sci: Mater Medicine 21:533–540.

Dubois P, Narayan R. 2003. Biodegradable compositions by reactive processing of aliphatic polyester/polysaccharide blends. Macromol Symp 198:233–243.

Dyrek K, Bidzińska E, Łabanowska M, Fortuna T, Przetaczek I, Pietrzyk S. 2007. ESR study of radicals generated in starch by microwaves or by conventional heating. Starch/Stärke 59:318–325.

Edi-Soetaredjo F. 2010. Numerical Simulation of Twin-screw Extrusion of Starch Based Material. Saarbrücken, Germany: LAP Lambert Academic Press.

Edwards RH, Berrios JD, Mossman AP, Takeoka GR, Wood DF, Mackey BE. 1998. Texture of jet cooked, high amylose corn starch-sucrose gels. Lebensm Wiss Technol 31:432–438.

Eskins K, Fanta GF. 1997. Non-separable starch-oil compositions. U.S. patent 6,676,994.

Eskins K, Fanta GF. 1999. Non-separable compositions of starch and water-immiscible organic materials. U.S. patent 5,882,713.

Eskins K, Fanta GF, Felker FC, Baker FL. 1996. Ultrastructural studies on microencapsulated oil droplets in aqueous gels and dried films of a new starch-oil composite. Carbohydr Polym 29:233–239.

Ezaki S, Hayashi R. 1992. High pressure effects on starch: Structural change and retrogradation. In: Balney C, Hayashi R, Heremans K, Rikimaru H (editors). High Pressure and Biotechnology. Montrouge, France: Colloque INSERM/John Libbey Eurotext Ltd. pp. 163–165.

Fanta GF. 1999. Starch graft copolymers. In: Salamone JC (editor). Concise Polymeric Materials Encyclopedia. Boca Raton, FL: CRC Press. pp. 1497–1499.

Fanta GF, Bagley EB. 1977. Starch, graft copolymers. In: Mark HF, Bikales NM (editors). Encyclopedia of Polymer Science and Technology, Supplement Vol. 2. New York, NY: John Wiley & Sons, Inc. pp. 665–699.

Fanta GF, Christianson DD. 1996. Starch-hydrocolloid composites prepared by steam jet cooking. Food Hydrocoll 10:173–178.

Fanta GF, Eskins K. 1995. Stable starch-lipid compositions prepared by steam jet cooking. Carbohydr Polym 28:171–175.

Fanta GF, Felker FC, Eskins K, Baker FL. 1999a. Aqueous starch-oil dispersions prepared by steam jet cooking. Starch films at the oil–water interface. Carbohydr Polym 38:25–35.
Fanta GF, Shogren RL, Salch JH. 1999b. Steam jet cooking of high amylose starch-fatty acid mixtures. An investigation of complex formation. Carbohydr Polym 38:1–6.
Fanta GF, Felker FC, Shogren RL, Knutson CA. 2001a. Starch-paraffin wax compositions prepared by steam jet cooking. Examination of starch adsorbed at the paraffin-water interface. Carbohydr Polym 46:29–38.
Fanta GF, Knutson CA, Eskins K, Felker FC. 2001b. Starch microcapsules for delivery of active agents. U.S. patent 6,238,677.
Fanta GF, Muijs HM, Eskins K, Felker FC, Erhan SM. 2002a. Starch-containing lubricant for oil filed applications. U.S. patent 6,461,999.
Fanta GF, Felker FC, Shogren RL, Salch JH. 2002b. Formation of hydrophilic starch coatings on polyethylene films. J Appl Polym Sci 84:1781–1788.
Fanta GF, Felker FC, Shogren RL. 2002c. Formation of crystalline aggregates in slowly-cooled starch solutions prepared by steam jet cooking. Carbohydr Polym 48:161–170.
Fanta GF, Felker FC, Salch JH. 2003. Graft polymerization of acrylonitrile onto starch-coated polyethylene film surfaces. J Appl Polym Sci 89:3323–3328.
Fanta GF, Felker FC, Shogren RL, Salch JH. 2006. Effect of fatty acid structure on the morphology of spherulites formed from jet cooked mixtures of fatty acids and defatted corn starch. Carbohydr Polym 66:60–70.
Fanta GF, Felker FC, Shogren RL, Salch JH. 2008. Preparation of spherulites from jet cooked mixtures of high amylose starch and fatty acids. Effect of preparative conditions on spherulite morphology and yield. Carbohydr Polym 71:253–262.
Fanta GF, Felker FC, Byars JA, Kenar JA, Shogren RL. 2009a. Starch-soybean oil composites with high oil:starch ratios prepared by steam jet cooking. Starch/Stärke 61:590–600.
Fanta GF, Salch JH, Felker FC, Shogren RL. 2009b. Adsorption of polyethylene from solution onto starch film surfaces. J Appl Polym Sci 114:1840–1847.
Fernandez-Martin F, Tabilo-Munizaga G, Barbosa-Cánovas GV. 2005. Starch and other polysaccharides under high pressure. Food Sci Technol 141:261–282.
Fernández-Martín F, Fernández-García M, Tabilo-Munizaga G. 2008. Pressurization of some starches compared to heating: Calorimetric, thermo-optical and X-ray examination. Food Res Internat 41:683–692.
Finkenstadt VL, Willett JL. 2005. Reactive extrusion of starch-polyacrylamide graft copolymers: Effects of monomer/starch ratio and moisture content. Macromol Chem Phys 206:1648–1652.
Flory PJ. 1953. Principles of Polymer Chemistry. New York: Cornell University Press.
Gabriel C, Gabriel S, Grant EH, Halstead BSJ, Mingos DMP. 1998. Dielectric parameters relevant to microwave heating. Chem Soc Rev 27:213–223.
Galema SA. 1997. Microwave chemistry. Chem Soc Rev 26:233–238.
Gallant D, Degrois M, Sterling C, Guilbot A. 1972. Microscopic effects of ultrasound on the structure of potato starch. Starch/Stärke 24:116–123.
Garzon GA, Gaines CS, Mohamed A, Palmquist DE. 2003a. Effect of oil content and pH on the physicochemical properties of corn starch-soybean oil composites. Cereal Chem 80:154–158.
Garzon GA, Gaines CS, Palmquist DE. 2003b. Use of wheat flour-lipid and waxy maize starch-lipid composites in wire-cut formula cookies. J Food Sci 68:654–659.
Garzon GA, McKeith FK, Gooding JP, Felker FC, Palmquist DE, Brewer MS. 2003c. Characteristics of low-fat beef patties formulated with carbohydrate-lipid composites. J Food Sci 68:2050–2056.
Garzon GA, Warner K, Felker FC, Palmquist DE, Eller F. 2004. Oxidation of soybean oil during storage of starch-oil composites. J Amer Oil Chem Soc 81:861–866.
Gogoi BK, Alavi SH, Rizvi SSH. 2000. Mechanical properties of protein-stabilized starch-based supercritical fluid extrudates. Internat J Food Prop 3:37–58.
Gray DA, Bowen SE, Farhat I, Hill SE. 2007. Lipid oxidation in glassy and rubbery state extrudates. Food Chem 106:227–234.
Green JW, Rubal MJ, Osman BM, *et al.* 2000. Silicon-organic hybrid polymers and composites prepared in supercritical carbon dioxide. Polym Advan Technol 11:820–825.
Grinberg O, Gedanken A. 2010. The development and characterization of starch microspheres prepared by a sonochemical method for the potential drug delivery of insulin. Macromol Chem Phys 211:924–931.
Gross RA, Kalra B. 2002. Biodegradable polymers for the environment. Science 297:803–807.
Guan J-J, Qiu A-Y, Liu X-Y, Hua Y-F, Ma Y-H. 2006. Microwave improvement of soy protein isolate-saccharide graft reactions. Food Chem 97:577–585.

Gugliemelli LA, Swanson CL, Baker FL, Doane WM, Russell CR. 1974. Cationic starch-polyacrylonitrile graft copolymer latexes. J Polym Sci, Polym Chem Edit 12:2683–2692.

Gugliemelli LA, Swanson CL, Doane WM, Russell CR. 1976a. Latexes of starch-based graft copolymers containing polymerized acrylonitrile. J Appl Polym Sci 20:3175–3183.

Gugliemelli, LA, Swanson CL, Doane WM, Russell CR. 1976b. Cationic starch graft-polychloroprenne latexes. J Polym Sci, Polym Lett Edit 14:215–218.

Gugliemelli LA, Doane WM, Russell CR. 1979. Preparation of soapless latexes by sonification of starch-based poly(isoprene-*co*-acrylonitrile) graft reaction mixtures. J Appl Polym Sci 23:635–644.

Gui-Jie M, Peng W, Xiang-Sheng M, Xing Z, Tong Z. 2006. Crosslinking of corn starch with sodium trimetaphosphate in solid state by microwave irradiation. J Appl Polym Sci 102:5854–5860.

Guy R. 2001. Extrusion Cooking: Technology and Applications. Boca Raton, FL: CRC Press.

Hanna MA, Fang Q. 2002. Starch esterification by reactive extrusion. Agro-Food-Industry Hi-Tech 13:33–36.

Harper JM. 1978. Food extrusion. CRC Crit Rev Food Sci Nutr 11:155–215.

Harris R, Jureller SH, Kerschner JL, Trzasko PT, Humphreys RWR. 27 January 1999. Derivatized polysaccharides and preparation thereof in a densified fluid. European Patent Application EP893451 A2.

Hasjim J, Jane J-L. 2009. Production of resistant starch by extrusion cooking of acid modified normal maize starch. J Food Sci 74:C556-C562.

Heinemann C, Escher F, Conde-Petit B. 2003. Structural features of starch-lactone inclusion complexes in aqueous potato starch dispersions: the role of amylose and amylopectin. Carbohydr Polym 51:159–168.

Hibi Y, Matsumoto T, Hagiwara S. 1993. Effect of high pressure on the crystalline structure of various starch granules. Cereal Chem 70:671–676.

Huang Q, Li L, Fu X. 2007. Ultrasound effects on the structure and chemical reactivity of Corn starch granules. Starch/Stärke 59:371–378.

Hubbard JD, Downing JM, Ram MS, Chung OK. 2004. Lipid extraction from wheat flour using supercritical fluid extraction. Cereal Chem 81:693–698.

Ida Y, Tuziuti T, Yasui K, Towata A, Kozuka T. 2008. Control of viscosity in starch and polysaccharide solutions with ultrasound after gelatinization. Innov Food Sci Emerg Technol 9:140–146.

Isono Y, Kumagai T, Watanabe T. 1994. Ultrasonic degradation of rice starch. Biosci Biotech Biochem 58:1799–1802.

Jackson DS. 1991. Solubility behavior of granular corn starches in methyl sulfoxide (DMSO) as measured by high performance size-exclusion chromatography. Starch/Stärke 43:422–427.

Jackson DS, Choto-Owen C, Waniska RD, Rooney LW. 1988. Characterization of starch cooked in alkali by aqueous high performance size-exclusion chromatography. Cereal Chem 65:493–496.

Jacobs H, Delcour JA. 1998. Hydrothermal modifications of granular starch with retention of the granular structure: A review. J Agric Food Chem 46:2895–2905.

Jambrak AR, Herceg Z, Šubarić D, *et al.* 2010. Ultrasound effect on physical properties of corn starch. Carbohydr Polym 79:91–100.

Jane J, Kasensuwan T, Chen JF, Juliano BO. 1996. Phosphorus in rice and other starches. Cereal Food World 41:827–832.

Janssen L, Moscicki L. 2009. Thermoplastic Starch. New York, NY: Wiley-VCH Verlag GmbH.

Jarowenko W. 1970. Starch. In: Mark HF, Gaylord NG (editors). Encyclopedia of Polymer Science and Technology, Vol. 12. New York, NY: John Wiley& Sons, Inc. pp. 787–862.

Jenkens PJ, Donald AM. 1998. Gelatinization of starch: a combined SAXS/WAXS/DSC and SANS study. Carbohydr Res 308:133–147.

Jyothi AN, Rajasekharan KN, Moorthy SN, Sreekumar J. 2005. Microwave-assisted synthesis and characterization of succinate derivatives of cassava (manihot esculenta crantz) starch. Starch/Stärke 57:556–563.

Jyothi A, Moorthy SN, Sreekumar JN, Rajasekharan KN. 2007. Studies on the properties of citrate derivatives of cassava (*Manihot esculenta* Crantz) starch synthesized by microwave technique. J Sci Food Agric 87:871–879.

Kalambur S, Rizvi SSH. 2006. An overview of starch-based plastic blends from reactive extrusion. J Plastic Film Sheeting 22:39–58.

Kardos N, Luche J-L. 2001. Sonochemistry of carbohydrate compounds. Carbohydr Res 332:115–131.

Kasica JJ, Eden JL. 1992. Continuous coupled jet-cooking/spray drying process and novel pregelatinized high amylose starches prepared thereby. U.S. patent 5,131,953.

Katopo H, Song Y, Jane J-L. 2002. Effect and mechanism of ultrahigh hydrostatic pressure on the structure and properties of starches. Carbohydr Polym 47:233–244.

Kawai K, Fukami K, Yamamoto K. 2007a. State diagram of potato starch-water mixtures treated with high hydrostatic pressure. Carbohydr Polym 67:530–535.

Kawai K, Fukami K, Yamamoto K. 2007b. Effects of treatment pressure, holding time, and starch content on gelatinization and retrogradation properties of potato starch-water mixtures treated with high hydrostatic pressure. Carbohydr Polym 69:590–596.

Kenar JA, Felker FC, Biresaw G, Kurth TL. 2009. Properties of dry film lubricants prepared by spray application of aqueous starch-oil composites. Ind. Crop. Prod. 29:45–52.

Khanal SK, Montalbo M, van Leeuwen J. 2007. Ultrasound enhanced glucose release from corn in ethanol plants. Biotechnol Bioeng 98:978–985.

Kim J-Y, Park D-J, Lim S-T. 2008. Fragmentation of waxy rice starch granules by enzymatic hydrolysis. Cereal Chem 85:182–187.

King JW. 2000. Advances in critical fluid technology for food processing. Food Sci Technol Today 14: 186–191.

Kitamura S, Yoneda S, Kuge T. 1984. Study on the retrogradation of starch. I. Particle size and its distribution of amylose retrograded from aqueous solutions. Carbohydr Polym 4:127–136.

Klem RE, Brogly DA. 1981. Methods for selecting the optimum starch binder preparation system. Pulp Paper 55:98–103.

Knutson CA, Eskins K, Fanta GF. 1996. Composition and oil-retaining capacity of jet-cooked starch-oil composites. Cereal Chem 73:185–188.

Kohlgruber K. 2007. Co-rotating Twin-screw Extruder. New York: Hanser Gardner.

Koroskenyi B, McCarthy SP. 2002. Microwave-assisted solvent-free or aqueous-based synthesis of biodegradable polymers. J Polym Environ 10:93–104.

Koxholt M, Altieri PA, Marentis RT, Trzasko PT. 3 December 2003. Process for purifying starches. European Patent Application EP 1 291 361 A1.

Kweon M, Slade L, Levine H. 2008a. Role of glassy and crystalline transitions in the responses of corn starches to heat and high pressure treatments: Prediction of solute-induced barostability from solute-induced thermostability. Carbohydr Polym 72:293–299.

Kweon M, Slade L, Levine H. 2008b. Effect of sodium chloride on glassy and crystalline melting transitions of wheat starch treated with high hydrostatic pressure: Prediction of solute-induced barostability from nonmonotonic solute-induced thermostability. Starch/Stärke 60:127–133.

Lai LS, Kokini JL. 1991. Physicochemical changes and rheological properties of starch during extrusion (a review). Biotechnol Prog 7:251–266.

Landerito NA, Wang Y-J. 2005. Preparation and properties of starch phosphates using waxy, common, and high-amylose corn starches. II. Reactive extrusion method. Cereal Chem 82:271–276.

Lee EY, Ryu G-H, Lim S-T. 1999. Effects of processing parameters on physical properties of corn starch extrudates expanded using supercritical CO_2 injection. Cereal Chem 76:63–69.

Lewandowicz G, Fornal F, Walkowski A. 1997. Effect of microwave radiation on physico-chemical properties and structure of potato and tapioca starches. Carbohydr Polym 34:213–220.

Lewandowicz G, Jankowski T, Fornal J. 2000. Effect of microwave radiation on physico-chemical properties and structure of cereal starches. Carbohydr Polym 42:193–199.

Lidström P, Tierney J, Wathey W, Westman J. 2002. Microwave assisted organic synthesis – a review. Tetrahedron 57:9225–9283.

Lima FF, Andrade CT. 2010. Effect of melt processing and ultrasonic treatment on physical properties of high-amylose maize starch. Ultrason Sonochem 17:637–641.

Lipatova IM, Losev NV, Yusova AA. 2002. Effect of ultrasonic field on the state of starch hydrogels. Russ J Appl Chem 75:526–530.

Liu Y, Selomulyo VO, Zhou W. 2008. Effect of high pressure on some physicochemical properties of several native starches. J Food Engin 88:126–136.

Liu H, Xie F, Yu L, Chen L, Li L. 2009. Thermal processing of starch-based polymers. Prog Polym Sci 34:1348–1368.

Liu W-C, Halley PJ, Gilbert RG. 2010. Mechanism of degradation of starch, a highly branched polymer, during extrusion. Macromol 43:2855–2864.

Luo Z, Fu X, He X, Luo X, Gao Q, Yu S. 2008. Effect of ultrasonic treatment on the physicochemical properties of maize starches differing in amylose content. Starch/Stärke 60:646–653.

Mariam I, Cho KY, Syed SH. 2008. Thermal properties of starch-based biodegradable foams produced using supercritical fluid extrusion (SCFX). Int J Food Prop 11:415–426.

Mercier C, Linko P, Harper JM. 1989. Extrusion Cooking. St. Paul, MN: American Association of Cereal Chemists.
Mongenot N, Charrier S, Chalier P. 2000. Effect of ultrasound emulsification on cheese aroma encapsulation. J Agric Food Chem 48:861–867.
Muhr AH, Blanshard JMV. 1982. Effect of hydrostatic pressure on starch gelatinization. Carbohydr Polym 2:61–74.
Muljana H, Picchioni F, Heeres HJ, Janssen LPBM. 2008. Green starch conversions: Studies on starch acetylation in supercritical CO_2. In: International Symposium on Chemical Reaction Engineering. Kyoto, Japan: Society of Chemical Engineers, Japan. pp. 26–28.
Muljana H, Picchioni F, Heeres HJ, Janssen LPBM. 2009. Supercritical carbon dioxide ($scCO_2$) induced gelatinization of potato starch. Carbohydr Polym 78:511–519.
Mulvaney SJ, Rizvi SSH. 1993. Extrusion processing with supercritical fluids. Food Technol 47(12):74–82.
Murua-Pagola B, Beristain-Guevara CI, Martinez-Bustos F. 2009. Preparation of starch derivatives using reactive extrusion and evaluation of modified starches as shell materials for encapsulation of flavoring agents by spray drying. J Food Engin 91:380–386.
Nabar Y, Narayan R, Schindler M. 2006. Twin-screw extrusion production and characterization of starch foam products for use in cushioning and insulation applications. Polym Engin Sci 46:438–451.
Narkrugsa W, Berghofer E, Camargo LC. 1992. Production of starch derivatives by cooking extrusion. Starch/Stärke 44:81–90.
Nikolić S, Mojović L, Rankin R, Pejin D, Pejin J. 2010. Ultrasound-assisted production of bioethanol by simultaneous saccharification and fermentation of corn meal. Food Chem 122:216–222.
Nitsch, E. 13 June 1995. Process for the production of starch degradation products with a narrow molecular weight distribution. U.S. patent 5,424,302.
Nordmark TS, Ziegler GR. 2002a. Spherulitic crystallization of gelatinized starch and its fractions. Carbohydr Polym 49:439–448.
Nordmark TS, Ziegler GR. 2002b. Structural features of non-granular spherulitic maize starch. Carbohydr Res 337:1467–1475.
Obanni M, BeMiller JN. 1995. Identification of starch from various maize endosperm mutants via ghost structures. Cereal Chem 72:436–442.
O'Brien S, Wang Y-J, Vervaet C, Remon JP. 2009. Starch phosphates prepared by reactive extrusion as a sustained release agent. Carbohydr Polym 76:557–566.
Oh HE, Hemar Y, Anema SG, Wong H, Pinder D.N. 2008a. Effect of high-pressure treatment on normal rice and waxy rice starch-in-water suspensions. Carbohydr Polym 73:332–348.
Oh HE, Pinder DN, Hemar Y, Anema SG, Wong H. 2008b. Effect of high pressure treatment on various starch-in-water suspensions. Food Hydrocoll 22:150–155.
Patel S, Venditti RA, Pawlsk JJ, Ayoub A, Rizvi SSH. 2009. Development of cross-linked starch microcellular foam by solvent exchange and reactive supercritical fluid extrusion. J Appl Polym Sci 111:2917–2929.
Pei-Ling L, Xiao-Song H, Qun S. 2010. Effect of high hydrostatic pressure on starches: A review. Starch/Stärke 62:615–628.
Peterson SC, Fanta GF, Adlof RO, Felker FC. 2005. Identification of complexed native lipids in crystalline aggregates formed from jet cooked corn starch. Carbohydr Polym 61:162–167.
Peterson SC, Eller FJ, Fanta GF, Felker FC, Shogren RL. 2007. Effects of critical fluid lipid extraction on the gelatinization and retrogradation of normal dent corn starch. Carbohydr Polym 67:390–397.
Peterson SC, Eller FJ, Fanta GF, Felker FC, Shogren RL. 2008. Comparison of the effects of critical fluid and reflux-extracted techniques on corn starch pasting properties. Carbohydr Polym 71:74–79.
Pushpadass HA, Babu GS, Weber RW, Hanna MA. 2008. Extrusion of starch-based loose fill packaging foams: effects of temperature, moisture and talc on physical properties. Packaging Technol Sci 21:171–183.
Qi ZH, Xu A. 1999. Starch-based ingredients for flavor encapsulation. Cereal Foods World 44:460–465.
Rankin JC, Phillips BS, Doane WM, Russell CR. 1976. Steam jet cooking of unmodified and modified cereal flours. Evaluation of paste properties, papermaking uses, and flocculation activity. Starch/Stärke 28:174–179.
Raquez J-M, Nabar Y, Srinivasan M, Shin B-Y, Narayan R, Dubois P. 2008. Maleated thermoplastic starch by reactive extrusion. Carbohydr Polym 74:159–169.
Rauwendaal C. 2001. Polymer Extrusion, Fourth Edition. New York: Hanser.
Repka MA, Battu SK, Upadhye SB, *et al.* 2007. Pharmaceutical applications of hot-melt extrusion: part II. Drug Devel Ind Pharm 33:1043–1057.

Riaz MN. 2000. Extruders in Food Applications. Boca Raton, FL: CRC Press.
Rizvi SSH, Mulvaney S. 9 June 1992. Extrusion processing with supercritical fluids. U.S. patent 5,120,559.
Rudnik E, Zukowska E. 2004. Studies on preparation of starch succinate by reactive extrusion. Polimery 49:132–134.
Rumpold BA, Knorr D. 2005. Effects of salts and sugars on pressure-induced gelatinization of wheat, tapioca and potato starches. Starch/Stärke 57:370–377.
Sajilata MG, Singhal RS, Kulkarni PR. 2006. Resistant starch – a review. Compr Rev Food Sci Food Safety 5:1–17.
Saka S, Ueno T. 1999. Chemical conversion of various celluloses to glucose and its derivatives in supercritical water. Cellulose 6:177–191.
San Martin-Martinez E, Aguilar-Mendez MA, Espinosa-Solares T, Pless RC, Quintana ZD. 2004. Starch phosphates produced by extrusion: Physical properties and influence on yogurt stability. Starch/Stärke 56:199–207.
Sánchez-Rivera MM, Flores-Ramírez I, Zamudio-Flores PB, González-Soto RA, Rodríguez-Ambríz SL, Bello-Pérez LA. 2010. Acetylation of banana (*Musa paradisiacal* L.) and maize (*Zea mays* L.) starches using a microwave heating procedure and iodine as catalyst: Partial characterization. Starch/Stärke 62:155–164.
Schittenhelm N, Kulicke W-M. 2000. Producing homologous series of molar masses for establishing structure–property relationships with the aid of ultrasonic degradation. Macromol Chem Phys 201:1976–1984.
Schuchmann HP, Danner T. 2000. Product engineering using the example of extruded instant powders. Chem Eng Technol 23:303–308.
Seker M, Sadikoglu H, Ozdemir M, Hanna MA. 2003. Cross-linking of starch with reactive extrusion and expansion of extrudates. Int J Food Prop 6:473–480.
Selmi B, Marion D, Perrier-Cornet JM, Douzals JP, Gervais P. 2000. Amyloglucosidase hydrolysis of high-pressure and thermally gelatinized corn and wheat starches. J Agric Food Chem 48:2629–2633.
Shahidi F, Han XQ. 1993. Encapsulation of food ingredients. Crit Rev Food Sci Nutr 33:501–547.
Shen L, Worrell E, Patel M. 2009. Present and future development in plastics from biomass. Biofuels Bioprod Bioref 4:25–40.
Shewale SD, Pandit AB. 2009. Enzymatic production of glucose from different qualities of grain sorghum and application of ultrasound to enhance the yield. Carbohydr Res 344:52–60.
Shogren RL. 1992. Effect of moisture content on the melting and subsequent physical aging of corn starch. Carbohydr Polym 19:83–90.
Shogren RL. 1993. Effects of moisture and various plasticizers on the mechanical properties of extruded starch. In: Ching C, Kaplan DL, Thomas EL (editors). Biodegradable Polymers and Packaging. Lancaster, PA: Technomic. pp. 141–150.
Shogren RL. 1994. Aging properties of extruded high-amylose starch. J Environ Polym Degrad 2:99–109.
Shogren RL. 2000. Modification of maize starch by thermal processing in glacial acetic acid. Carbohydr Polym 43:309–315.
Shogren RL, Biswas A. 2006. Preparation of water-soluble and water swellable starch acetates using microwave heating. Carbohydr Polym 64:16–21.
Shogren RL, Swanson CL, Thompson AR. 1992. Extrudates of corn starch with urea and glycols: structure/mechanical property relations. Starch/Stärke 44:335–338.
Shogren RL, Fanta GF, Doane WM. 1993. Development of starch based plastics – a reexamination of selected polymer systems in historical perspective. Starch/Stärke 45:276–280.
Shogren RL, Fanta GF, Felker FC. 2006. X-ray diffraction study of crystal transformations in spherulitic amylose/lipid complexes from jet-cooked starch. Carbohydr Polym 64:444–451.
Shrestha AK, Ng CS, Lopez-Rubio A, Blazek J, Gilbert EP, Gidley MJ. 2010. Enzyme resistance and structural organization in extruded high amylose maize starch. Carbohydr Polym 80:699–710.
Shroyer JA. 1997. Understanding the basics of steam injection heating. Chem Eng Prog 93:52–55.
Siemion P, Jabłońska J, Kapuśniak J, Kozioł JJ. 2004. Solid state reactions of potato starch with urea and biuret. J Polym Environ 12:247–255.
Siemion P, Kapuśniak J, Kozioł JJ. 2005a. Thermally induced reaction of potato starch with thiourea. J Polym Environ 13:19–27.
Siemion P, Kapuśniak J, Kozioł JJ. 2005b. Solid state thermal reactions of starch with semicarbazide hydrochloride. Cationic starches of a new generation. Carbohydr Polym 62:182–186.
Siemion P, Kapuśniak, J, Kozioł JJ. 2006. Solid state reaction of starch with thiosemicarbazide. Carbohydr Polym 66:104–109.

Sifferman TR, Muijs HM, Fanta GF, Felker FC, Erhan SM. 2003. Starch-lubricant compositions for improved lubricity and fluid loss in water-based drilling muds. In: SPE 80213, SPE Symposium on Oilfield Chemistry. Richardson, TX: Society of Petroleum Engineers. pp. 1–5.
Simonin H, Marzouki S, Guyon C, Orlowska M, Le-Bail A, de Lamballerie M. 2009. Pasting properties of high-pressure-treated starch suspensions. High Press Res 29:726–731.
Singh M. 2008. Particle size characteristics of starch-lipid composites. J. Food Proc Preserv 32:404–415.
Singh M, Byars JA. 2009. Starch-lipid composites in plain set yogurt. Internat J Food Sci Technol 44: 106–110.
Singh M, Kim S. 2009. Yogurt fermentation in the presence of starch-lipid composite. J Food Sci 74(2): C85–C89.
Singh V, Tiwari A. 2008. Microwave-accelerated methylation of starch. Carbohydr Res 343:151–154.
Singh B, Rizvi SSH, Harriot, P. 1996. Measurement of diffusivity and solubility of carbon dioxide in gelatinized starch at elevated pressures. Indust Eng Chem Res 35:4457–4463.
Singh V, Tiwari A, Pandey S, Singh SK. 2006. Microwave accelerated synthesis and characterization of potato starch-*g*-poly(acrylamide). Starch/Stärke 58:536–543.
Singh J, Dartois A, Kaur L. 2010. Starch digestibility in food matrix: a review. Trends Food Sci Technol 21:168–180.
Sjoqvist M, Boldizar A, Rigdahl M. 2009. Processing and properties of expanded starch materials. J Cellular Plast 45:51–66.
Sreeram KJ, Nidhin M, Nair BU. 2008. Microwave assisted template synthesis of silver nanoparticles. Bull Mater Sci 31:937–942.
Staroszczyk H, Janas P. 2010. Microwave-assisted synthesis of zinc derivatives of potato starch. Carbohydr Polym 80:962–969.
Staroszczyk H, Fiedorowicz M, Zhong W, Janas P, Tomasik P. 2007a. Microwave assisted solid-state sulphation of starch. e-Polymers, number 140.
Staroszczyk H, Tomasik P, Janas P, Poreda A. 2007b. Esterification of starch with sodium selenite and selenate. Carbohydr Polym 69:299–304.
Stepto RFT, Tomka I. 1987. Injection moulding of natural hydrophilic polymers in the presence of water. Chimia 41:76–81.
Stute R. 1992. Hydrothermal modification of starches: The difference between annealing and heat/moisture treatment. Starch/Stärke 44:205–214.
Stute R, Klingler RW,Boguslawski S, Eshtiaghi MN, Knorr D. 1996. Effects of high pressure treatment on starches. Starch/Stärke 48:399–408.
Suguchi M, Higasa T, Mori T. 1994. Study of wheat starch structures by sonication treatment. Cereal Chem 71:636–639.
Suslick KS. 1990. Sonochemistry. Science 247:1439–1445.
Suslick KS, Flannigan DJ. 2009. Inside a collapsing bubble: sonoluminescence and the conditions during cavitation. Ann Rev Phys Chem 59:659–83.
Svihus B, Uhlen AK, Harstad OM. 2005. Effect of starch granule structure, associated components and processing on nutritive value of cereal starch: a review. Animal Feed Sci Technol 122:303–320.
Tan F-J, Dai W-T, Hsu K-C. 2009. Changes in gelatinization and rheological characteristics of japonica rice starch induced by pressure/heat combinations. J Cereal Sci 49:285–289.
Tang H, Kitagawa K. 2005. Supercritical water gasification of biomass: Thermodynamic analysis with direct Gibbs free energy minimization. Chem Eng J 106:261–267.
Tester RF, Debon SJJ. 2000. Annealing of starch – a review. Internat J Biol Macromol 27:1–12.
Tharanathan RN. 2005. Starch-value addition by modification. Crit Rev Food Sci Nutr 45:371–384.
Tomasik P, Schilling CH. 2004. Chemical modification of starch. Adv Carbohydr Chem Biochem 59:175–403.
Tomasik P, Zaranyika MF. 1995. Nonconventional methods of modification of starch. In: Horton D (editor). Advances in Carbohydrate Chemistry and Biochemistry, Vol. 51. New York: Academic Press. p. 243–320.
Tomasik P, Wang YJ, Jane J-L. 1995. Facile route to anionic starches. Succinylation, maleination and phthalation of corn starch on extrusion. Starch/Stärke 47:96–99.
Tomka I, inventor. 1990. Thermoplastically processable starch and a process for producing it. PCT International Application WO 90/05161.
Trivedi NR, Rajan MG, Johnson JR, Shukla AJ. 2007. Pharmaceutical approaches to preparing pelletized dosage forms using the extrusion-spheronization process. Crit Rev Therapeut Drug Carrier Sys 24:1–40.
Vallons KJR, Arendt EK. 2009a. Effects of high pressure and temperature on buckwheat starch characteristics. Eur Food Res Technol 230:343–351.

Vallons KJR, Arendt EK. 2009b. Effects of high pressure and temperature on the structural and rheological properties of sorghum starch. Innov Food Sci Emerg Technol 10:449–456.

Van Den Einde RM, Van Der Goot AJ, Boom RM. 2003. Understanding molecular weight reduction of starch during heating-shearing processes. J Food Sci 68:2396–2404.

Wang L, Wang Y-J. 2004a. Application of high-intensity ultrasound and surfactants in rice starch isolation. Cereal Chem 81:140–144.

Wang L, Wang Y-J. 2004b. Rice starch isolation by neutral protease and high intensity ultrasound. J Cereal Sci 39:291–296.

Wang L, Shogren RL, Willett JL. 1997. Preparation of starch succinates by reactive extrusion. Starch/Stärke 49:116–120.

Wang, B., Li, D., Wang, L.-j., Chiu, Y.L., Chen, X.D. & Mao, Z.-h. 2008. Effect of high- pressure homogenization on the structure and thermal properties of maize starch. J Food Engineer 87:436–444.

Warner K, Eskins K, Fanta GF, Nelsen TC, Rocke J. 2001. Use of starch-lipid composites in low-fat ground beef products. Food Technol 55:36–41.

Weaver MO. 1988. Cationic starch graft copolymer latexes for stabilizing soil to water erosion: cationic charge on starch vs cationic charge on grafted branches. J Polym Mater 5:109–113.

Weaver MO. 1989a. Starch-*g*-poly(methyl acrylate) latexes for stabilizing soil to water erosion: Varying the starch moiety. J Polym Mater 6:271–6.

Weaver MO. 1989b. Starch-*g*-poly(methyl acrylate) latexes for stabilizing soil to water erosion: extending the range of polymer add-on. Starch/Stärke 41:106–110.

Weaver MO, Fanta GF. 1987. Cationic starch graft copolymer latexes as soil stabilizers. J Polym Mater 4:51–65.

Whistler RL, Daniel JR. 1997. Starch. In: Howe-Grant M (editor). Kirk-Othmer Encyclopedia of Chemical Technology, 4th Ed., Vol. 22. New York, NY: John Wiley & Sons, Inc. pp. 699–719.

Willett JL, Shogren RL. 2002. Processing and properties of extruded starch-polymer foams. Polymer 43:5935–5947.

Willett JL, Millard MM, Jasberg BK. 1997. Extrusion of waxy maize starch: melt rheology and molecular weight degradation of amylopectin. Polymer 38:5983–5989.

Williams JR, Clifford AA, Al-Saidi SHR. 2002. Supercritical fluids and their applications in biotechnology and related areas. Molec Biotechnol 22:263–286.

Wilpiszewska K, Spychaj T. 2008. Chemical modification of starch by reactive extrusion. Polimery/Polymers 53:268–275.

Winfrey VL, Black WC. 1964. Method of treating starch with steam. U.S. patent 3,133,836.

Wing RE, Willett JL. 1997. Water soluble oxidized starches by peroxide reactive extrusion. Ind Crops Prod 7:45–52.

Wolf B. 2010. Polysaccharide functionality through extrusion processing. Curr Opinion Colloid Interface Sci 15:50–54.

Wurzburg OB. 1986. Modified Starches: Properties and Uses. Boca Raton, FL: CRC Press.

Xie F, Liu H, Chen P, Xue T, Chen L, Yu L, Corrigan P. 2006a. Starch gelatinization under shearless and shear conditions. Int J Food Engin 2:article No. 6.

Xie F, Yu L, Liu H, Chen L. 2006b. Starch modification using reactive extrusion. Starch/Stärke 58:131–139.

Xing G-X, Zhang S-F, Ju B-Z, Yang J-Z. 2006. Microwave-assisted synthesis of starch maleate by dry method. Starch/Stärke 58:464–467.

Yalpani M. 1993. Supercritical fluids: Puissant media for the modification of polymers and biopolymers. Polymer 34:1102–1105.

Yamamoto K, Handschin S, Conde-Petit B, Escher F. 2002. Starch-amphiphile complex formed by high pressure. Trends High Press Biosci Biotechnol 19:405–409.

Yoon KJ, Carr ME, Bagley EB. 1994. Reactive extrusion vs. batch preparation of starch-g-polyacrylonitrile. J Appl Polym Sci 45:1093–1100.

Yu L, Dean K, Li L. 2006. Polymer blends and composites from renewable resources. Prog Polym Sci 31:576–602.

Zavareze ER, Dias ARG. 2011. Impact of heat–moisture treatment and annealing in starches: A review. Carbohydr Polym 83:317–328.

Zhan X, Bean SR, Mo X, Sun XS, Boyle D. 2006. Ethanol production from supercritical-fluid-extracted cooked sorghum. Indust Crops Prod 23:304–310.

Zhang Z, Feng H, Niu Y, Eckhoff SR. 2005a. Starch recovery from degermed corn flour and hominy feed using power ultrasound. Cereal Chem 82:447–449.

Zhang Z, Niu Y, Eckhoff SR, Feng H. 2005b. Sonication enhanced corn starch separation. Starch/Stärke 57:240–245.

Zhang L, Gu S, Liu Y, Gao W. 2008. Preparation of yam starch acetates by using ultrasonic-assisted method and characterization of its products. Shipin Keji 11:247–252.

Ziegler GR, Nordmark TS, Woodling SE. 2003. Spherulitic crystallization of starch: influence of botanical origin and extent of thermal treatment. Food Hydrocoll 17:487–494.

Ziegler GR, Creek JA, Runt J. 2005. Spherulitic crystallization in starch as a model for starch granule initiation. Biomacromolecules 6:1547–1554.

Zobel HF. 1988. Starch crystal transformations and their industrial importance. Starch/Stärke 40:1–7.

3 Protein Processing in Food and Bioproduct Manufacturing and Techniques for Analysis

Joyce Irene Boye and Chockry Barbana

3.1 INTRODUCTION

Proteins and amino acids are the building blocks of life. They play a critical role as components of blood, muscle, enzymes, hormones and hormone receptors. In addition to their biological role in sustaining the functions of living organisms, proteins also play important roles in foods providing taste, texture and flavor, which are essential criteria for food selection. Due to their versatile functionality and complex molecular structure, proteins have also been explored in many industrial applications. Some recent examples include adhesives, protein plastics, gels, coating, additives and biomaterials. Protein scaffolds and cross-linked networks also hold much promise for use in the pharmaceutical industry.

In nature, proteins exist as components of biological matrices with other compounds such as lipids, carbohydrates, minerals, and other minor components. This composite existence is ideal in foods, as it allows various sources of nutrients to be supplied at the same time. In some instances, however, there are distinct advantages in separating protein fractions in order to obtain enriched or purified fractions for specific food, nutraceutical or industrial application. Thus, for sports nutrition as an example, there may be the need for highly enriched protein beverages that quickly help to build and restore muscle. Engineering of protein scaffolds for pharmaceutical applications may require the use of highly purified proteins as raw materials for production. Preparation of simulated meat products using plant proteins as meat analogs may also require the use of highly purified protein sources.

This chapter provides an overview of the properties of food proteins and details some of the techniques currently used for the extraction and purification of proteins from different food sources. Protein processing for human food use is emphasized. However, most of the described processes are also applicable for feed and other industrial and pharmaceutical applications. As the purity and quality of extracted proteins have to be frequently assessed to determine their suitability for different applications, the chapter also provides a summary of some of the principal techniques used for analyzing the properties of proteins in food and bioproduct manufacturing.

Food and Industrial Bioproducts and Bioprocessing, First Edition. Edited by Nurhan Turgut Dunford.

3.2 GENERAL PROPERTIES OF PROTEINS

Proteins are comprised of amino acids linked by peptide bonds (Figure 3.1). A list of the 20 amino acids most commonly found in nature is presented in Table 3.1. Other rare amino acids found in nature include ornithine, citrulline and cycloleucine. Amino acid exists in the D or L conformation depending on their rotatory optical activity (Figure 3.2). Most amino acids found in nature exist in the L conformation.

Proteins have complex molecular structures. The linear sequence of the amino acids comprising a protein is classified as its primary structure. In different proteins, these linear sequences assume conserved structures along the axis of the polypeptide in the form of alpha-helixes, 3_{10}-helix, beta sheets, or random coils, turns which are described as the secondary structure of the protein. These secondary structures are stabilized primarily by hydrogen bonds. For thermodynamic stability, proteins rearrange themselves into tertiary structures comprising several secondary structures stabilized by van der Waal's, electrostatic, or hydrophobic interactions, hydrogen bonding, as well as disulfide cross-links. Some proteins have a fourth structural level called the quaternary structure in which two or more

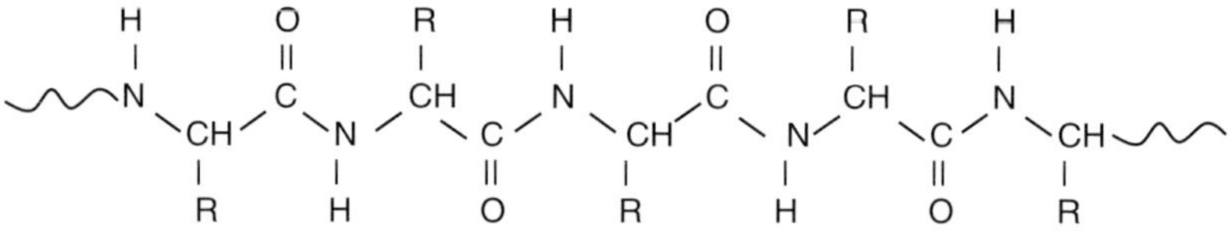

Figure 3.1 Amino acids linked by peptide bonds.

Table 3.1 The 20 amino acids most commonly found in nature and their characteristics.

Amino Acid	Abbrev.	Letter	Type	MW
Alanine	ala	A	Aliphatic	89.1
Arginine	arg	R	Basic	174.2
Asparagine	asn	N	Amidic	132.1
Aspartic acid	asp	D	Acidic	133.1
Cysteine	cys	C	Sulfur containing	121.1
Glutamine	gln	Q	Amidic	146.1
Glutamic acid	glu	E	Acidic	147.1
Glycine	gly	G	Aliphatic	75.1
Histidine	his	H	Basic	155.2
Isoleucine	ile	I	Aliphatic	131.2
Leucine	leu	L	Aliphatic	131.2
Lysine	lys	K	Basic	146.2
Methionine	met	M	Sulfur containing	149.2
Phenylalanine	phe	F	Aromatic	165.2
Proline	pro	P	Aliphatic	115.1
Serine	ser	S	Hydroxylic	105.1
Threonine	thr	T	Hydroxylic	119.1
Tryptophan	trp	W	Aromatic	204.2
Tyrosine	tyr	Y	Aromatic	181.2
Valine	val	V	Aliphatic	117.1

D-amino acid L-amino acid

Figure 3.2 Conformation of D and L amino acids.

monomers, sometimes called "subunits", associate to form complex structures stabilized by non-covalent bonds.

Amino acids may be classified as acidic, basic, sulfur containing, hydroxylic, aliphatic, hydrophobic or amidic (Table 3.1). The number and type of amino acids in a given protein determine, respectively, its size and net charge at different pHs. Proteins vary in molecular mass, ranging from <10 000 Da to >1 million Da, depending on the number of amino acids contained within the protein. Environmental conditions, such as pH and the presence of salts, can alter the charge state of amino acid residues and, consequently, the net charge of the protein. The isoelectric point of a protein is defined as the pH at which the net charge on the protein is zero. These two properties of proteins (i.e. charge and size) play a critical role in the technologies and processing conditions used for their separation.

Another property of food proteins that is of interest in processing is their solubility. Osborne (1924) classified proteins based on their solubility as follows: water soluble proteins (albumins), salt soluble proteins (globulins), alcohol soluble proteins (prolamins) and acid and alkali soluble proteins (glutenins). The solubility of proteins under aqueous conditions can be further modified by heat treatment, hydrolysis and the presence of protein structure perturbing chemical reagents. Depending on the extent of heat treatment, proteins can unfold, exposing more hydrophilic amino acids, which can enhance solubilization. In contrast, harsh heat treatment can result in extensive denaturation, association and irreversible aggregation, which can reduce protein solubility. Hydrolysis of proteins using chemical agents, such acid or alkali or enzymes, can break up proteins into smaller fragments, which can increase their solubility.

3.3 PROTEIN SEPARATION PROCESSES IN FOOD AND BIOPRODUCT MANUFACTURING

Proteins can be processed to obtain enriched flours, concentrates or isolates for various food and bioproduct applications. In this chapter, the term bioproduct is used to indicate functional foods, nutraceutical products, cosmetic products and other industrial (non-food) products.

In general, protein flours may contain up to 65% w/w protein on a dry basis (db), whereas concentrates and isolates contain >65% (w/w, db) and >80–90% (w/w, db) protein, respectively. Microbiological and quality standards will vary for different applications.

Proteins used in the food and bioproduct sectors are derived from either animal or plant sources, which vary significantly in their lipid, carbohydrate and protein content. Proteins of interest may be separated using dry or wet processing techniques. Dry processing primarily involves air classification and is frequently applied to plant materials such as cereals (e.g., wheat) and grain legumes (e.g., peas, chickpeas, lentils) containing high amounts of starch and protein. Wet processing has several processing steps. Some of the major unit

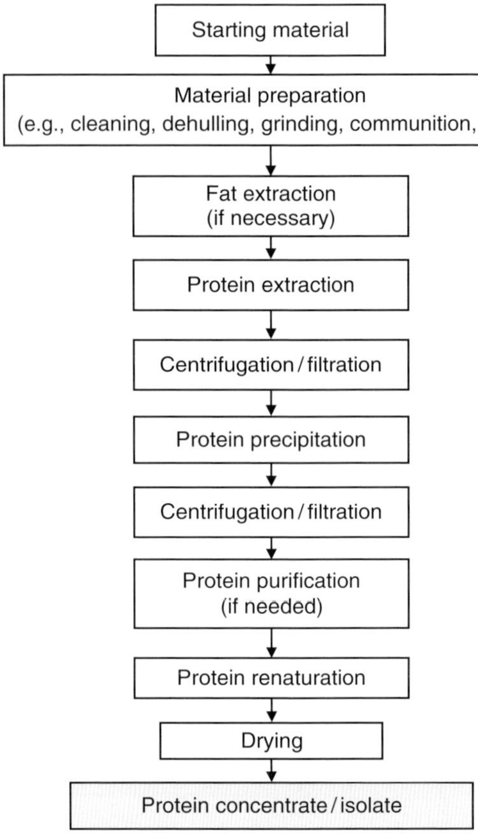

Figure 3.3 General schematic showing some key unit operations for the wet processing of proteins.

operations used for the wet processing of proteins are presented in Figure 3.3. Deviations to these operations occur depending on the starting material and the desired end product. The material preparation step for wet processing will, therefore, vary depending on the source material.

3.3.1 Dry processing

3.3.1.1 *Air classification*

Air classification is a milling technique that allows the fractionation of plant grains/seeds into high starch and high protein flours. The milling process yields flours having particles of two discrete sizes and densities. Air classification exploits this phenomenon to separate the light fine fraction (protein) from the heavy coarse fraction (starch).

During air classification, whole or de-hulled seed is ground into very fine flour, and the flour is subsequently classified in a spiral air stream to separate the starch from the protein. Repeating the process a few times improves separation efficiency, as protein bodies can still adhere to the surface of starch granules after the initial run. Agglomerates present in this initial starch fraction consist of starch granules embedded in a protein matrix, but by repeated pin milling and air classification, further purification can be obtained (Gueguen *et al.*, 1984).

For optimal separation efficiency, the cotyledons of seeds must be finely ground to achieve complete cellular disruption and maximum protein and starch separation during air classification (Tyler and Panchuk, 1982). The milling technique used must, therefore, be capable of producing a very fine grind, yet selective enough to break up cells and cell fragments without severely damaging the starch granules (Jones *et al.*, 1959). A portion of the protein derived from the membranes and stroma of the choloroplasts in which the starch granules developed cannot be milled free of the starch granules despite repeated milling (Tyler, 1984). The purities of the coarse (starch) and the fine (protein) fractions obtained by air classification are, therefore, often lower than what may be obtained by wet extraction.

Amongst others, high protein flours (up to 75% purity in some cases) have been successfully produced from wheat, soybean, beans, lentils, chickpeas and peas using air classification (Wu and Strongfellow, 1979, 1981; Tyler and Panchuk, 1982; Wolf *et al.*, 2002).

3.3.2 Wet processing

3.3.2.1 *Material preparation*

Animal by-products

Animal by-products (e.g., blood, skin, bones and offal) may serve as useful sources of biomass for harvesting high-value proteins. Prior to extraction, proteins are solubilized by disruption of the cells and tissues retaining them. Protein solubilization is a major critical step affecting yield and quality of the extract. Methods available for tissue disruption include grinding, homogenization and sonication. Mechanical disruption using colloidal mills, in which the biomass is fed through a rotor-stator, high speed dispersion mills or bead mills, which break down cell walls through their tumbling action, may also be used. In some instances enzymes (e.g., zymolase, lysozyme, and lysostaphin) may be added to facilitate tissue disruption followed by homogenization, sonication or vigorous vortexing.

Equipment selection will depend on the specific material being processed and the scale (analytical, pilot or industrial). As animal materials frequently contain proteases that can hydrolyze proteins of interest and decrease their functionality, appropriate precautions are required to slow down or prevent these reactions (e.g., thermal inactivation or processing at cold temperatures). After the appropriate level of communition has been attained, a filtration step can be used to recuperate the supernatant containing the desired proteins for further downstream processing.

Plant by-products

Plants contain high amounts of fiber and cellulosic material that needs to be removed prior to protein extraction. Some plant materials are also high in fat, making a defatting step necessary prior to protein extraction. Effectively disintegrating plant materials to obtain particle sizes that allow for maximum fat and protein extraction is an important first step in processing. This is best done by beginning the processing with a milling step. Examples of milling equipment used in research and in industrial settings include centrifugal mills, hammer mills, ball mills, roller mills and disc attrition mills. De-hulling and milling can be done as a single unit operation or as two separate unit operations depending on the type of equipment used and the material being processed (i.e., ease of de-hulling).

Materials such as soybeans seeds, which are high in fat (17–23%), may require a defatting step. The techniques most commonly used are solvent extraction using hexane, mechanical extraction and aqueous extraction. Details of these processes are summarized elsewhere

(Mustakas, 1980; Serrato, 1981). The material obtained for further downstream processing will vary depending on the process used and, in the case of soybean, could be a full fat soybean meal or flour, defatted meal (untoasted, mildly toasted or fully toasted). The toasting process refers to the heat treatment applied to remove residual solvent after hexane extraction. Both mechanical and solvent extraction techniques have their limitations (Russin *et al.*, 2010). Mechanical extraction, or pressing, is limited in its applicability, particularly with low oil content oilseeds such as soybean. Elevated temperatures employed during pressing can also have a deleterious effect on the quality of the extracted oil and residual meal (Sugarman, 1956; Nelson *et al.*, 1987). Hexane extraction has many economic, environmental and safety limitations. Economically, one of the main concerns is the stability of both hexane supply and price due to fluctuation in the fossil fuel market (Friedrich and List, 1982; Lusas *et al.*, 1990; Gandhi *et al.*, 2003; Russin *et al.*, 2010). Concerns also exist about the environmental impacts of hexane use and its toxicity.

As an alternative, aqueous techniques for fat extraction are being explored. In this case, after communition the full fat soy material is solubilized to perform a solid–liquid extraction/separation. During this step, insoluble compounds are removed, leaving a liquid solution containing both proteins and lipids. This solution is further separated by three-phase centrifugation to yield a solid, an aqueous and an oil/emulsion phase, each of which can be further processed downstream (Russin *et al.*, 2010). Enzymes (e.g., lipases, cellulases) may be added during this process to breakdown fat and carbohydrate components to facilitate protein extraction (Russin *et al.*, 2010). The main principle of enzyme-assisted extraction is the use of enzymes which damage and/or degrade plant cell walls, so increasing the permeability of the oil in the oilseed (Domínguez *et al.*, 1994). The two main approaches include the use of single and mixed enzymatic systems. The latter has increased utility, given that the mixed systems allow for various enzymes to simultaneously act on the cellular structures, leading to a more effective release of oil (Fullbrook, 1984; Domínguez *et al.*, 1993; Russin *et al.*, 2010). The use of lipases or phospholipases to breakdown fats, particularly in high fat aqueous extraction systems where emulsions are likely to occur, is of interest; however, adequate care must be taken to minimize oxidation during processing, as this could result in the generation of off-flavors in the finished product.

3.3.2.2 *Protein extraction*

Alkaline extraction

Aqueous alkaline extraction is one of the most commonly used techniques for protein extraction; it takes advantage of the solubility of proteins at alkaline pH. In this process the prepared biomass, which may be full fat, partially defatted, or fully defatted, and in the case of plant materials may or may not contain fibrous materials such as hulls, is dispersed in water using flour:water ratios ranging from 1:5 to 1:20. The pH of the mixture is adjusted to alkaline (pH 8–11) and the mixture is continuously stirred for 30 to 180 min to maximize protein solubilization. During this time the pH is maintained at the desired value and the temperature may be elevated (up to 55–65 °C) to further enhance protein solubilization and extraction. The mixture is subsequently filtered to remove any insoluble material to yield a supernatant containing the extracted proteins. Some extraction processes call for a second extraction of the precipitate using similar pH as in the first extraction or higher in order to extract any remaining proteins in the precipitate and increase protein recovery (Boye *et al.*, 2010a).

Lawhon and co-workers (1981a) found the extractability of protein in aqueous medium (water) from full fat soy flour to be higher when using higher flour to water ratios (i.e. 1:30

or 1:25 soy flour to water ratio by weight) than when using lower ratios (i.e., 1:12 or using a double extraction at 1:10 followed by 1:6). In the former case, extraction at 60 °C for 30 min at pH 6.6, 8.0 and 9.0 gave nitrogen recoveries of 91.4, 94.2 and 89.3%, respectively. In the latter case (1:12 flour to water ratio) the extraction was done at pH 9.0 and a recovery of 80.4% nitrogen was obtained, whereas for the double extraction (1:10 followed by 1:6) at pH 2.5, 81.8% nitrogen recovery was obtained. The authors indicated that water ratios influenced nitrogen extractability more significantly than pH.

pH and temperatures used during alkaline extraction must be carefully chosen to avoid extensive denaturation as well as the development of by-products, such as lysinoalanine (N6-(DL-2-amino-2-carboxyethyl)-L-lysine), an unusual amino acid implicated as a renal toxic factor in rats. Lysinoalanine has been found in proteins of home-cooked and commercial foods and ingredients and was initially thought to occur in both edible and non-food proteins only after alkali treatment. However, some reports have shown that it can be generated in various proteins when heated under non-alkaline conditions (Sternberg *et al.*, 1975).

Acid extraction

The solubility of some proteins increases under acidic conditions (i.e., pH <4). This low pH range can, therefore, be used to solubilize proteins prior to their recovery. The principle of acid extraction is similar to that of alkaline extraction, except that the initial protein extraction is conducted under acidic conditions. The acid extraction technique is generally used less frequently than the alkaline extraction technique and, as with alkaline extraction, processing conditions can influence the yield and purity of the finished product. Using this technique, Alli and co-workers (1993) extracted a bipyramidal crystalline protein preparation from dried seeds of white kidney bean (*Phaseolus vulgaris*) by extracting ground seeds with citric acid solution (0.4 N, pH 4.0) followed by refrigeration (4 °C, 18 h) to precipitate the protein material. The isolate obtained contained 95.7% protein.

Salt extraction

The selective extraction of proteins in aqueous solutions having different ionic strengths can be used for their fractionation and separation. The process is based on the salting-in and salting-out phenomenon of food proteins.

At low molarities (0.5–1 M), ions of neutral salts promote the solubilization of proteins ("salting in"). Interactions between the ions and charges of proteins reduce electrostatic attractions between protein molecules enhancing their solubility. Additionally, the hydration of the ions increases the solvation of the proteins contributing to increased solubility. At higher salt concentrations (>1 M), competition between salts and proteins for available water forces the proteins to precipitate ("salting out"). This phenomenon can, therefore, be used for protein recovery as explained later.

For salt soluble globulins as per the Osborne classification, addition of salts to the extracting medium facilitates their solubilization. After extraction, extensive dilution of the solution can cause these globulins to precipitate out of solution especially at low temperatures leading to their fractionation.

Foam fractionation of proteins

Foam fractionation (or separation) is an adsorptive bubble separation technique in which soluble, surface-active substances can be removed from solution by preferential adsorption at the gas–liquid interface (Wang and Liu, 2003). Proteins contain both hydrophilic and hydrophobic amino acid residues that are surface active. During foam formation, bubbles

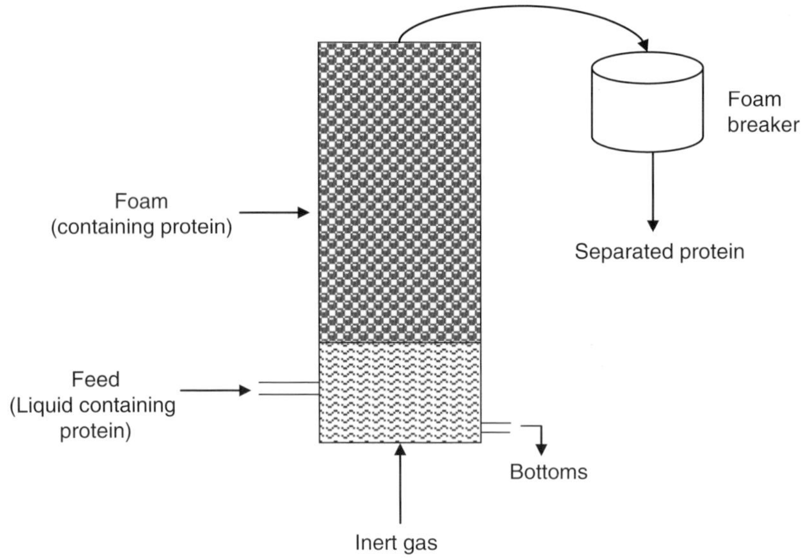

Figure 3.4 Schematic process for foam fractionation.

are created as a result of the passage of gas through a liquid medium. Proteins with greater surface activity orientate themselves and are capable of being adsorbed at the gas–liquid interface created during bubble formation. A schematic of a typical foam fractionation set up is presented in Figure 3.4. Bubbles leaving the surface of the feed solution carry both adsorbed compounds and bulk liquid into the foam; as the interstitial liquid drains back along the lamella of the bubbles, non-adsorbed solutes are returned to the retentate leaving an enriched foam (Wang *et al.*, 2009). The remaining liquid contained in the gas phase is, thus, selectively enriched with protein having greater surface activity compared to the proteins retained in solution.

Foam fractionation is a relatively inexpensive technique for protein separation. Most studies in the literature are, however, experimental and very few report on industrial applications. Among other applications, foam fractionation has been used to separate wheat flour proteins, ovalbumin, lysozyme, egg albumin, milk proteins (e.g., beta-casein, bovine lactoferrin, bovine serum albumin, alpha-lactalbumin, and beta-lactoglobulin) and potato protein from potato juice waste water after starch extraction (Weijenberg *et al.*, 1978; Keller *et al.*, 1997; Hossain and Fenton, 1998; Brown *et al.*, 1999; Wang and Liu, 2003; Wang *et al.*, 2009).

Parameters influencing efficiency of protein separation include bubble size distribution, bubble rupture and coalescence, fluid drainage and rate of bubble formation. Separation ratio (SR) defines the ratio between the protein concentration in the foam and that in the starting solution (feed). Separation efficiency (SE), on the other hand, is the product of the flow rate of the foam phase and protein concentration in the foam divided by the product of the flow rate of the liquid effluent and protein concentration of the effluent. Separation ratio is affected by the feed protein concentration (FPC), feed flow rate (FFR), gas velocity (GV) and the height of sampling (HS). For a given superficial GV, SR decreases with an increase of FPC (Wang and Liu, 2003). Additionally, the averaged protein concentration in the foam decreases with increasing GV; SR decreases with increasing FFR. Increasing the gas to feed flow rate ratio (G/F) decreases SR up to a critical G/F beyond which a plateau is reached.

SE on the other hand increases with increasing G/F as it enhances the amount of protein molecules transferred from the liquid to the foam phase. At high gas to feed flow rate ratios, however, there is greater likelihood of bubble burst. Additionally, Hossain and Fenton (1998) reported higher enrichment and protein recoveries with continuous processing compared to semi-batch mode.

Wang and co-workers (2009) successfully used foam fractionation to separate lysozyme from chicken eggs and reported that separation efficiency was significantly influenced by pH and sodium chloride concentration. Activity recovery and protein recoveries were highest (40% and 60%, respectively) at pH 7 and 0.3 M NaCl concentration. Addition of sodium dodecyl sulfate (SDS), cetyltrimethylammonium bromide, Triton-X and Tween 80, resulted in protein recoveries of 93%, 88%, 68% and 72% and activity recoveries of 30%, 20%, 55% and 58%, respectively.

Gas flow requirements to produce stable foams were found to be lower for bovine serum albumin and beta-lactoglobulin, which had greater surface activity compared to beta-casein and alpha-lactalbumin, and the enrichment factor and percentage volume loss in foam were also higher for the former proteins (Hossain and Fenton, 1998).

Matouq (2008) compared the separation of proteins from yogurt whey and cheese whey. Less foam formation occurred with cheese whey compared to yogurt whey and higher enrichment values were obtained with yogurt whey, although the cheese whey contained higher protein content. As higher concentrations of fat tend to decrease foam formation, the author concluded that the higher fat content of the cheese whey may explain the decreased foam formation and stability.

Various workers have explored ways to enhance the SE of the foam fractionation by using adsorptive ligands, such as Cibacron blue of a triazine dye bound with a polyethylene glycol, kaolin, iron oxide, graphite and SDS (Keiichi *et al.*, 1998; Yoshihiro and Toshiro, 2000; Suzuki *et al.*, 2002).

Yoichiro (1986) proposed continuous countercurrent foam separation equipment in which samples are introduced into the middle portion of a gas–liquid dual countercurrent flow system. Material having an affinity to the foam is carried with the foam stream whereas other materials are carried with the liquid stream.

As conditions used during foam fractionation can change the molecular structure of target proteins and, consequently, protein function, it is important that these conditions be selected carefully. Changes can occur due to the direct effects of ligands used, unfolding of proteins at the gas–liquid interface, high shear stress rates and chemical damage due to oxidation (Maruyama *et al.*, 2007).

Extraction using reverse micelles

Proteins can be extracted from solution using liquid–liquid separation techniques, also known as reverse micellization. The technique involves the use of a biphasic liquid system comprising an aqueous solution of the protein and an organic micellar solution containing monodisperse aggregates of surface-active molecules, usually ranging in diameter from 1 to 10 nm (Asenjo and Chaudhuri, 1996). Shaking of the two solutions results in partitioning of proteins from the aqueous into the micellar phase (Figure 3.5). Synthetic protein chaperones can be added to protect proteins from extensive denaturation and to aid in renaturation especially for pharmaceutical type applications.

A solid state extraction version of the method involves suspension of protein powder in the micellar phase followed by gentle stirring to allow the selective separation of the target proteins into the organic micellar solution.

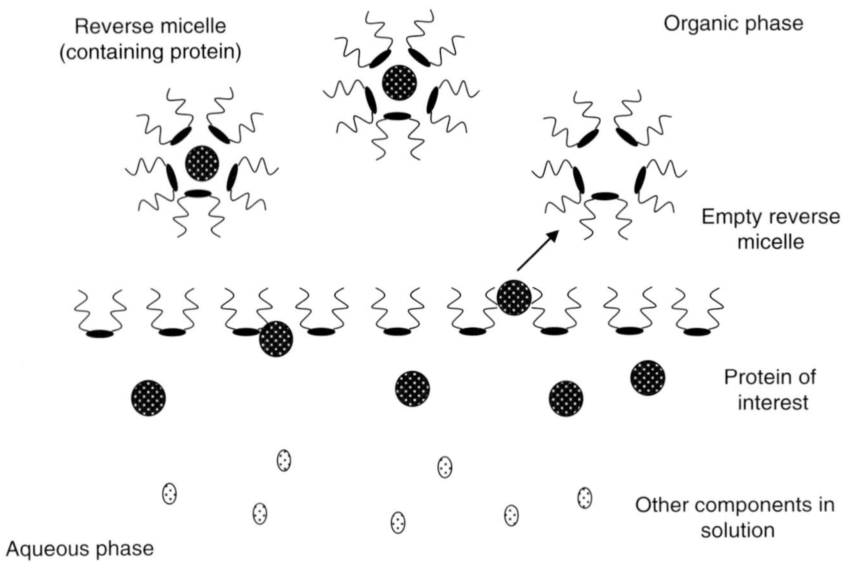

Figure 3.5 Schematic representation of the extraction of proteins using reverse micellization.

Protein solubilized in the reverse micellar solution can be transferred back into an aqueous solution by contacting the micellar solution with an aqueous solution containing a high concentration of salt (e.g., KCl, $CaCl_2$), which has the capability to exchange with protein in the micelles (Asenjo and Chaudhuri, 1996). Reverse micellization has been successfully used to separate a variety of proteins including enzymes (e.g., lysozyme, trypsin and ribonucease).

3.3.2.3 *Protein recovery*

Isoelectric precipitation

Most proteins precipitate at pH values close to their isoelectric point. This property can be used to selectively precipitate different proteins from solution. As most food proteins have their isoelectric point in the range pH 4–5, this pH range is frequently used for protein recovery in the food processing industry. Typically, after alkaline, acid or salt extraction, the pH of the protein extract is adjusted to the desired isoelectric point to induce protein precipitation, followed by centrifugation to recover the protein, washing to remove salts, neutralization and drying.

Salting out

Water from the hydrated protein is removed at high enough neutral salt concentrations (>1 M), leaving the exposed proteins to interact with each other through hydrophobic interactions. This results in protein aggregation and precipitation, a phenomenon known as "salting out". The Hofmeister series promote salting out, aggregation and stabilization or unfolding, dissociation and salting in depending on the ascending or descending order of the ions: $SO_4^{2-} < F^- < CH_3COO^- < Cl^- < Br^- < NO_3^- < I^- < ClO_4^- < SCN^-$, $NH_4^+ < K^+ < Na^+ < Li^+ < Mg^+ < Ca_2^+$ (Cheftel *et al.*, 1985). Ammonium sulfate is frequently used for the selective precipitation of proteins because it is relatively inexpensive and has high solubility. Other salts used in protein applications are magnesium chloride ($MgCl_2$) and calcium chloride ($CaCl_2$).

Achouri and co-workers (2010) reported that initial extraction of soy proteins using isoelectric precipitation at pH 6.8 followed by cryo-precipitation yielded 4.2% product recovery for the 11 S soybean protein fraction with 98% protein purity for a control extracted with sodium hydroxide, and average yields of 4.4% and 5.17%, respectively, when sodium sulfate (Na_2SO_4) and ammonium sulfate [$(NH_4)_2SO_4$] were used. Addition of calcium chloride ($CaCl_2$) doubled the extraction yield to 9%.

As some salts may be toxic, an appropriate salt must be selected for the target end-product application (industrial, food, pharmaceutical). The type of proteins precipitated in a given solution will also vary as a function of the salt concentration used.

In a modified version of the salting out process, salt proteins may be preferentially separated out of solution by dilution of the protein extract. In this case, after extraction of protein using an appropriate salt solution at desired ionic strength, the solution is extensively diluted to decrease the solubility of the salt soluble proteins, inducing protein precipitation.

Paredes-López and co-workers (1991) extracted protein from a 10% (w/v) solution of defatted chickpea with sodium chloride (0.5 M, pH 7.0) and obtained a chickpea protein isolate containing 87.8% protein using this method. After concentration of the extract by ultrafiltration, protein was flocculated by the addition of water (4 °C, pH 7, 1:4 v/v ratio of protein extract: water). Márquez and co-workers (1996) reported protein contents ranging from 74.7 to 84.2% for common bean protein extract using similar methods.

Microfiltation and ultrafiltration

Microfiltation (MF) and ultrafiltration (UF) are pressure-driven filtration processes that utilize a porous membrane to selectively retain compounds larger than a nominal molar mass (retentate) while allowing particles of lower molar mass to pass through the membrane (permeate). It is a frequently used alternative to isoelectric precipitation.

During this process, solutions containing extracted or dissolved proteins are subjected to MF and/or UF to concentrate the proteins. MF membranes retain larger molecular weight compounds and are better suited for retaining particles with a molecular mass greater than 300 000 and in the 0.02–10 micron size range. They are, therefore, more useful for the separation of cellular material, such as in biological, biochemical, pharmaceutical and nutraceutical applications. A useful application of MF in food processing is the removal of yeasts and bacteria without the requirement for thermal treatment. In protein purification work, MF could be used to specifically retain larger molecular weight proteins of interest, or to remove these interfering proteins. The permeate can be further processed to obtain a higher purity protein extract.

UF membranes retain smaller molecular mass particles (300–300 000) with sizes ranging from 0.001 to 0.02 microns and are frequently used in protein processing. UF membranes are typically constructed from regenerated cellulose, cellulose acetate, ceramic composites, polysulfone, polyethersulfone, polyamide, polyacrylonitrile or polyvinyl alcohol. The molecular weight cut-off of ultrafiltration membranes is diffuse by about one order of magnitude, which makes it difficult to achieve absolute retention. Thus, a UF membrane with a molecular weight cut-off of 200 000 could retain particles as small as 20 000 molecular mass. In addition to molecular mass the shape of particles will also influence their retention or permeation. Membranes with specific molecular weight cut-offs must be carefully selected in order to retain the desired protein of interest. Other factors affecting efficient separation of proteins by microfiltration or UF include the type of membrane, the molecular weight cut-off, the volume concentration ratio and diafiltration conditions. A major problem with membrane separation processes such as MF and UF is fouling of the membranes during processing, as this can affect membrane selectivity.

MF and UF have been used extensively to separate proteins from milk, soybeans, peas, chickpeas, lentils amongst others (Romero-Baranzini *et al.*, 1995; Cheryan, 1998; Fredrikson *et al.*, 2001; Boye *et al.*, 2010a).

Cryo-precipitation

The solubility of some proteins decreases significantly as the temperature is reduced below room temperature, a property that can be selectively used for their recovery. Cryo-precipitation refers to the method of protein precipitation under refrigerated conditions or from freezing and cold thawing.

Melcher and Fraij (1980) successfully used cryo-precipitation at –20 °C to separate a zein II extract obtained from corn meal into a methionine-rich polypeptide fraction (cryo-precipitate) and a cryo-supernatant containing predominantly polypeptides with amino acid composition similar to that of the zein 1 proteins. The cryo-precipitate was readily dissolved in 70% ethanol containing mercaptoethanol and precipitated again on cooling, indicating that the cryo-precipitaion was reversible.

In medical terms, a cryo-precipitate is defined as a concentrate of high molecular weight plasma proteins which precipitate in the cold (Pantanowitz *et al.*, 2003). Various researchers have used the technique to recover cryoglobulins, a heterogenous group of immunoglobulins (monoclonal and polyclonal) which precipitate or gel at reduced temperature (e.g., 4 °C) and redissolve on heating to 37 °C (Weber and Clem, 1981; Wang, 1988). The purified proteins exhibit crystalline, amorphous or gelatinous structure as a result of their differing primary structure and physicochemical properties. All major immunoglobulin classes (IgM, IgG, IgA) and their subclasses have been found in cryo-precipitates, including monoclonal immunoglobluins, mixed immunoglobulins and polyclonal immunoglobulins. Other proteins found in plasma that are insoluble at cold temperatures include cryofibriniogen, heparin precipitable protein and aggregates of C-reactive protein and albumin (Stathakis *et al.*, 1978; Shanbrom, 1980; Wang, 1988; Coelho and Wolf, 1991).

Francis and co-workers (2000) found that frozen saliva samples contained variable amounts of precipitate on thawing. The cryo-precipitates comprised of low molecular weight proteins (<14 KDa), which were either absent or present at reduced concentrations in the supernatant.

Cryo-precipitation is, therefore, a promising technique for the recovery of high value proteins and is a technique that is likely to be increasingly explored in the food, nutraceutical and pharmaceutical sectors.

Electrodialysis using bipolar membranes

Proteins can be preferentially precipitated from solution using a technique known as electrodialysis coupled with the use of bipolar membranes. The application is based on the specific properties of electrolysis (redox reactions at the electrodes) coupled with membrane action (Bazinet *et al.*, 1998) resulting in either electro-acidification or electro-alkalinization. A bipolar membrane is a composite membrane consisting of a cation exchange layer and an anion exchange layer which can split solvents (water and methanol, so far) into H^+ and OH^-/CH_3O^- at the interface under reverse potential bias (Xu and Huang, 2008).

For a solution circulated into an electrodialysis cell on the cationic side of the bipolar membrane, where the H^+ are generated, the pH of the solution will decrease. Similarly, a solution circulated on the anionic side of the bipolar membrane, where OH^- are generated, will experience an increase in pH (Bazinet *et al.*, 1998). During processing, protons generated by the bipolar membrane come into contact with proteins circulating in the electrodialysis cells on the cationic side of the bipolar membrane. At the isoelectric point when the net charge on the protein is zero, the proteins aggregate/precipitate allowing for their selective separation. Proteins can be subsequently recovered by centrifugation or filtration as is done in other conventional processes for protein separation.

Using this technique, Bazinet and co-workers (1998) lowered the pH of a soy protein solution from 8.0 to 4.5 in a cell of $100\,cm^2$ effective electrode surface, at a constant current of $25\,mA/cm^2$ and obtained a precipitate with 95% protein purity. Mondor and co-workers (2004a, 2004b) also successfully applied bipolar membrane electrodialysis to separate soy proteins.

Application of bipolar electrodialysis for the simultaneous removal of inhibitory acetate and pH control during *E. coli* fermentation was investigated by Wong and co-workers (2010). The final biomass and recombinant protein concentrations obtained increased by up to 37 and 20%, respectively.

There has been limited application of bipolar membrane electrodialysis for protein separation at the commercial scale due to cost, complexity of use and lack of availability of units for large scale processing.

Protein precipitation using organic solvents

Organic solvents such as acetone and ethanol can be used to induce protein precipitation. Similar to the salting-out phenomenon, addition of organic solvents removes water from the hydration spheres of the protein allowing electrostatic forces to bring oppositely charged regions of the protein together. Water is, thus, removed both by bulk replacement by the organic solvent and by structuring of the water around the organic molecules (Singh, 1995). Acetone, ethanol, acetone–methanol, chloroform–methanol, tricholoroacetic acid–ethanol are examples of solvents and solvent mixtures frequently used for protein precipitation. Modifications in protein functionality, safety of solvent handling and miscibility are some of the challenges associated with organic solvent protein precipitation.

3.3.2.4 *Protein Purification further downstream*

Washing

After recovery, proteins can be further purified by repeated washing with an appropriate solvent followed by centrifugation and/or filtration. This extra unit operation allows the removal of non-protein soluble components trapped within the interstitial spaces of the precipitate. The process can be further optimized to remove contaminating soluble proteins prior to drying (e.g., by selecting an appropriate pH of the water used for washing).

By washing the acid precipitated soy protein with pH 4.5 wash water, then resolubilizing and washing at pH 9.0 culminating with a final pH 4.5 precipitation, fat content was further decreased from 7–9% to 3–6%, while increasing the protein content from ~81–85% to ~90–92% without the added washing step (Mattil *et al.*, 1979).

Lawhon and co-workers (1981b) also reported that solubilizing full fat soy flour in water under controlled conditions (1:12 w/w solid:water ratio, pH 9 at 60 °C for 30 min), separating the aqueous phase from the slurry by centrifugation and washing the residue with water (1:5 solid:water ratio) at pH 9.0, precipitating the combined aqueous phase at pH 4.5. Washing the curd with acid water (pH 4.5) then decreased the fat content to 3.2%, while ensuring a protein content approaching 90%.

The addition of extra washing steps ultimately increases the efficiency of oil and protein extraction. To maintain desired functionality, the pH of the precipitate must be appropriately adjusted prior to drying.

Dialysis

Dialysis removes small molecular weight contaminants by selective and passive diffusion through a semi-permeable membrane. At the analytical scale, proteins can be purified using dialysis membranes with frequent water changes and constant stirring to drive the removal of contaminating smaller molecular weight compounds. Temperature, pore size, surface area, stirring speed and concentration gradient are factors influencing the rate of microsolute removal.

In general, dialysis proceeds faster at room temperature than at 4 °C. However, depending on the dialysis time, it may be prudent to conduct dialysis at lower temperature to prevent spoilage and loss of functionality. Thermal stability of the molecule should also be considered in temperature selection. Pore size of dialysis membranes must be carefully selected in accordance with the protein of interest. Although dialysis is an efficient way to increase protein purity at the analytical scale, it is less practical at the industrial scale, and therefore other techniques, such as diafiltration and chromatography, may be of greater interest.

Diafiltration

Diafiltration is typically performed as part of the membrane separation process described above. It refers to the process of adding water to the retentate during MF and/or UF and continuing the elimination of membrane-permeating species along with the water removed in the permeate (Cheryan, 1998). This allows removal of (contaminants) unwanted soluble materials such as smaller molecular weight sugars, alcohols, acids (e.g., phytic acid), and minerals from solution resulting in further protein purification.

Diafiltration can be performed in the continuous or discontinuous (batch) mode. In the discontinuous mode, the volume of the retentate is reduced, followed by redilution with water and repeating the UF in consecutive steps. In the continuous mode, water at the processing temperature and pH is added to the feed tank at the same rate as the permeate flux, resulting in a constant feed volume.

To enhance microsolute removal, UF/diafiltration can be carried out in a dialysis mode where water is pumped into the permeating membrane at three to ten times greater flow rate than the flux, while keeping the net pressure in the permeate channel lower than the feed pressure (Cheryan, 1998). The significant decrease in the concentration of microsolutes in permeate compared to the feed establishes a concentration gradient which enhances microsolute permeation. Reverse osmosis, a filtration technique which allows the removal of very low molecular weight compounds, can be used to subsequently concentrate the solute while producing the water for the UF dialysis.

3.3.2.5 *Chromatographic techniques*

Chromatography offers a wide variety of techniques that could be used to analyze as well as purify proteins. Many of these techniques can be used at relatively large scales to purify and collect proteins from biological mixtures for further analysis or for final use. In contrast, analytical chromatography is primarily used for qualitative and quantitative characterization purposes. Short descriptions of the most widely used methods are detailed in the following sections.

Gel filtration chromatography

Gel filtration chromatography has been widely used to purify biomolecules due to its simplicity and efficiency (Pastorello and Trambaioli, 2001). The technique separates proteins on the basis of their molecular size independent of the composition of the buffer used as eluent. Different media are used to perform the molecular exclusion in packed columns. The inertness and physicochemical stability of the chosen media are key requirements in column selection. Details of the technique can be found elsewhere (Amersham, 2002). Gel filtration is generally used in combination with the other chromatographic techniques described (Neyestani *et al.*, 2003).

Ion exchange chromatography

Ion exchange chromatography separates biomolecules according to their net chemical charge under determined experimental conditions. The degree of interaction between the charged proteins and the opposite charges of the ionic groups of the ion exchange medium, as well as differences in the charge properties of the proteins being separated, are key factors influencing separation (Ishihara and Yamamoto, 2005). The elution method generally involves a gradient of at least two buffers, such that the elution of the proteins is performed after the disruption of the electrostatic bonds between the biomolecule and the ionic groups of the chromatographic medium resulting from an increase in ionic strength or pH change (Levison, 2003). Two ion exchangers are commonly used: anionic exchange columns are generally used to separate acidic proteins, whereas cationic columns are more widely utilized for the purification of basic proteins (Pastorello and Trambaioli, 2001).

Reversed phase chromatography

Reversed phase chromatography is a widely used chromatographic technique which offers high resolution in separating proteins on the basis of their hydrophobicity (Pastorello and Trambaioli, 2001). The stationary phase is generally composed of an *n*-alkyl hydrocarbon that is able to interact with the proteins through hydrophobic interactions. The separation is often performed using a gradient elution. Thus, after adsorption to the immobilized *n*-alkyl hydrocarbon, the injected proteins are eluted due to the disruption of their hydrophobic bonds with the stationary phase after decreasing the polarity of the mobile phase (Pastorello and Trambaioli, 2001). Reversed phase chromatography has been extensively used in analytical analysis as well as at the preparative scale for the purification of food and recombinant proteins (Olson *et al.*, 1994).

Affinity chromatography

Affinity chromatography separates proteins according to their specific reversible interactions with a chemical ligand attached to the chromatographic matrix; it might be a substrate analogue, enzymatic inhibitor or activator, or a specific ligand of the targeted protein (Chaiken, 1986). Hence, proteins could be separated based on their biological function and specific structure.

Immunochromatography refers specifically to the immobilization of antibodies onto the gel matrix for affinity chromatography and is a technique that has been widely used for the isolation of high purity proteins (Riggin and Sportsman, 1993).

Proteins separated by chromatography using any of these techniques can be detected using UV spectroscopy, refractive index, evaporative light scattering, fluorescence or mass spectrometry detection.

3.3.2.6 *Drying Technologies*

Drying represents one of the most important costs in protein processing. Three techniques most commonly used are freeze drying (lyophilization), spray drying or drum drying.

Freeze drying

Freeze drying involves the removal of water from frozen materials by sublimation under high vacuum. The technique consists of three main stages: freezing, sublimation and desorption. The sample to be dried (i.e., example wet protein precipitate), is first frozen at a temperature below 0 °C, which should be sufficient to completely solidify the sample. The frozen sample is then placed under vacuum with pressures lower than 610 Pa at a high enough temperature to stimulate the sublimation of the free ice crystals. Water vapor is mechanically removed from the environment by vacuum pumps or steam jet ejectors (Mujumdar and Devahastin, 2000). To facilitate desorption of remaining water adsorbed by the sample, the temperature of the sample is raised at the latter stages of freeze drying to between 40 and 60 °C, which is sufficient to break any remaining bonds between the water molecules and the sample.

Freeze drying is costly and requires considerable time and energy. However, it has several advantages (Baker, 1997). It has a very low impact on the quality of the processed products and enables the creation of a powder with a quick rehydration rate. Freeze drying may be best suited for proteins targeted for the pharmaceutical and nutraceutical sectors, and perhaps for high end food applications.

Spray drying

Spray drying is one of the most widely used industrial drying processes in the food industry. It can transform low concentration emulsions, solutions or suspensions into powder and is frequently used in the agri-food industry because it is quick and can be completed in a single step. The product to be dried is pumped into the spray dryer and passes through an atomizer, which sprays the suspension in tiny droplets. There are many atomizer models on the market, each having specific advantages and disadvantages (e.g., vanes, nozzles, etc.). Hot air fed into the drying chamber dries the atomized droplets. On contact with the air, the water present in the droplets is quickly vaporized and the powder that forms drops out of the drying chamber. More recent spray dryers offer the possibility of lowering the pressure in the drying chamber, which allows products to be exposed to lower temperatures. This minimizes the impact of temperature on heat-sensitive products, such as proteins, which can denature at relatively low temperatures (80 °C) (Mujumdar and Devahastin, 2000).

Drum drying

Prior to the development of spray dryers, drum drying was one of the most commonly used techniques for drying. Drum drying can be used to dry a variety of liquid materials and is particularly effective for drying viscous liquids. Drum dryers consist of one or two horizontally mounted hollow cylinder(s) made of high-grade cast iron or stainless steel, a supporting frame, a product feeding system, a scraper and auxiliaries (Tang *et al.*, 2003). During processing, steam at very high temperatures (up to 200 °C) heats up the interior of the drums while the liquid material to be dried is applied uniformly as a thin layer onto the exterior surface(s) of the drum(s). The heat evaporates the water from the material leaving a dry product that is scraped off as the drum rotates towards the scrapers. The rate of drying and final moisture content depend on the type of material being dried, residence time, feed rate, thickness of material, steam pressure and temperature, and roll speed of the drum. Drum dryers have been successfully used to dry a variety of protein extracts.

Although drum dryers are easy to operate and have high energy efficiency, they have relatively low throughput compared to spray dryers. Materials may also be unevenly dried and become scorched, resulting in the generation of unacceptable colors or flavors, or the loss of functionality. Thus, drum dryers may be better suited for the drying of lower cost protein extracts (e.g., food and feed applications) than for pharmaceutical or nutraceutical products.

3.4 CALCULATING PROTEIN YIELDS AND RECOVERY

Protein extraction efficiency is an important measure of the effectiveness of protein separation. Calculating the amount of protein recovered is, therefore, important to assess the efficiency of the technique used. Protein recovery may be defined as the ratio of the total mass of protein recovered in the dried extract divided by the total mass of protein present in the starting material as shown in Equation 3.1 below:

$$\text{Protein recovery (\%)} = \frac{\text{Mass of protein extracted (g)}}{\text{Total mass of protein in starting material (g)}} \times 100 \qquad (3.1)$$

Thus, the closer the mass of the protein extracted to the original amount present in the starting material, the more efficient is the protein recovery.

Protein extraction efficiency is sometimes simply reported as protein yield defined as the ratio of the total mass of protein recovered in the dried extract divided by the total mass of the starting material (Equation 3.2):

$$\text{Protein yield (\%)} = \frac{\text{Mass of protein extracted (g)}}{\text{Total mass of starting material (g)}} \times 100 \qquad (3.2)$$

Another frequently used measure of protein extraction efficiency is protein purity, defined as the concentration of protein in the final extract as measured by the Kjedhal or Dumas methods or a method analyzing for true protein (e.g., Lowry, Bradford or BCA methods). Protein purity may, therefore, be defined as the ratio of the mass of protein in the dried recovered extract divided by the total mass of the dried extract as shown in Equation 3.3:

$$\text{Protein purity (\%)} = \frac{\text{Mass of protein in extract extracted (g)}}{\text{Total mass of recovered extract (g)}} \times 100 \qquad (3.3)$$

3.5 PROCESSING EFFECTS ON YIELD AND PROTEIN QUALITY

Protein purity, yield and quality are easily affected by processing conditions (e.g. temperature, time, flour to solvent ratio, condition and protein solubility of the starting material, type of equipment and process used, *g* forces used for centrifugation, laboratory vs. pilot scale extraction, batch vs. continuous extraction, etc.).

Russin and co-workers (2007) reported that by decreasing the average particle diameter of the starting raw material (i.e. soy flour) from 223.4 to 89.5 μm increased total solids recovery from 23 to 32%, while protein recovery increased from 40 to 52%. Final protein content (i.e. purity) of the soy protein isolates was, however, not significantly impacted by average particle size. The results demonstrate that protein recovery can be significantly increased by decreasing the average particle size of the starting raw material (i.e. defatted soy flour), without having any detrimental effect on the purity of the final soy protein isolate.

In the processing of pulse proteins, protein contents of fractions extracted using different techniques have been variable, probably due to differences in the processing conditions used (Boye *et al.*, 2010a). Chakraborty and co-workers (1979) extracted pulse proteins at pH 8.5 (30 min, 40 °C) followed by precipitation at pH 4.5 and obtained isolates containing 88.3–90.5% protein for chickpea, pea, Great Northern bean, lentil protein, lima bean, and mung bean isolates. Paredes-López and co-workers (1991) used similar conditions and obtained a chickpea isolate containing 84.8% protein. In the latter study, chickpea protein isolates were obtained from defatted chickpea flour using an initial aqueous extraction at 10% (w/v) solution and pH 8.5, followed by precipitation at pH 4.5, centrifugation and freeze drying. In other studies, de-hulled faba bean flour and pea flour were dispersed in water (1:5 w/v ratio, pH 9, 20 min, room temperature) followed by centrifugation and protein precipitation from the supernatant (pH 4, room temperature, 20 min). The extracts obtained contained 81.2 and 84.9% protein, respectively (Fernández-Quintela *et al.*, 1997).

In addition to yield and purity, conditions used during processing (pH, temperature, extraction times, solvents) can modify protein structure and affect protein thermal stability and functionality. These conditions, therefore, need to be properly selected taking the final product application into consideration.

3.5.1 Protein characterization

3.5.1.1 *Biochemical methods*

Slab gel electrophoresis

Electrophoresis is widely applied for protein separation and characterization. The technique involves the separation of proteins on the basis of their mobility in an electric field. Proteins move with different speeds depending on their charge, shape and size. Polyacrylamide gels (PAG), which are formed by the copolymerization of acrylamide monomers with a cross-linking agent such as bisacrylamide, are generally used as the matrix for electrophoresis (Dunn, 1989). Different types of gels containing a linear or non-linear concentration gradient of polyacrylamide can be used to optimize protein separation (Dunn, 1989).

A range of electrophoretic techniques are available which can separate proteins on the basis of their properties, such as size, net charge and relative hydrophobicity. However, the most used electrophoresis techniques for the characterization of food and bioproduct proteins are: native PAGE, SDS-PAGE, immunoelectrophoresis and isoelectric focusing. Protein detection for these different methods is often done by staining with Coomassie Brilliant Blue in a fixative solution or, after fixation, with silver, gold and fluorescence or chemiluminescence probes. More detailed information on protein staining can be found elsewhere (Amersham, 1999a).

Electrophoresis separation can be monodimensional (1D) or bidimensional (2D). In 1D electrophoresis, proteins are separated in one dimension on the basis of only one property, such as charge (isoelectric point) or size. 2D electrophoresis combines the 1D electrophoresis with a second property in the orthogonal direction (Pastorello and Trambaioli, 2001). 2D electrophoresis is a powerful method for the analysis of complex protein mixtures extracted from cells, tissues, or other biological samples, and can resolve hundreds to thousands of individual protein species detected as spots or bands after staining (Friedman *et al.*, 2009).

Native Polyacrylamide Gel Electrophoresis (PAGE)
PAGE under native conditions has been used to separate soluble proteins which retain their biological and enzymatic properties. Factors affecting the migration of proteins in native PAGE are: size, shape, and native charge (Dunn, 1989). A derivative method from native PAGE developed by Schägger and Von Jagow (1991), known as Blue Native PAGE, was originally used for the qualitative and quantitative analysis of mitochondrial protein complexes and proteins ranging in molecular weights from 10 to 10000 kDa (Dresler *et al.*, 2010).

SDS-PAGE
Sodium dodecyl sulfate (SDS) is an anionic detergent that denatures proteins through specific binding to the hydrophobic tail around the polypeptide backbone, giving a net uniform negative charge per molecule (Dunn, 1989). The binding of SDS unfolds the protein as a consequence of hydrogen bond cleavage, and blockage of hydrophobic interactions. Consequently, electrophoretic separation is only dependent on the molecular weight and proteins migrate in the anodal direction. Moreover, proteins can be denatured by the addition of strong reducing agents, such as dithiothreitol (DTT) or β-mercaptoethanol, which disrupt disulfide bonds between cysteine residues, allowing a more detailed characterization of proteins (Dunn, 1989).

Lately, the discontinuous Laemmli system buffer (containing 0.1% w/v SDS) is most commonly used for SDS-PAGE. The method has been widely used for the characterization of the protein profiles of processed whey, meat, cereal, legume, and seed proteins (Roy *et al.*, 2007; Miyamoto *et al.*, 2009; Boye *et al.*, 2010b; Sikes *et al.*, 2010; Warchalewski and Gralik, 2010) as well as the analysis of membrane proteins (Braun *et al.*, 2009). Further information on SDS-PAGE electrophoresis can be found elsewhere (Amersham, 1999a).

Immunoelectrophoresis
Immunoelectrophoresis is a specific method characterized by the presence of polyclonal monospecific antibodies against a specific protein in the agarose gel. The running buffer used in this method should have a pH corresponding to the isoelectric point of the antibody. The fixation of the latter in the gel leads to the formation of a precipitate comprised of the antigen and the specific antibody, which may be visualized by using a Coomassie blue stain or some other suitable technique (Westermeier and Scheibe, 2009). Immunoelectrophoresis has been used for the qualitative and quantitative analysis of individual proteins in complex mixture (Hansen and Larsen, 2008).

Isoelectric focusing
Isoelectric focusing (IEF) is an electrophoretic method that separates proteins based on their isoelectric points (pI, neutral charge state). The migration is performed in the presence of a continuous pH gradient generated using carrier ampholytes or immobilized pH gradients gels (Dunn, 1989; Amersham, 1999a). After migration according to their charge, proteins attain a steady sate when they reach the pH value corresponding to their pI. The technique is commonly used in combination with SDS-PAGE in 2D electrophoresis. IEF can also be performed under native conditions (Dunn, 1989; Friedman *et al.*, 2009).

Capillary electrophoresis

New developments in the field of capillary electrophoresis (CE) offer a powerful tool for the rapid separation of proteins (Mattiasson and Hatti-Kaul, 2003; Huck and Bonn, 2008). Separation of proteins by CE is based on electrophoretic migration driven by an electroosmotic flow (EOF) and occurs in capillary tubes, filled with background electrolytes (BGE) (Tagliaro *et al.*, 1998) following similar principles as for standard electrophoresis. CE is characterized by high efficiency, short time of analysis, low consumption of samples and running buffers, and possible connectivity to detectors originally designated for liquid chromatography, including mass spectrometry. Examples of CE use in protein analysis include successful quantification of 7 S and 11 S fractions in several soybean varieties (Blazek and Caldwell, 2009), characterization of cereal proteins (Piergiovanni, 2007; Salmanowicz, 2008; Salmanowicz *et al.*, 2008; Di Luccia *et al.*, 2009), and the analysis of casein/caseinate addition in processed cheeses (Miralles *et al.*, 2006). Further information on the use of CE to purify proteins and study their conformations can be found in the work by Mattiasson and Hatti-Kaul (2003).

Thermal analysis by differential scanning calorimetry

Heat treatment is one of the most commonly used techniques in food processing, and is frequently responsible for the transition of proteins from the native folded state to the denatured unfolded state. Thermal denaturation of proteins involves a structural change which affects the nutritional quality of foods (Plum, 2009).

Differential scanning calorimetry (DSC) is a suitable method for the characterization of thermal and thermodynamic stabilities of the protein. DSC provides several tools for the study of the thermal properties of proteins under controlled heating and cooling rates and can be used to determine the apparent specific heat of proteins (O'Brien and Haq, 2004).

In brief, a DSC instrument comprises two cells fixed in an adiabatic chamber. One cell contains the sample to be tested, the second cell contains a reference solution or an empty DSC pan. The adiabatic chamber is maintained under pressure to avoid the evaporation of the sample (Plum, 2009). A DSC-thermogram represents the plot of heat capacity difference ΔC_p (between the sample and the reference) as a function of temperature. Thermodynamic parameters, such as T_m, ΔH and ΔS, could be determined by the DSC curve analysis. T_m is the temperature at which the concentration of denatured and native forms of the protein are equal. This specific temperature is also called the midpoint of the thermal transition. ΔH represents the enthalpy of thermal transition determined from the integration of the DSC curve. The entropy (ΔS) of the thermodynamic transition of the protein may be calculated from the integrated area under the curve of $\Delta C_p/T$ vs. T. The free energy (ΔG), which gives an indication of the protein stability, can also be determined at any temperature from the values of ΔH and ΔS (O'Brien and Haq, 2004; Plum, 2009). Thermal and thermodynamic properties of proteins analyzed by DSC are greatly affected by the experimental conditions used, such as pH, salts, alcohols, and the presence of other food components (e.g., lipids, polysaccharides) (Grinberg *et al.*, 2009).

3.5.1.2 *Biophysical methods*

Biophysics offers a variety of accurate methods that enable kinetic resolution in the submillisecond time scale as well as high protein structural resolution. This section summarizes some of the most frequently used techniques for the biophysical analysis of proteins.

Fourier transform infrared spectroscopy

Fourier transform infrared spectroscopy (FTIR) is a powerful tool used to monitor changes in protein and polypeptide secondary structure during processing. After exposure of a protein to infrared light, its secondary structure can be determined from the spectra obtained from the absorption of different wavelengths corresponding to specific vibration frequencies of the amide bonds (Jackson and Mantsch, 1995a).

Two kinds of the vibrations, known as amide I and amide II vibrations, are directly correlated to the molecular structure of proteins and polypeptides. The stretching of the carbonyl C=O double bonds is linked to the amide I vibration, and the deformation of the N–H bonds causes the amide II vibration. FTIR determines the amount of light absorbed corresponding to each of the vibrations over a wide range of frequencies (Kumosinski and Farrell, 1993). Empirical correlations between the frequency of the amide I and amide II absorptions of the protein and the secondary structure composition in helix, extended β-sheets, loops and unordered structures have been determined (Jackson and Mantsch, 1995).

The amide I absorption region (1700–1600 cm^{-1}) in the FTIR is widely used for the determination of the secondary structure of proteins (Álvarez *et al.*, 2008). Parallel and antiparallel β-sheets are attributed to bands at 1622 and 1632 cm^{-1}, respectively, whereas, α-helix structures correspond to the band at ~1655 cm^{-1}. β-turns and unordered structures can be detected respectively at band positions of 1692 and 1645 cm^{-1} (Jackson and Mantsch, 1995).

Circular dichroism

Circular dichroism (CD) is a powerful method for the study of protein structure and folding in solution under various conditions. The technique has been widely utilized to monitor structural transitions of food and bioproduct proteins resulting from changes in different processing conditions (Martin and Schilstra, 2008). CD is a spectroscopic technique based on the difference in protein interactions with left (L) and right (R) circularly polarized light (Sreerama and Woody, 2004). Hence, CD analysis of proteins is based on the measurements of the difference in absorbance between L and R circularly polarized components expressed as ellipticity (θ) in degrees (Kelly *et al.*, 2005). CD spectra thus result from the chirality of some structures, such as the carbon atom bound to four different substituents, the C–S–S–C structure and molecular asymmetric environments (Kelly *et al.*, 2005). Table 3.2 shows the spectral region and contributing protein chromophores. The determination of the relative proportions of α-helix, β-sheets, β-turns and random coil structures, as well as the overall features of the secondary structure, could be determined by CD measurements in the far UV region (Sreerama and Woody, 2004). All α-helix rich proteins, such as α-lactalbumin and lysozyme, are characterized by an intense negative band with two peaks at 208 and 222 nm, and a strong positive band at 191–193 nm (Barbana *et al.*, 2006), whereas the β sheets rich proteins are distinguished by a negative band at 210–225 nm and a stronger positive band at

Table 3.2 Contributing chromophores in CD spectra of proteins.

Chromophore	CD spectra region
Backbone amides	<250 nm (far UV)
Aromatic groups	250–300 nm (near UV)
Extrinsic groups	Above 300 nm (near UV–visible region)

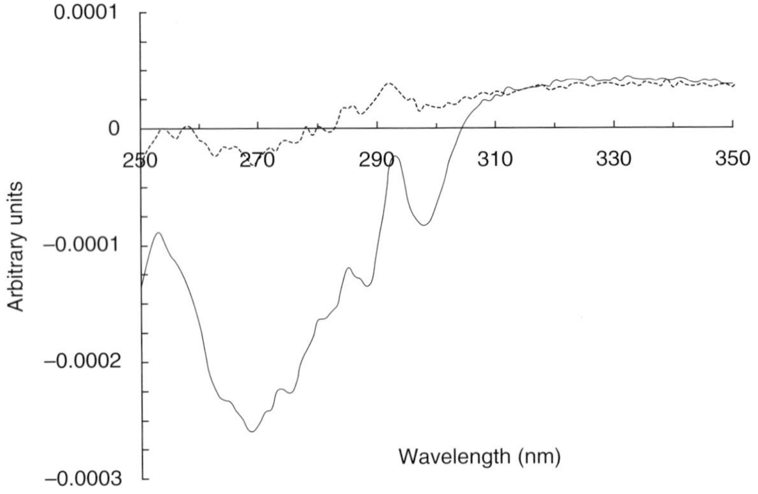

Figure 3.6 Circular dichroism spectra in the near UV region of bovine holo- (solid line) and apo- (dashed line) α-lactalbumin. Circular dichroism spectra were recorded at a protein concentration of 70 mM in 5 mM Tris buffer, pH 7.0, at 20 °C. (Reproduced from Barbana *et al.*, copyright 2006, with permission from Elsevier.)

190–200 nm. Unordered proteins are generally characterized by a negative band at 195–200 nm, and a much weaker signal between 215 and 230 nm (Martin and Schilstra, 2008).

CD spectra analysis at the near UV region provides important information about protein tertiary structure. Spectra at the near UV region arise from the aromatic amino acids, and depend on their number, mobility, spatial disposition and their environment (Kelly *et al.*, 2005). CD signals could also be useful for the study of the conformational changes in proteins caused by ligand binding (Barbana *et al.*, 2006, 2008). Figure 3.6 shows the near UV CD spectra of bovine α-lactalbumin in its holo and apo forms. As observed, apo-α-lactalbumin has very low signal in the near UV region of the CD spectrum, indicative of loss of the native tertiary structure with rapidly rotating aromatic side chains after the release of Ca^{2+}.

Fluorescence spectroscopy

Fluorescence is the light emitted subsequent to absorption of UV or visible light due to electronic state transitions from excited singlet to several vibrational levels (Lakowicz, 1999; Hof *et al.*, 2005). The emission spectrum provides information for both qualitative and quantitative analysis.

Fluorescence spectroscopy is a widely used biophysical tool for the study of the protein structure based mainly on tryptophan fluorescence, which can only be excited by light of specific wavelengths. The fluorophore that is able to emit fluorescence in the trypotophan structure is the indole group. Generally, the fluorescence spectra recorded between 300 and 400 nm after excitation of the tryptophan fluorescence at 298 nm could be very useful for understanding the effect of processing on protein structure. A shift in the maximum of tryptophan fluorescence emission suggests an exposition of the tryptophanyl residues to a polar environment. However, an increase in fluorescence intensity corresponds to an increase in solvent polarity and reduced quenching of some tryptophanyl residues (Barbana *et al.*, 2006, 2008). Figure 3.7 shows the fluorescence spectra of α-lactalbumin before (holo) and after (apo) the removal of Ca^{2+} by chelating agent. The fluorescence intensity of the spectrum of

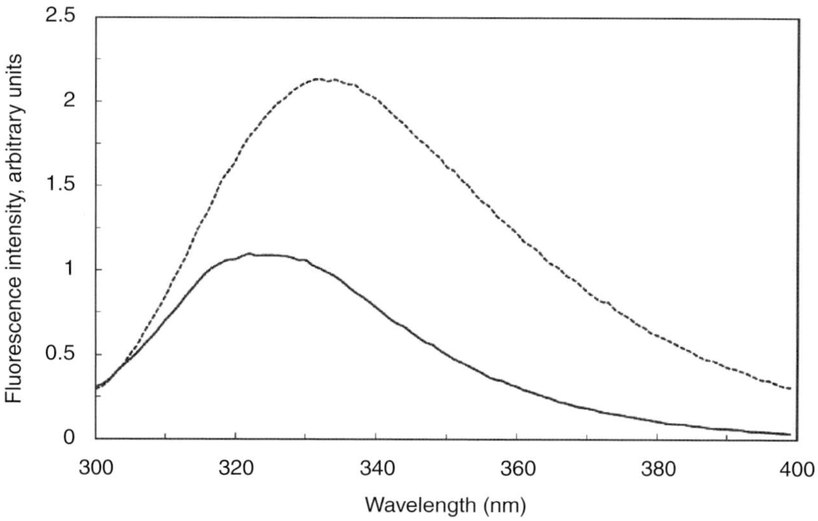

Figure 3.7 Fluorescence spectra of human holo- (solid line) and apo- (dashed line) α-lactalbumin recorded for 20 mM protein solution in 10 mM Tris buffer, pH 8.5, at 20°C. Fluorescence excitation was at 287 nm (Barbana *et al.*, 2008). Reproduced with kind permission from Springer Science and Business Media.

holo-α-lactalbumin reaches its maximum at 325 nm. The removal of calcium bound to the protein causes an increase in fluorescence intensity and shifting of the wavelength maximum to a significantly higher value (336 nm). These results indicated that tryptophan residues are more exposed to solvent in the apo form, reflecting a conformational change in the protein caused by the release of bound calcium (Barbana *et al.*, 2008). Fluorescence properties of proteins, however, depend on experimental conditions, such as temperature, pH, concentration, polarity of the environment and quenching (intra- or inter-molecular interactions) (Andersen *et al.*, 2008).

3.5.1.3 *Immunochemical methods*

Immunochemical methods are highly sensitive techniques for the detection and identification of protein targets by antigen–antibody specific reactions. Radial immunodiffusion and western blotting are among the most used techniques for the quantitative estimation of processed food and bioproduct proteins based on their immunoreactivity with specific antibodies.

Radial immunodiffusion

Radial immunodiffusion is used extensively for the quantitative estimation of antigens. In a uniformly thin layer of agar containing a specific antibody, the area of the radial diffusion of the corresponding antigen from wells cut in the gel is directly proportional to the concentration of the antigen employed. The method is very accurate and sensitive, with lower detection limits corresponding to an antigen concentration of 1.25 μg/ml (Mancini *et al.*, 1965). This technique could be used for the determination of antigens and evaluation of the effect of processing on the denaturation of food proteins (Wehbi *et al.*, 2005).

Western Blotting

After PAGE electrophoretic separation, protein bands can be transferred to an activated thin support membrane, generally nitrocellulose membrane, by electro-transfer techniques.

The second step involves the immobilization of the bands to the surface of the membrane in the same pattern as in the original PAGE gel. Subsequently, the selected proteins could be analyzed by immunoassays after the treatment of the membrane with specific antibodies which bind to the antigens in the membrane and can be detected by a variety of techniques involving treatments with secondary antibodies and staining (Phillips, 1992). Prior to antibody binding and staining the remaining protein binding sites on the membrane must be blocked (e.g., by bovine serum albumin (BSA) or ovalbumin) to avoid non-specific interaction with the antibodies. Additionally, the removal of unbound reagents by a series of washes between each binding step is necessary to avoid non-specific binding.

The detection of the antigen–antibody complexes on the membrane could be performed by a variety of techniques. The most common method is the application of a secondary antibody which binds specifically to the primary antigen–antibody complex. Hence, radiochemicals, fluorescent compounds, colloidal gold, enzymes, or biotin labels could be easily bound to the secondary antibody, and protein detection could be deduced from the intensity of the colored, fluorescent or chemiluminescent end product. Further details on blotting can be found elsewhere (Amersham, 1999).

3.6 CONCLUSION

Proteins play a key role in food, feed, and in many health and industrial applications. Due to their versatility, the market for semi-purified and highly purified proteins is likely to grow. There has been significant effort to identify novel sources of healthy and functional proteins. Examples of such new sources of proteins include canola and flax proteins that are currently in industrial production for the food, feed and pharmaceutical industries.

As consumers and the food and industrial products sectors become more environmentally conscious, techniques that are more energy efficient, sustainable and have a lower carbon foot print will be required. In this regard novel non-thermal technologies, such as high pressure processing, pulse electric field and ultrasonication, will be preferred for protein extraction and/or improving protein functionality. Since the most expensive unit operation in protein processing is drying, new innovative drying techniques that are cheaper, easy to use and do not adversely alter protein functionality need to be developed. Additionally, optimization of the processing conditions for the current techniques and the development of new processes that could enhance protein functionality while increasing its health benefits and nutritional properties would be extremely useful.

REFERENCES

Achouri A, Boye JI, Belanger D, Chiron T, Yaylayan, VA, Yeboah FK 2010. Functional and molecular properties of calcium precipitated soy glycinin and the effect of glycation with kappa-carrageenan. Food Res Int 43(5):1494–1504.

Alli I, Gibbs BF, Okoniewska MK, Konishi Y, Dumas F 1993. Identification and characterization of phaseolin polypeptides in a crystalline protein isolated from white kidney beans (*Phaseolus vulgaris*). J Agric Food Chem 41:1830–1834.

Álvarez PA, Ramaswamy HS, Ismail AA 2008. High pressure gelation of soy proteins: Effect of concentration, pH and additives. J Food Eng 88:331–340.

Amersham 1999a. Protein Electrophoresis Technical Manual. Amersham Biosciences.

Amersham 1999b. Blot Processing with the Processor Plus Protocol Guide for Western Blotting. Amersham Biosciences.

Amersham 2002. Gel Filtration Principle and Methods. Amersham Biosciences.

Andersen CM, Wold JP, Engelsen SB 2008. Autofluorescence Spectroscopy in Food Analysis In: Ötleş S (editor). Handbook of Food Analysis Instruments. Boca Raton, FL:CRC Press Taylor & Francis Group. pp. 347–363.

Asenjo JA, Chaudhuri JB 1996. Innovative separation methods in bioprocessing. In:Grandison AS and MJ Lewis (editors). Separation processes in the food and biotechnology industries. Cambridge, UK: Woodhead Publishing Ltd. pp. 179–206.

Baker CGJ 1997. Industrial drying of foods. First edition. London, UK:Blackie Academic & Professional. pp. 90–114 and 134–155.

Barbana C, Pérez MD, Pocoví C, Sánchez L, Wehbi Z 2008. Interaction of human α-lactalbumin with fatty acids:determination of the binding parameters. Biochemistry (Mosc) 73:711–716.

Barbana C, Pérez MD, Sánchez L, *et al.* 2006. Interaction of bovine α-lactalbumin with fatty acids as determined by partition equilibrium and fluorescence spectroscopy. Int Dairy J 16:18–25.

Bazinet L, Lamarche F, Ippersiel D 1998. Bipolar membrane electrodialysis:Applications of electrodialysis in the food industry. Trends Food Sci Technol 9(3):107–113.

Blazek V, Caldwell RA 2009. Comparison of SDS gel capillary electrophoresis with microfluidic lab-on-a-chip technology to quantify relative amounts of 7S and 11S proteins from 20 soybean cultivars. Int J Food Sci Technol 44:2127–2134.

Boye JI, Aksay S, Roufik S, *et al.* 2010a. Comparison of the functional properties of pea, chickpea and lentil protein concentrates processed using ultrafiltration and isoelectric precipitation techniques. Food Res Int 43(2):537–546.

Boye JI, Roufik S, Pesta N, Barbana C 2010b. Angiotensin I-converting enzyme inhibitory properties and SDS-PAGE of red lentil protein hydrolysates. LWT – Food Sci Technol 43(6):987–991.

Braun RJ, Kinkl N, Zischka H, Ueffing M 2009. 16-BAC/SDS-PAGE analysis of membrane proteins of yeast mitochondria purified by free flow electrophoresis. Methods Mol Biol 528:83–107.

Brown AK, Kaul A, Varley J 1999. Continuous foaming for protein recovery: Part II. Selective recovery of proteins from binary mixtures. Biotechnol Bioeng 62(3):291–300.

Chaiken IM 1986. Analytical affinity chromatography in studies of molecular recognition in biology:A review. J Chromatogr B 376:11–32.

Chakraborty P, Sosulsi F, Bose A 1979. Ultracentrifugation of salt soluble proteins in ten legume species. J Sci Food Agric 30:766–771.

Cheftel JC, Cuq JL, Lorient D 1985. Amino acids, peptides, and proteins. In: Fennema OR (editor). Food Chemistry. Basel, Switzerland:Marcel Dekker Inc. pp. 245–369.

Cheryan M 1998. Ultrafiltration and microfiltration handbook. Pennsylvania, USA: Technomic Publishing Co. Inc.

Coelho PH, Wolf T 1991. Device for fractionating constituent components of a substances using cryoprecipitation. US Patent 5261255.

Di Luccia A, Lamacchia C, Mamone G, *et al.* 2009. Application of capillary electrophoresis to determine the technological properties of wheat flours by a glutenin Index. J Food Sci 74(4):C307–C311.

Domínguez H, Nuñez MJ, Lema JM 1993. Oil extractability from enzymatically treated soybean and sunflower:range of operational variables. Food Chem 46(3):277–284.

Domínguez H, Nuñez MJ, Lema JM 1994. Enzymatic pretreatment to enhance oil extraction from fruits and oilseeds:a review. Food Chem 49(3):271–286.

Dresler J, Klimentova J, Stulik J 2010. Bacterial protein complexes investigation using blue native PAGE. Microbiol Res. DOI:10.1016/j.micres.2010.01.005 (Available online).

Dunn MJ 1989. Electrophoretic analysis Methods. In:Harris ELV and Angal S (editors). Protein Purification Applications:A Practical Approach. Oxford, UK:ORL Press. pp. 18–39.

Fernández-Quintela A, Macarulla MT, Del Barrio AS, Martínez JA 1997. Composition and functional properties of protein isolates obtained from commercial legumes grown in northern Spain. Plant Foods Hum Nutr 51:331–342.

Francis CA, Hector MP, Proctor GB 2000. Precipitation of specific proteins by freeze-thawing of human saliva. Arch Oral Biol 45(7):601–606.

Fredrikson M, Biot P, Alminger ML, Carlsson NG, Sandberg AS 2001. Production process for high-quality pea-protein isolate with low content of oligosaccharides and phytate. J Agric Food Chem 49:1208–1212.

Friedman DB, Hoving S, Westermeier R 2009. Isoelectric focusing and two-dimensional gel electrophoresis. Methods Enzymology 463:515–540.

Friedrich JP, List GR 1982. Characterization of soybean oil extracted by supercritical carbon dioxide and hexane. J Agric Food Chem 30(1):192–193.

Fullbrook PD 1984. Extraction of vegetable oils. UK Patent 2127425.

Gandhi AP, Joshi KC, Krishna J, *et al.* 2003. Studies on alternative solvents for the extraction of oil-I soybean. Int J Food Sci Technol 38(3):369–375.

Grinberg VY, Burova TV, Tolstoguzov VB 2009. Thermal Analysis of Denaturation and Aggregation of Proteins and Protein Interactions in a Real Food System. In:Kaletunç G (editor). Calorimetry in Food Processing:Analysis and Design of Food Systems. Ames, IA:Wiley-Blackwell. pp. 67–85.

Gueguen J, Vu AT, Schaeffer F 1984. Large-scale purification and characterisation of pea globulins. J Sci Food Agric 35:1024–1033.

Hansen GN, Larsen JN 2008. Immunoelectrophoresis for the characterization of allergen extracts. Methods Mol Med 138:147–165.

Hof M, Fidler V, Hutterer R 2005. Basics of Fluorescence Spectroscopy in Biosciences. In:Hof MR, Hutterer R and V Fidler (editors). Fluorescence Spectroscopy in Biology Advanced Methods and their Applications to Membranes, Proteins, DNA, and Cells (Wolfbeis OS, Series editor). Springer Series on Fluorescence Methods and Applications. Berlin, Germany:Springer. pp. 3–29.

Hossain M, Fenton G 1998. Application of Foam Separation Processes for Protein Extraction/Stripping. Australas Biotechnol 8:289–295.

Huck CW, Bonn GK 2008. Analysis of Proteins by Capillary Electrophoresis. In:Schmitt-Kopplin P (editor). Capillary electrophoresis methods and protocols. Methods in molecular biology (Walker JM, series editor). Totowa, USA:Humana Press. pp. 507–540.

Ishihara T, Yamamoto S 2005. Optimization of monoclonal antibody purification by ion-exchange chromatography:Application of simple methods with linear gradient elution experimental data. J Chromatogr A 1069(1):99–106.

Jackson M, Mantsch HH 1995. The use and misuse of FTIR spectroscopy in the determination of protein structure. Crit Rev Biochem Mol Biol 30(2):95–120.

Jones RW, Taylor NW, Senti FR 1959. Electrophoresis and fractionation of wheat gluten. Arch Biochem Biophys 84:363–376.

Keiichi K, Yuichiro O, Atsushi T 1998. New selective protein-separation using triazine dye by a foam separation. Separation between BSA and hemoglobin. Nippon Kagakkai Baiotekunoroji Bukai Shinpojiumu Koen Yoshishu 3:9.

Keller RCA, Orsel R, Hamer RJ 1997. Competitive adsorption behaviour of wheat flour components and emulsifiers at an air-water interface. J Cereal Sci 25(2):175–183.

Kelly SM, Jess TJ, Price NC 2005. How to study proteins by circular dichroism. Biochim Biophys Acta 1751:119–139.

Kumosinski TF, Farrell HM 1993. Determination of the global secondary structure of proteins by Fourier transform infrared (FTIR) spectroscopy. Trends Food Sci Technol 4:169–175.

Lakowicz JR 1999. Introduction to Fluorescence. In: Lakowicz JR)(editor). Principles of Fluorescence Spectroscopy, 2nd Edn. New York, USA: Kluwer Academic/Plenum Publishers. pp. 1–24.

Lawhon JT, Manak LJ, Rhee KC, Rhee KS, Lusas EW 1981a. Combining aqueous extraction and membrane isolation techniques to recover protein and oil from soybeans. J Food Sci 46(3):912–919.

Lawhon JT, Rhee KC, Lusas EW 1981b. Soya protein ingredients prepared by new processes – aqueous processing and industrial membrane isolation. J Am Oil Chem Soc 58(3):377–384.

Levison PR 2003. Large-scale ion-exchange column chromatography of proteins:Comparison of different formats. J Chromatogr B 790:17–33.

Lusas EW, Watkins LR, Rhee KC 1990. Separation of fats and oils by solvent extraction: non-traditional methods. In:Erickson DR (editor). Edible fats and oils processing:basic principles and modern practices. Champaign, IL:AOCS Press. pp. 56–78.

Mancini G, Carbonara AO, Heremans JF 1965. Immunochemical quantitation of antigens by single radial immunodiffusion. Immunochem 2:235–254.

Marquez UML, Barros RMC, Lajolo FM 1996. Chemically determined total and available methionine in beans (*Phaseolus vulgaris* L.) and isolated protein fractions. Food Chem 55:179–184.

Martin SR, Schilstra MJ 2008. Circular Dichroism and Its Application to the Study of Biomolecules. Methods Cell Biol 84:263–293.

Maruyama H, Seki H, Suzuki A, Inoue N 2007. Batch foam separation of a soluble protein. Water Res 41(3):710–718.

Matouq M 2008. Investigation of the bubble foam separation technique to extract protein from whey. Am J Appl Sci 5(5):468–472.

Mattiasson B, Hatti-Kaul R 2003. Process monitoring during protein purification. In:Hatti-Kaul R and B Mattiasson (editors). Isolation and purification of proteins. New York:Marcel Dekker Inc. pp. 515–548.

Mattil KF, Rhee KC, Cater CM 1979. Soybean protein extract. US Patent 4151310.

Melcher U, Fraij B 1980. Methionine-rich protein fraction prepared by cryo-precipitation from extracts of corn meal. J Agric Food Chem 28(6):1334–1336.

Miralles B, Krause I, Ramos M, Amigo L 2006. Comparison of capillary electrophoresis and isoelectric focusing for analysis of casein/caseinate addition in processed cheeses. Int Dairy J 16:1448–1453.

Miyamoto Y, Matsumiya K, Kubouchi H, Noda M, Nishimura K, Matsumura Y 2009. Effects of heating conditions on physicochemical properties of skim milk powder during production process. Food Sci Technol Res 15(6):631–638.

Mondor M, Ippersiel D, Lamarche F, Boye JI 2004a. Effect of electro-acidification treatment and ionic environment on soy protein extract particle size distribution and ultrafiltration permeate flux. J Membr Sci 231:169–179.

Mondor M, Ippersiel D, Lamarche F, Boye JI 2004b. Production of soy protein concentrates using a combination of electroacidification and ultrafiltration. J Agric Food Chem 52:6991–6996.

Mujumdar, AS, Devahastin, S 2000. Mujumdar's practical guide to industrial drying. In:Devahastin S (editor). Brossard, Canada:Exergex Corporation. pp. 41–50.

Mustakas GC 1980. Recovery of oil from soybeans. In:Erickson DR, Pryde EH, Brekke OL, Mounts TL, and Falb RA (editors). Handbook of Soy Oil Processing and Utilization. Champaign, IL:American Soybean Association and AOCS Press. pp. 49–65.

Nelson AI, Wijeratne WB, Yeh SW, Wei TM, Wei LS 1987. Dry extrusion as an aid to mechanical expelling of oil from soybeans. J Amer Oil Chem Soc 64(9):1341–1347.

Neyestani TR, Djalali M, Pezeshki M 2003. Isolation of α-lactalbumin, β-lactoglobulin, and bovine serum albumin from cow's milk using gel filtration and anion-exchange chromatography including evaluation of their antigenicity. Protein Expr Purif 29:202–208.

O'Brien R, Haq I 2004. Applications of Calorimetry in the Biological Sciences. In:Ladbury JE and LM Doyle (editors). Biocalorimetry 2. Chichester, UK: John Wiley & Sons, Ltd., pp. 3–34.

Olson CV, Reifsnyder DH, Canova-Davis E, Ling VT, Builder, SE 1994. Preparative isolation of recombinant human insulin-like growth factor 1 by reversed-phase high-performance liquid chromatography. J Chromatogr A 675:101–112.

Osborne TB 1924. The vegetable proteins, 2nd edn. Monographs on Biochemistry. London, UK: Longmans, Green and Co.

Pantanowitz L, Kruskall MS, Uhl L 2003. Cryoprecipitate. Patterns of use. Am J Clin Pathol 119(6):874–881.

Paredes-López O, Ordorica-Falomir C, Olivares-Vazquez MR 1991. Chickpea protein isolates: Physicochemical, functional and nutritional characterization. J Food Sci 56:726–729.

Pastorello EA, Trambaioli C 2001. Isolation of food allergens. J Chromatogr B 756:71–84.

Phillips TM 1992. Elution and blotting techniques. In:Phillips TM (editor). Analytical techniques in immunochemistry. New York, USA:Marcel Dekker. pp. 211–234.

Piergiovanni AR 2007. Extraction and separation of water-soluble proteins from different wheat species by acidic capillary electrophoresis. J Agric Food Chem 55:3850–3856.

Plum GE 2009. Calorimetry of proteins in dilute solutions. In:Kaletunç G (editor). Calorimetry in Food Processing:Analysis and Design of Food Systems. Ames, IA: Blackwell Publishing Ltd., pp. 67–85.

Riggin A, Sportsman JR 1993. Immunochromatographic analysis of proteins Identification, characterization and purity determination. J Chromatogr 632:37–44.

Romero-Baranzini AL, Yanez-Farias GL, Barron-Hoyos JM 1995. A high protein product from chickpeas *(Cicer arietinum* L.) by ultrafiltration. Preparation and functional properties. J Food Process Preserv 19:319–329.

Roy A, Moktan B, Sarkar PK 2007. Characteristics of *Bacillus cereus* isolates from legume-based Indian fermented foods. Food Control 18(12):1555–1564.

Russin T, Arcand Y, Boye JI 2007. Raw Material Particle Size Effect on Yield and Purity of Soy Protein Isolates. J Food Process Preserv 31:308–319.

Russin TA, Boye JI, Rajamohamed SH 2010. Alternative techniques for defatting soy: A practical review. Food Bioprocess Technol. DOI:10.1007/S11947-010-0367-8. (Available Online).

Salmanowicz BP 2008. Detection of high molecular weight glutenin subunits in triticale (× Triticosecale Wittm.) cultivars by capillary zone electrophoresis. J Agric Food Chem 56:9355–9361.
Salmanowicz BP, Surma M, Adamski T, Rębarz M 2008. Effects of amounts of HMW glutenin subunits determined by capillary electrophoresis on technological properties in wheat doubled haploids. J Sci Food Agric 88:1716–1725.
Schägger H, von Jagow G 1991. Blue native electrophoresis for isolation of membrane protein complexes in enzyma- tically active form. Anal Biochem 199(2):223–231.
Serrato AG 1981. Extraction of oil from soybeans. J Am Oil Chem Soc 58(3):157–159.
Shanbrom E 1980. Method of selectively increasing yield and purity of certain cryoprecipitate proteins by heating. US Patent 4305871.
Sikes A, Tornberg E, Tume R 2010. A proposed mechanism of tenderising post-rigor beef using high pressure–heat treatment. Meat Sci 84(3):390–399.
Singh R 1995. Precipitation of biomolecules. In:Singh RK and Rizvi SSH (editors). Bioseparation Processes in Foods. New York, USA:Marcel Dekker Inc. pp. 139–174.
Sreerama N, Woody RW 2004. Computation and Analysis of Protein Circular Dichroism Spectra. Methods Enzymology 383:318–351.
Stathakis NE, Mosesson MW, Chen AB, Galanakis DK 1978. Cryoprecipitation of fibrin-fibrinogen complexes induced by the cold-insoluble globulin of plasma. Blood 51(6):1211–1222.
Sternberg M, Kim CY, Schwende FJ 1975. Lysinoalanine:presence in foods and food ingredients. Science 190(4218):992–994.
Sugarman N 1956. Process for simultaneously extracting oil and protein from oleaginous materials. US Patent No 2762820.
Suzuki A, Yasuhara K, Seki H, Maruyama H 2002. Selective foam separation of binary protein solution by SDS complexation method. J Colloid Interface Sci 253(2):402–8.
Tagliaro F, Deyl Z, Mikšík I, Ulfelder KJ 1998. Concepts and principles of high performance capillary electrophoresis. In:Rossomando EF (editor). HPLC in enzymatic analysis, second edition. New York, John Wiley & Sons, Inc. pp. 41–63.
Tang J, Feng H, Shen GQ (2003). Drum drying. In:Heldman DR (editor). Encyclopedia of Agricultural, Food, and Biological Engineering. New York, USA:Marcel Dekker. pp. 211–214.
Tyler RT 1984. Impact milling quality of grain legumes. J Food Sci 49:925–930.
Tyler RT, Panchuk BD 1982. Effect of seed moisture content on the air classification of field peas and faba beans. Cereal Chem 59:31–33.
Wang AC 1988. Molecular basis for cryoprecipitation. Springer Seminars in Immunopathology 10:21–34.
Wang SSS, Liu HS 2003. A study on protein concentration by foam fractionation. J Chin Inst Chem Eng 34(3):291–297.
Wang W, Yue H, Yuan YQ 2009. A primary study on partial purification of lysozyme from chicken egg white using foam separation method. Biotechnol Biotechnol Equip 23(2):1237–1241.
Warchalewski JR, Gralik J 2010. Influence of microwave heating on biological activities and electrophoretic pattern of albumin fraction of wheat grain. Cereal Chem 87(1):35–41.
Weber RJ, Clem LW 1981. The molecular mechanism of cryoprecipitation and cold agglutination of an IgM Waldenstrôm macroglobulinemia with anti-Gd specificity: sedimentation analysis and localization of interacting sites. J Immunol 127:300–305.
Wehbi Z, Pérez MD, Sánchez L, Pocoví C, Barbana C, Calvo M 2005. Effect of heat treatment on denaturation of bovine α-lactalbumin:determination of kinetic and thermodynamic parameters. J Agric Food Chem 53:9730–9736.
Weijenberg DC, Mulder JJ, Drinkenburg AAH, Stemerding S 1978. The recovery of protein from potato juice waste water by foam separation. Ind Eng Chem Process Des Devel 17(2):209–213.
Westermeier R, Scheibe B 2009. Gel Electrophoresis. In: Ötleş S (editor). Handbook of Food Analysis Instruments. Boca Raton, FL:CRC Press, Taylor & Francis Group. pp. 423–438.
Wolf WJ, Sessa DJ, Wu YV, Thompson AR 2002. Air classification of pin milled soybean hulls. Cereal Chem 79(3):439–444.
Wong M, Woodley JM, Lye GJ 2010. Application of bipolar electrodialysis to *E. Coli* fermentation for simultaneous acetate removal and pH control. Biotechnol Lett 32(8):1053–1057.
Wu YV, Strongfellow AC 1979. Protein concentrate from air classification of high protein soft wheat flours. J Food Sci 44:453–455.
Wu YV, Strongfellow AC 1981. Protein concentrate from air classification of flour and horny endosperm from high-lysine sorghum. J Food Sci 46:304–305.

Xu T, Huang C 2008. Electrodialysis-based separation technologies: A critical review. Am Inst Chem Eng 54(12):3147–3159.

Yoichiro I 1986. Methods for continuous countercurrent foam separation. US Patent 4615805.

Yoshihiro S, Toshiro M 2000. Adsorption behaviour of proteins on suspended particles, and suitable protein for coagulation and foam separation. Proc Env Eng Res 37:237–245.

4 Advancements in Oil and Oilseed Processing

Nurhan Turgut Dunford

4.1 INTRODUCTION

Oil and oilseeds are important feedstocks for bioproduct manufacturing. Selection of the appropriate oil and oilseed processing techniques is influenced by the planned production capacity and product end use. For example, supercritical fluid technology might be a viable option for high value specialty oil production but not for commodity products because of the high capital cost of this technique. Both chemical composition and quality of feedstock and type of processing techniques used for production have significant effect on final product quality and cost. Traditionally, edible oil quality was mainly defined by organoleptic properties, such as taste, odor, and color. Today, nutritional quality and functionality are the most important criteria for selection of oils for edible and industrial applications.

Oilseed and oil processing are designed to obtain high quality oil with minimal undesirable components, achieve high extraction yields and produce high value meal. Over the years advances in processing equipment and design have led to significant reduction in costs and product losses. The trend toward reducing the environmental impact of the oil processing industry has been motivated by consumer pressure, government regulations, and corporate responsibility. Today, many oil and oilseed processors are making business decisions based on environmental sustainability of feedstocks, processes, and products as well as the economics of the entire system. Since the 1980s, physical refining processes have been gradually introduced to replace chemical ones in an effort to reduce waste streams and meet consumer demand for "natural or no chemical" products. Advanced processing techniques, such as supercritical fluid technology and enzyme-aided processing, are receiving attention from oil and oilseed processors.

This chapter covers both conventional and advanced processing techniques used for oil and oilseed processing and their environmental impacts. Effect of processing on final product properties, and by-product and waste generation, will also be highlighted. It is important to note that this chapter is not intended to cover the details of each unit operation. The references cited are excellent resources for detailed information on the processes discussed here.

Food and Industrial Bioproducts and Bioprocessing, First Edition. Edited by Nurhan Turgut Dunford.

4.2 OILSEED PRETREATMENT

4.2.1 Handling and storage

Oilseeds are shipped in railroad cars, trucks or barges for ship transportation. Seeds are received and stored at elevators. The physical properties of the seeds, such as size, shape, bulk density and flow characteristics, affect the design of oilseed handling facilities. For example, canola, rapeseed, delinted cottonseed, soybean and sunflower have very good flow properties. They are usually stored in vertical storage bins. Copra and undelinted cottonseed do not flow well. Hence, vertical bins are not suitable for these types of seeds. Copra is stored in large flat warehouses. Storage bins can be built of stainless steel, concrete, brick or other materials. Stainless steel bins are preferred where ambient temperature is low while concrete bins are built at places where average outside temperature is relatively high (35–45 °C) (Berger, 1985). Some oilseeds are abrasive, and abrasion-resistant materials need to be used on seed contact surfaces.

Proper handling and storage of oil-containing materials are very important to minimize deterioration and maintain good quality of both contained oil and meal. Whole, intact low moisture oilseeds (about 8–10% moisture) may be stored for extended time under suitable conditions. Deterioration of oilseeds is accompanied by respiration or carbon dioxide evolution and heat generation by oxidation reactions. Damaged and high moisture seeds may release higher amounts of carbon dioxide than sound and intact seeds (5 vs. 0.1 cubic centimeter or more of carbon dioxide per gram of seed per day) (Rukunudin *et al.*, 2004; Mattil *et al.*, 1964). Furthermore, heat generation by oxidation reactions may increase the temperature of the stored seed, accelerating deterioration even to the point of charring the seeds.

Activity of native enzymes present in oilseeds, infestation by insects and mites along with microbial activity during storage are also important factors affecting oil and meal quality. If microbial growth occurs it is accompanied by production of metabolites and mycotoxins, particularly aflatoxins. Aflatoxins produced by strains of Aspergillus *flavus* emerge partly in the expressed oil. In such a case oil requires further refining for aflatoxin removal. Aflatoxin-containing meal is also unsafe as animal feed.

In general, high moisture content in seeds (above 14–15% moisture) and high relative humidity conditions accelerate lipolytic enzyme reactions and increase the free fatty acid (FFA) content in oils. Oil in mature seeds may contain about 0.5% FFA. However, if seeds are damaged mechanically or by frost or become wet during harvest, handling and storage, the acidity of oil can be much higher. Sprouting, which is considered a damage factor, is another important issue during storage of high moisture seeds. Sprouted seeds may have lower oil and higher FFA content compared to sound seeds (Mostafa *et al.*, 1987).

Today, air dryers are essential components of modern storage facilities to maintain oilseed quality. Most oilseed storage bins are equipped with aeration ducts and ventilation blowers to cool the seeds. Since oxidation reactions are aerobic processes, a low oxygen atmosphere in storage bins helps to slow down oxidation and quality deterioration. Mature seeds can be stored longer than immature seeds because of the lower activity of oil-splitting enzymes in mature seed. Proper storage of harvested oilseeds may also contribute to desirable changes. For example, oil extraction yields from fresh soybeans can be lower than for soybeans stored more than five months. The storage of soybeans also decreases chlorophyll content of green beans (Gomes *et al.*, 2003; Mattil *et al.*, 1964).

Logistical implications of transportation and handling of genetically modified oilseeds are particularly important. Bioengineered oilseeds with enhanced quality attributes need to be segregated from traditional crops to preserve their identity. Otherwise enhanced quality will be lost. Furthermore, consumer acceptance of genetically modified crops is still a controversial issue and labeling of bioengineered crops has been mandated in some international markets. Therefore, separate handling systems for bioengineered and traditional products may need to be created.

4.2.2 Preparation of seeds for oil extraction

Unit operations for preparation of seeds for oil extraction vary slightly depending on the physical properties and oil content. However, most oilseeds go through cleaning, drying, de-hulling, size reduction, flaking and cooking processes.

Oilseeds need to be cleaned to remove plant stems, sticks, leaves and foreign material before storage. Such material may decompose and generate heat in the stored seed mass, diminishing oil and meal quality. Foreign materials in seeds are typically separated out by a combination of rotating or vibrating coarse screens, reels and aspiration. This process is commonly referred to as scalping. Sand and dirt are also removed by fine screening. Magnetic devices, plate, drum or electromagnets, installed over conveyor belts remove metal contaminants from seeds. Some oilseeds, such as peanuts, may contain stones which are similar in size to the seeds. Hence they cannot be separated by screening. Seeds are separated from stones by gravity in destoners.

The moisture content of oilseeds often has to be reduced to minimize degradation in storage and to improve the effectiveness of downstream processing. For example, soybeans are often received at 13% moisture or higher and need to be dried to 10% moisture to facilitate efficient hull removal. Depending on the regulations of the country in which oilseeds are traded, seeds with high moisture contents may be accepted with some financial penalty (Sutherland and Ghaly, 1982). High moisture seeds are dried with heated air either on the farm or at the elevators. Fixed-bed-batch (both parallel and radial flow), recirculating-batch and continuous-flow dryers are used for drying oilseeds. The design of oilseed dryers requires basic knowledge of drying parameters, including resistance of seeds to air flow, moisture sorption properties and drying rates, and also safe drying temperatures. Because of the high oil content in oilseeds, safe drying air temperatures and fire risk are very important factors that have to be taken into account during dryer design. A study carried out with sunflower and rapeseed (16% initial moisture content) showed that the maximum inlet drying air temperature to preserve germination was 60 °C, which was also the maximum safe drying air temperature in a fluidized-bed dryer (Sutherland and Ghaly, 1982). It was also indicated in the latter study that higher inlet air temperatures could be used provided that seed temperature at any point within the dryer does not exceed the stated safe value. Inlet air temperatures higher than maximum safe drying temperature was not recommended for fixed-bed batch dryers, as seeds near the air inlet could approach the inlet air temperature, but for recirculating batch and continuous-flow dryers a slight increase in inlet air temperature could be allowable.

The amount of hull on oilseeds varies significantly. The percentage of hulls for cotton seed, sunflower seeds and soybean are 36, 24 and 8%, respectively (Clark *et al.*, 1980; Pierce, 1970). Hulls and shells of oilseeds do not contain a significant amount of oil (less than 1%). Most oilseeds need to be separated from their outer husk or shell prior to oil extraction. De-hulling or shelling or decorticating increases oil production efficiency and capacity of the extraction equipment, and reduces wear in the extractor as the husks are abrasive. If not removed, hulls reduce the total oil yield by absorbing or retaining oil in the pressed cake. Furthermore, wax and color compounds present in the hulls are extracted along with oil. These compounds are not desirable in edible oils and need to be removed during the refining process. De-hulling reduces fiber and increases protein content of the meal. Knife, disk and impact type de-hullers are widely used. An impact de-huller, which is commonly used for sunflower seeds, consists of a rotating blade that drives seeds onto a hard wall outside the diameter of the blades. The force of impact causes the hulls to break. The seeds are then separated using shaking conveyor belts, multiple sifting screens or vacuum (Miller *et al.*, 1986). De-hulling efficiency is often measured by the residual fiber content in the meal and the residual oil content in the hulls. The industry standards for oil and fiber contents of soybean meal are: minimum 0.5% oil and maximum 3.3–3.5% fiber for (minimum) 47.5–49% protein meal, and maximum 7% fiber and minimum 0.5% oil in 44% protein meal (12% moisture content) (National Oilseed Processors Association, 2010).

4.2.3 Size reduction and flaking

Most oilseeds are reduced in size to facilitate hull removal, heating, drying and flaking prior to oil extraction. Canola, rapeseed and corn germ do not require size reduction as they are already sufficiently small. Cracking mills are used for seed size reduction. High capacity cracking mills can process up to 1000 metric tons (tonnes) per day of oilseeds. In general, oilseeds are flaked prior to solvent extraction. A flaking mill has two large diameter rolls turning in opposite direction and forced together by hydraulic cylinders. As the seeds are pulled through the flaking mill, they are stretched and flattened. Unfortunately, literature on the mechanical aspects of oilseed flaking, specifically technical information on deformations and forces that seeds experience as they move through the flaking rolls, is very limited (Levine, 1993). Such information is very important for designing flaking rolls for oilseeds. Flaking ruptures seed cellular structure and reduces the distance that solvent has to travel to contact oil in the cells. Typical flake thickness is in the range of 0.01–0.015 inch or 0.25–0.37 mm (Singha *et al.*, 1999). Flaking of oilseeds also increases surface area for increased contact between solvent and seed during the solvent extraction process. Oil from the cracked or flaked seeds should be extracted as quickly as possible (within 24 h) to minimize meal and oil quality deterioration. Flaking mills which can process 300–500 tonnes of seeds per day are available.

4.2.4 Cooking/Tempering

Oilseeds are cooked or tempered to denature proteins, release oil from the cells and inactivate enzymes. For example, rapeseed contains the enzyme myrosinase. This enzyme catalyzes hydrolysis of glucosinolates, which are naturally present in rapeseed. During the hydrolysis process undesirable compounds such as isothiocyanates and nitriles form. These compounds are soluble in oil and lower the quality of oil. Rapeseed is cooked in multistage cookers to keep the glucosinolates intact and inactivate the myrosinase (Christian, 2010). Rapeseed is preheated to 20–50 °C in less than five minutes and contacted with direct steam at 120 °C. The cooking temperature for canola is lower (less than 100 °C) than that of conventional rapeseed varieties because of the lower glucosinolate content of canola (Gunstone, 2004). Cooking also gives seeds elasticity for efficient pressing. Oil needs to be extracted from cooked seeds immediately. Cooking at high temperature is not necessary for sunflower seeds (Pierce, 1970). Cracked and de-hulled soybeans are conditioned or tempered by increasing the temperature to 65 °C and adjusting the moisture by using live steam.

Conditioning is done in rotating drums with indirect steam heating. Tempering improves flaking performance and extraction efficiency.

4.3 OIL EXTRACTION

Two common oilseed extraction processes are solvent extraction and mechanical extraction using a screw press. Solvent extraction with hexane is the standard practice in today's modern oilseed-processing facilities. Solvent extraction plant capacities range from 100 to 9000 metric tonnes/day. High oil content seeds, that is canola and sunflower, are processed by using a combination of mechanical pressing and solvent extraction. Seeds are first pressed to reduce the oil content to about 20% and the remaining cake is then solvent extracted. Use of the screw press without solvent extraction is preferred by small processors because of its low capital cost. Supercritical fluid, water and enzyme-aided water extraction processes are of interest for specialty and gourmet oils production.

4.3.1 Solvent extraction

Solvent extraction is a separation process by which solutes are transferred from a solid or liquid mixture into a solvent. Oilseed extraction refers to preferential dissolution of oil by contacting oilseeds with a liquid solvent. Solvent extraction is the most efficient technique to recover oil from oilseeds. The process efficiency depends on the oilseed preparation prior to extraction, temperature, mode of operation (batch vs. continuous and co-current vs. countercurrent operations), equipment design and, most importantly, solvent type. Residual oil in the meal is expected to be less than 1% after commercial solvent extraction.

4.3.1.1 *Solvent selection*

There is no ideal solvent but there are numerous physical and chemical properties that have to be considered while choosing a solvent for oil extraction from oilseeds: availability, cost, oil solubility, toxicity, flammability, selectivity for the desired oil components, specific heat, latent heat of vaporization, boiling point, viscosity, specific gravity, stability to heat and light, inertness to oil, meal and equipment (Johnson and Lusas, 1983). Light paraffinic petroleum fractions – pentane (boiling point 31–36 °C), hexane (boiling point 63–69 °C), heptanes (boiling point 90–99 °C) and octane (boiling point 102–129 °C) – can be used for oil extraction. Currently hexane is the solvent of choice for commodity vegetable oil extraction. Toxicological and environmental concerns, several tragic explosions and fires at hexane extraction plants have motivated searches for alternative solvents. In 1990, the US Environmental Protection Agency Clean Air Act listed hexane as a hazardous air pollutant (EPA, 1990). Flammability of hexane has been a major issue, limiting its use in small processing plants. Only large centralized facilities can afford the capital investment and the highly skilled labor required to operate a hexane extraction plant.

A number of alternative solvents have been examined to replace hexane. For example, trichloroethylene, which is a non-flammable and non-explosive compound, received some attention as an alternative solvent to hexane during the period 1930–1955. Trichloroethylene has a high solvent capacity (oil solubility) but it is less selective and extracts more pigments than hexane. Yet, oil can be bleached without excessive losses during refining (Duncan, 1948). Oilseed extraction plants using trichloroethylene were converted to hexane after only one year of operation because ruminants fed on trichloroethylene extracted meal died of "bloody nose disease". The exact mechanism of trichloroethylene extracted meal toxicity was never identified but it was hypothesized that trichloroethylene reacted with sulfhydryl groups of the amino acid cysteine, because S-(trans-dichloro-vinyl)-L-cysteine produced the same symptoms when fed to animals (Seto *et al.*, 1958).

Ethanol has also been utilized for oil extraction. Oil solubility in ethanol varies with temperature and water content. Soybean oil is completely miscible with absolute ethanol above 70 °C (Johnson and Lusas, 1983). As ethanol concentration decreases and water content increases, oil solubility is significantly reduced in the mixture. The higher cost and latent heat of vaporization are the major disadvantages of ethanol as a solvent for oilseed extraction. Recent developments in bioethanol production may reduce the cost of ethanol, making it a viable alternative to hexane. Solvent mixtures can also be used to extract oil. Hexane/alcohol azeotropes have been used for extraction of residual lipids from hexane-extracted meals to improve flavor and odor, specifically from soybean and peanuts. "Grassy" and "beany" flavors in oilseeds are associated with the presence of phosphatides, which can be easily extracted with hexane/alcohol mixtures. Similarly, hexane/alcohol azeotropes, specifically hexane/methanol, are very effective in extracting aflatoxin from meal.

Dichloromethane or methylene chloride is an excellent solvent for oil extraction because of its low boiling point (39.8 °C), which makes desolventization of oil and meal easy. Furthermore, it is non-flammable and has low specific heat, latent heat of vaporization and low solubility of water. Utilization of dichloromethane for oil extraction was first demonstrated in the 1940s but the process was not economically feasible at the time because of the relatively high cost of dichloromethane. In 1986, the feasibility of cottonseed oil extraction by using dichloromethane was demonstrated at a pilot scale study (Johnson *et al.*, 1986). Residual oil content in the meal was lower than typically achieved with hexane extraction. Cottonseed meal produced during the process was suitable for use in poultry feed formulations, because gossypols present in cottonseed were extracted with oil and removed from meal. No residual aflatoxin was detected in alkali-refined oil.

Acetone, isopropyl alcohol, methyl and ethyl acetate esters, ethylene glycol monomethyl ether (methyl cellosolve), ethylene glycol monoethyl ether (ethylene cellosolve) and amines have also been examined to determine their suitability for oil extraction from oilseeds (Johnson and Lusas, 1983) but hexane remains to the choice of solvent for large oilseed extraction operations today.

4.3.1.2 *Extraction process*

There are three major steps in a traditional solvent extraction system: oil extraction, oil desolventizing and meal toasting. In general, residual oil in solvent-extracted oil would be less than 1%. Oil extraction is a diffusion controlled process that involves diffusion of solvent into a solid matrix to solubilize oil then diffusion of the oil–solvent mixture into a liquid phase. Mathematical models have been developed to describe the solvent extraction process (Abraham *et al.*, 1988; Becker, 1978; Mattil *et al.*, 1964). It has been shown that the oil extraction rate for hexane extraction of flaked soybeans is proportional to the 3.97 power of the flake thickness and 3.5 power of the residual oil (Mattil *et al.*, 1964). In general, oil solubility in a solvent increases with increasing extraction temperature. High temperature has also a positive effect on viscosity and diffusivity of oil. Viscosity decreases while diffusivity increases as the extraction temperature increases, resulting in shorter extraction times. Energy required for solvent recovery decreases when higher operating temperature is employed in an extractor. However, high temperatures may cause deterioration and denaturation of some oil and meal components. Hence, determination of appropriate extraction temperature is based on the solvent properties and required specifications of the final product.

Expanders are used to form oilseeds into porous pellets, which are commonly known as collets. There are two types of extruders used in the oilseed processing industry, expanders and dry extruders. Dry extruders are discussed in the mechanical expression section of this chapter. Expanders consist of an un-segmented barrel housing a worm screw with interrupted flights and needle valves installed at the flight interruptions to inject water or steam onto the product stream, flaked seeds. Conversion of flaked seeds into extruded collets improves solvent extraction efficiency. Collets are larger and stronger and have larger bulk density than flakes. Better solvent flow and micelle (oil+solvent) drainage are achieved in the solvent extractor when collects are used for solvent extraction. Because of the larger bulk density of collets less space is required for the same weight oilseed in the extractor. Production capacity in a plant is increased when collets, rather than flaked seeds, are used for solvent extraction. High oil content seeds such as sunflower, safflower, peanut, canola and rapeseed have too much oil to produce good flakes and collets. Hence, they are pre-pressed to reduce meal oil content to 16–18% before solvent extraction. Soybean does not require pre-pressing because of its relatively lower oil content (18–20%). A new screw press design combines pre-pressing and extruded collets formation into a single processing unit. A drainage cage and a pressing screw are included into the barrel of an expander, so there is no need for purchasing a separate extruder and pre-press.

In the 1800s small batch type oil extraction systems were common in Europe. The same vessel was used for both extraction and meal desolventizing. Today, only extraction of specialty and high value oils and recovery of oil from spent bleaching clay are carried out in batch extractors. A rotating drum and a vertical cylindrical kettle equipped with a vapor tight cover and a low-speed agitator are two popular batch type extractor designs.

Use of large scale continuous processes began in the early 1920s. Today, immersion and percolation type continuous extractors are used for commercial solvent extraction of oilseeds (Bernardini, 1976; Becker, 1978; Christensen, 1983; Woerfel, 1995). In immersion type extractors flaked oilseeds are completely submerged in solvent. The material is gently agitated by a screw stirrer. Micella is collected from the top of the extractor. As the extracted material comes out of the extractor bottom it leaves the solvent bath and is conveyed to an elevator. While moving up the elevator in a series of baskets with perforated bottoms, the solvent is drained off and finally residual meal is delivered into a desolventizer (Becker, 1978). This type of extractor is suitable for very fine ground materials, which cause poor drainage through the fines in other type of extractors unless the material is pelletized.

Early extractor designs based on solvent percolation were basket-type extractors in which flaked seeds were placed in baskets with perforated bottoms. These systems looked like an enclosed bucket elevator. The baskets were supported by endless chains in a sealed housing and continuously raised and lowered at a slow rate (1 revolution/h). Each basket was filled with flaked seeds by an automatic feed hopper at the top. As the basket started descending solvent is sprayed over the baskets. The spent flakes in baskets ascended to the top of the housing on the opposite side of the feed hopper. At the top baskets were automatically inverted and spent seeds were discharged into a hopper, from which they were transferred to a meal desolventizer on a conveyor belt. Basket type extractors were bulky and hard to maintain. The newer extractor designs are horizontal and rotary type. The design principle for horizontal extractors is similar to the basket-type extractors but the baskets rotate in a single horizontal plane rather than vertically (Becker, 1978). In percolation-type extractors, flaked seeds are formed into a fixed bed usually 50 cm deep. Under these conditions solvent channeling or micella flow through the flakes is minimized; consequently complete extraction is achieved (Bernardini, 1976). The particle size of the seeds needs to be large enough to allow a good solvent flow rate through the seed bed. Fine particles in the bed block the solvent flow, causing flooding of the bed and consequently lowering extraction efficiency.

Combination of immersion and percolation extractors has also been used for oil extraction from oilseed. This process involves partial extraction of the oil in a percolation extractor (until there is 10–15% residual oil in the seeds) followed by wet flaking of the partially defatted seed as it comes out of the percolation extractor. This operation is carried out in a special solvent-tight flaker operating in a solvent vapor-saturated atmosphere. Finally, flaked seeds are extracted in an immersion-type extractor. The advantage of this process is that seeds are extracted at a low temperature (maximum 50 °C) and the subsequent meal desolventizing is carried out lower than 105 °C, producing a meal suitable for protein concentrates and isolates manufacturing (Bernardini, 1976).

4.3.1.3 *Meal and oil desolventizing*

After oil extraction, residual solvent in meal (spent seeds) is removed in a desolventizer-toaster (DT), which may have three sections for pre-desolventizing, desolventizing and toasting. Meal is heated to a temperature that evaporates solvent, but does not deteriorate the nutritional value of the meal. A DT consists of trays mounted along a vertical shaft with sweep arms attached to move the meal around. Meal from a DT may still contain some moisture, which is removed in a rotary steam tube dryer and then cooled by air before being sent to storage. Desolventizer toasters are commonly used to produce toasted meal for ruminant and poultry feeds. In some cases, desolventizing and drying is done in the same equipment, referred to as a desolventizer-toaster-dryer-cooler (Becker, 1978). The drying is done by hot air and the cooling is accomplished by blowing ambient air through the meal. Another system, "Vapor Desolventizer-Deodorizer", allows time, temperature and moisture in the system to be adjusted during operation. The solvent in the meal is removed by superheated vapor in the desolventizer; then meal passes to the deodorizer through a lock (Becker, 1978). The steam pressure in the deodorizer can be regulated between 0.5 and 2 bars. By regulating steam pressure and minor tunings of the height of the discharge dam, the operators have the flexibility to adjust meal properties, that is the protein dispersible index (PDI) over a range of 10 to 90.

Solvent recovery from micella (oil + solvent) is carried out in single or dual effect evaporators. Flashing and low-pressure steam stripping remove hexane from the oil down to less than 0.15% (w/w). Solvent vapors from distillation and stripping columns are condensed and recycled to the extractor. After this, crude oil goes through a refining process, which removes undesirable components in the oil.

4.3.2 Mechanical oil expression

Mechanical pressing of liquid out of liquid-containing solids is referred to as "expression". The conventional theories about oil expression from oilseeds were based on the assumption that oilseed cell walls were non-porous; hence, they had to be raptured by mechanical and thermal treatment before substantial oil expression could be achieved (Mattil *et al.*, 1964). A later theory has suggested that the pores (plasmodesmata) of certain biological cell walls might be the channels through which oil flows out of the seed during expression (Diekert and Diekert, 1976). Microstructure studies of rapeseed and cashew using transmission electron microscopy confirmed the porous structure of oilseed cell walls, indicating that oil expression could occur through a porous microstructure without cell wall rupture (Mrema and McNult, 1984). A mathematical model based on the latter theory was developed to describe the mechanical oil expression process (Mrema and McNulty, 1985). It was assumed that the process involved three phases: (1) oil flow through the cell wall pores, (2) oil flow in the inter-kernel voids and (3) consolidation of the oilseed cake (reduction of cake volume). Three fundamental equations – the Hagen Poiseulle equation for flow of fluids in pipes to describe the flow of oil through the pores on the cell wall; Darcy's law of fluid flow through porous media to describe the flow of oil through the inter-kernel voids; and a modified form of Terzaghi's equation for the consolidation of saturated soils – were used to develop a general equation to describe the oil expression process. The general equation was solved for two loading regimes: a linearly increasing load regime to simulate the load application phase in hydraulic presses and the overall screw expeller process; and a constant load to simulate the hydraulic pressing. Experimental and predicted data obtained from the model were in good agreement. The flow of oil across cell walls in the seed was the rate determining step (Mrema and McNulty, 1985).

Screw presses, roll presses and mills, collapsible-plate and frame-filter presses, disc mills, rack and cloth presses and hydraulic presses are examples of the wide variety of equipment available for expression processing (Khan and Hanna, 1983; Bredeson, 1977; Bockisch, 1998). Lever and wedge presses were the early mechanical oil expression equipment used for oilseed processing. The first cottonseed oil mill built in the United States in the 1920s utilized a hydraulic press. Seeds in filter bags were manually loaded into perforated, horizontal boxes between the head block and the ram of the press. Boxes were pressed together by applying hydraulic pressure on the ram. Oil was pressed through the filter bag. Then the filter bag containing spent cake was manually removed from the hydraulic press (Khan and Hanna, 1983). Later versions of the hydraulic presses used cages instead of filter bags. Hydraulic presses were in use until the 1950s. They were replaced with continuous screw presses and continuous solvent extraction plants, which are less labor intensive. The olive oil industry still utilizes hydraulic presses today.

The mechanical screw press consists of a vertical feeder and a horizontal screw with increasing body diameter to exert pressure on the oilseeds as it advances along the length of the press. The barrel surrounding the screw has slots along its length to allow it to expel air and then drain the oil. Oil is collected in a trough under the screw and the de-oiled cake discharged at the end of the screw. The main advantages of the screw press are that large quantities of oilseeds can be processed with minimal labor and it allows continuous oil extraction (Ward, 1976). The maintenance of screw presses is challenging. The moving parts and high pressure and temperature generated in the press increase wear. The primary improvement in mechanical screw press design has been developing materials of construction that extend the life of the parts that wear the most. Equipment parts that used to last three months before requiring replacement may now last up to two years. Today, much larger screw presses have been built. Former capacities of five tonnes per day are now up to over 100 tonnes per day for full pressing and over 800 tonnes per day for pre-pressing applications.

A growing number of mini crushing mills, which employ extruding-expelling (E-E) technology, have been built as farmer-owned cooperatives or as on-farm operations to process locally produced soybeans or other oilseeds. E-E uses a dry extruder in which heat is generated internally by friction and dissipation of mechanical energy prior to screw pressing, replacing more capital intensive use of steam-heated dryers and cookers and associated steam generation equipment. These are high-shear extruders running at screw speeds of about 600 rpm (Wijeratnea *et al.*, 2004). The partially defatted meal produced by E-E, typically with 6–7% residual oil content, is used in livestock rations.

4.3.3 Aqueous extraction

Aqueous extraction (AE) is traditionally used in many developing countries. The process, called water flotation, has been used for extracting coconut and palm oil. It involves heating oily material, grinding with or without water and boiling with water to liberate the oil. The oil which appears on the surface is collected and heated to remove moisture. Oil extraction yields of around 50% are generally considered to be satisfactory for traditional non-commercial processes. This process can be used to extract oil as well as high quality proteins. However, there are certain limitations to the process. Finely ground seeds, which cause dusting, may lead to an explosion if the processing area is not ventilated well. Other disadvantages for aqueous oilseed include the need for breaking emulsion (that might form during the process) to separate oil and water phases and the costs of the treatment of aqueous effluents and wet meal.

4.3.4 Enzyme and surfactant-aided oil extraction

The objective of enzyme-aided oil extraction is to improve aqueous oil extraction yields. In general the process involves the following steps: (1) oilseed grinding (wet or dry), (2) incubation of the feed material with aqueous enzyme solution, (3) separation of liquid and solid phases by centrifugation or filtration and (4) recovery of oil from liquid phase. Additional unit operations may be needed depending on the chemical composition of the specific oil-bearing material. For example, corn germ was hydrotreated to inactivate the intrinsic lipase and increase the quality of oil prior to the incubation of corn germ with enzyme (Bocevska *et al.*, 1993; Domínguez *et al.*, 1994; Moreau *et al.*, 2004; Dickey *et al.*, 2008). Similarly, rapeseed or canola seeds need to be cooked prior to enzyme incubation to inactivate native enzyme myrosinase, which can produce toxic compounds, such as nitriles and isothiocyanate, during the extraction process (Rask *et al.*, 2000; Zhang *et al.*, 2007; Latif *et al.*, 2008).

Unlike in the organic solvent extraction where oil is recovered as free oil, enzymatic extraction produces oil in three forms: free oil, emulsified oil and oil in skim. The presence of all three forms of oil during enzymatic extraction of soybean has been reported (Rosenthal *et al.*, 2001). Both emulsified and free oil were obtained from corn germ (Moreau *et al.*, 2004), rapeseed (Zhang *et al.*, 2007) and coconut (Chen Man *et al.*, 1996; Sant'Anna *et al.*, 2003). Enzymatic extraction of sunflower and Moringa oleifera seeds resulted in only the emulsified form of oil (Domínguez *et al.*, 1995; Abdulkarim *et al.*, 2006). The forms of oil obtained from enzymatic extraction depend on the compositions of the seeds as well as the processing parameters, such as particle size of the comminuted material, liquid to solid ratio, enzyme concentration, incubation time and mixing rate (Sharma *et al.*, 2002).

The role of enzymes in oil extraction from oilseeds is to hydrolyze the structural components of plant cells and facilitate the release of oil. Hence, enzyme selection depends on the composition of the cell wall and membrane that encapsulates oil bodies and the cytoplasmic network where oil bodies are imbedded in the oilseed. Carbohydrases such as cellulase, hemicellulase and pectinase and protease are effective in breaking up the cell structure. Single enzyme or mixtures of several enzymes have been tested for their efficacy to release oil from oilseeds. In rapeseed oil extraction, pectinase was more effective than cellulase, β-glucanase and xylanase, but was less effective than the mixture of these four enzymes (Zhang *et al.*, 2007). A similar trend was observed when protease, pectinase and mixtures of cell wall-degrading enzymes were used for oil extraction from canola seeds. The use of protease for soybean oil extraction resulted in higher yields than those obtained by using cellulase (Lamsal *et al.*, 2006). However, in corn germ oil extraction protease resulted in lower oil yield than that obtained by cellulase, indicating the importance of choosing the right enzyme for different oilseeds (Moreau *et al.*, 2004). Not only do the types of enzymes, but also sources of the same type of enzyme may have different impacts on oil extraction yield. For example, protease isolated from fungus was superior to protease from papaya (papain) (Hanmoungjai *et al.*, 2002). Similarly, cellulase secreted by *Trichoderma reesei* showed higher efficacy in extracting oil than those produced by *Trichoderma viride* and *Aspergillus niger* (Moreau *et al.*, 2004).

Smaller particle size of feed material is an indication of efficient grinding, which mechanically breaks apart the cell structure of materials (Rosenthal *et al.*, 1996). In addition, smaller particle size results in larger surface area, which allows better contact between oil-bearing material and solvent and reduces resistance to the diffusion of oil, as well as other components, from the solid matrix to the aqueous medium. The rate of enzyme diffusion into the solid substrate is also improved with smaller particle size. Although the importance of smaller particle size has been well recognized, only a few studies have been carried out to investigate its effect on oil extraction yield. A study on enzymatic soybean oil extraction showed that the oil extraction yield increased with decreasing particle size from 1200 μm to less than 200 μm (Rosenthal *et al.*, 2001).

The enzyme to be used for the process dictates the pH and temperature of the extraction medium. Optimum temperature and pH for enzyme activity are usually the conditions used for extraction process. The pH of the medium could actually change the solubility of some cell components, especially that of proteins. In enzymatic extraction, protein solubility is closely associated with oil extraction yield. In general, a high oil extraction yield is obtained when the solubility of protein is high (Rosenthal *et al.*, 1996, 1998; Hanmoungjai *et al.*, 2002). High protein solubility is achieved when pH is away from the isoelectric point of the protein (pI) (Rosenthal *et al.*, 1996, 1998; Hanmoungjai *et al.*, 2002). The amount of solvent used for recovery of target substances is an important parameter that must be optimized to achieve high extraction yields and minimize solvent use, which leads to lower energy consumption for residual meal drying and waste processing. It has been shown that during the enzymatic extraction of soybeans oil yield increased as the liquid to solid ratio increased from 5:1 to 10:1 (de Moura and Johnson, 2009). Unfortunately, an optimization study was not conducted to minimize water use.

The concentration of enzyme in the extraction medium greatly affects the enzymatic reaction rate and the oil extraction yield. It has been demonstrated that an increase in enzyme concentration from 0.2 to 2.5% (v/w) resulted in significant improvement in oil recovery, but further increase to 5% did not bring additional enhancement in rapeseed oil extraction yield (Zhang *et al.*, 2007). The amount of enzyme needed to achieve acceptable oil yields needs to be determined to optimize yield and the cost of enzyme for economic feasibility. Action of enzyme on oilseeds and mass transfer between the aqueous medium and the seed material are time dependent processes. In most cases an increase in extraction time results in an improvement in the oil extraction yield, but the improvement slows down and eventually stops as time prolongs (Sharma *et al.*, 2002). Hence, extraction time is a variable that needs to be included in process optimization studies. Several studies have reported the presence of significant interactions among extraction time, enzyme concentration and water to solid ratio (Xie *et al.*, 2011; Zhang *et al.*, 2007). The optimum combination of these factors needs to be investigated by employing an appropriate experimental design such as factorial design or response surface methodology (RSM) in order to achieve the maximum oil extraction yield.

Microemulsion-based aqueous extraction of oilseeds by using surfactants has also been examined (Ugolini *et al.*, 2008; Naksuk *et al.*, 2009; Do and Sabatini, 2010). The role of the surfactant in oil extraction is to lower interfacial tension between the aqueous solvent and the oil vegetable seeds. As a result, oil can be extracted from the seeds. Application of this technology to oil extraction from oilseeds is still at research phase. The main disadvantages of the surfactant-based oil extraction techniques include: (1) production of wet meal, (2) requirement for small particle size that might cause significant problems in large oilseed production operations, (3) high energy cost for solvent removal and (4) in some cases presence of surfactant in water or oil phase might not be desirable or acceptable.

4.3.5 Supercritical fluid technology

Supercritical fluids (SCF) have been studied as alternative solvents to conventional hexane for oilseed extraction. The fundamentals of SCF technology and its utilization for oilseed processing have been reviewed elsewhere (Dunford *et al.*, 2003; Dunford, 1995; King and List, 1996; McHugh and Krukonis, 1994; Stahl *et al.*, 1980; Stahl *et al.*, 1987). Supercritical carbon dioxide (SC-CO_2), which refers to CO_2 above its critical pressure (7.3 MPa) and temperature (31 °C), has been the preferred solvent for the edible applications of SCF. However, nitrous oxide and some organic compounds, such as ethane, propane and butane, as well as halogenated hydrocarbons, such as chlorotrifluoromethane, have also been employed as SCF for extraction of natural materials. The advantage of SC-CO_2 is the easy removal of solvent from the extract. When pressure is released, CO_2 returns to the gas phase and oil precipitates out of the CO_2–oil mixture, then CO_2 is recycled to the system. Hence, CO_2 release from the system is not an environmental issue. Density and selectivity of SCF can be regulated by adjusting pressure and temperature of the system. This property makes the selective extraction of desired compounds from a complex matrix possible by using SCF.

Oil extraction with SCF was first reported in patent literature in the 1950s (Dickinson, 1947; Palmer and Fanwood, 1950). Processes for the SC-CO_2 extraction of copra, sunflower seeds, soybeans and groundnuts were patented later (Vitzthum and Hubert, 1971; Zosel, 1964). Some of the oil and oilseeds successfully extracted with SC-CO_2 are soybean (Stahl *et al.*, 1983; Friedrich, 1984; Snyder *et al.*, 1984), jojoba oil (Stahl *et al.*, 1983; Friedrich *et al.*, 1998), cottonseed and corn (Friedrich and Pryde, 1984), sunflower and rapeseed (Stahl *et al.*, 1984), palm (Kalra *et al.*, 1987), evening primrose (Favati *et al.*, 1991) and canola (Dunford, 1995; Dunford and Temelli, 1996, 1997; Fattori *et al.*, 1988). The SC-CO_2 technique has not been widely used because of the high costs of high-pressure equipment and high-pressure processing. A study carried out with soybeans demonstrated that a substantial reduction in the operating costs could be obtained by optimization of heat transfer and recovery systems. In such a case the operating cost of oil extraction with SC-CO_2 could be in the same range as conventional hexane extraction (Reverchon and Osséo, 1994).

In gas-assisted mechanical expression, liquid CO_2 is introduced into the screw press under high pressure. This process takes advantage of the high solubility of dense CO_2 in oil to enhance the extraction yield. Oilseed cell structure swells and eventually ruptures because of the dissolved CO_2 in oil, consequently releasing the oil. The oil yield is expected to increase because of the presence of free oil in seed and lower viscosity of CO_2-saturated oil (Venter *et al.*, 2006). However, during the actual operation, when the pressure is released at the nozzle of the press, CO_2 turns to the gas phase and oil is entrained in the meal (Dijkstra, 2009).

Although the industrial applications of SC-CO_2 technology have been growing steadily, currently most of the applications involve high value products. The high investment cost of SCF systems remains to be a drawback for this technology. However, cost saving can be realized from simplified separations and the lower solvent cost. The non-toxicity of SC-CO_2 and significantly reduced waste streams from the process makes this technology unique in terms of meeting increasingly stringent health and environmental regulations.

4.4 OIL REFINING

Crude oil obtained by both solvent extraction and mechanical pressing contains both desirable and undesirable compounds. Triacylglycerides (TAGs) (neutral lipids) and health beneficial compounds such as tocopherols and phytosterols are the desirable compounds in the oil. Free fatty acids (FFAs), phospholipids (PLs), also referred to as gums, and lipid oxidation products are removed during oil refining because of their adverse effects on oil quality. There are several unit operations in a crude oil refining operation. Degumming, deacidification/refining, bleaching, deodorization and winterization are commonly used for edible oil production.

4.4.1 Degumming

Phospholipids are essential components of oilseed cell structure. They are oil soluble and extracted along with oil. PLs are not desirable in oil because they settle out of the oil during shipping and storage and have adverse effects on the color and flavor of oil. The presence of PLs in oil may create problems during processing, for example frying, because of their emulsifying properties (they reduce interfacial tension between immiscible liquids, i.e. water/oil). There are two types of PLs, hydratable and non-hydratable. PLs containing the functional groups choline and ethanolamine have higher affinity for water than that of the PLs with acids, acid salts (calcium, magnesium, and iron) and inositol groups. Phosphatidic acid and the salts of phosphatic acid are commonly known as "non-hydratable phospholipids". The amount of non-hydratable PLs in crude oil varies significantly depending on the quality and type of the seed and the oilseed handling and processing conditions.

4.4.1.1 *Water and acid degumming*

Degumming refers to the unit operation during which PLs are removed from crude oil. Several reviews on oil degumming have already been published (Xu and Diosady, 2004; Gupta, 2008; Mattil *et al.*, 1964). Oil degumming is usually carried out at the crushing or extraction plant. Hydratable PLs can be removed from the oil by water degumming. Hot deionized water (at 65–80 °C) or steam is injected into the warm oil. The amount of water/steam added depends on the amount of hydratable PLs present in the oil. As a rule of thumb, about 2% water is added to oil and mixed for 30–60 min during a batch operation. Continuous degumming processes utilize an on-line mixer for mixing oil and the residence time is usually 10–15 min. During this process PLs absorb water and lose their lipophilic characteristics (affinity to lipids), become oil insoluble and agglomerate into a gum phase. Gum is separated by centrifugation and added back to meal or further processed to produce lecithin, which is used as an emulsifier in food and feed applications. In industry the PL concentration of oil is expressed as parts per million (ppm or mg/kg oil) phosphorous, which can be converted to phospholipids by using the relationship:

$$\%\text{PLs} = [(\text{Phosphorous in oil, ppm}) \times 31.7)] \times 10^4$$

The residual phosphorous level in degummed oil is usually around 100 mg/kg oil after water degumming. PL content of the oil can be further decreased (30–50 mg/kg) in a process called super-degumming by adding organic acid (1500–2500 mg/kg) into the oil at 40–55 °C (Segers, 1982). A newer process, "Ultra Degumming" utilizes an ultra-sheer mixer to attain shear forces around 100 000 sec^{-1} to enhance oil–acid–water contact and reduces the phosphorous content of the oil to 5 ppm (Gupta, 2008). Degummed oil is cooled to below 40 °C before entering a feed tank for further refining.

The "Total Degumming (TOP)" process is similar to the "Ultra Degumming" and utilizes a high-shear mixer to bring previously water degummed oil–acid–water mixture into close contact. An average acid droplet size well below 10 μm, equivalent to at least 10 million acid droplets per gram oil, is attained by high shear mixing. After a sufficient contact time, a base is added and mixed into the acid-in-oil dispersion to adjust the pH of the aqueous phase between 2.5 and 7.0 (Xu and Diosady, 2004). An emulsion forms with minimal soap formation. During the process the acid initially dissociates metal/phosphatidic acid (PA) complexes into insoluble metal salts and PA (in acid form). PA is then hydrated by partial neutralization with the base added, and removed from the oil by centrifugation. Another version of the same process utilizes two centrifuges to remove the hydrated PLs with a high efficiency and minimal losses. The first centrifuge removes the bulk of the gum phase. Since the gums are sticky and very viscous, some PLs still remain in the oil after first centrifugation. The second centrifuge removes the remaining gums in the oil. The gums recovered from the second centrifuge are recycled to the process because of their high oil content. After degumming, the oil is washed with water to remove residual phosphorus salts and acid. The process produces a low phosphorous (about 5 ppm) and iron (<0.1 ppm) oil, but when a water-degummed oil has a higher Ca/Mg content, TOP becomes less effective (Xu and Diosady, 2004). The investment cost for the TOP process is high because of the use of two centrifuges. The process has been combined with alkali refining and named "integrated TOP", where another base is added to the oil immediately after the degumming stage but before water washing, to extract fatty acids and save costs.

The "SOFT degumming" process removes non-hydratable PLs by using a chelating agent (i.e. EDTA), which forms complexes with Ca^{2+} and Mg^{2+} salts of PA and improves interfacial exchange properties (Choukri *et al.*, 2001). This process produces degummed oil with low residual phosphorous (<5 ppm) and iron (<0.05 ppm) contents.

Ultrasonic or acoustic cavitation occurs when a liquid is subjected to high intensity sound waves (20–10 MHz). This causes the formation, growth and rapid recompression of vapor bubbles in the liquid. The implosive bubble collapse generates localized heat and pressure and produces intimate contact among liquid components. The effect of ultrasonic cavitation on PL removal was examined in a continuous degumming system. It was found that phosphorous removal varied between 90 and 99% depending on the ultrasonic power applied. Low power (22 watts) resulted in higher PL removal. The weight of the gum collected from the process increased as the ultrasonic power increased possibly due to larger amount of neutral oil entrainment in the gum (Moulton and Mounts, 1990).

4.4.1.2 *Membrane degumming*

Membrane degumming of crude oils has also been reported (Moulton and Mounts, 1990; Young *et al.*, 1994). Membrane processing is simple, consumes less energy than other degumming techniques, can be carried out at ambient temperature and retains nutrients and other desirable compounds in the degummed oil (Raman *et al.*, 1996). Utilization of ultrafiltration systems equipped with hexane-resistant membranes for degumming hexane–oil miscella has been studied (Iwama, 1987; Gupta, 1977). Two membranes (DS-7 and AN03) were evaluated for their flux and rejection properties. A 99.6% rejection of PLs was achieved at 20.4 Bar and 40 °C by using DS-7 membrane; significant reduction in the color of the oil was also observed after membrane processing (Lin *et al.*, 1997). A number of ultrafiltration membranes prepared in the laboratory were tested for their PLs rejection. A pilot scale study demonstrated that 93% PLs rejection was possible when vegetable oils were processed without adding a solvent to reduce viscosity (Zhang *et al.*, 1996). The microfiltration membranes made of PTFE and PVDF (polyvinylidenefluoride) have also been shown to be effective in rejecting PLs. Although, in general, hydrophobic membranes are preferred for oil processing, hydrophilic PTFE membranes with 100 nm pore size were also successfully used for oil degumming without any compromise in flux rates (Subramanian *et al.*, 1999). Non-porous composite polymeric membranes have selectively rejected PLs (97.4–99.9% PLs rejection), resulting in less than 240 ppm phosphorous in the permeate without any pretreatment or dilution of crude oil with organic solvent. It was also reported that the permeate flux needed improvement before industrial adoption of non-porous membranes for oil degumming (Subramanian and Nakajima, 1997).

An interesting approach, surfactant-aided membrane degumming, has been reported to improve membrane degumming of soybean and rapeseed oils (Subramanian *et al.*, 1999). The hydration rates of different phospholipids, phosphatidylcholine (PC), phosphatidylinositol (PI), phosphatidylethanolamine (PE) and phosphatidic acid (PA), were reported to be in the magnitude of 100, 44, 16, and 8.5 on an arbitrary scale of 100 (Segers and van der Sande, 1990). Calcium salts of PE and PA had very poor hydration rates of less than one on the above scale. It was also reported that fast-hydrating PLs such as PC had the ability to encapsulate other phospholipids. The effect of soybean lecithin addition into crude oil prior to degumming by using a porous polymeric membrane has been evaluated. A PLs reduction of 85.8–92.8% in soybean oil was achieved during membrane processing. The phosphorus content in the permeate varied between 20 and 58 ppm depending on the processing conditions (Subramanian *et al.*, 1999).

High temperature durability, mechanical strength, chemical inertness, organic solvent resistance and unique surface characteristics are some of the advantages of ceramic membranes as compared to polymeric membranes. These membranes are also resistant to microbial growth and steam sterilization. Utilization of ceramic membranes for oil degumming has been examined (De *et al.*, 1998; Ribeiro *et al.*, 2008). Soybean oil degumming in oil/hexane miscella by using a multichannel ceramic membrane resulted in 99.7% PLs retention and 2.2 ppm phosphorous in the permeate (Ribeiro *et al.*, 2008).

The main limitations of conventional membrane processing are short membrane life, low temperature and pressure ranges for polymeric membranes, sensitivity of membranes to chlorine, and instability of membranes at high and low pH ranges. Membrane oil degumming

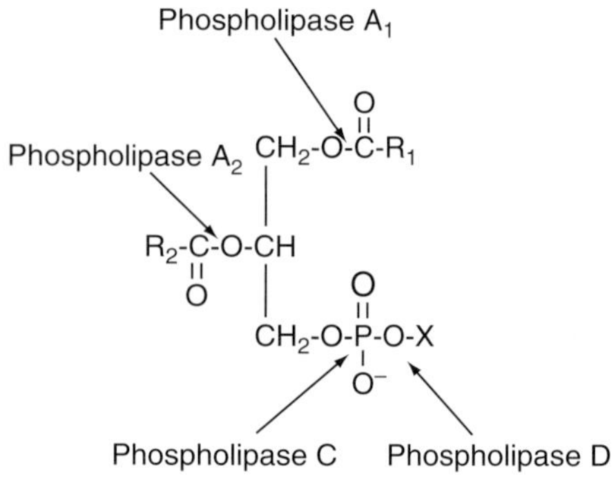

Figure 4.1 Indication of bonds on a phospholipid molecule that are hydrolyzed by various phospholipase enzymes. X = H, choline, ethanolamine, inositol.

processes could be a good alternative to conventional oil degumming processes. However, acceptance of this technology by oil refiners requires improvements in flux rates, membrane life and membrane availability at lower costs.

4.4.1.3 *Enzymatic degumming*

Although the idea of using enzymes to remove PLs from crude oil has been around since the 1980s, it is attracting a lot of attention just recently. Oil degumming can be accomplished by using certain phospholipases, which can hydrolyze the ester bonds of phospholipids. Lysophospholipids that are hydrophilic and insoluble in oil are released from the glyceride backbone of PLs by the action of enzymes, then these insoluble compounds (gums) are separated by centrifugation. The main phospholipase types include A1, A2, C and D (Figure 4.1). The first enzymatic degumming process, EnzyMax®, used Lecitase® 10L (porcine phospholipase A2) (Dahlke and Eichelsbacher, 1998). The process was not very successful for the following reasons: (1) the production capacity of Lecitase® 10L was very limited as the enzyme was extracted from porcine pancreas, (2) the optimal pH for Lecitase® 10L activity was too high for some processors, (3) the enzyme did not meet the requirements for a kosher grade product because of its source (Clausen, 2001). Later, microbial enzymes, the phospholipases A1 (Lecitase Novo and Ultra) and, more recently, a phospholipase C (Purifine) and a lipid acyl transferase (LysoMax a phospholipase A2), became commercially available. Purifine only acts on PC and PE but does not catalyze the hydrolysis of PI or PA and its non-hydratable salts (Dijkstra, 2009). Diacylglycerols (DAG) formed during the degumming process using Purifine remain in the oil, consequently minimizing oil losses. Unlike Purifine, Lecitase Ultra produces free fatty acids and contributes to the oil loss during the degumming process (Dijkstra, 2009). LysoMax catalyzes the transfer of a fatty acid moiety from a phosphatide to free sterols or stanols present in the oil being degummed. During LysoMax treatment lysophosphatides are formed and removed with the water phase. Hence, this process does not form any FFA unless there is insufficient sterol or stanol substrate present to receive the FFA (Dijkstra, 2009).

Response surface methodology was used to optimize enzymatic rapeseed oil degumming process parameters (Yang *et al.*, 2006). Lecitase Ultra was used to degum rapeseed oil. The optimal conditions were determined as follows: enzyme dosage 39.6 mg/kg, temperature

48.3 °C and pH 4.9. The process produced rapeseed oil with 3.1 ppm phosphorous content. The process was then scaled up to 400 tonnes oil/day production. Degumming process efficiency was significantly affected by pH at the large scale. When the pH was maintained between 4.6 and 5.1, phosphorous content in the degummed rapeseed oil could be reduced to less than 10 ppm.

In general, enzymatic degumming minimizes oil retained in the gum, resulting in improved oil yields. The enzymatic hydrolysis reaction typically requires 5–6 hours and utilizes less water than water and acid degumming processes. Reductions in enzyme production costs and enhanced enzyme activity and stability could make enzymatic degumming process very attractive for oil refining industry in the near future.

4.4.2 Deacidification/Refining

Good quality oil contains over 95% neutral lipids, triacylglycerides (TAGs). FFA content of commercial crude oils is usually about 1–3% while high quality crude oils can have only 0.5% or less FFAs. However, palm, olive, rice bran and some specialty oils, such as wheat germ and rice bran oils, may contain 20% or more FFAs. As an industry rule, the FFA content of refined edible oils should be less than 0.05%. Although most of the long chain FFAs do not significantly impair the taste of the oil, the short chain FFAs may have a soapy and rancid flavor. Furthermore, FFAs accelerate oxidation reactions, consequently reducing the oxidative stability of the oils. Hence, removal of FFAs is essential to obtain high quality oil.

4.4.2.1 *Chemical refining*

Crude oils are traditionally deacidified or refined by chemical methods. During chemical refining a heavy soapstock (sodium or potassium salts of fatty acids) is formed. Soapstock is separated from refined oil by gravity settling, filtration or centrifugation. Sodium hydroxide, also referred to as caustic or lye, is widely used for chemical oil refining. The proper strength and amount of lye is critical for achieving high FFA removal with minimal TAG loss and degradation. The presence of color and surface-active compounds in oil interfere with the reaction of FFAs with lye (Mattil *et al.*, 1964). The degummed oil at 30–40 °C is mixed with the required amount of lye and pumped through a high-shear mixer. The mixing time is usually 5–10 min. The oil is then heated to 75 °C and centrifuged to remove soapstock. Soda ash or sodium carbonate can also be used to remove FFAs from crude oil. However, carbon dioxide released during refining causes foaming. In addition, entrainment of gas in the soapstock prevents proper settling. The deacidification with alkali refining process has a number of drawbacks: (1) high oil losses due to oil entrainment in soapstock, (2) production of low value soapstock, (3) high water requirement to wash oil after caustic treatment which leads high waste water volume and disposal costs.

4.4.2.2 *Miscella refining*

In cottonseed, gossypol, a complex polyphenolic compound, contributes to oil toxicity and dark color and is regarded as an undesirable component. During cottonseed processing, gossypol must be removed to produce edible oil and animal feed. Gossypol in crude cottonseed oil is typically removed in miscella (mixture of oil + hexane) before oil desolventation in the extraction plant. The crude oil–hexane mixture (45–65% oil : 35–55% hexane) is filtered to remove any meal, scale or insoluble impurities that may be carried from the extraction process. Next the crude miscella is pumped to a reaction vessel where lye is added and mixed thoroughly until the impurities in the crude oil precipitate in the soap phase. Then the light colored refined miscella is separated from the dark and gummy soapstock by using an explosion-proof centrifuge. The light yellow miscella is pumped to a stripper to recover hexane. Leaving the stripper at 104 °C, the refined oil passes to a pressure leaf-type filter to remove the last traces of soap and any impurities before cooling and entering the storage tank. During miscella refining FFAs and PLs are also removed along with gossypol. The major advantage of miscella refining over conventional refining is that crude miscella is refined immediately after solvent extraction before the solvent stripping has set the color (Hendrix, 1984).

4.4.2.3 *Steam refining*

Steam refining, also referred to as physical refining, is the removal of fatty acids and other volatiles by superheated steam at 200–270 °C at low pressure after preliminary degumming and bleaching steps. Physical refining, which has been adapted by oil refining industry as a viable alternative to the caustic/chemical refining process, is based on the higher volatility of FFAs than TAGs at high temperatures and low pressures. During the process, volatile compounds including FFAs are volatilized and neutral oil droplets are entrained within the stripping steam. The final FFA content in the refined oil can be reduced to 0.005% when physical refining is used. Utilization of a packed column with 4–5 m height and 250 m^2 of packing surface area/m^3 was suggested for steam refining of palm oil (about 30% FFA content) after degumming (Bernardini, 1993). The advantages of the steam refining process include: elimination of objectionable soapstock production, consequently reducing oil losses, production of higher quality FFA and simplified operations. However, during high temperature processing heat labile oil components might be destroyed. It is important to note that phosphorous content of oil has a significant effect on physical refining. Oils with high non-hydratable PLs are not suitable for physical refining. This process is energy intensive because of the high temperatures involved.

4.4.2.4 *Selective removal of FFAs*

Selective removal of FFAs from oil by microorganisms, specifically the Pseudomonas strain (BG1), has been reported. (Cho *et al.*, 1990). BG1 assimilates lauric, myristic, palmitic, stearic, and oleic acids as carbon sources without secreting extracellular lipases. However, short chain fatty acids with less than 12 carbon atoms and linoleic acid are not utilized and sometimes inhibited the growth of BG1. Solubility of fatty acids in water had a significant effect on the rate of FFA removal. Maximum biomass was obtained when BG1 was grown on oleic acid. Even though butyric, valeric, caproic, caprylic and capric acids all have higher solubility in water than oleic acid, they were not utilized by BG1, probably due to the toxicity of short chain fatty acids to microorganisms. Although removal of FFAs by microorganisms is an interesting approach, this technique remains at the research level.

Although it is not widely used, selective solvent extraction is practiced by small operations to neutralize very high FFA content oils. Selective solvent or liquid–liquid extraction of FFAs from TAGs is based on the differences in solubility of FFA and TAG in various organic solvents. Phase separation and selective extraction of FFAs (0–50%) in peanut oil could be achieved by use of aqueous isopropyl alcohol (IPA) (75% and 80%). Phase separation was found to be depended on the fatty acid content in the mixture and the concentration of aqueous IPA (Shah and Venkatesan, 1989). This process can be carried out at room temperature and atmospheric pressure, reducing the energy consumption for oil refining. Because of the large difference between the boiling points of the solvent and lipids, solvent removal from refined oil and extract stream can be easily achieved by evaporation or distillation at relatively low temperatures (less than 80 °C), which could be further reduced to 55 °C by performing the solvent separation under moderate vacuum. A continuous deacidification system using aqueous ethanol (6% water) reduced FFA content of corn oil from 3.5% to less than 0.3% (Pina and Meirelles, 2000). The TAG loss was less than 5%, which is significantly lower than the values reported for alkali or physical refining of corn oil. Selective solvent extraction is a potential alternative partial deacidification process. However, use of a second solvent such as ethanol or isopropyl alcohol besides the hexane used for extraction of oils from oilseeds may require higher capital and energy costs.

4.4.2.5 *Membrane refining*

There have been a number of attempts to utilize ultrafiltration and nanofiltration for deacidification of oils. The molecular weights of FFAs and TAGs are less than 300 Da and larger than 800 Da, respectively. Ideally, hydrophobic nanofiltration membranes with a molecular weight cut-off around 500–600 Da could be used to separate FFAs from TAG. However, in practice the difference in molecular weights of FFAs and TAGs is too narrow for utilization of nanofiltration membranes to separate these two groups of compounds. There have been some efforts to use non-porous denser polymeric membranes for oil refining. FFA rejection of non-porous membranes was 8–27% when sunflower oil used for the experiments (Subramanian *et al.*, 2001). In another study, simultaneous desolventizing and deacidification by using membrane technology was examined. About 40% reduction in FFA content and greater than 50% savings in the energy needed for hexane evaporation have been reported while processing a model system containing 20% soybean oil with 2% FFA (Raman *et al.*, 1996). It appears that the selectivity of the current membranes is not high enough for industrial adoption of membrane deacidification process.

4.4.2.6 *Esterification of FFA*

Identification of microbial lipases capable of synthesizing a TAG molecule from a fatty acid and glycerol has kindled the research on the utilization of these enzymes for oil deacidification. The approach was to esterify FFAs present in crude oil with added glycerol or mono or diacylglycerides (DAG) to produce TAGs. Indeed, the FFA content of rice bran oil was reduced from 30 to 3.6% by esterification of the FFA with added glycerol using a 1,3-specific lipase (Mucor miehei) (Bhattacharyya and Bhattacharyya, 1987). Esterification of crude palm olein FFAs with DAG using a lipase from *Pseudomonas fluorescens* has also been reported (Bhosle and Subramanian, 2005). The extent of FFA esterification was higher when glycerol was replaced with DAG. This result was attributed to the better solubility of DAG in oil. Enzymatic esterification of FFAs with monoglycerides has also been demonstrated in the laboratory. Although enzymatic esterification of FFAs neutralizes oil, increases TAG concentrations, minimizes oil loss and produces a high quality product, low conversion rates and the high cost of enzymes remain an issue for commercialization of this process.

The esterification of FFA can also be done chemically. The reaction is performed at high temperature, usually over 200 °C, with or without a catalyst. Zinc and tin chlorides are suitable catalyst for FFA esterification. FFA concentration in very high FFA content oils (about 40%) could be significantly reduced (to less than 10%, usually 1–3%) by chemical esterification (Bhattacharyya and Bhattacharyya, 1987). Although this process is technically feasible, it is not cost competitive with physical refining process and has not been commercialized yet (De and Bhattacharyya, 1999; Bhosle and Subramanian, 2005).

4.4.2.7 *Removal of FFAs with supercritical fluids*

Supercritical CO_2 has been used to remove FFAs from oils. Selectivity of SC-CO_2 for FFAs is higher than that for TAGs at lower pressures. Based on this concept, different processing schemes have been proposed to extract FFAs and refine oils using SC-CO_2. One approach involves a two-step process during which FFAs, unsaponifiable matter and tocopherols are extracted at low pressure before the TAGs remaining in the feed material are recovered with SC-CO_2 at a higher pressure (Zhao *et al.*, 2000). The disadvantage of this process is that health beneficial bioactive compounds such as tocopherols end up in FFAs rather than TAGs which is the product used for edible applications.

A patented process describes a continuous countercurrent SC-CO_2 fractionation technique for deacidification of crude vegetable oils using a packed column (Dunford and King, 2004). Fractionation at low pressure (138 Bar) and high temperature (80 °C) was effective in removing FFAs from crude rice bran oil without any oryzanol loss in the extract fraction (Dunford and King, 2000, 2001; Dunford *et al.*, 2002). A columnar fractionation technique was also successful in removing FFAs from crude wheat germ oil and concentrating health beneficial bioactive compounds in the TAGs fraction (Eisenmenger *et al.*, 2006). Deacidification of soybean oil deodorizer distillate could be achieved using SC-CO_2 technology (King and Dunford, 2001). Lampante olive oil refining (Bondioli *et al.*, 1992) and deacidification of roasted peanut (Ziegler and Liaw, 1993) and olive oil (Brunetti *et al.*, 1989) using the SC-CO_2 technique have also been demonstrated. These studies showed that FFA solubility in SC-CO_2 was directly related to the degree of unsaturation and deodorization was mass transfer controlled, while deacidification was thermodynamically constrained.

Supercritical CO_2 is an excellent solvent to remove FFAs from crude oil. The capital and processing costs of a SC-CO_2 deacidification process would be relatively lower than those for oil extraction because of the higher solubility and selectivity of FFAs in SC-CO_2 at lower pressures than those of TAGs. Certainly this technology has a lot of potential for oil refining, specifically for high value specialty products.

4.4.2.8 *Removal of FFA by adsorption*

Adsorption processes have also been examined to remove FFAs from oils. A process which utilizes magnesium oxide as adsorbent to remove FFAs from oils has been patented (Munson *et al.*, 1995). Aluminum hydroxide gel is also effective for removing FFAs. Mechanically expressed soybean oil was successfully deacidified by using Magnesol®, a commercial adsorbent designed to remove FFA from oils. The FFA content of the oil was reduced to 0.04% by adsorption with 3% Magnesol®. This material also adsorbed primary and secondary oil oxidation products. The Magnesol refining process was much milder than conventional refining, as evidenced by very small amount of primary and secondary lipid oxidation product formation and less loss of tocopherols (Wang and Johnson, 2001).

4.4.3 Bleaching

Chlorophyll or pheophytin and their degradation products give oil a greenish color and carotenoids contribute to yellow and red tints that may be undesirable. Thus, oils are usually bleached after deacidification/refining and before deodorization. Originally bleaching was used to remove color compounds. Today, bleaching is designed to remove undesirable oil components, including peroxides, aldehydes, ketones, phosphatides, oxidative trace metals, soaps and other contaminants, such as pesticides and polycyclic aromatic hydrocarbons.

Clays used for bleaching are commonly called "Bentonites". Activated carbon, alumina, silicic acid, aluminum- and magnesium-silicate, silica gel and synthetic silicates are also used to adsorb impurities from refined oil (Neuman and Dunford, 2004). Although natural clays do not remove color, they are very effective in cleaning up residual soap from alkali refining and gums from degumming, and normally retain lower amount of oil as compared to activated clays (Rini, 1960). Bleaching is normally carried out under vacuum (20–30 mm Hg) to minimize oxidation reactions and control moisture levels. Preheated oil (90 °C) is pumped into a slurry tank and adsorbent is added to the tank simultaneously. After mixing, the clay/oil system is fed into a vacuum bleacher. Bleaching process takes 15–30 min in a temperature range of 80–120 °C. Although high temperature increases adsorption efficiency, bleaching at very high temperatures is not recommended because it promotes undesirable reactions. Temperature should be high enough to maintain a low oil viscosity, which improves diffusion and mass transfer rates.

Wet bleaching is practiced when processing oils containing PLs. Water will act as a carrier for PLs into the bleaching clay particles. The optimal amount of water used for wet bleaching is about 50–100% of the adsorbent used for the process. Initially, oil (about 0.5% moisture) is treated with water and adsorbent (8–15% moisture) at 70–90 °C for 20 min under atmospheric conditions. Then bleaching is carried out under vacuum for 15–30 min. The amount of adsorbent required for bleaching depends on the types of adsorbent and the oil and its pretreatment. The adsorbent dosage range varies significantly, usually 0.1–2.0% (of oil processed) depending on the oil quality and type of adsorbent used. In some cases it can be as high as 5%. Physically refined oils require a higher amount of adsorbent than chemically refined oils. After bleaching, oil is filtered and separated from the adsorbent.

Currently bleaching is performed in slurry tanks. However, fixed-bed bleaching using granular clays is an option. At present, there are a few applications of fixed-bed post-bleach "polishing" before storage or shipment. Fixed-bed bleaching requires highly purified feed streams, that is, no soaps, no phospholipids and low particulate content. Development of regenerable bleaching clays could reduce the impact of oil refining on the environment. Patents do exist covering bleaching clay regeneration technologies but currently these technologies are not economically feasible.

There are several relatively new techniques being examined to replace conventional bleaching or at least to combine bleaching with other processing steps. In one study the bleaching step was incorporated into a SC-CO_2 extraction process by integrating an adsorbent-containing high-pressure vessel after the oil extractors in the process cycle (Woerlee and Pellikaan, 2003; Ambrogi *et al.*, 2003). Although the contact time between oil and the adsorbent in the reactor was very short, a significant bleaching effect was achieved using this process. Furthermore, the used adsorbent did not retain oil. During the process the peroxide value of the oil was significantly reduced, indicating that catalytic decomposition of hydroperoxides on bleaching earth takes place under high pressure. Ketones, aldehydes and FFAs were not adsorbed and they were detected in the bleached oil. Oil extracted with this method needs to be further refined to remove FFAs. Currently this technology would be suitable for specialty oils due to the high cost of high-pressure equipment and low throughputs.

Application of an electrostatic field through a non-conducting liquid such as oil facilitates agglomeration of electrically charged bleaching earth on one electrode and oil leaves the electrofilter completely clear (Transfeld, 1998). Finely ground bleaching earth is found to be more effective than coarser bleaching earth in electrofiltration. Removal of finely ground clay from oil is not feasible by the current filtration techniques but bleaching earth can be removed as sludge by discontinuing the electric field in the bleacher. It has been reported that this technology reduces the amount of bleaching clay used and power requirement for filtration (Transfeld, 1998). Other advantages of this technique are: easier filter cleaning and reduction of pre-coat filter area because of the formation of a highly porous filter cake by the agglomerated bleaching clay. An electrofilter/agglomerator can be fitted in a conventional bleaching unit.

Activation of bleaching clays with microwave treatment has been investigated (Boukerroui and Ouali, 2002). Clay was impregnated with ammonium chloride then activated in a microwave oven. Optimum conditions for activation were achieved at 3 M ammonium chloride concentration and 15 min heating time. Microwave activation improved chlorophyll removal efficiency from the clay as compared to strong acid activation. This technique presents two important advantages over the strong acid activation method: shorter activation time and elimination of sulfuric acid, which is a toxic waste.

4.4.4 Deodorization

The presence of FFAs in oil correlates well with flavor and odor. The practical rule is that if an oil has 0.1% FFA it will have an odor which could be eliminated by reducing the FFA content to 0.01–0.03% (zero peroxide value) (Gavin, 1978). Deodorization is a steam distillation process during which volatile and odoriferous compounds are stripped off with steam. The objective of deodorization is to produce a bland and stable product by removing FFAs, aldehydes, ketones, and peroxides from bleached oil. The amount of FFA removed from the oil is inversely proportional to the system pressure and directly proportional to the vapor pressure of the FFA and the sparge steam rate. Temperature plays a critical role during deodorization. The rate at which odor compounds are removed could be tripled when deodorization temperature increased from 178 to 205 °C. If the temperature is further raised to 232 °C, that rate can be expected to triple again. This means higher deodorization temperature reduces processing time (Dudrow, 1983). However, high temperatures cause development of undesirable polymers. Hence, optimization of time and temperature is necessary for a given process. High vacuum is desirable for deodorization because it inhibits oil hydrolysis. The volume of stripping steam needed in steam deodorizers is affected by vacuum. For example, a deodorizer operating at 12 mm Hg pressure would require twice the stripping steam of a unit operated at 6 mm Hg. Currently, 6 mm Hg vacuum is commonly used for vegetable oil deodorizers. A relatively new approach is to use nitrogen instead of steam for sparging. It has been demonstrated that this process improved oil quality.

A patent describes a continuous process for deodorizing oils with carbon dioxide at 150–250 °C and 100–250 Bars (Zosel, 1979). Impurities extracted with carbon dioxide are removed by a solid adsorbent, such as activated charcoal, before the carbon dioxide is recycled to the deodorizer column. The FFA content of soybean oil could be lowered from 0.4 to 0.02% by using this technology.

A recently identified food contaminant, 3-chloropropane-1,2-diol (3-MCPD-esters) is formed from glycerol or acylglycerols and chloride ions. For refined vegetable oils the average content of 3-MCPD-esters increases in the order rapeseed oil, soybean oil, sunflower oil, safflower oil, walnut oil to palm oil. Deodorization was clearly identified as the critical step in the formation of 3-MCPD-esters (Weißhaar, 2008). It appears that a significant reduction of 3-MCPD-esters in foods is not simple but current research and development efforts are expected to bring solutions to this issue.

4.4.5 Winterization

Winterization is a separation process by which higher melting point acylglycerides and waxes, which are responsible for the turbidity of some edible oils at cold temperatures, are crystallized and removed (Leibovitz and Ruckenstein, 1984). Composition of the oil, rate of cooling, temperature of crystallization and mobility of TAG molecules in the oil are critical factors affecting efficiency of winterization. These factors play a significant role both in formation of the solid phase and then separation of the solids from the liquid portion. The edible oil industry utilizes the liquid fraction to make high quality salad oils whereas the solid fraction is used in shortening or margarine formulations. During the winterization process, the oil is cooled from room temperature to a predetermined temperature of crystallization. The cooled oil is kept at this temperature for a certain period prior to the separation of solid phase from the liquid oil by filtration of the oil-solid fat slurry. In a winterization process, cooling rate and temperature of crystallization are extremely important. Too low a temperature and high cooling rates will result in high viscosity and reduce crystal growth rate. A mild agitation is recommended to provide a gentle motion to the crystals to enhance their growth rate and keep the temperature and composition uniform in the bulk oil. The agitator design should be such that no shear to break the crystals is generated. In commercial winterization operations crystal modifiers or an appropriate solvent are used to facilitate filtration of solid phase from the liquid oil.

The presence of high concentrations of waxes and mucilagenous materials in the oil makes winterization of some oils difficult. In such cases filtration of the oil is hindered by the high viscosity of the cold oil, the lack of uniformity in crystal formation of the waxes, and the presence of mucilaginous materials which coat the wax crystals (Morrison and Robertson, 1975). Washing crude oil with emulsifiers removes waxes and the majority of PLs and helps with the subsequent winterization process. Solvent winterization from acetone or an 85/15 acetone–hexane mixture is shown to be effective in peanut and sunflower oil winterization by reducing viscosity. In addition, small amounts of diatomaceous earth addition increased the filtration rate of cold sunflower seed oil by a factor of four.

4.5 CONCLUSIONS

The oil and oilseed processing industry has gone through many changes over the last two decades. Major developments in the industry have been technical improvements in unit operations and equipment. Changes in technology have been in reaction to the more strict environmental regulations imposed on processors, industry consolidation, consumer awareness and agricultural and biotechnical advances. Liquid or solid waste generated during oil and oilseed processing has always been a major concern. Today solvent emissions and waste disposal, that is spent bleaching clay, waste water, soapstock and deodorizer distillate, have been strictly regulated. The general spills and leaks of products in the plant are now considered as a serious ground water pollutant in connection with storm water run-off. Many facilities have redesigned their drainage systems and installed on-site water treatment and oil recovery systems.

Use of expanders to prepare oilseeds for extraction, new developments in screw press design, installation of efficient heat recovery systems and integration of automated filters, self-cleaning centrifuges and electronic systems with programmable logic controllers (PLCs) for automation of plant operations in processing plants have improved process efficiencies considerably. The thin film deodorizers first used for palm oil processing is being slowly adapted for processing other commodity oils. The new countercurrent desolventizers use significantly less steam than the conventional ones and reduced hexane concentration in the meal from 500–700 ppm to 100–250 ppm.

Today protection of oil from oxidation and deterioration during processing is a priority. This has been achieved by cooling oils between processes and storing them under nitrogen. Although this has resulted in increased costs in energy and cooling water, installation of improved heat and cooling water recovery systems in processing plants offset these costs (Carlson and Scott, 1991). Minimization of tocopherol losses during processing also helped to protect oil because of the antioxidant properties of tocopherols.

Consumer concerns about the use of chemicals for oil and oilseed processing encouraged processors to look at the alternative extraction and refining techniques, such as enzyme-aided processing and physical refining techniques, more closely. Integration of extraction and refining facilities reduced overall operating costs, created more opportunities for energy and waste management and effluent recycling and allowed immediate refining freshly extracted oil at the same plant reducing transportation costs and product losses, consequently improving product quality. Technical, economic, social and environmental sustainability of oil and oilseed processing operations is an ongoing dialogue. More and more companies are implementing sustainable practices in processing plants and making product development, feedstocks purchasing and other business decisions based on sustainability and offering healthier products to meet consumer demands. It is expected that advancements in biotechnology and bioprocessing techniques will have a significant positive impact on the industry profit margins and the types and quality of the products offered to the consumers.

REFERENCES

Abdulkarim SM, Lai OM, Muhammad SKS, Long K, Ghazali HM. 2006. Use of Enzymes to Enhance Oil Recovery During Aqueous Extraction of Moringa Oleifera Seed Oil. J. Food Lipids 13:113–130.

Abraham G, Hron R, Koltun S. 1988. Modeling the Solvent Extraction of Oilseeds. J. Am. Oil Chem. Soc. 65(1):129–135.

Ambrogi A, Cardarelli DA, Eggers R. 2003. Separation of Natural Colorants Using a Combined High Pressure Extraction-Adsorption Process. Lat. Am. Appl. Res. 33(3):323–326.

Becker W. 1978. Solvent Extraction of Soybeans. J. Am. Oil Chem. Soc. 55(11):754–761.

Berger KG. 1985. Quality Control in Storage and Transport of Edible Oils. J. Am. Oil Chem. Soc. 62(2):438–442.

Bernardini E. 1976. Batch and Continuous Solvent Extraction. J. Am. Oil Chem. Soc. 53(6):275–278.

Bernardini M. 1993. Deodorization. In: Applewhite, T.A. (editor). Proceedings of the World Conference on Oilseed Technology and Utilization. Budapest, Hungary. Champaign, IL: AOCS Press. pp. 186–193.

Bhattacharyya AC, Bhattacharyya DK. 1987. Deacidification of High FFA Rice Bran Oil by Reesterification and Alkali Neutralization. J. Am. Oil Chem. Soc. 64:128–131.

Bhosle BM, Subramanian R. 2005. New Approaches in Deacidification of Edible Oils – a Review. J. Food Eng. 69(4):481–494.

Bocevska M, Kariovic D, Turkulov J, Pericin D. 1993. Quality of Corn Germ Oil Obtained by Aqueous Enzymatic Extraction. J. Am. Oil Chem. Soc. 70(12):1273–1277.

Bockisch M. 1998. Fats and Oils Handbook. Champaign, IL: AOCS Press.

Bondioli P, Mariani C, Lanzani A, Fedeli E, Mossa A, Muller A. 1992. Lampante Olive Oil Refining with Supercritical Carbon Dioxide. J. Am. Oil Chem. Soc. 69(5):477–480.

Boukerroui A, Ouali M. 2002. Edible Oil Bleaching with Bentonite Activated by Micro Wave Irradiation. Ann. Chim. Sci. Mat. 27(4):73–81.

Bredeson D. 1977. Mechanical Pressing. J. Am. Oil Chem. Soc. 54(6):a489–a490.

Brunetti LA, Daghetta E, Fedeli E, Kikic I, Zanderighi L. 1989. Deacidification of Olive Oils by Supercritical Carbon Dioxide. J. Am. Oil Chem. Soc. 66(2):209–212.

Carlson KF, Scott JD. 1991. Recent Developments and Trends Processing of Oilseeds, Fats an Oils. Inform 2(11):1034–1060.

Chen Man YB, Suhardiyono AAB, Azudin MN, Wei LS. 1996. Aqueous Enzymatic Extraction of Coconut Oil. J. Am. Oil Chem. Soc. 73:683–686.

Cho SY, Kwon TW, Yoon SH. 1990. Selective Removal of Free Fatty Acids in Oils Using a Microorganism. J. Am. Oil Chem. Soc. 67:558–560.

Choukri A, Kinany M, Gibon V, Tirtiaux A, Jamil S. 2001. Improved Oil Treatment Conditions for Soft Degumming. J. Am. Oil Chem. Soc. 78(11):1157–1160.

Christensen P. 1983. Solvent Extraction: Recent Developments. J. Am. Oil Chem. Soc. 60(2):214–215.

Christian M. 2010. Rapeseed/Canola. Biotechnology in Agriculture and Forestry 64:409–422.

Clark SP, Baker GW, Wan PJ, Matlock SW. 1980. Separating Kernels and Hulls from Oilseed Type Sunflower Seed. Trans. ASABE 23(1):20–0024.

Clausen K. 2001. Enzymatic Oil-Degumming by a Novel Microbial Phospholipase. Eur. J. Lipid Sci. Technol. 103(6):333–340.

Dahlke K, Eichelsbacher M. 1998. Enzymax® and Alcon® – Lurgi's Route to Physical Refining. In: Koseoglu, S.S., Rhee, K.C. and Wilson, R.F. (editors). Emerging Technologies, Current Practices, Quality Control, Technology Transfer, and Environmental Issues. Champaign, IL: AOCS Press. pp. 53–59.

De BK, Bhattacharyya DK. 1999. Deacidification of High-Acid Rice Bran Oil by Reesterification with Monoglyceride. J. Am. Oil Chem. Soc. 76:1243–1246.

De BK, Das R, Dutta BK, Bhattacharyya DK. 1998. Membrane Degumming and Dewaxing of Rice Bran Oil and Its Refining. Lipid/Fett 100(9):416–421.

Dickey LC, Kurantz MJ, Parris N. 2008. Oil Separation from Wet-Milled Corn Germ Dispersions by Aqueous Oil Extraction and Aqueous Enzymatic Oil Extraction. Industrial Crops and Products 27:303–307.

Dickinson JT. 1947. Fractionation of Fatty Materials. US Patent 2 660 590.

Diekert JW, Diekert MC. 1976. The Chemistry and Cell Biology of the Vacuolar Proteins of Seeds. J. Food Sci. 41:475–482.

Dijkstra AJ. 2009. Recent Developments in Edible Oil Processing. Eur. J. Lipid Sci. Technol. 111: 857–864.

Do L, Sabatini D. 2010. Aqueous Extended-Surfactant Based Method for Vegetable Oil Extraction: Proof of Concept. J. Am. Oil Chem. Soc. 87(10):1211–1220.

Domínguez H, Núñez MJ, Lema JM. 1994. Enzymatic Pretreatment to Enhance Oil Extraction from Fruits and Oilseeds: A Review. Food Chemistry 49:271–286.

Domínguez H, Núñez MJ, Lema JM. 1995. Aqueous Processing of Sunflower Kernel with Enzymatic Technology. Food Chemistry 53:427–434.

Dudrow F. 1983. Deodorization of Edible Oil. J. Am. Oil Chem. Soc. 60(2):272–274.

Duncan I. 1948. Commercial Extraction of Vegetable Oils by Means of Trichlorethylene. J. Am. Oil Chem. Soc. 25(8):277–278.

Dunford NT. 1995. Use of Supercritical Carbon Dioxide for Edible Oil Processing. Department of Agricultural, Food and Nutritional Science. Edmonton, AB, Canada: University of Alberta.

Dunford NT, King JW. 2000. Phytosterol Enrichment of Rice Bran Oil by a Supercritical Carbon Dioxide Fractionation Technique. J. Food Sci. 65:1395–1399.

Dunford NT, King JW. 2001. Thermal Gradient Deacidification of Crude Rice Bran Oil Utilizing Supercritical Carbon Dioxide. J. Am. Oil Chem. Soc. 78:121–125.

Dunford NT, King JW. 13 January 2004. Supercritical Fluid Fractionation Process for Phytosterol Ester Enrichment in Vegetable Oils. US Patent 6,677,469 B1.

Dunford NT, Temelli F. 1996. Effect of Supercritical CO_2 on Myrosinase Activity and Glucosinolate Degradation in Canola. J. Agric. Food Chem. 44:2372–2376.

Dunford NT, Temelli F. 1997. Extraction Conditions and Moisture Content of Canola Flakes as Related to Lipid Composition of Supercritical CO_2 Extracts. Journal of Food Science 62(1):155–159.

Dunford NT, Teel JA, King JW. 2002. A Continuous Countercurrent Supercritical Fluid Deacidification Process for Phytosterol Ester Fortification in Rice Bran Oil. Food Research Intenational 36(2):175–181.

Dunford NT, King JW, List GR. 2003. Supercritical Fluid Extraction in Food Engineering. In: Tzia, C. and Liadakis, G. (editors). Extraction Optimization in Food Engineering. NY: Marcel Dekker, Inc. pp. 57–93.

Eisenmenger M, Dunford NT, Eller F, Taylor S, Martinez J. 2006. Pilot Scale Supercritical Carbon Dioxide Extraction and Fractionation of Wheat Germ Oil. J. Am. Chem. Soc. 10(2)(2):863–868.

EPA. 1990. Modifications to the 112(B)1 Hazardous Air Pollutants. Office of Air Quality Planning & Standards. http://www.epa.gov/ttn/atw/atwsmod.html (accessed 8 November 2011).

Fattori M, Bulley RN, Meisen A. 1988. Carbon Dioxide Extraction of Canola Seed: Oil Solubility and Effect of Seed Treatment. J. Am. Oil Chem. Soc. 65(6):968–974.

Favati F, King JW, Mazzanti M. 1991. Supercritical Carbon Dioxide Extraction of Evening Primrose Oil. J. Am. Oil Chem. Soc. 68(6):422–427.

Friedrich JP. 1984. Supercritical CO_2 Extraction of Lipids from Lipid-Containing Materials,. US Patent 4,466,923.

Friedrich JP, Pryde EH. 1984. Supercritical CO_2 Extraction of Lipid-Bearing Materials and Characterization of the Products. J. Am. Oil Chem. Soc. 61(2):223–228.

Friedrich JP, List GR, Spencer GG. 1998. Semicontinuous Supercritical CO_2 System for Rapid Extraction of Jojoba and Other Oilseeds. In: Baldwin, A.R. (editor). 7th International Conference on Jojoba and Its Uses. Champaign, IL: AOCS. pp. 165–172.

Gavin A. 1978. Edible Oil Deodorization. J. Am. Oil Chem. Soc. 55(11):783–791.

Gomes MSO, Sinnecker P, Tanaka RT, Lanfer-Marquez UM. 2003. Effect of Harvesting and Drying Conditions on Chlorophyll Levels of Soybean (*Glycine Max* L. Merr). J. Agric. Food Chem. 51(6):1634–1639.

Gunstone FD. 2004. Rapeseed and Canola Oil: Production, Processing, Properties and Uses. Boca Raton, FL: CRC Press LLC.

Gupta AKS. 1977. Process for Refining Crude Glyceride Oils by Membrane Filtration. US Patent 4,062,882.

Gupta M. 2008. Practical Guide for Vegetable Oil Processing. Urbana, IL: AOCS Press.

Hanmoungjai P, Pyle DL, Niranjan K. 2002. Enzyme-Assisted Water-Extraction of Oil and Protein from Rice Bran. J. Chem. Technol. Biotech. 77:771–776.

Hendrix WM. 1984. Current Practices in Continuous Cottonseed Miscella Refining. J. Am. Oil Chem. Soc. 61(8):1369–1372.

Iwama A. 1987 New Process for Purifying Soybean Oil by Membrane Separation and Economical Evaluation of the Process. J. Am. Oil Chem. Soc. 64:1258.

Johnson LA, Lusas EW. 1983. Comparison of Alternative Solvents for Oils Extraction. J. Am. Oil Chem. Soc. 60(2):229–242.

Johnson L, Farnsworth J, Sadek N, Chamkasem N, Lusas E, Reid B. 1986. Pilot Plant Studies on Extracting Cottonseed with Methylene Chloride. J. Am. Oil Chem. Soc. 63(5):647–652.

Kalra H, Chung SYK, Chen CJ. 1987. Phase Equilibrium Data for Supercritical Extraction of Lemon Flavors and Palm Oils with CO_2. Fluid Phase Equil. 36:263–278.

Khan LM, Hanna MA. 1983. Expression of Oil from Oilseeds – a Review. Journal of Agricultural Engineering Research 28(6):495–503.

King JW, Dunford NT. 2001. Phytosterol-Enriched Triglyceride Fractions from Vegetable Oil Deodorizer Distillates Utilizing Supercritical Fluid Fractionation Technology. Separation Science and Technology 37(2):451–462.

King JW, List GR. 1996. Supercritical Fluid Technology in Oil and Lipid Chemistry. Champaign, IL: AOCS Press.

Lamsal B, Murphy P, Johnson L. 2006. Flaking and Extrusion as Mechanical Treatments for Enzyme-Assisted Aqueous Extraction of Oil from Soybeans. J. Am. Oil Chem. Soc. 83:973–979.

Latif S, Diosady LL, Anwar F. 2008. Enzyme-Assisted Aqueous Extraction of Oil and Protein from Canola (Brassica Napus L.) Seeds. Eur. J. Lipid Sci. Technol. 110:887–892.

Leibovitz Z, Ruckenstein C. 1984. Winterization of Sunflower Oil. J. Am. Oil Chem. Soc. 61(5):870–872.

Levine L. 1993. Musing on the Mechanics of Flaking Rolls. Cereal Foods World 38:872–874.

Lin L, Rhee KC, Koseoglu SS. 1997. Bench-Scale Membrane Degumming of Crude Vegetable Oil: Process Optimization. Journal of Membrane Science 134(1):101–108.

Mattil KF, Norris FA, Swern D. 1964. Bailey's Industrial Oil and Fat Products, 3rd edn. New York: Interscience Publishers.

McHugh MA, Krukonis VJ. 1994. Supercritical Fluid Extraction Principles and Practice, 2nd edn. Boston, MA: Butterworth-Heinemann.

Miller N, de Villiers JBM, Peulen JMJ. 1986. A Process for the Dehulling of High Oil-Content Sunflower Seeds. Fette Seifen Anstrich. 7:268–271.

Moreau R, Johnston D, Powell M, Hicks K. 2004. A Comparison of Commercial Enzymes for the Aqueous Enzymatic Extraction of Corn Oil from Corn Germ. J. Am. Oil Chem. Soc. 81:1071–1075.

Morrison W, Robertson J. 1975. Solvent Winterization of Sunflower Seed Oil. J. Am. Oil Chem. Soc. 52(5):148–150.

Mostafa MM, Rahma EH, Rady AH. 1987. Chemical and Nutritional Changes in Soybean During Germination. Food Chemistry 23(4):257–275.

Moulton K, Mounts T. 1990. Continuous Ultrasonic Degumming of Crude Soybean Oil. J. Am. Oil Chem. Soc. 67(1):33–38.

de Moura J, Johnson L. 2009. Two-Stage Countercurrent Enzyme-Assisted Aqueous Extraction Processing of Oil and Protein from Soybeans. J. Am. Oil Chem. Soc. 86(3):283–289.

Mrema GC, McNult PB. 1984. Microstructure of Rapeseed and Cashew as Related to Mechanical Oil Expression. Ir. J. Fd Sci Technol. 8:59–66.

Mrema GC, McNulty PB. 1985. Mathematical Model of Mechanical Oil Expression from Oilseeds. J. Agric. Eng. Res. 31:361–370.

Munson JR, Bertram BL, Caldwell JD (The Dallas Group of America, Inc.). 1995. Treatment of Cooking Oils and Fats with Magnesium Silicate and Alkali Materials. US Patent 5597600.

Naksuk A, Sabatini DA, Tongcumpou C. 2009. Microemulsion-Based Palm Kernel Oil Extraction Using Mixed Surfactant Solutions. Industrial Crops and Products 30(2):194–198.

National Oilseed Processors Association. 2010. National Oilseed Processors Association Yearbook and Trading Rules. http://www.nopa.org/content/trading/NOPA%20SBM%20Trading%20Rules_amended%20Feb%2006_updated%20labs%20Nov%202010.pdf (accessed 16 February 2011).

Neuman T, Dunford NT. 2004. Edible Oil Bleaching. In: Dunford, N.T. and Dunford, H.B. (editors). Nutritionally Enhanced Oil and Oilseed Processing. Champaign, IL: AOCS Press. pp. 148–160.

Palmer GH, Fanwood NJ. 1950. Refining Fatty Oils. US Patent 2 658 907.

Pierce RM. 1970. Sunflower Processing Techniques. J. Am. Oil Chem. Soc. 47:248A–269A.

Pina CG, Meirelles AJA. 2000. Deacidification of Corn Oil by Solvent Extraction in a Perforated Rotating Disc Column. J. Am. Oil Chem. Soc. 77:553–559.

Raman L, Cheryan M, Rajagopalan N. 1996. Deacidification of Soybean Oil by Membrane Technology. J. Am. Oil Chem. Soc. 73(2):219–224.

Rask L, Andréasson E, Ekbom B, Eriksson S, Pontoppidan B, Meijer J. 2000. Myrosinase: Gene Family Evolution and Herbivore Defense in Brassicaceae. Plant Molecular Biology 42:93–114.

Reverchon E, Osséo L. 1994. Comparison of Processes for the Supercritical Carbon Dioxide Extraction of Oil from Soybean Seeds. J. Am. Oil Chem. Soc. 71(9):1007–1012.

Ribeiro APB, Bei N, Goncalves LAG, Petrus JCC, Viotto LA. 2008. The Optimisation of Soybean Oil Degumming on a Pilot Plant Scale Using a Ceramic Membrane. J. Food Eng. 87:514–521.

Rini SJ. 1960. Refining, Bleaching, Stabilization, Deodorization, Plasticization of Fats, Oils, and Shortening. J. Am. Oil Chem. Soc. 37:512–520.

Rosenthal A, Pyle DL, Niranjan K. 1996. Aqueous and Enzymatic Processes for Edible Oil Extraction. Enzyme Microb. Technol. 19:402–420.

Rosenthal A, Pyle DL, Niranjan K. 1998. Simultaneous Aqueous Extraction of Oil and Protein from Soybean: Mechanisms for Process Design. Food Bioprod. Process. 76:224–230.

Rosenthal A, Pyle DL, Niranjan K, Gilmour S, Trinca L. 2001. Combined Effect of Operational Variables and Enzyme Activity on Aqueous Enzymatic Extraction of Oil and Protein from Soybean. Enzyme and Microbial Technology 28:499–509.

Rukunudin IH, Bern CJ, Misra MK, Bailey TB. 2004. Carbon Dioxide Evolution from Fresh and Preserved Soybeans. Trans. ASABE 47(3):827–833.

Sant'Anna BPM, Freitas SP, Coelho MAZ. 2003. Enzymatic Aqueous Technology for Simultaneous Coconut Protein and Oil Extraction. Grasas Y Aceites 54:77–80.

Segers JC. 1982. Superdegumming, a New Degumming Process and its Effect on the Effluent Problems of Edible Oil Refining. Fette Seifen, Anstrich. 84(S1):543–546.

Segers JC, van der Sande RLKM. 1990. Degumming–Theory and Practice. Proceedings of the World Conference on Edible Fats and Oils Processing: Basic Principles and Modern Practices. Champaign, IL: AOCS. pp. 88–93.

Seto TA, Schultze MO, Perman V, Bates FW, Sautter JH. 1958. Toxicity in Oil Meal, Properties of the the Toxic Factor in Trichloroethylene-Extracted Soybean Oil Meal. J. Agric. Food Chem. 6(1):49–54.

Shah KJ, Venkatesan TK. 1989. Aqueous Isopropyl Alcohol for Extraction of Free Fatty Acids from Oils. J. Am. Oil Chem. Soc. 66:783–787.

Sharma A, Khare SK, Gupta MN. 2002. Enzyme-Assisted Aqueous Extraction of Peanut Oil. J. Am. Oil Chem. Soc. 79:215–218.

Singha PP, Maiera DE, Okosa MR, Cattanachb E, Trumbleb KP. 1999. Effects of Physical Properties and Operating Parameters on Soybean Flaking. J. Am. Oil Chem. Soc. 76(8).

Snyder M, Friedrich JP, Christianson DD. 1984. Effect of Moisture and Particle Size on the Extractability of Oils from Seeds with Supercritical CO_2. J. Am. Oil Chem. Soc. 61(12):1851–1856.

Stahl E, Schutz E, Mangold HK. 1980. Extraction of Seed Oils with Liquid and Supercritical Carbon Dioxide. J. Agric. Food Chem. 23:1153–1157.

Stahl E, Quirin K-W, Gerard D. 1983. Solubilities of Soybean Oil, Jojoba Oil and Cuticular Wax in Dense Carbon Dioxide. Fette Seifen Anstrich. 85:458–463.

Stahl E, Quirin K-W, Blagrove RJ. 1984. Extraction of Seed Oils with Supercritical Carbon Dioxide: Effect on Residual Proteins. J. Agric. Food Chem. 32:938–940.

Stahl E, Quirin K-W, Gerard D. 1987. Dense Gasses for Extraction and Refining. Berlin-Heidelberg: Springer-Verlag.

Subramanian R, Nakajima M. 1997. Membrane Degumming of Crude Soybean and Rapeseed Oils. J. Am. Oil Chem. Soc. 74(8):971–975.

Subramanian R, Nakajima M, Yasui A, Nabetani H, Kimura T, Maekawa T. 1999. Evaluation of Surfactant-Aided Degumming of Vegetable Oils by Membrane Technology. J. Am. Oil Chem. Soc. 76(10):1247–1253.

Subramanian R, Raghavarao KSMS, Nabetani H, Nakajima M, Kimura T, Maekawa T. 2001. Differential Permeation of Oil Constituents in Nonporous Denser Polymeric Membranes. J. Membr. Sci. 187:57–69.

Sutherland JW, Ghaly TF. 1982. Heated-Air Drying of Oilseeds. Journal of Stored Products Research 18(2):43–54.

Transfeld P. 1998. Electrofiltration Cuts Bleaching Earth Use. Inform 9:502–506.

Ugolini L, De Nicola G, Palmieri S. 2008. Use of Reverse Micelles for the Simultaneous Extraction of Oil, Proteins, and Glucosinolates from Cruciferous Oilseeds. J. Agric. Food Chem. 56(5):1595–1601.

Venter MJ, Willems P, Kuipers NJM, Haan ABd. 2006. Gas Assisted Mechanical Expression of Cocoa Butter from Cocoa Nibs and Edible Oils from Oilseeds. The Journal of Supercritical Fluids 37(3):350–358.

Vitzthum O, Hubert P (HAG AG). 1971. German Patent 21 27 596.

Wang T, Johnson L. 2001. Natural Refining of Extruded-Expelled Soybean Oils Having Various Fatty Acid Compositions. J. Am. Oil Chem. Soc. 78(5):461–466.

Ward J. 1976. Processing High Oil Content Seeds in Continuous Screw Presses. J. Am. Oil Chem. Soc. 53(6):261–264.

Weißhaar R. 2008. 3-MCPD-Esters in Edible Fats and Oils – a New and Worldwide Problem. Eur. J. Lipid Sci. Technol. 110(8):671–672.

Wijeratnea WB, Wang T, Johnson LA. 2004. Extrusion-Based Oilseed Processing Methods. In: Dunford, N.T. and Dunford, H.B. (editors). Nutritionally Enhanced Oil and Oilseed Processing. Champaign, IL: AOCS Press. pp. 71–88.

Woerfel JB. 1995. Extraction. In: Erickson, D.R. (editor). Practical Handbook of Soybean Processing and Utilization. Champaign, IL: AOCS Press. pp. 65–92.

Woerlee GF, Pellikaan HC. 2003. Bleaching of Lipid Materials Using Carbon Dioxide as Moderator. In: Brunner, G. and Perrut, M. (editors). 6th Internation Symposium on Supercritical Fluids. Versailles, France: International Society for Advancement of Supercritical Fluids (ISASF). pp. 261–264.

Xie M, Dunford NT, Goad C. 2011. Enzymatic Extraction of Wheat Germ Oil. J. Am. Oil Chem. Soc. DOI: 10.1007/s11746-011-1861-5.

Xu L, Diosady LL. 2004. Degumming. In: Dunford, N.T. and Dunford, H.B. (editors). Nutritionally Enhanced Edible Oil and Oilseed Processing. Champaign, IL: AOCS Press. pp. 117–147.

Yang B, Wang Y-H, Yang J-G. 2006. Optimization of Enzymatic Degumming Process for Rapeseed Oil. J. Am. Oil Chem. Soc. 83(7):653–658.

Young FVK, Poot C, Biernoth E, Krog N, Davidson GJ, Gunstone FD. 1994. Processing of Fats and Oils. In: Gunstone, F.D., Harwood, J.L. and Padley, F.B. (editors). The Lipid Handbook, 2nd edn. London: Chapman and Hall. pp. 249–276.

Zhang SQ, Kutowy O, Kumar A. 1996. Application of Ultrafiltration Membranes in Refining of Crude Vegetable Oils. Proceedings of the 8th Annual Meeting of North American Membrane Society. Ottawa, Canada. p. 95.

Zhang S, Wang Z, Xu S. 2007. Optimization of the Aqueous Enzymatic Extraction of Rapeseed Oil and Protein Hydrolysates. J. Am. Oil Chem. Soc. 84:97–105.

Zhao Y, Sheng G, Wang D. 2000. Pilot-Scale Isolation of Tocopherols and Phytosterols from Soybean Sludge in a Packed Column Using Supercritical Carbon Dioxide. 5th International Symposium on Supercritical Fluids (ISSF). Atlanta, Georgia.

Ziegler GR, Liaw YJ. 1993. Deodorization and Deacidification of Edible Oils with Dense Carbon Dioxide. J. Am. Oil Chem. Soc. 70(10):947–953.

Zosel K. 1964. German Patent 1493190.

Zosel K. (Studiengesellschaft Kohle mbH). 1979. Process for Deodorizing Fats and Oils. US Patent 4156688.

5 Food-grade Microemulsions As Nano-scale Controlled Delivery Vehicles

Natasha Berry, Rickey Yada and Dérick Rousseau

5.1 INTRODUCTION

Microemulsions were first recognized in the 1940s with the pioneering work of Hoar and Schulman (Hoar and Schulman, 1943; Friberg and Venable, 1983; Karamustafa and Celebi, 2008) who referred to them as oleopathic hydro-micelles. In 1959, the term microemulsion was coined by Schulman *et al.* (Schulman *et al.*, 1959; Friberg and Venable, 1983). Microemulsions are now recognized as self-assembled nano-scale entities (Ruckenstein and Chi, 1975; Friberg and Venable, 1983; Moulik and Paul, 1998; Flanagan and Singh, 2006; Polizelli *et al.*, 2006; Karamustafa and Celebi, 2008). On the basis of size alone, however, the term microemulsion is a misnomer, and has led to confusion. *Vis-à-vis* standard emulsions, which normally consist of dispersed droplet sizes in the low micron range (1–20 μm), microemulsion dispersed domains are normally well below 100 nanometres (nm) in diameter. More recently, nano-emulsions have gained attention as vehicles for controlled release and delivery. Despite the fact that these also consist of nano-scale droplets (typically 50–200 nm), they- lack the thermodynamic stability of microemulsions, which suggests the need for kinetic stability (though practice often dictates otherwise).

Given their inherent stability and nano-sized domains, microemulsions have a number of advantages for use in food, pharmaceuticals and other applications (e.g., cosmetics and petrochemicals), including ease of preparation, a clear appearance promoting consumer acceptance and patient compliance, the improved bioavailability of incorporated compounds and a possible shelf life of many years (Oberdorster *et al.*, 2005; Chau *et al.*, 2007). As a result, microemulsions have been investigated as delivery systems for pharmaceuticals, nutraceuticals and ingredients, as they can improve the solubility of poorly-soluble compounds (e.g., sterols) (Garti *et al.*, 2004), increase bioavailability (e.g., of oil-soluble vitamins) (Rozner *et al.*, 2010) and protect labile compounds against light and/or enzyme-induced breakdown (e.g., oxidation of omega-3 fatty acids) (Swenson and Curatolo, 1992; Flanagan and Singh, 2006; Sanguansri and Augustin, 2006). Given their high surface-to-volume ratio (Gaonkar and Bagwe, 2002), microemulsions have also been used as nano-scale "bioreactors" for the efficient synthesis of numerous compounds, including lysophospholipids, free fatty acids (Morgado *et al.*, 1996) and 2-furfurylthiol, a key product associated with Maillard browning (Vauthey *et al.*, 2000).

Food and Industrial Bioproducts and Bioprocessing, First Edition. Edited by Nurhan Turgut Dunford.

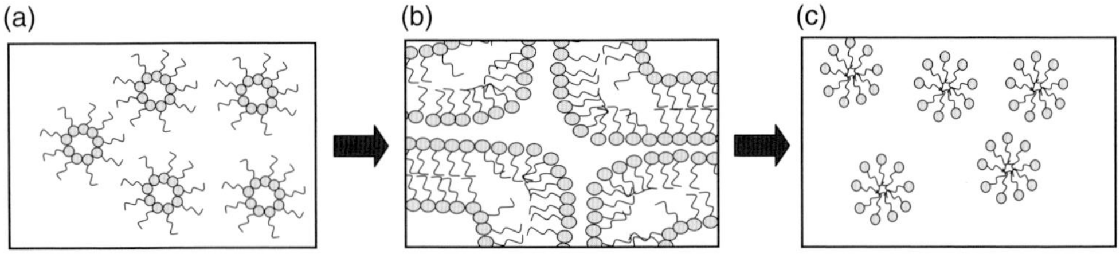

Figure 5.1 Possible nanostructures present within microemulsions: (a) water-in-oil, (b) bicontinuous and (c) oil-in-water (adapted from Garti *et al.*, 2005).

The preparation of microemulsions requires the delicate admixture of an aqueous phase, an oil phase and a surfactant phase. Often, co-surfactants are added to aid in formation and to extend the range of compositions leading to microemulsion formation for a given set of ingredients. Figure 5.1 shows the three known types of microemulsions: (a) water-in-oil (w/o) (L_2 phase), (b) bicontinuous (comparable amounts of oil and water) and (c) oil-in-water (o/w) (L_1 phase). As shown in Figure 5.1, a duplex interfacial film separates the oil and aqueous phases. The formation, properties and stability of a microemulsion strongly depend on the number of extent of interactions between the functional groups present in the components solubilized in the oil, aqueous and interfacial regions. Other structures, such as liquid crystals and gels, may also exist, either alone or in combination with true microemulsion phases (Hamdan *et al.*, 1995; von Corswant and Soderman, 1998; Kunieda *et al.*, 1999; Garti, 2003; Yaghmur *et al.*, 2005).

Though they are seeing increasing use in both food and other applications, notable limitations to microemulsion usage exist, particularly for food-compliant applications. These include:

(1) difficulty in solubilizing long-chain triglycerides due to their relative lack of molecular flexibility;
(2) the limited choice of food-grade surfactants and co-surfactants due to toxicity concerns;
(3) the amount of surfactant required for microemulsion formation is often higher than permitted by legislation;
(4) non-dilutable microemulsions have limited usage in food products;
(5) elimination and substitution of alcohols with other acceptable food-grade surfactants/co-surfactants whilst maintaining microemulsion integrity can be challenging, although progress has been made in developing alcohol-free microemulsions (Constantinides and Scalart, 1997; Polizelli *et al.*, 2006).

5.2 WINSOR CLASSIFICATION/PHASE BEHAVIOR

Winsor (1948) developed an approach to classify equilibrium systems consisting of mixed water, oil and surfactants (Figure 5.2), which is still used to this day. The four categories are:

(1) Type I: o/w microemulsion in equilibrium with excess oil.
(2) Type II: w/o microemulsion in equilibrium with excess water.
(3) Type III: bicontinuous structure in equilibrium with excess oil and water.
(4) Type IV: single phase microemulsion (o/w or w/o) with no excess oil or water.

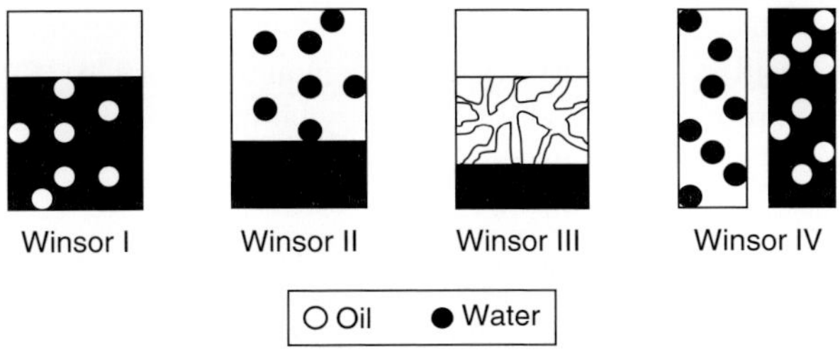

Figure 5.2 Mixed water–oil–surfactant structures as defined by Winsor. (Adapted from: Winsor, 1948 by permission of the Royal Society of Chemistry; Moulik and Paul, copyright 1998, with permission from Elsevier.)

5.3 THEORIES OF MICROEMULSION FORMATION

Three key theories to explain microemulsion formation of have been proposed: the mixed film, solubilization and thermodynamic theories. As described, these theories are not mutually exclusive, as elements from each can contribute to an understanding of microemulsion formation and stability.

5.3.1 Mixed film theory

An amphiphilic film at the oil–water interface exists as a duplex film influenced by the properties of the oil and water phases (Schulman *et al.*, 1959; Flanagan and Singh, 2006). When oil molecules have the ability to associate with those within the interfacial film, a microemulsion readily forms. For instance, a mixed system consisting of cetyl alcohol, benzene and 2-amino-2-methyl-1-propanol (AMP) oleate does not form microemulsions due to the lack of complementarity between these compounds. However, if benzene is replaced by a straight-chained hydrocarbon chain (7–18 C), a microemulsion readily forms due to the ability of the oil molecules to associate with the alcohol and AMP oleate at the interface (Schulman *et al.*, 1959). It is normally assumed that microemulsion formation requires the presence of a duplex film with differing tensions on either side of the interface, which impart curvature and thus the possibility of forming dispersed domains or bicontinuous structures (Prince, 1969).

5.3.2 Solubilization theory

This lesser-known theory states that microemulsions exist as swollen micellar systems. Inverse (reverse) micelles can integrate water within their hydrophilic core and grow in size beyond micellar dimensions, resulting in w/o microemulsions. In a similar manner, normal micelles can solubilize oil and grow in size resulting in o/w microemulsions (Rance and Friberg, 1977; Moulik and Paul, 1998; Flanagan and Singh, 2006).

5.3.3 Thermodynamic theory

Ruckenstein (Ruckenstein and Chi, 1975; Friberg and Venable, 1983; Flanagan and Singh, 2006) explored the stability and size of microemulsion domains based on a thermodynamic approach where interactions (Van der Waals forces, electric double layer potential, and free energy and entropy of formation) were considered. He studied the dependence of the free energy change of mixing as a function of droplet radius and determined whether the free energy change was positive or negative (Ruckenstein and Chi, 1975). Reiss (1975) studied the contribution of entropy in generating thermodynamically stable microemulsions, and distinguished between spontaneous dispersions formed due to entropy effects from micellar solutions formed due to energetic considerations.

5.4 WHAT MAKES MICROEMULSIONS THERMODYNAMICALLY STABLE?

Every molecular assembly theoretically organizes itself such that the Gibbs free energy (ΔG) of the system is minimized (Anton, *et al.*, 2008). The Gibbs free energy depends on interfacial energy ($\gamma\Delta A$) and the change in entropy of the system upon droplet formation ($T\Delta S$):

$$\Delta G = \gamma \Delta A - T \Delta S \tag{5.1}$$

where γ is the interfacial tension, ΔA is the surface area of the droplets, T is the temperature and ΔS is the entropy of the system. In emulsions, ΔG is always positive whereas in microemulsions ΔG is negative or near zero, due to the interfacial tension being extremely small given the high proportions of surfactant(s)/co-surfactant(s) (Anton *et al.*, 2008). According to Kumar and Mittal (1999), microemulsions will form when the oil–water interfacial tension is below 10^{-2} mN/m. However, as food-grade surfactants alone cannot normally lower γ to such low levels, co-surfactants such as alcohols are often used for further reduction. Thus, microemulsions form nearly spontaneously, in large part due to the negative ΔG resulting from the entropy ($T\Delta S$) of mixing and the formation of nano-sized droplets (Anton *et al.*, 2008).

5.5 METHODS OF MICROEMULSION FORMATION

There are numerous ways to generate microemulsions, including the low-energy and phase inversion methods (Flanagan and Singh, 2006). With the former, microemulsions can be formed either via: (1) dilution of an oil–surfactant mixture with water; (2) dilution of a water–surfactant mixture with oil; or (3) mixing all components at once. In some systems, the order of ingredient addition may determine whether a microemulsion forms or not. For instance, Flanagan *et al.* (2006) established that the order of ingredient addition was crucial in systems with soybean oil, ethoxylated mono- and di-glycerides as surfactants and a mixture of sucrose and ethanol as the aqueous phase. Transparent microemulsions resulted from dilution of the oil–surfactant mixtures with water along several regions in the pseudo-ternary phase diagram. However, upon dilution of aqueous–surfactant mixtures with oil, microemulsions only resulted when the aqueous phase:surfactant ratio was 3:2. The authors did not explain the reason for such behavior and only mentioned the existence of a kinetic energy barrier that inhibited microemulsion formation.

With the phase inversion method, when an aqueous solution consisting of a non-ionic surfactant and solubilized oil is heated to, or above, the cloud point, the system separates into a water-rich phase and a surfactant-rich phase (Saito and Shinoda, 1967, 1970; Friberg and Venable, 1983). As the temperature is increased further, the surfactant phase coalesces to form a hydrocarbon solution with solubilized water. Phase reversal takes place at the Phase Inversion Temperature (PIT), that is, the temperature range in which an o/w microemulsion inverts to a w/o type (Flanagan and Singh, 2006).

5.6 POLYDISPERSITY

Polydispersity is a measure of the uniformity of a droplet size distribution and typically varies from 0.0 to 1.0 (unit-less) (Constantinides and Yiv, 1995), where values of 0.000 to 0.02 indicate a monodisperse or nearly monodisperse distribution, values of 0.02–0.08 are common for narrowly-distributed droplet sizes, and values higher than 0.08 indicate broad distributions. Despite microemulsions being polydispersed, they are frequently interpreted as being monodispersed to simplify analysis.

Polydispersity can be measured using techniques such as dynamic light scattering (DLS) and time-resolved fluorescence quenching (TRFQ). Also, small angle light scattering has been used in various systems, for example, water–AOT–*n*-hexane microemulsions (Ricka *et al.*, 1991). Understanding the importance of polydispersity can provide insight into the overall stability of microemulsions. Ideally, microemulsions should be monodispersed, with a narrow size distribution. An increase in breadth usually implies reduced stability, and is often observed when the composition of a microemulsion system is altered, for example, with the addition of a bioactive compound to the microemulsion formulation or with changes in environmental (pH, temperature, ionic strength, etc.) conditions. For example, Constantinides and co-workers (Constantinides and Yiv, 1995) studied the influence of temperature and dilution of w/o microemulsions on polydispersity and mean particle diameter using DLS, reporting a significant change in these parameters above 60 °C, which was used as an indicator of the impact of storage conditions on system stability.

5.7 COMPOSITION

Most microemulsions consist of an organic phase (usually oil), an aqueous phase, surfactant and a co-surfactant. However, in some studies, co-surfactants can be eliminated by the judicious combination of high and low HLB (hydrophilic lipophilic balance) surfactants that are structurally complementary (Constantinides and Scalart, 1997; Polizelli *et al.*, 2006). For instance, surfactants with bulkier head groups [e.g., polysorbates or polyoxyethylene lauryl ether (von Corswant and Soderman, 1998)] will tend to form o/w microemulsions whereas surfactants with bulkier tails [e.g., AOT (sodium bis(2-ethylhexyl) sulfosuccinate)] will tend towards w/o microemulsion formation (Gaonkar and Bagwe, 2002). Thus, combining such surfactants can promote a wide range of compositions leading to microemulsion formation.

5.7.1 Organic phase

Many food-based oils, including mineral, soy, corn, cottonseed and sunflower, have been used to generate microemulsions, although not necessarily for food applications (Hamdan *et al.*, 1995; von Corswant *et al.*, 1997; Gaonkar and Bagwe, 2002; Garti, 2003; Flanagan *et al.*, 2006). Another commonly used oil is mineral oil (Hamdan *et al.*, 1995; Garti, 2003), given its ability to form well-structured dispersed phases (Gaonkar and Bagwe, 2002).

Long-chain triglycerides (e.g., with 18 carbons) are difficult to solubilize given their high molecular weight, which limits their flexibility at the interface and restricts film penetration and formation (Gaonkar and Bagwe, 2002). These have been often substituted by medium-chain triglycerides (e.g., tricaprylin), where the smaller fatty acid chain lengths permit greater flexibility at the interface (Constantinides and Scalart, 1997; Fanun *et al.*, 2001b). Also, medium-chain triglycerides have been used to enhance absorption and bioavailability of water-soluble compounds such as calcein and peptides (Constantinides *et al.*, 1995, 1996; Kamm *et al.*, 2000).

5.7.2 Aqueous phase

Water is an essential component of microemulsions. In o/w microemulsions, water acts as the continuous phase. Water's immiscibility with hydrocarbon oils combined with its ability to hydrogen bond with surfactants results in the formation of oil droplets when it is present as the major component (although this depends on the properties of the surfactant). Formation of microemulsions with small droplets usually requires the presence of a water-soluble co-surfactant (e.g., short chain polyols such as propylene glycol), as discussed later.

5.7.3 Surfactants

Surfactants are surface-active compounds that consist of a hydrophilic head group attached to a hydrophobic tail (usually a long alkyl chain). They have a high affinity for water or oil depending on the dominant moiety (Walz, 1998; McClements, 2004). When present in sufficiently high concentrations, surfactants form a monolayer at the interface between the oil and water, with the hydrophobic tails of the surfactant orientated towards the oil phase and the hydrophilic head groups towards the aqueous phase. There are four categories of surfactants available in the food industry: ionic, non-ionic, zwitterionic and cationic.

5.7.3.1 *Ionic surfactants*

Ionic surfactants have been used to formulate microemulsions. However, their toxic nature at concentrations normally encountered in microemulsions limits their application in foods (Alander and Warnheim, 1989; Flanagan *et al.*, 2006).

5.7.3.2 *Non-ionic surfactants*

Surfactants from this group are commonly used to formulate microemulsions, due to their low toxicity, lack of irritability and capacity to easily form microemulsions (Yaghmur *et al.*, 2004). Examples include sugar ester surfactants (e.g., sorbitan monooleate), polyoxyethylene ether surfactants such as Brij 96 [used in the detergent industry (Gaonkar and Bagwe, 2002; Garti, 2003)] and ethoxylated sorbitan esters (polysorbates) such as Tween 60 (polyoxyethylene sorbitan monostearate) and Tween 80 (polyoxyethylene sorbitan monooleate) (Gaonkar and Bagwe, 2002; Garti, 2003; Flanagan *et al.*, 2006). Mixtures of non-ionic surfactants are far more effective than the other class of surfactants for microemulsion formation, as these potentially eliminate the need for a co-surfactant (Friberg and Venable, 1983).

5.7.3.3 *Zwitterionic surfactants*

Phospholipids are perhaps the common zwitterionic surfactants used in microemulsions. However, they are too lipophilic to be used alone and have a fairly high critical packing parameter, making their incorporation into microemulsions difficult (Moreno *et al.*, 2003). If used, co-surfactants are also required to increase the fluidity of the interface (Gaonkar and Bagwe, 2002; Patel *et al.*, 2006).

5.7.3.4 *Cationic surfactants*

Presently, no cationic surfactants (e.g., cetyl trimethylammonium bromide) are used in the manufacture of food-grade microemulsions. This class of surfactants is mostly used in industrial chemistry applications (Joshi and Mukherjee, 2003; Zuev *et al.*, 2004).

5.7.4 Co-surfactants

A co-surfactant (e.g., propylene glycol or ethanol) can play many roles in microemulsion formation: (a) by reducing the interfacial tension to almost zero (Attwood, 1994; Garti *et al.*, 2002; Yaghmur *et al.*, 2002, 2004; de Campo *et al.*, 2004a; Garti and Aserin, 2006); (b) by increasing the flexibility and fluidity of the interface by positioning itself between the surfactant tails (Dimeglio *et al.*, 1895; Garti *et al.*, 2001; de Campo *et al.*, 2004a), which alters the solvent properties of both the dispersed and continuous microemulsion phases; (c) by lowering overall viscosity, which restricts formation of more rigid structures such as gels and liquid crystals (Leung and Shah, 1987a, 1987b; Martino and Kaler, 1995; Iwanaga *et al.*, 1998; Nagarajan and Wang, 2000; Garti *et al.*, 2001; Fanun, 2007); (d) being often soluble in both organic and aqueous phases, co-surfactants help solubilize poorly-soluble compounds (e.g., peptides, vitamins, etc.) (Sottmann and Strey, 1996; Krafft and Riess, 1998; Yaghmur *et al.*, 2002; Garti, 2003; de Campo *et al.*, 2004); (e) use of co-surfactants can result in fully dilutable microemulsions (Martino and Kaler, 1995; Kunieda *et al.*, 1999; Yaghmur *et al.*, 2002, 2005); and (f) they aid in "connecting" the w/o and o/w regions via a bicontinuous region (Friberg *et al.*, 1994; Fanun, 2007), hence the transition from one regime to another occurs without phase separation.

The inherent properties of alcohol co-surfactants will impact on usage. In particular, alcohol chain length has a significant effect on the ionization of the surfactant and disorder at the oil/water interface. Thus, shorter-chained alcohols will increase interfacial disorder (flexibility) (Bansal *et al.*, 1979; Tabony *et al.*, 1983). However, toxicity concerns limit the use of effective co-surfactants such as 1-butanol, 2-butanol and *tert*-butanol, which are toxic to humans (Attwood, 1994). The most commonly used food-safe co-surfactant is ethanol, though its use in foods limits product marketing due to regulatory requirements, which differing from country to country.

5.8 FACTORS AFFECTING PHASE BEHAVIOR

Microemulsion phase behavior and microstructure depend on two interfacial parameters, namely the spontaneous film curvature (the optimal curvature that the surfactant film attains) and the elasticity (bending modulus) of the surfactant film (Attwood, 1994; Sottmann and

Strey, 1996; Gaonkar and Bagwe, 2002). The free energy per unit area for curvature deformation is (Strey, 1994; Sottmann and Strey, 1996; Gradzielski, 1998):

$$f = 2k(H - H_0)^2 + \bar{k}K \tag{5.2}$$

where $H = 1/2(c_1 + c_2)$is the mean film curvature, in which c_1 and c_2 are the principal curvatures of a regular surface at each point, K is the Gaussian curvature, H_0 is the spontaneous film curvature, and k and $\bar{k}$ describe the elastic properties of the surfactant film (bending modulus), which refers to the deformability of the surfactant film. For $H_0 > 0$, the surfactant film is convex toward water (o/w microemulsion) whereas with H_0 values near zero, bicontinuous microemulsions or lamellar liquid crystalline phases are formed (Strey, 1994; Leitao *et al.*, 1996; Sottmann and Strey, 1996). If the surfactant film is convex toward the oil phase, $H_0 < 0$, and a w/o microemulsion will form. Values of H_0 may be affected by the nature of the surfactant and the composition of the polar and apolar phases (Sottmann and Strey, 1996; Gradzielski, 1998; Gaonkar and Bagwe, 2002). Co-surfactants are also used as "tuning parameters" to modify H_0 and to increase the flexibility of the surfactant film. These compounds may affect the spontaneous curvature of the interface by changing the polarity of the polar and apolar phases.

5.9 PARAMETERS THAT MODIFY MICROEMULSION STRUCTURE

In developing controlled-release matrices, the solubilization capacity of microemulsions should be optimized, which depends on various factors, such as the critical micelle concentration, the surfactant's HLB value, the critical packing parameter and molecular compatibility between the oil, surfactant and co-surfactant (Friberg *et al.*, 1994; Aboofazeli *et al.*, 1995; von Corswant and Soderman, 1998).

5.9.1 Critical micelle concentration

Surfactants at low concentrations in solution are predominantly dispersed as monomers. Above the critical micelle concentration (CMC), they spontaneously associate into (reverse) micelles, either through strong interaction of their hydrophobic tails (resulting in reverse micelles, w/o) or by hydrophilic interaction of the polar head groups (micelles, o/w). Most micellar solutions consist of particles in the 2–5 nm range. Surfactant monomers have the highest surface activity, yet this activity is greatly diminished once micelles are formed, given that either the tails or heads are buried and thus unavailable to participate in surface/interfacial tension reduction (Walz, 1998; Flanagan and Singh, 2006).

5.9.2 Critical packing parameter

The critical packing parameter (CPP) is a measure of the preferred geometry adopted by a surfactant in solution:

$$p = v / la_o \tag{5.3}$$

where p is the packing parameter, v and l are the volume and length of the hydrophobic tail and a is the optimal head group area. In solution, when the hydrophilic head group volume

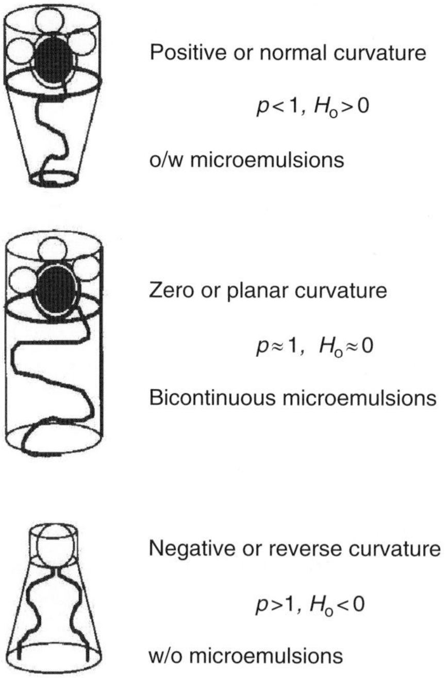

Figure 5.3 Formation of w/o, bicontinuous and o/w microemulsions, and their relationship to p and H_0 (adapted from: Israelachvili, 1994; Lawrence and Rees, 2000).

is greater than that of its hydrophobic tail, o/w microemulsions will form. When the head group has a smaller volume than its hydrophobic tail, w/o microemulsions are preferentially formed. Bicontinuous structures will prevail when these volumes are similar, resulting in $p \approx 1$ (Babak and Stebe, 2002; Malmsten, 2002) (Figure 5.3).

5.9.3 Hydrophile–lipophile balance

The hydrophile–lipophile balance (HLB) of a surfactant typically ranges from 1 to 20 and is an indication of the relative affinity of a surfactant molecule for the oil or aqueous phase. A widely used semi-empirical method of calculating the HLB value is:

$$\text{HLB} = 7 + \sum(\text{hydrophilic group numbers}) - \sum(\text{hydrophobic group numbers}) \quad (5.4)$$

Each surfactant has a specific HLB value, which depends on its chemical structure, with a higher HLB number indicating a higher ratio of hydrophilic to hydrophobic groups and vice versa (Strey, 1994; McClements, 2004). Surfactants with a low HLB values (about 3–8) promote w/o microemulsion formation whereas o/w microemulsions will be formed by surfactants with higher HLB values (about 8–18).

5.9.4 Ingredient compatibility

There is a strong relationship between microemulsion monophasic total area, molecular complementarity and the interface's flexibility. Compatibility between the alkyl chains of the surfactant and oil will influence microemulsion formation by strongly affecting the bending modulus of the interfacial layer (Walz, 1998; Yaghmur *et al.*, 2002), with increased molecular compatibility reducing the bending modulus of the interfacial layer (Yaghmur *et al.*, 2002).

An increase in the alkyl chain length of the compound in the oil will typically decrease its solubility in water (Anton *et al.*, 2008), and consequently reduce the total area of the microemulsion monophasic region. Using the BSO (Bansal–Shah–O'Connell) equation, the maximum solubilization for ionic microemulsions occurs when (Bansal *et al.*, 1980; Garti *et al.*, 1995, 2001; Shiao *et al.*, 1998; Yaghmur *et al.*, 2002):

$$l_o + l_a = l_s \tag{5.5}$$

where l_o = oil alkyl chain length, l_a = alcohol alkyl chain length and l_s = surfactant alkyl chain length.

However, this equation should be used with caution. Shiao and co-workers (1998) reported that the BSO equation failed if the microemulsion contained components with branched alkyl chains, as these decreased the packing order of the hydrocarbons in the interface compared to linear alkyl chains. Garti and co-workers (1995, 2000a) showed that the BSO equation was valid for non-ionic surfactants in five-component microemulsions only when the surfactant (alcohol) exhibited limited solubility in both the aqueous and oil phases.

5.10 CHARACTERIZATION TECHNIQUES

As a first approximation, the presence of microemulsions in a mixed system is often gauged using ternary phase diagrams (TPDs) combined with visual observations. Further characterization can be performed with a number of techniques, including DLS, cryo-transmission electron microscopy (cryo-TEM), small angle X-ray scattering (SAXS), small angle neutron scattering (SANS), nuclear magnetic resonance (NMR) and TRFQ. Others include conductance, viscosity, thermal conductivity, ultrasonic absorption, transient electrical birefringence and vibrational spectroscopy (infrared and Raman).

5.10.1 Ternary phase diagrams

TPDs are used to identify and characterize microemulsion regions (Figure 5.4). The three components (an aqueous phase, an organic phase and a surfactant phase) that compose the system are located at the triangle's apexes, where their corresponding volume fraction is 100%. Moving away from that apex reduces its volume fraction and increases the volume fraction of one or both of the other two components. Thus, each point within the triangle represents a unique combination of the three components, unless positioned along an axis where the mixture is binary. Varying the composition leads to the determination of phase boundaries, and hence phase behavior. As with the surfactant phase, the aqueous and organic phases may also be mixtures of multiple compounds. Where four or more components are used, pseudo-ternary phase diagrams (PTPDs) are used with one or more apexes represent a defined mixture of two or more components (Flanagan and Singh, 2006).

Within a putative TPD (Figure 5.4), any point within the triangle represents a three-component mixture of the apexes. For point "X", the proportion of water (A) can be calculated by drawing a line through "X", parallel to the surfactant–oil (BC) axis. As apex A in the TPD represents 100% water, point X shows a water concentration of 70%. By

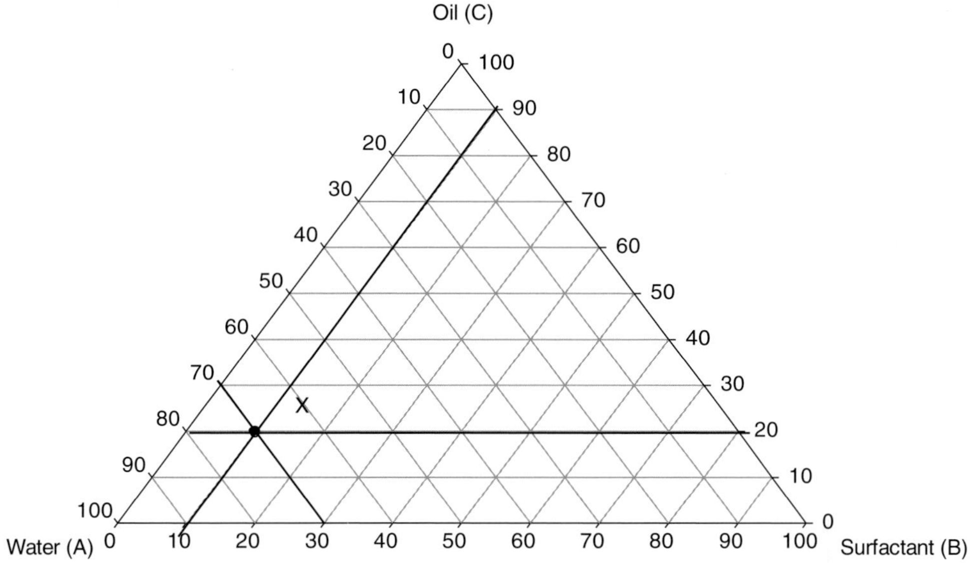

Figure 5.4 Putative ternary phase diagram.

applying the same logic, the proportions of the surfactant (B) (10%) and oil phases (C) (20%) can be determined. Any point along the axes consists of two components only.

A key property dictating the functionality of microemulsions will be whether these can be infinitely diluted by the continuous phase without loss of structure. Such fully dilutable ("U-type") microemulsions yield large isotropic regions extending from the oil-rich to the water-rich apex, without any phase separation (Garti, 2003). Garti and co-workers were the first to develop dilutable microemulsions, in which the L_2 microemulsions were dilutable with aqueous phase (Garti *et al.*, 2005). This was achieved by addition of co-surfactants to the system, thereby increasing the flexibility of the interface. These dilutable microemulsions had a high solubilization capacity and could be used for the transport of hydrophobic guest molecules (nutraceuticals, drugs, etc.), which are normally water insoluble.

5.10.2 Small angle scattering techniques

Small angle scattering techniques, such as SAXS and SANS, are often used to provide information on microemulsion droplet shape and size (Moulik and Paul, 1998; Garti and Aserin, 2006). In these methods, the intensity of the light, scattered at low angles, provides structural information about the scattering species (e.g., oil or water droplets in a microemulsion) (Moulik and Paul, 1998). The scattering intensity depends on the scattering length density of the droplets and the continuous phase (Garti and Aserin, 2006). In both techniques, there is a sharp change of scattering length density across the amphiphilic film, which separates the water and oil regions in a microemulsion (Regev *et al.*, 1996).

5.10.2.1 *Small angle X-ray scattering*

This technique is quite commonly used to characterize microemulsions, providing information on domain size and spatial arrangement. Regev and co-workers (1996) characterized octaethylene glycol mono *n*-dodecyl ether–dodecane–pentanol–water systems as a fire-resistant hydraulic fluid using SAXS and established the change in scattering behavior as a function of the percentage of water, noting microstructural changes associated with water solubilizing capacity. Using SAXS to analyze monolinolein–water–Pluronic F127 systems, de Campo and co-workers (2004b) reported the existence of an intact inner w/o structure when dispersed to form an emulsified w/o microemulsion – an o/w emulsion where water is microemulsified within the inner oil droplets. These authors also reported the transition of an emulsified cubic phase via an emulsified hexagonal phase to emulsified L_2 phase, establishing that temperature reversibly influenced the internal structure of the dispersed particles.

Polizelli and co-workers (2006) used SAXS to distinguish structurally different regions L_2 (w/o microemulsions) and L_2'(dry lamellar structures) within the PTPD consisting of soybean oil–water–monoolein:AOT. Fanun and co-workers (2008a) reported an increase in the periodicity of microemulsions, and hence an increase in domain size, upon increasing the volume fraction of water in water–sucrose laurate–ethoxylated mono-di-glyceride-R(+)-limonene systems. Along with NMR results, they elucidated the structural transitions in the microemulsions from water-in-oil via bicontinuous to oil-in-water dispersions.

5.10.2.2 *Small angle neutron scattering*

Small angle neutron scattering has been effectively used to study droplet shape, size and polydispersity of microemulsions. For the purpose of simplifying structural analysis in dispersions, droplets are considered spherical and monodispersed (Kotlarchyk and Chen, 1983; Toprakcioglu *et al.*, 1984). Eastoe and co-workers (1996) established that the alkyl chain structure of the surfactant was critical to the penetration of cyclohexane/n-heptane (oil phase) at the interface of o/w microemulsions. The study screened various surfactants (phosphatidylcholines, dialkyldimethyl ammonium bromides and AOT) to establish their effect on the oil penetration at the oil–water interface. Kellay and co-workers (1993) studied interfacial film structure in AOT–alkane–brine mixtures at low AOT concentrations as a function of brine salinity, noting the lamellar arrangement of octane and decane films. At the highest salinity, they reported the existence of an oil/brine bicontinuous phase. Finally, Full and Kaler (1994) studied the structure of styrene–dodecyltrimethylammoniumbromide–brine microemulsions, noting the presence of swollen micelles upon the addition of styrene.

5.10.3 Cryo-transmission electron microscopy

Cryo-transmission electron microscopy is widely used to study the microstructure of micelles (Clausen *et al.*, 1992; Zana and Talmon, 1993; Kaplun *et al.*, 1994; Danino *et al.*, 2000) and microemulsions (Jahn and Strey, 1988; Strey, 1994). This technique is unique in providing direct images of the structure (shape and size) at a resolution of ~1 nm (Jahn and Strey, 1988; Danino *et al.*, 2000; Garti and Aserin, 2006). The procedure for studying microemulsions using cryo-TEM involves four steps: sample preparation, freeze–fracturing, replication, and viewing under the microscope (Jahn and Strey, 1988; Strey, 1994). Figure 5.5 shows the structure of a representative microemulsion characterized via cryo-TEM, along with the corresponding particle size distribution obtained via DLS.

One of the first attempts at distinguishing water and oil regions in microemulsions using cryo-TEM was undertaken by Jahn and Strey (1988), who reported the first evidence of bicontinuous structures, using water–*n*-octane–*n*-dodecylpentaethylene glycol micro-emulsions as a model. Regev *et al.* (1996) reported the presence of different structures based on the amount of water in microemulsions with cryo-TEM, including spherical

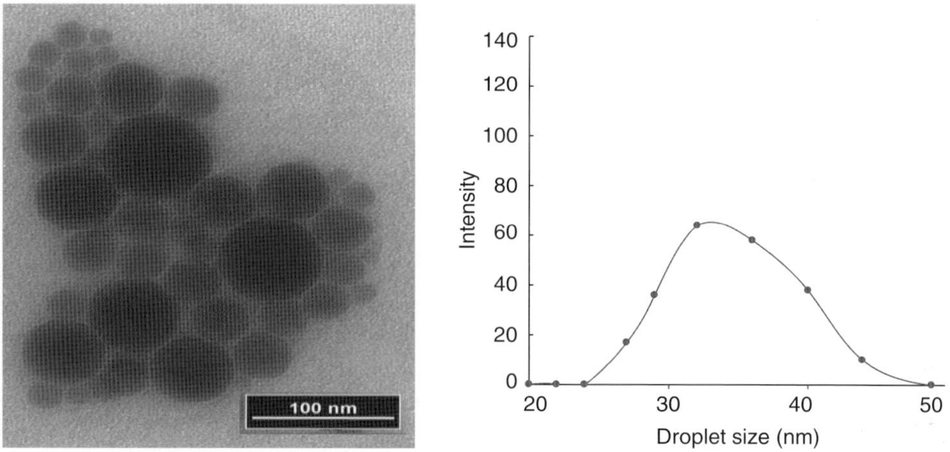

Figure 5.5 Cryo-TEM of a microemulsion consisting of soybean oil, glycerol monooleate, polysorbate 80 and water (left) and the corresponding size distribution obtained via DLS (right) (unpublished results).

micelles and striated (lamellar) structures. The authors also noted that samples were susceptible to electron beam damage.

Fast Fourier Transform cryo-TEM has been found to be a suitable replacement to SAXS to study the nano-structure of microemulsions (Sagalowicz *et al.*, 2006). This technique offers the advantage of being able to study individual particles as well the co-existence of particles with varying structural details in the same sample (Leser *et al.*, 2006; Sagalowicz *et al.*, 2006).

Finally, a novel technique is freeze–fracture direct imaging, where sensitive samples such as microemulsions can be studied without the need for a replica. This technique does not have a blotting step (as per cryo-TEM) that can damage microemulsion structures (Belkoura *et al.*, 2004; Gradzielski, 2008).

5.10.4 Dynamic light scattering

Dynamic light scattering (DLS) is a commonly used technique to determine droplet size and polydispersity in microemulsions. When a coherent light beam interacts with colloidal particles undergoing Brownian motion, the particles scatter light. The original measurement is a time correlation function of the scattered intensity of the particles within the microemulsion. The decrease of this correlation function with time (lag time) is used to extract the diffusion coefficient of a particle or droplet in solution (Goddeeris *et al.*, 2006). The measured diffusion coefficient can be used to calculate a hydrodynamic radius (R_h) of the droplet using the Stokes–Einstein equation:

$$R_h = kT / 6\pi\eta D \tag{5.6}$$

where k is the Boltzmann constant, T is the absolute temperature in Kelvin, η is the viscosity (cP) of the continuous phase, and D (cm^2/s) is the diffusion coefficient (Provencher, 1979; Goddeeris *et al.*, 2006). For proper interpretation of results, it is very important to establish whether the R_h values are based on unimodal or multimodal distributions.

DLS can be used to study structural changes associated with the volume fraction of the dispersed phase in microemulsions, where changes are often the result of modified droplet interactions. These interactions increase when droplet size increases, alcohol (co-surfactant) chain length decreases and salinity decreases (Cazabat and Langevin, 1980; Bellocq *et al.*, 1984; Cazabat, 1985; Hou *et al.*, 1988). Furthermore, it can be used to confirm domain diameter, as per Figure 5.5.

5.10.5 Nuclear magnetic resonance

This approach is widely used for the determination of the diffusion coefficients of the various components in microemulsions, namely the oil, water and surfactant(s). These measurements are useful in elucidating the structural changes occurring in microemulsions with changes in composition or environmental conditions. It is also helpful in establishing the presence of long-range order and particle anisotropy (Moulik and Paul, 1998).

Numerous authors have established structural transitions and diffusion coefficients in microemulsions using NMR. Fanun (2008b) reported structural transitions in water:propylene glycol–sucrose esters–benzaldehyde:ethanol systems upon increasing volume fraction of water. The diffusion measurements suggested a transition from water-in-oil microemulsions via bicontinuous microstructure to oil-in-water microemulsions. In a follow-up study, Fanun (2008a) suggested structural transitions occurring in water–sucrose laurate–ethoxylated mono-di-glyceride-R(+)-limonene systems. The diffusion coefficient of water increased and that of oil decreased upon increasing the volume fraction of water. Determination of self-diffusion coefficients in water–octylglucoside–pentanol–decane microemulsions led Parker and co-workers (1993) to report a gradual decrease in the mean curvature of the interfacial film upon water addition at a fixed oil content. These changes suggested a phase transition from o/w via bicontinuous to w/o microemulsions.

5.11 APPLICATIONS

Microemulsions have been attracting considerable attention due to properties such as ease of formation, thermodynamic stability, transparency and high solubilization capacity. Microemulsions are excellent delivery systems for nutraceuticals (Rozner *et al.*, 2007, 2008) and as vehicles for chemical reactions, such as acid autocatalysis (McIlwaine *et al.*, 2008). They have also been used for the separation and purification of proteins, metal extraction and as drug delivery systems (Hatton, 1989; Pileni, 1989; Garti *et al.*, 2006; Spernath and Aserin, 2006; Kogan *et al.*, 2007).

5.11.1 Solubilization of poorly-soluble drugs

A key benefit of microemulsions is their ability to disperse food ingredients, such as bioactives, flavors and preservatives, that are poorly water soluble in aqueous systems, as these compounds can be conveniently solubilized into the oil droplets of o/w microemulsions (Weiss *et al.*, 2006). For example, Garti and co-workers (2004) studied the solubilization of water-insoluble and poorly oil-soluble nutrients – phytosterol, lutein and lycopene – in U-type microemulsions and showed an increase in the solubility of these nutrients by factors of 15, 11 and 5, respectively, by incorporating polysorbate 80 (or polysorbate 60) in a limonene–ethanol mixture.

5.11.2 Emulsified microemulsions

Few formulations based on double emulsions (in which droplets of the dispersed phase further contain smaller dispersed droplets) are on the market as these consist of large, polydispersed droplets (10–100 μm) that are thermodynamically unstable. Controlling the release of compounds incorporated in double emulsions is normally achieved by stabilizing the interface of the inner emulsion, careful composition of the oil and aqueous phases, and/or stabilization of the secondary interface (Davis *et al.*, 1985; Garti, 1997). However, most double emulsions tend to release entrapped compounds in an uncontrolled manner. A novel possibility for the stabilization of the inner emulsion is to use microemulsions. Pilman and co-workers (1980) reported the use of L_2 microemulsions dispersed within the inner phase of water-in-oil-in-water (w/o/w) duplex emulsions, and stabilized in a sodium caseinate solution resulting in emulsified microemulsions. In principle, these emulsified microemulsions may be used as controlled delivery system for nutraceuticals and pharmaceuticals.

Lutz and co-workers (2007) formulated a w/o/w emulsified microemulsions based on a two-step emulsification process whereby the w/o microemulsion was dispersed in a continuous aqueous phase containing a steric stabilization amphiphile – graft copolymer Pluronic F127. Water up to 15 wt-% in the inner microemulsion and 30 wt-% of microemulsion in the double emulsion was entrapped, in comparison to only 5 wt-% obtained in previous studies carried out by Yaghmur and co-workers (2005), which suggested the possibility of enhanced solubilization of entrapped nutrients.

5.11.3 Protection against oxidation/light

Microemulsions are known to enhance the efficiency of antioxidants (Flanagan and Singh, 2006). Moberger and co-workers (1987) studied triglyceride oxidation in microemulsions consisting of soybean oil, sunflower oil monoglycerides and water. Soybean oil was stable against oxidation for an additional 50 days upon incorporation of 5% (w/w) ascorbic acid into the water phase compared to similar microemulsions without added ascorbic acid.

Components solubilized within microemulsions are generally protected from degradation as they are localized within dispersed domains that, being thermodynamically stable, do not break down with time, thereby limiting exposure. For example, Garti and co-workers (2003) found that lycopene, which is highly photosensitive, could be stabilized via solubilization in dilutable U-type microemulsions.

5.11.4 Controlled release delivery systems

Controlled release and delivery involves delivering a drug/nutrient at a controlled (desirable) rate within the body. Along with film/polymer coatings, microencapsulation is one of the most common techniques used in the development of controlled-release matrices. In microemulsions, the location of an incorporated bioactive compound, whether in the oil, water or interface, will affect its release pattern (Leser *et al.*, 2006).

As controlled-release matrices, microemulsions offer several advantages over emulsions, namely (Tadros, 1993; Garti and Aserin, 1996; Aungst, 2000; Lawrence and Rees, 2000; Guo *et al.*, 2006; Kogan and Garti, 2006; Spernath and Aserin, 2006; Mehta *et al.*, 2007; Gupta and Moulik , 2008):

(1) the possibility of tailored absorption (rate and predictability) and enhanced bioavailability of poorly-soluble bioactive compounds;
(2) higher solubilization capacity and/or entrapment rate of lipophilic bioactives;
(3) aqueous dosage form for hydrophobic compounds, given the much smaller domain size;
(4) better protection of labile compounds (e.g., against oxidation) and flavors/aromas;
(5) improved transdermal penetration compared to other matrices, given the much smaller domain size;
(6) straightforward preparation and scaling up for industrial production, though this also applies to emulsions.

Generally, the poor aqueous solubility of most drugs and nutraceuticals can result in low absorption after *in vivo* administration, thus limiting biological efficacy. Enhancement in biological efficacy and lowering of toxicity may be achieved through judicious encapsulation and delivery of the bioactive compound in aqueous- or oil-based microemulsion delivery systems (Garti and Aserin, 1996).

5.11.5 Microemulsions as nano-reactors

Properties such as large interfacial area and an ability to solubilize both oil-soluble and water-soluble reactants in a single phase system makes microemulsions ideal as reaction media (Flanagan and Singh, 2006; Gaonkar and Bagwe, 2002). For example, Morgado and co-workers (1996) used a continuous reversed micellar system to synthesize lysophospholipids and free fatty acids from lecithin hydrolysis, with applications to the food, pharmaceutical and chemical industries. Hydrolysis was catalyzed by porcine pancreatic phospholipase A_2. Carvalho and Cabral (2000) reviewed the use of reversed micellar systems as reactors to carry out lipase-catalyzed esterification, biocatalysis, transesterification, and hydrolysis reactions.

The Maillard reaction is a reaction between a reducing sugar and an amino acid at high temperatures which is responsible for many color and flavor compounds in foods. Microemulsions have served as nano-reactors for such reactions given their ability to enhance reaction yield and/or pathway (Leser *et al.*, 2006). Vauthey and co-workers (2000) studied the Maillard reaction between furfural and L-cysteine at 100 °C in microemulsions and cubic phases. The systems comprised monoglycerides and water where the production of 2-furfurylthiol, a compound with distinct sensory properties, was monitored. The authors reported a significant increase in the reaction yield of this compound compared to the reaction in the phosphate buffer at pH 5 with the use of microemulsions. Garti and co-workers (Garti *et al.*, 2000b; Fanun *et al.*, 2001a) studied the same reaction and found similar results, that is, enhanced formation of 2-furfurylthiol in sucrose stearate–dodecane–butanol w/o microemulsions at 60 °C. They noted that the partitioning of the co-surfactant butanol between the water phase, the water/oil interface and the oil phase significantly altered the overall reaction rate.

5.12 CONCLUSIONS

There has recently been tremendous progress made in the development of microemulsion formulations for food use. However, there remains a limited understanding of how these systems interact within the body and if they are toxic, given currently insufficient safety testing and assessment in food environments. With the ever increasing development of nano-scale matrices for food use, it is essential that they meet safety regulations and be acceptable to consumers. In concert, it is important to understand the mechanism of how nutrients (bioactive peptides, vitamins, etc.) solubilized within microemulsions are delivered and absorbed in the body. Part of this challenge lies in understanding the changes in phase behavior upon dilution or addition of a nutrient into microemulsions.

For detailed structural studies, a multi-pronged approach combining scattering (DLS, SAXS and SANS), imaging (electron microscopy, cryo-TEM) and other techniques (NMR, conductance, viscosity, etc.) has proven highly beneficial for the characterization of microemulsion properties. As well, the development of novel techniques has helped to better elucidate microemulsions structure; for instance, freeze–fracture direct imaging is now a viable alternative to standard cryo-TEM. As well, computer simulation is now providing new knowledge regarding the interactions between the components present in microemulsions.

As microemulsions have been well studied and applied in the pharmaceutical industry, an inter-disciplinary (pharma–food) approach would highly benefit the development of novel foods. Lastly, continued in-depth research on microemulsions beyond the laboratory benchtop is crucial for the development of new and improved delivery systems to be used as health-romoting functional foods.

REFERENCES

Aboofazeli R, Patel N, Thomas M, and Lawrence MJ. 1995. Investigations into the formation and characterization of phospholipid microemulsions .4. Pseudo-ternary phase-diagrams of systems containing water-lecithin-alcohol and oil – the influence of oil. International Journal of Pharmaceutics: 107–116.

Alander J, and Warnheim T. 1989. Model microemulsions containing vegetable oils .2. Ionic surfactant systems. Journal of the American Oil Chemists Society: 1661–1665.

Anton N, Benoit J, and Saulnier P. 2008. Design and production of nanoparticles formulated from nano-emulsion templates – a review. Journal of Controlled Release: 185–199.

Attwood D. 1994. *Microemulsions as drug delivery systems*. Marcel Dekker, Inc., New York, 1994.

Aungst B. 2002. Intestinal permeation enhancers. Journal of Pharmaceutical Sciences: 429–442.

Babak V and Stebe M. 2002. Highly concentrated emulsions: Physicochemical principles of formulation. Journal of Dispersion Science and Technology: 1–22.

Bansal V, Chinnaswamy K, Ramachandran C and Shah D. 1979. Structural aspects of microemulsions using dielectric-relaxation and spin label techniques. Journal of Colloid and Interface Science: 524–357.

Bansal V, Shah D and O'Connell J. 1980. Influence of alkyl chain-length compatibility on micro-emulsion structure and solubilization. Journal of Colloid and Interface Science: 462–475.

Belkoura, L, Stubenrauch C and Strey R. 2004. Freeze fracture direct imaging: A new freeze fracture method for specimen preparation in cryo-transmission electron microscopy. Langmuir: 4391–4399.

Bellocq, A, Biais J, Bothorel P, *et al.* 1984. Microemulsions. Advances in Colloid and Interface Science: 167–272.

Carvalho C and Cabral J. 2000. Reverse micelles as reaction media for lipases. Biochimie: 1063–1085.

Cazabat, AM. 1985. Probes of microemulsion structure. In: V Degiorgio and M Corti (Eds), *Physics of Amphiphiles, Micelles, Vesicles and Microemulsions*. North Holland, Amsterdam, The Netherlands. pp. 723–753.

Cazabat A and Langevin D. 1980. Light scattering by water in oil microemulsions. In: V Degiorgio, M Corti and M Gilio (Eds). *Light Scattering in Liquids and Macromolecular Solutions*. Plenum Press, New York. pp. 139–155.

Chau C, Wu S and Yen G. 2007. The development of regulations for food nanotechnology. Trends in Food Science & Technology: 269–280.

Clausen T, Vinson P, Minter J, Davis H, Talmon Y and Miller W. 1993 Viscoelastic micellar solutions – microscopy and rheology. Journal of Physical Chemistry: 474–484.

Constantinides P and Scalart J. 1997. Formulation and physical characterization of water-in-oil microemulsions containing long- versus medium-chain glycerides. International Journal of Pharmaceutics: 57–68.

Constantinides P and Yiv S. 1995. Particle-size determination of phase-inverted water-in-oil microemulsions under different dilution and storage conditions. International Journal of Phamaceutics: 225–34.

Constantinides P, Lancaster C, Marcello J, *et al.* 1995. Enhanced intestinal-absorption of an rgd peptide from water-in-oil microemulsions of different composition and particle size. Journal of Controlled release: 109–116.

Constantinides P, Welzel G, Ellens H, *et al.* 1996. Water-in-oil microemulsions containing medium-chain fatty acids salts: formulation and intestinal absorption enhancement evaluation. Pharmaceutical Research: 210–215.

Danino D, Talmon Y and Zana R. 2000. Cryo-TEM of thread-like micelles: on-the-grid microstructural transformations induced during specimen preparation. Colloids and Surfaces A –Physicochemical and Engineering Aspects: 67–73.

Davis S, Hadgraft J and Palin K. 1985. Medical and Pharmaceutical Applications of Emulsions. In: P Becher (Ed.), *Encyclopedia of Emulsion Technology, Applications*, Vol. 2. Marcel Dekker, Inc., New York. pp. 159–238.

de Campo L, Yaghmur A, Garti N, Leser M, Folmer B and Glatter O. 2004a. Five-component food-grade microemulsions: structural characterization by SANS. Journal of Colloid and Interface Science: 251–267.

de Campo L, Yaghmur A, Sagalowicz L, Leser M, Watzke H and Glatter O. 2004b. Reversible phase transitions in emulsified nanostructured lipid systems. Langmuir: 5254–5261.

Dimeglio J, Dvolaitzky M and Taupin C. 1985. Determination of the rigidity constant of the amphiphilic film in birefringent microemulsions – the role of the cosurfactant. Journal of Physical Chemistry: 871–874.

Eastoe J, Hetherington K, Sharpe D, Dong J, Heenan R and Steytler D. 1996. Mixing of alkanes with surfactant monolayers in microemulsions. Langmuir: 3876–3880.

Fanun M. 2007. Conductivity, viscosity, NMR and diclofenac solubilization capacity studies of mixed nonionic surfactants microemulsions. Journal of Molecular Liquids: 5–13.

Fanun M. 2008a. A study of the properties of mixed nonionic surfactants microemulsions by NMR, SAXS, viscosity and conductivity. Journal of Molecular Liquids: 103–110.

Fanun M. 2008b. Phase behavior, transport, diffusion and structural parameters of nonionic surfactants microemulsions. Journal of Molecular Liquids: 14–22.

Fanun M, Leser M, Aserin A and Garti N. 2001a. Sucrose ester microemulsions as microreactors for model Maillard reaction. Colloids and Surfaces A – Physicochemical and Engineering Aspects: 175–187.

Fanun M, Wachtel E, Antalek B, Aserin A and Garti N. 2001b. A study of the microstructure of four-component sucrose ester microemulsions by SAXS and NMR. Colloids and Surfaces A – Physicochemical and Engineering Aspects: 173–186.

Flanagan J and Singh H. 2006. Microemulsions: A potential delivery system for bioactives in food. Critical Reviews in Food Science and Nutrition: 221–237.

Flanagan J, Kortegaard K, Pinder D, Rades T and Singh H. 2006. Solubilisation of soybean oil in microemulsions using various surfactants. Food Hydrocolloids: 253–260.

Friberg S and Venable R. 1983. Microemulsions. In: P Becher (Ed.), *Encyclopedia of Emulsion Technology*, Vol. 1. Marcel Dekker, Inc., New York. pp. 287.

Friberg S, Brancewicz C, and Morrison D. 1994. O/W microemulsions and hydrotropes – the coupling action of a hydrotrope. Langmuir: 2945–2949.

Full A and Kaler E. 1994. Structure and dynamics of polymerizable microemulsions. Langmuir: 2929–2938.

Gaonkar A and Bagwe R. 2003. Microemulsions in foods: Challenges and applications. Adsorption and Aggregation of Surfactants in Solution: 407–430.

Garti N. 1997. Progress in stabilization and transport phenomena of double emulsions in food applications. Food Science and Technology – Lebensmittel-Wissenschaft & Technologie: 222–235.

Garti N. 2003. Microemulsions as microreactors for food applications. Current Opinion in Colloid & Interface Science: 197–211.

Garti N and Aserin A. 1996. Pharmaceutical emulsions, double emulsions and microemulsions. In: S Benita (Ed.), Microencapsulation:Methods and Industrial Applications. Informa Health Care, Drugs and Pharmaceutical Sciences. Marcel Dekker, Inc., New York. pp. 411–534.

Garti N and Aserin A. 2006. Microemulsions for solubilization and delivery of nutraceuticals and drugs, 2nd edn. CRC Press, New York.

Garti N, Aserin A, Ezrahi S and Wachtel E. 1995. Water solubilization and chain-length compatibility in nonionic microemulsions. Journal of Colloid and Interface Science: 428–436.

Garti N, Aserin A and Fanun M. 2000a. Non-ionic sucrose esters microemulsions for food applications. Part 1. Water solubilization. Colloids and Surfaces A – Physicochemical and Engineering Aspects: 27–38.

Garti N, Clement V, Fanun M and Leser M. 2000b. Some characteristics of sugar ester nonionic microemulsions in view of possible food applications. Journal of Agricultural and Food Chemistry: 3945–3956.

Garti, N, A Yaghmur, ME Leser, V Clement and HJ Watzke. 2001. Improved oil solubilization in oil/water food grade microemulsions in the presence of polyols and ethanol. Journal of Agricultural and Food Chemistry: 2552–2562.

Garti N, Yaghmur A, Aserin A, Spernath A, Elfakess R and Ezrahi S. 2003. Solubilization of active molecules in microemulsions for improved environmental protection. Colloids and Surfaces A – Physicochemical and Engineering Aspects: 183–90.

Garti N, Amar-Yuli I, Spernath A and Hoffman R. 2004. Transitions and loci of solubilization of nutraceuticals in U-type nonionic microemulsions studied by self-diffusion NMR. Physical Chemistry Chemical Physics: 2968–2976.

Garti N, Spernath A, Aserin A and Lutz R. 2005. Nano-sized self-assemblies of nonionic surfactants as solubilization reservoirs and microreactors for food systems. Soft Matter: 206–218.

Garti N, Avrahami M and Aserin A. 2006. Improved solubilization of Celecoxib in U-type nonionic microemulsions and their structural transitions with progressive aqueous dilution. Journal of Colloid and Interface Science: 352–365.

Gaonkar A and Bagwe R. 2002. Microemulsions in foods: Challenges and applications. In: O Dinesh, K Shah and L Mittal (eds), Adsorption and Aggregation of Surfactants in Solution. CRC Press, baton Rouge, FL. 2002. pp. 407–430.

Goddeeris C, Cuppo F, Reynaers H, Bouwman W, and Van den Mooter G. 2006. Light scattering measurements on microemulsions: estimation of droplet sizes. International Journal of Pharmaceutics: 187–195.

Gradzielski M. 1998. Effect of the cosurfactant structure on the bending elasticity in nonionic oil-in-water microemulsions. Langmuir: 6037–6044.

Gradzielski M. 2008. Recent developments in the characterisation of microemulsions. Current Opinion in Colloid & Interface Science: 263–269.

Guo R, Qian S, Zhu J and Qian J. 2006. The release of cephanone in Ctab/N-$C_5H_{11}OH/H_2O$ system. Colloid and Polymer Science: 468–474.

Gupta S and Moulik S. 2008. Biocompatible microemulsions and their prospective uses in drug delivery. Journal of Pharmaceutical Sciences: 22–45.

Hamdan S, Lizana R and Laili C. 1995. Aqueous and nonaqueous microemulsion systems with a palm oil-base emollient. Journal of the American Oil Chemists Society: 151–155.

Hatton, T. 1989. Reverse micellar extraction of proteins. In: JF Scamehorn and JH Harwell (Eds) *Surfactant-Based Separation Processes*. Marcel Dekker, Inc., New York, pp. 55–90.

Hoar T and Schulman J. 1943. Transparent water-in-oil dispersions the oleopathic hydro-micelle. Nature: 102–103.

Hou M, Kim M and Shah D. 1988. A light-scattering study on the droplet size and interdroplet interaction in microemulsions of AOT-oil-water system. Journal of Colloid and Interface Science: 398–412.

Israelachvili J. 1994. The science and applications of emulsions – an overview. Colloids and Surfaces A – Physicochemical and Engineering Aspects: 1–8.

Iwanaga T, Suzuki M and Kunieda H. 1998. Effect of added salts or polyols on the liquid crystalline structures of polyoxyethylene-type nonionic surfactants. Langmuir: 5775–5781.

Jahn W and Strey R. 1988. Microstructure of microemulsions by freeze-fracture electron-microscopy. Journal of Physical Chemistry: 2294–2301.

Joshi R and Mukherjee T. 2003. Hydrated electrons in water-in-oil microemulsion: a pulse radiolysis study. Radiation Physics and Chemistry: 397–402.

Kamm W, Jonczyk A, Jung T, Luckenbach G, Raddatz P and Kissel T. 2000. Evaluation of absorption enhancement for a potent cyclopeptidic alpha(nu)beta(3)-antagonist in a human intestinal cell line (Caco-2). European Journal of Pharmaceutical Sciences: 205–214.

Kaplun A, Talmon Y, Konikoff F, Rubin M, Eitan A, Tadmor M and Lichtenberg D. 1994. Direct visualization of lipid aggregates in native human bile by light-transmission and cryo-transmission electron-microscopy. FEBS Letters: 78–82.

Karamustafa F and Celebi N. 2008. Development of an oral microemulsion formulation of alendronate: Effects of oil and co-surfactant type on phase behaviour. Journal of Microencapsulation: 315–323.

Kellay H, Hendrikx Y, Meunier J, Binks B and Lee L. 1993. Local properties of an AOT monolayer at the oil–water interface – neutron-scattering experiments. Journal de physique II: 1747–1757.

Kogan A and Garti N. 2006. Microemulsions as transdermal drug delivery vehicles. Advances in Colloid and Interface Science: 369–385.

Kogan A, Aserin A and Garti N. 2007. Improved solubilization of carbamazepine and structural transitions in nonionic microemulsions upon aqueous phase dilution. Journal of Colloid and Interface Science: 637–647.

Kotlarchyk M and Chen S. 1983. Analysis of small-angle neutron-scattering spectra from polydisperse interacting colloids. Journal of Chemical Physics: 2461–2469.

Krafft M and Riess J. 1998. Highly fluorinated amphiphiles and colloidal systems, and their applications in the biomedical field. A contribution. Biochimie: 489–514.

Kumar P and Mittal K (Eds). 1999. Handbook of Microemulsions Science and Technology. Marcel Dekker, Inc., New York.

Kunieda H, Umizu G and Yamaguchi Y. 1999. Mixing effect of polyoxyethylene-type nonionic surfactants on the liquid crystalline structures. Journal of Colloid and Interface Science: 88–96.

Lawrence M and Rees G. 2000. Microemulsion-based media as novel drug delivery systems. Advanced Drug Delivery Reviews: 89–121.

Leitao H, Somoza A, daGama M, Sottmann T and Strey R. 1996. Scaling of the interfacial tension of microemulsions: A phenomenological description. Journal of Chemical Physics: 2875–2883.

Leser M, Sagalowicz L, Michel M and Watzke H. 2006. Self-assembly of polar food lipids. Advances in Colloid and Interface Science: 125–136.

Leung R and Shah D. 1987a. Solubilization and phase-equilibria of water-in-oil microemulsions. 1. Effects of spontaneous curvature and elasticity of interfacial films. Journal of Colloid and Interface Science: 320–329.

Leung R and Shah D. 1987b. Solubilization and phase-equilibria of water-in-oil microemulsions. 2. Effects of alcohols, oils, and salinity on single-chain surfactant systems. Journal of Colloid and Interface Science: 330–344.

Lutz R, Aserin A, Wachtel E, Ben-Shoshan E, Danino D and Garti N. 2007. A study of the emulsified microemulsion by SAXS, Cryo-TEM, SD-NMR, and electrical conductivity. Journal Of Dispersion Science And Technology: 1149–1157.

Malmsten M. 2002. Liquid crystalline phases. In: *Surfactants and Polymers in Drug Delivery*. Marcel Dekker, Inc., New York. pp. 51–86.

Martino A and Kaler E. 1995. Phase-behavior and microstructure of nonaqueous microemulsions. 2. Langmuir: 779–784.

McClements D. 2004. *Food Emulsions, Principle, Practices, and Techniques*, 2nd edn. CRC Press, New York.

McIlwaine R, Fenton H, Scott S and Taylor A. 2008. Acid autocatalysis and front propagation in water-in-oil microemulsions. Journal of Physical Chemistry C: 2499–2505.

Mehta S, Kaur G and Bhasin K. 2007. Analysis of Tween based microemulsion in the presence of TB drug rifampicin. Colloids and Surfaces B – Biointerfaces: 95–104.

Moberger L, Llarsson K, Buchheim W and Timmen H. 1987. A study of fat oxidation in a microemulsion system. Journal of Dispersion Science and Technology: 207–215.

Moreno M, Ballesteros M and Frutos P. 2003. Lecithin-based oil-in-water microemulsions for parenteral use: Pseudoternary phase diagrams, characterization and toxicity studies. Journal of Pharmaceutical Sciences: 1428–1437.

Morgado M, Cabral J and Prazeres D. 1996. Phospholipase A(2)-catalyzed hydrolysis of lecithin in a continuous reversed-micellar membrane bioreactor. Journal of the American Oil Chemists Society: 337–346.

Moulik S and Paul B. 1998. Structure, dynamics and transport properties of microemulsions. Advances in Colloid and Interface Science: 99–195.

Nagarajan R and Wang C. 2000. Theory of surfactant aggregation in water/ethylene glycol mixed solvents. Langmuir: 5242–5251.

Oberdorster G, Oberdorster E and Oberdorster J. 2005. Nanotoxicology: An emerging discipline evolving from studies of ultrafine particles. Environmental Health Perspectives: 823–839.

Parker W, Genova C and Carignano G. 1993. Study of micellar solutions and microemulsions of an alkyl oligoglucoside via nmr-spectroscopy. Colloids and Surfaces A – Physicochemical and Engineering Aspects: 275–284.

Patel N, Schmid U and Lawrence M. 2006. Phospholipid-based microemulsions suitable for use in foods. Journal of Agricultural and Food Chemistry: 7817–7824.

Pileni MP (Ed.). 1989. *Structure and Reactivity in Reverse Micelles*. Elsevier, Amsterdam, The Netherlands.

Pilman E, Larsson K and Tornberg E. 1980. Inverse micellar phases in ternary-systems of polar lipids-fat-water and protein emulsification of such phases to w-o-w-microemulsion-emulsions. Journal of Dispersion Science and Technology:267–281.

Polizelli M, Telis V, Amaral L and Feitosa E. 2006. Formation and characterization of soy bean oil/surfactant/ water microemulsions. Colloids and Surfaces A – Physicochemical and Engineering Aspects: 230–236.

Prince L. 1969. A Theory of aqueous emulsions. 2. Mechanism of film curvature at oil/water interface. Journal of Colloid and Interface Science: 216–221.

Provencher S. 1979. Inverse Problems in Polymer Characterization – Direct Analysis of Polydispersity with Photon Correlation Spectroscopy." Makromolekulare Chemie – Macromolecular Chemistry and Physics: 201–209.

Rance D and Friberg S. 1977. Micellar solutions versus microemulsions. Journal of Colloid and Interface Science: 207–209.

Regev O, Ezrahi S, Aserin A, *et al.* 1996. A study of the microstructure of a four-component nonionic microemulsion by cryo-TEM, NMR, SAXS, and SANS. Langmuir: 668–674.

Reiss, H. 1975. Entropy-Induced Dispersion of Bulk Liquids. Journal of Colloid and Interface Science: 61–70.

Ricka J, Borkovec M and Hofmeier U. 1991. Coated droplet model of microemulsions – optical matching and polydispersity. Journal of Chemical Physics: 8503–8509.

Rozner S, Aserin A, Wachtel E and Garti N. 2007. Competitive solubilization of cholesterol and phytosterols in nonionic microemulsions. Journal of Colloid and Interface Science: 718–726.

Rozner S, Aserin A and Garti N. 2008. Competitive solubilization of cholesterol and phytosterols in nonionic microemulsions studied by pulse gradient spin-echo NMR. Journal of Colloid and Interface Science: 418–425.

Rozner S, Shalev D, Shames A, Ottaviani M, Aserin A and Garti N. 2010. Do food microemulsions and dietary mixed micelles interact?" Colloids and Surfaces B – Biointerfaces: 22–30.

Ruckenstein E and Chi J. 1975. Stability of microemulsions. Journal of the Chemical Society – Faraday Transactions II: 1690–1707.

Sagalowicz L, Michel M, Adrian M, *et al.* 2006. Crystallography of dispersed liquid crystalline phases studded by cryo-transmission on electron microscopy. Journal Of Microscopy: 110–121.

Saito H and Shinoda K. 1967. Solubilization of hydrocarbons in aqueous solutions of nonionic surfactants. Journal of Colloid and Interface Science 24(1): 10–16.

Saito H and Shinoda K. 1970. Stability of w/o type emulsions as a function of temperature and of hydrophilic chain length of emulsifier. Journal of Colloid and Interface Science 32(4): 647–651.

Sanguansri P and Augustin M. 2006. Nanoscale materials development – a food industry perspective. Trends in Food Science & Technology: 547–556.

Schulman J, Stoeckenius W and Prince L. 1959. Mechanism of formation and structure of micro emulsions by electron microscopy. Journal of Physical Chemistry: 1677–1680.

Shiao S, Chhabra V, Patist A, *et al.* 1998. Chain length compatibility effects in mixed surfactant systems for technological applications. Advances in Colloid and Interface Science: 1–29.

Sottmann T and Strey R. 1996. Shape similarities of ultra-low interfacial tension curves in ternary microemulsion systems of the water-alkane-C(i)E(j) type. Berichte Der Bunsen-Gesellschaft – Physical Chemistry Chemical Physics: 237–241.

Spernath A and Aserin A. 2006. Microemulsions as carriers for drugs and nutraceuticals. Advances in Colloid and Interface Science: 47–64.

Strey R. 1994. Microemulsion microstructure and interfacial curvature. Colloid and Polymer Science: 1005–1019.

Swenson E and Curatolo W. 1992. Intestinal permeability enhancement for proteins, peptides and other polar drugs – mechanisms and potential toxicity. 2. Advanced Drug Delivery Reviews: 39–92.

Tabony J, Llor A and Drifford M. 1983. Quasielastic neutron-scattering measurements of monomer molecular motions in micellar aggregates. Colloid and Polymer Science: 938–946.

Tadros T. 1993. Industrial Applications of Dispersions. Advances in Colloid and Interface Science: 1–47.

Toprakcioglu C, Dore JC, Robinson BH, Howe A and Chieux P. 1984. Small-angle neutron-scattering studies of microemulsions stabilized by aerosol-OT. 2. Critical scattering and phase-stability. Journal of the Chemical Society – Faraday Transactions I: 413–422.

Vauthey S, Milo C, Frossard P, Garti N, Leser M and Watzke H. 2000. Structured fluids as microreactors for flavor formation by the Maillard reaction. Journal of Agricultural and Food Chemistry: 4808–4816.

von Corswant C and Soderman O. 1998. Effect of adding isopropyl myristate to microemulsions based on soybean phosphatidylcholine and triglycerides. Langmuir: 3506–3511.

von Corswant C, Engstrom S and Soderman O. 1997. Microemulsions based oil soybean phosphatidylcholine and triglycerides. Phase behavior and microstructure. Langmuir: 5061–5070.

Walz J. 1998. The effect of surface heterogeneities on colloidal forces. Advances In Colloid and Interface Science: 119–168.

Weiss J, Takhistov P and McClements J. 2006. Functional materials in food nanotechnology. Journal of Food Science: R107–R116.

Winsor P. 1948. Hydrotropy, Solubilisation and Related Emulsification Processes. 1. to .4. Transactions of the Faraday Society: 376–398.

Yaghmur A, Aserin A and Garti N. 2002. Phase behavior of microemulsions based on food-grade nonionic surfactants: effect of polyols and short-chain alcohols. Colloids and Surfaces A – Physicochemical and Engineering Aspects: 71–81.

Yaghmur A, de Campo L, Aserin A, Garti N and Glatter O. 2004. Structural characterization of five-component food grade oil-in-water nonionic microemulsions. Physical Chemistry Chemical Physics: 1524–1533.

Yaghmur A, de Campo L, Sagalowicz L, Leser M and Glatter O. 2005. Emulsified microemulsions and oil-containing liquid crystalline phases. Langmuir: 569–577.

Zana R and Talmon Y. 1993. Dependence of aggregate morphology on structure of dimeric surfactants. Nature: 228–230.

Zuev Y, Mirgorodskaya A, Kudryavtseva L, Idiyatullin B and Kharnidullin R. 2004. Influence of the structure of cetyltrimethylammonium bromide based microemulsions on base hydrolysis of carboxylic acid esters. Russian Journal of General Chemistry: 1051–1056.

6 Emulsions, Nanoemulsions and Solid Lipid Nanoparticles as Delivery Systems in Foods

Umut Yucel, Ryan J. Elias and John N. Coupland

6.1 DELIVERY SYSTEMS IN FOODS

The functional properties of foods depend to a great extent on the presence of relatively low concentrations of certain endogenous or exogenous small molecules. These molecules may be classified by their function, such as flavors (e.g., limonene, citral), colorants (e.g., β-carotene), and increasingly phytochemicals (e.g., tea catechins, β-carotene, highly polyunsaturated fats), but many of them share the common property of producing a biological reaction in the consumer (i.e., are bioactive). Most of these molecules are lipophilic and often labile to oxidation, elevated temperatures or highly acidic environments. They are amongst the most expensive ingredients in the food and their addition is critical to the economics of the product while their loss can limit shelf life. There is, therefore, a broad need for methods to allow these bioactive lipophilic ingredients (BLI) to be used effectively in aqueous foods. Most of these approaches involve some form of encapsulation.

McClements and co-workers (2007) suggested an effective delivery system for BLI should be easy to manufacture, physically stable, capable of binding useful quantities of the molecule of interest while maintaining its chemical stability yet allowing release when necessary (to ensure bioavailability). The delivery system should, of course, be compatible with the food matrix and not unduly change the properties of the food. Based on these recommendations and on the nature of BLI molecules, some inferences to the nature of suitable delivery systems can be made. Firstly, the encapsulation should be supported by non-covalent interactions and/or physical entrapment; covalent bonds are too strong to allow release of the ingredient and would render it non-functional. As BLI ingredients are non-polar, hydrophobic interactions will likely provide the basis for at least some of the binding necessary; the encapsulation system must provide a hydrophobic domain in a hydrophilic media. If the hydrophobic domain is to be dispersed in an aqueous food, then the encapsulation system must allow bulk physical stability (e.g., have some hydrophilic regions). The encapsulation system must also be small enough or share sensory properties common to the native product, so as not to be perceived in the food. Given these constraints, the number of delivery systems available for use with BLI in food is relatively limited and, in practice, can be classified into three main groups.

In the first group of delivery systems, the BLI is bound directly by a food polymer (polysaccharide or protein) and then dispersed within the food. It has been shown that the

Food and Industrial Bioproducts and Bioprocessing, First Edition. Edited by Nurhan Turgut Dunford.

bioavailability of lycopene entrapped within whey proteins was similar to that provided by tomato paste (Richelle *et al.*, 2002) and much greater than in unprocessed tomato. In another work, casein micelles were introduced as a potentially useful delivery system for the lipid soluble vitamin D2 (Semo *et al.*, 2007). Moreover, spray-dried starch-based delivery systems were studied in order to encapsulate phytochemicals, such as carotenoids from carrots (Wagner and Warthesen, 1995).

In the second group of delivery systems, the BLI is incorporated into a thermodynamically stable surfactant self-association structure (e.g., a microemulsion). For example, Amar and co-workers (2003) used microemulsions of different surfactant types to improve the solubilization of lutein and lutein esters, which are otherwise insoluble in water and have limited solubility in food-grade oils. In the final group of delivery systems, the BLI can be dissolved in a lipid carrier and subsequently homogenized to form fine emulsion droplets. This final group is known as emulsion-based delivery systems (EBDS) and is the focus of this chapter.

Emulsions are attractive candidates for delivery systems, as they are already widespread in foods and utilize commonly available ingredients and process technologies. There is a parallel between the need to encapsulate BLI in foods with the need to incorporate poorly-soluble drugs in a form that can be ingested, injected or applied topically. Indeed, in the pharmaceutical industry lipophilic drugs are delivered parenterally as commercial EBDS products (Mehnert and Mader, 2001; Muller *et al.*, 2000). The lipid phase provides a reservoir for the BLI, which can then be uniformly dispersed in an aqueous food by homogenization and stabilized with emulsifiers. It is possible to tune the performance of the delivery system by altering the size or number of the lipid droplets, or by altering the interfacial and/or lipid composition. Work in pharmaceutical EBDS has pioneered the use of solid lipid nanoparticles (SLN) and nano-structured lipid carriers (NLC), fine crystalline emulsion droplets, as delivery systems (Bonacucina *et al.*, 2009; Mehnert and Mader, 2001; Muller *et al.*, 2000). Several workers have considered parallel applications in foods (McClements *et al.*, 2007, 2009; Velikov and Pelan, 2008; Weiss *et al.*, 2008). In this chapter, the relationships between structure of emulsions and crystalline emulsions, both micro- and nano-scale, and their functionality as delivery systems in foods are examined.

6.2 STRUCTURE OF EMULSIONS

An emulsion is a mixture of two immiscible liquids, one of which is uniformly dispersed within the other as small droplets (i.e., droplet diameter in the range 0.1–100 µm) (McClements, 2005). In this chapter, the concern is exclusively with dispersions of oil in water, that is, oil-in-water emulsions. Advances in homogenizer technology have allowed the development of nanoemulsions with droplets far smaller and with more uniform size distributions than commonly seen in manufactured foods (Weiss *et al.*, 2008). There is not a commonly accepted size cut-off for nanoemulsions and different researchers have used different definitions: below 1000 nm (Muller *et al.*, 2000), 500 nm (Anton *et al.*, 2008), 200 nm (Higami *et al.*, 2003; Solans *et al.*, 2005), and 100 nm (Luykx *et al.*, 2008). Emulsions with nano-scale crystalline droplets are sometimes referred to as solid lipid nanoparticles (SLN).

Emulsions are formed either by using mechanical energy to reduce the particle size of a coarse mixture or, less frequently, by a controlled phase separation (e.g., on dilution of raki or ouzo). Various technologies are used for particle size reduction in emulsions; the most common are high-pressure valve homogenizers, although other devices, including high intensity ultrasound, membrane homogenizers, and microfluidic devices, are also capable of generating very fine particles with uniform droplet size distributions (Solans *et al.*, 2005). Although the efficiency of droplet formation (i.e., size and uniformity of the droplets) depends on many factors, in general there is an approximately linear inverse relationship between the homogenization pressure and particle size (McClements, 2005).

Emulsions are thermodynamically unstable structures given a degree of kinetic stability by an adsorbed interfacial layer of amphiphilic emulsifiers. The emulsifiers serve to lower the interfacial tension and provide some inter-droplet repulsive forces to stabilize the dispersions (e.g., steric and electrostatic). The interfacial layer is typically between about 1 and 10 nm thick for food grade emulsifiers, such as surfactants, phospholipids, proteins, or polysaccharides, and the interfacial concentration is in the order of a few mg per square meter of surface (McClements, 2005; McClements and Decker, 2000).

The various mechanisms for emulsion destabilization include flocculation, coalescence (in fluid particles), partial coalescence (in semi-crystalline particles), Ostwald ripening and gravitational separation (creaming or sedimentation) (Claesson *et al.*, 2004; Dickinson, 1992; McClements, 2005; Vanapalli and Coupland, 2004). In general, the instability processes are slower for fine particles (e.g., the rate of creaming by Stokes Law is inversely proportional to the square of particle diameter). The major exception is destabilization via Ostwald ripening. Ostwald ripening is the growth of large particles at the expense of fine particles due to diffusion of molecules from one droplet to another. Ostwald ripening is driven by differences in surface curvature and is, therefore, faster in very fine particles. Ripening processes are particularly important when molecules in the lipid phase have some solubility in the aqueous phase, so while triacylglycerols (TAGs) typically ripen slowly, the transport of more polar emulsified BLI may be faster. Aqueous surfactant micelles can increase the solubility of lipids in the aqueous phase and thus increase the rate of ripening (Weiss *et al.*, 1996).

6.3 LOCALIZATION OF BLI IN EMULSIONS

An emulsion can be envisaged as consisting of at least three distinct microenvironments for added BLI molecules (Figure 6.1): the continuous aqueous phase, the lipid core of the droplets, and the surface of the droplets as the amphiphilic region. The relative volume of the aqueous and lipid phases can be altered by changing the formulation of the emulsion, while the surface to lipid volume ratios can be altered by changing the particle size (Figure 6.2). BLI localized in the lipid core or aqueous phases will be predominantly surrounded by TAG and water molecules, respectively, while BLI molecules in the interfacial region will interact with both to some extent as well as with the high concentration of emulsifiers on the surface.

BLI molecules will partition between the different microenvironments depending on their chemical structure and intermolecular interactions. As they are lipophilic molecules, the aqueous concentration will be very low. Molecular lipophilicity is most commonly expressed in terms of the log P value, where P is the ratio of equilibrium concentrations in octanol and water. Large log P values suggest the molecule will exist predominantly inside the emulsion droplets (Table 6.1). One weakness of the log P approach is that it neglects the

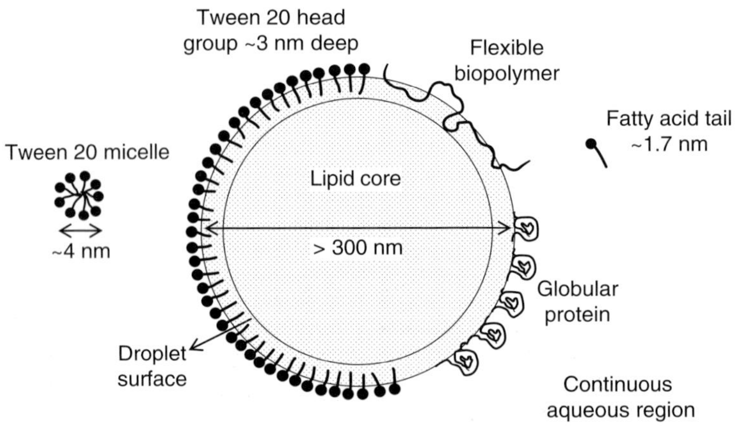

Figure 6.1 Regions in a conventional emulsion stabilized by different emulsifiers, i.e., Tween 20, a flexible biopolymer, and a globular protein (not to scale; adapted from Povey, 2001).

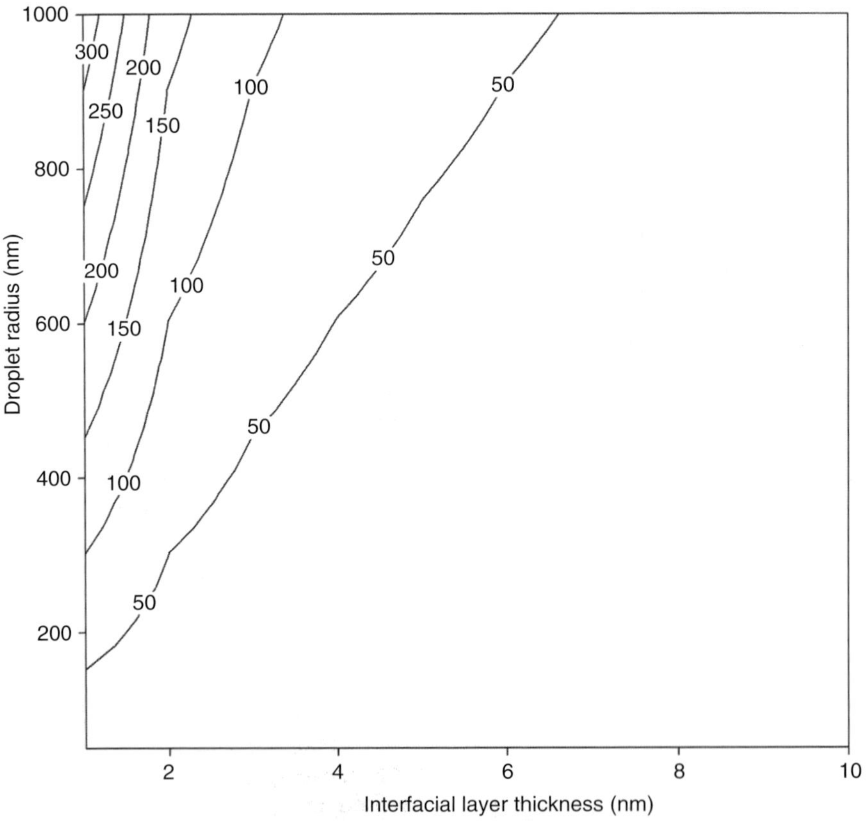

Figure 6.2 Ratio of the internal phase volume to interfacial phase volume as a function of droplet size and the thickness of the interfacial layer.

Table 6.1 Calculated log P values for selected BLI molecules (from Yannai, 2004, unless otherwise indicated).

Compound	Functional role	log P
Benzaldehyde	flavor, antioxidant	1.5
Hexanal	flavor, antioxidant	1.8 (Reineccius, 2006)
Benzoic acid	antimicrobial	1.88
Curcumin	flavor, antioxidant	2.05
Eugenol	flavor, antioxidant	2.4 2.73 (Reineccius, 2006)
Apiole	flavor, antioxidant	2.5
Quinine	flavor, antioxidant	2.93
Octanoic acid (caprylic acid)	antimicrobial	2.94
Anethole	flavor, antioxidant	3.31 3.39 (Reineccius, 2006)
Thymol	flavor, antioxidant	3.4
Capsaicin	flavor, antioxidant	3.5
Hexyresorcinol	flavor, antioxidant	3.95
Valerenic acid	flavor, antioxidant	4.74
limonene	flavor, antioxidant	4.83 (Reineccius, 2006)
narasin A	flavor, antioxidant	5.54
bullatacinone	flavor, antioxidant	8
vitamin D2	vitamin	9.39
vitamin D3	vitamin	9.48
vitamin K	vitamin	11.76
α-tocopherol; ac (vitamin E acetate)	vitamin	12.15
β-carotene	colorant, antioxidant	15.23

Table 6.2 Experimental (from Rothwell *et al.*, 2005) vs. calculated (from Yannai, 2004) log P values for selected molecules.

Compound	Experimental	Calculated
genistein	2.51+0.06	0.98
kaempferol	3.11+0.54	0.31
luteolin	3.22+0.08	0.38
naringenin	2.60+0.03	1.8
quercetin	1.82+0.32	0.29

presence of the surface region, which, as has been noted, enjoys properties distinct from the lipid core and aqueous phase. Many BLI have at least some amphiphilic character with a few polar groups located on an otherwise non-polar structure. Such molecules will have a tendency to align at interfaces, losing some rotational entropy to optimize the chemical environments of its parts. A second weakness in applying log P values to predict BLI partitioning in very fine droplets is that the BLI solubility tends to increase as particle size decreases due to the Laplace pressure effects. Finally, many of the log P values reported in the literature are values calculated from the chemical structure rather than experimentally measured numbers; these calculations can be unreliable (Table 6.2).

Measurement of the distribution of BLI (or model analog molecules) between microenvironments in an emulsion is challenging. The three main groups of approaches are:

(1) Separation and direct analysis in one of the phases. If the aqueous phase can be separated and analyzed it is possible to measure the proportion of BLI associated with the droplets (surface and core) by difference. Unfortunately, fine particles are hard to filter or separate by sedimentation. More aggressive separation procedures (e.g., solvent extraction) often destroy the equilibrium under investigation (especially surface and Laplace pressure effects). Alternatively, a volatile BLI can be determined in the headspace allowing its activity coefficient in the emulsion to be determined non-destructively (Ghosh *et al.*, 2006).

(2) Chemical degradation kinetics. An indirect method to determine the distribution of BLI is to measure their chemical degradation kinetics in response to an aqueous oxidizing agent, radical source, or acid. The assumption in these studies is usually that BLI at the interface will degrade faster than those in bulk. Studies of BLI stability are obviously valuable in their own right. However, care must be taken in inferring distribution between microenvironments, as the assumption that molecules at the surface are most vulnerable to degradation is rarely confirmed by other methods. For example, Boon and co-workers (2008) compared lycopene oxidation kinetics in lipid nanoemulsions of different lipid and surfactant types. They showed that the oxidative stability was affected by surface properties, and higher in emulsions stabilized with cationic and non-ionic surfactants than in emulsions stabilized with anionic surfactants. In another study, Jenning and Gohla (2001) compared the chemical stability of selected drug ingredients (i.e., retinoids) with different hydrophobicity in SLN and lipid nanoemulsions and argued that crystallization of the lipid droplets resulted in concentration of drug molecules on the droplet surface while encapsulation efficiency was improved by increased lattice defects. In contrast, Helgason and co-workers (2009) used this approach to argue that high melting lecithin prevented the accumulation of β-carotene at the surface of a SLN.

(3) Spectroscopic methods. The spectroscopic signal of many molecules is sensitive to their immediate chemical environment. If the spectroscopic method is capable of analyzing the BLI (or more likely a labeled analog) *in situ* in the emulsion then there will be distinguishable signals for the proportion in each microenvironment. For example, Arboleya and co-workers (2005) prepared a palm oil-in-water emulsion containing a stable free radical spin probe in the lipid phase. The probes were monitored using electron paramagnetic resonance (EPR) spectroscopy and the investigators were able to deconvolute the spectra and determine the proportion in the lipid and aqueous phases. The molecular mobility of the probe as a function of lipid crystallization was also determined. Nuclear magnetic resonance (NMR) spectroscopy is a similar technique to EPR, and can similarly be used to probe the properties of different compartments. For example, proton NMR (H^1-NMR) was used to investigate the mobility and distribution of a liquid lipid (medium chain TAG) loaded into solid lipid droplets (long chain TAG) (Jenning *et al.*, 2000a, 2000b).

6.4 EMULSIONS AS DELIVERY SYSTEMS

Emulsion-based delivery systems (EBDS) has been introduced as successful carrier systems for BLI. For example, ω-3 fatty acid-rich TAGs were incorporated as emulsions into different food products, such as milk, yogurts, ice cream, and meat patties (Chee *et al.*, 2005, 2007; Lee *et al.*, 2005, 2006a, 2006b; McClements and Decker, 2000; Sharma, 2005).

Other BLI have been dissolved in carrier lipids prior to emulsification, for example lycopene (Ribeiro *et al.*, 2006; Tyssandier *et al.*, 2001), astaxanthin (Ribeiro *et al.*, 2005, 2006), lutein (Losso *et al.*, 2005; Santipanichwong and Suphantharika, 2007), β-carotene (Santipanichwong and Suphantharika, 2007), and plant sterols (Sharma, 2005). As noted above, the delivery system must chemically stabilize the BLI without impairing bioavailability.

Many emulsified lipids and encapsulated BLI are vulnerable to chemical destruction; most commonly via oxidative reactions or acid hydrolysis. The structure of the emulsion and the microlocalization of the BLI affect the rate of chemical degradation. It is also important to consider the relative localization of any other reagents (e.g., oxygen, protons), catalysts (e.g., transition metals), and inhibitors of oxidation reactions (e.g., antioxidants) involved in the BLI degradation reaction.

The presence of an emulsifier layer at the oil/water interface can affect the distribution of small molecules in the emulsion. It has been demonstrated that the electrostatic properties of emulsion droplets can be modified in order to repel prooxidative metals for the purpose of inhibiting lipid oxidation (McClements and Decker, 2000). Emulsifiers can modify the electrical charge on the droplet surface, which can affect the rate of exchange of ions between the aqueous phase and the droplet surface. For example, the degradation of an omega-3 enriched oil-in-water emulsion by oxidation reactions can be inhibited by conferring cationic properties to emulsion droplets (Donnelly *et al.*, 1998). This was accomplished by lowering the pH of the aqueous continuous phase below that of the emulsifying protein's isoelectric point (pI), thereby establishing a net positive charge on the droplet interface. Under such conditions, positively charged transition metals are electrostatically repelled from the emulsified lipid, and oxidation reactions are inhibited. An opposite trend can be seen at pH values above the pI of the emulsifier, in which case the droplets carry a net negative charge that attracts transition metal catalysts to emulsion interfaces, accelerating lipid oxidation reactions.

Other studies have demonstrated the importance that droplet charge has on lipid oxidation kinetics in oil-in-water emulsions (McClements and Decker, 2000; Hu *et al.*, 2003; Klinkesorn *et al.*, 2005; Mei *et al.*, 1998a, 1998b). In a similar study, Boon and co-workers (2008) showed that the oxidative stability of lycopene emulsions is higher when stabilized with cationic and non-ionic surfactants than in emulsions stabilized with anionic surfactants.

The interfacial thickness of emulsion droplets is an important parameter affecting lipid oxidation reaction rates. Increasing interfacial membrane thickness can conceivably hinder the physical interaction between aqueous phase prooxidants (e.g., transition metals) and emulsified lipids(Chaiyasit *et al.*, 2000; Silvestre *et al.*, 2000). For example, Silvestre and co-workers (2000) showed that iron-catalyzed cumenehydroperoxide reduction, as well as salmon oil-in-water emulsion oxidation, was slower when Brij 700 was used in place of Brij 76. Brij 700 and 76 are small molecule surfactants with identical hydrophobic tail group lengths ($CH_3(CH_2)_{17}$–), but vary only with respect to the size of their polar head groups: Brij 700 and Brij 76 consist of 100 and 10 oxyethylene head groups, respectively. Lower hydroperoxide decomposition and lipid oxidation rates in Brij 700-stabilized emulsions suggest that a thicker interfacial layer was able to act as a physical barrier to decrease lipid–prooxidant interactions (Silvestre *et al.*, 2000).

A similar effect can be seen in protein-stabilized oil-in-water emulsions. For example, casein is known to form a thicker interfacial layer (10 nm) than whey proteins (1–2 nm) (Dalgleish *et al.*, 1995), which may explain, at least in part, why emulsions stabilized with casein tend to have greater oxidative stability than those prepared with whey protein. This is despite the fact that whey protein-stabilized emulsions droplets are more positively charged (+55.9 mV) than their casein counterparts (+29.9 mV) at the low pH studied (Chaiyasit *et al.*, 2007).

Smaller droplets (larger interfacial area) increase the exposure of the encapsulated BLI to reactive molecules, so would be expected to reduce their chemical stability. Despite this, comparisons of BLI degradation rates at different particle sizes are relatively few and the postulated relationship is not always seen. For example, Kiokias and co-workers (2007) did not report any effects of droplet size (d_{32} = 0.5–2 μm) on the rate of lipid oxidation in model food emulsions as a function of temperature, while Let and co-workers (2007), and Nakazawa and co-workers (2008) showed fine ω3 fatty acid-rich oil and methyl linolenate emulsions oxidized more slowly than coarse ones (d_{32} = 0.5–1.5 μm and median = 0.02–1 μm, respectively). Droplet size distribution affects not only the chemical stability of BLI and the physical stability of the EBDS but also the bioavailability of BLI within the EBDS.

Bioavailability, the amount of bioactive ingredient metabolized at the appropriate part of the living body, is believed to increase with decreasing particle size in the EBDS. Acosta (2009) reviewed various bioavailability studies and showed that, in general, there was an increased bioavailability in emulsions with droplets smaller than 500 nm; this was even greater for droplets less than 100 nm (Acosta, 2009). Various mechanisms have been suggested to explain this effect and perhaps the most universal is the increased solubility of BLI in aqueous phase around the very fine particles (i.e., Laplace pressure effects). However, it has also been shown that residence time of the carrier particles in the intestine increases with decreasing particle size, which in turn increases the amount absorbed (Kreuter, 1991). The shorter diffusion paths in fine particles have been suggested as an alternative reason for increased bioavailability (Acosta, 2009; McClements *et al.*, 2007). However, while the diffusion paths in the lipid phase are certainly reduced, it is unclear if they are rate limiting in the overall process of absorption. Finally, Jani and co-workers (1990) proposed that very fine particles ($d < 100$ nm) could be adsorbed whole by living cells.

6.5 CRYSTALLIZATION IN EMULSIONS

The equilibrium solid fat content (SFC) of a lipid is determined by the temperature and the type of molecules present (e.g., longer saturated or trans-unsaturated fats have higher melting points). The vegetable oils most commonly used in conventional food emulsion preparation are liquid at both storage and use temperatures. Other food lipids, notably milk fat, partially hydrogenated vegetable oil and some tropical fats are semi-crystalline at room temperature and below. Finally, hard fats (e.g., stearin fractions from tropical oils and animal fats, fully hydrogenated vegetable oils, certain waxes) are (almost) completely crystalline at room temperatures. The physical properties of lipids can be manipulated by changing the TAG composition by either chemical (e.g., hydrogenation or inter-esterification) or physical means (e.g., fractionation or blending) (Haumann, 1994). In all cases, commercial food lipids are a complex mixture of different TAG molecules.

TAG molecules can crystallize typically as one of three polymorphic crystal structures, α, β' and β, each with different subcell structures (i.e., hexagonal, orthorhombic-perpendicular, and triclinic-parallel, respectively) and with increasing melting points and enthalpies of fusion (Sato, 1999). If a liquid oil is cooled quickly, it will tend to nucleate into a less stable polymorphic form and then possibly transform into a more stable polymorphic form with time, particularly if stored at higher temperatures (Himawan *et al.*, 2006). In some cases, different yet physically similar TAG molecules can co-crystallize to form a compound crystal (Sato, 2001). If a lipid is processed into colloidal size droplets, the crystallization process is considerably different, affecting both the kinetics of crystallization, crystalline microstructure, and the structure of the fat crystal matrix (Coupland, 2002).

6.5.1 Kinetics of crystallization in fine droplets

The process of crystallization proceeds via two distinct processes: crystal nucleation and growth (Garside, 1985). The nucleation kinetics in fine droplets is often different from nucleation in the same liquid in bulk. In a fine emulsion, the number of droplets exceeds the number of potential nucleation catalysts (impurities) present in the liquid oil. Thus, a proportion of the lipid is effectively catalyst free and must nucleate by other mechanisms. This may either be completely spontaneous homogeneous nucleation or, more probably, some catalytic effect of the droplet surface (Coupland, 2002). In either case, the crystallization temperature is greatly reduced below the melting point and depends both on particle size and on the nature of the emulsifier selected. For example, Higami and co-workers (2003) showed that the crystallization temperature of trilaurin molecules decreased from 18.9 °C, the crystallization temperature of the bulk lipid, to –9.5 °C when emulsified into droplets smaller than 100 nm.

In a fine droplet, droplets crystallization is most often governed by surface effects, i.e., surface heterogeneous nucleation. The most generally accepted model for surface heterogeneous nucleation is based on an ordering of lipid molecules by the hydrophobic portion of the emulsifier at the interface. If the hydrophobic tail of the surfactant molecule has a similar molecular structure to the lipid, it may catalyze the surface nucleation event (Coupland, 2002). A good example of surfactant-directed interfacial heterogeneous nucleation is provided by Awad and Sato (2002) and Kaneko and co-workers (1999), who showed that hydrophobic sucrose oligoesters of stearic acid and palmitic acid accelerated the crystallization rate in emulsified palm kernel oil and hexadecane, respectively. They argue the hydrophobic surfactant absorbs to the interior surface of the droplet, where it provides a template for crystal formation.

The ordering effect preceding interfacial heterogeneous nucleation is believed to take place via the formation of rotator phase; a transient structure corresponding to TAG molecules orientated along their long axis but retaining some rotational mobility (Sirota, 1998). This phase is believed to act as an interfacial frozen monolayer, which in turn acts as a template for crystallization of the remainder of the lipid droplet (Wu *et al.*, 1993; Sirota, 1998). However, Gulseren and Coupland (2008) argue that the region of lipid affected by the surfactant is much greater than a monolayer. Some recent and elegant support for this mechanism is provided by Shinohara and co-workers (2008), who used a focused X-ray beam to study hexadecane emulsions stabilized with different polysorbate surfactants. They showed that when the size and shape of the hydrophobic tail of the surfactant is similar to hexadecane, then the rotator of the lipid was detectable before crystallization and produces a radial distribution of crystals after crystallization is complete (Figure 6.3).

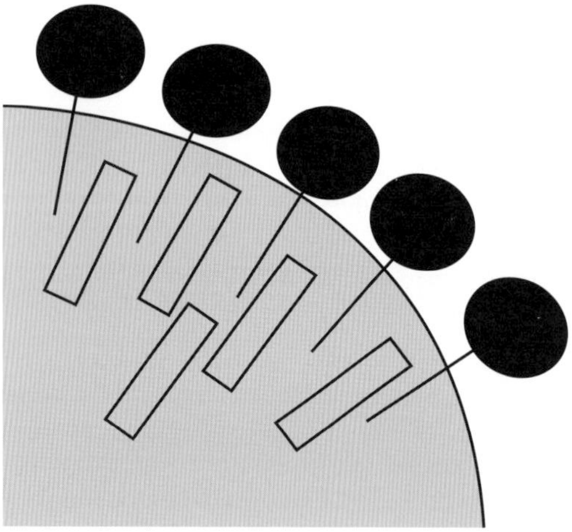

Figure 6.3 Alkane molecules (open rectangles) align with surfactant at the interface and produce a template for crystal nucleation (adapted from Awad and Sato, 2002).

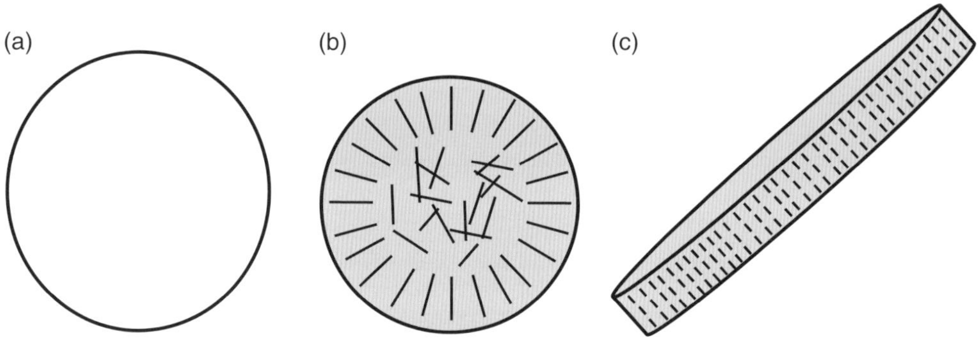

Figure 6.4 Proposed structures of (a) liquid nanoemulsions and SLN from TAG in the (b) α and (c) β crystal form (adapted from Bunjes *et al.*, 2007). Straight lines show the alignment of selected molecular backbones in the crystal lattice. The α crystals are aligned near the droplet surface and the more disordered structure in the core. The β crystals are aligned in lamellae as shown. (Highly schematic and not to scale.)

The second effect of emulsification on fat crystallization is thermodynamic in origin and arises from the elevation in internal droplet pressure due to the curvature of the surface. The lipid is pressurized by the interfacial tension and thus the formation of the more dense crystal phase is favored. This effect is only really important for very small particles (<100 nm). For example, Higami and co-workers (2003) showed that the melting temperature of very fine trilaurin droplets (β-polymorph, bulk melting point 46.7 °C) was reduced to as low as 26.5 °C. In some very fine particles, the different lamellae of the lipid may have distinct melting points. For example, Bunjes and co-workers (2000) showed that as the particle size of TAG droplets was decreased the melting range broadened and in DSC showed a number of distinct peaks, which were attributed to the melting of distinct layers of the crystal.

6.5.2 Structure of crystalline fat droplets

The type of TAG polymorph is determined by many factors and affects the crystalline structure within a droplet. Emulsions prepared from pure TAG usually also nucleate into the α-polymorph upon cooling, followed by a solid–solid transformation into a β' or β polymorph on storage, or reheating (Awad *et al.*, 2008; Bunjes and Unruh, 2007; Higami *et al.*, 2003; Weiss *et al.*, 2008). Bunjes and co-workers (2003) reported an unusual polymorphic form in rapidly cooled SLN corresponding to α-polymorph without any apparent long range order (i.e., no SAXS pattern). Moreover, the α-crystals are aligned with the surface (Figure 6.4) in a similar manner seen in X-ray microbeam measurements on larger droplets (Shinohara *et al.*, 2008), suggesting a potential role of the surface emulsifier in directing crystallization.

In general, the α form is preferred over β form for SLN, firstly because BLI can be more readily incorporated in the less-dense crystals and, secondly, because the formation of the β form is associated with a change in droplet shape and subsequent gelation of the suspension (see below). The rate of polymorphic transformation in fine particles is:

(1) greater for fine droplets and becomes especially significant at the nano-scale (Awad *et al.*, 2008; Bunjes and Koch, 2005; Bunjes *et al.*, 2003, 2007; Helgason *et al.*, 2008; Higami *et al.*, 2003; Illing *et al.*, 2004; Unruh *et al.*, 2002; Westesen and Siekmann, 1997). The increased α to β transformation tendency in SLN can be attributed to "the relaxation of lattice strain developing during the transition is facilitated in smaller crystalline domains due to the high surface-to-volume ratio, as suggested for bulk material by Dafler (1977)" (Bunjes and Westesen, 2001).
(2) greater for shorter chain TAGs than longer ones as the molecular mobility of small molecules is greater (Bunjes *et al.*, 1996).
(3) lower at lower temperatures. Helgason and co-workers (2008) showed that the polymorphic transformation rates of tripalmitin nanoparticles (r ~150 nm) increased by increasing the temperature from 1 to 10 °C. Similarly, Awad and co-workers (2008) showed that faster cooling rates retard the polymorphic transition.
(4) dependent on the surfactant selected (Awad and Sato, 2002; Bunjes *et al.*, 2002, 2003). Although the mechanism of surfactant action is not understood thoroughly, Bunjes and co-workers (2002, 2003) suggested that, since the selection of hydrophilic surfactants can significantly alter the transformation kinetics, the polymorphic transition starts at the surface rather than the lipid core. This observation is reasonable assuming that TAG molecules in the surface region interacting with surfactant molecules have higher free energy and mobility. The capacity of different surfactants to stabilize the α form varies considerably. For example, Bunjes and Koch (2005) showed that saturated long chain phospholipids decreased the rate of polymorphic transitions compared to unsaturated soy bean phospholipids. Awad and Sato (2002) saw similar results in larger droplets with other hydrophobic emulsifiers. Bunjes and co-workers (2003) found that combining bile salts with saturated long chain phospholipids stabilized the α form during recrystallization. One of the most striking examples published to date is a nine month shelf life in refrigerated saturated TAG nanoparticles stabilized solely with polyvinyl alcohol (Rosenblatt and Bunjes, 2009).

In very fine droplets, the recrystallization event from less stable to more stable polymorphic form is often associated with a change in droplet shape from spherical to plate-

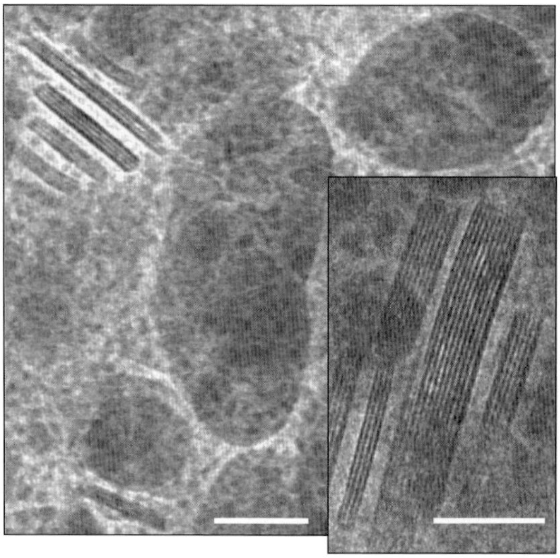

Figure 6.5 Cryo-TEM micrograph of tristearin nanoparticles (scale bar = 100 nm) showing a characteristic plate-like morphology in the β polymorphic form. (Reproduced with permission from Langmuir, Bunjes *et al.*, 2007, copyright 2007, the American Chemical Society.)

like (Bunjes *et al.*, 2007) (Figure 6.4), and subsequent gelation of the dispersion. For example, tristearin droplets with particle sizes around 100–200 nm crystallized in the α form were spheriodal in shape with some possible elongation and also with concentric layers of TAG molecules. When these droplets converted to the β modification they were plate-like nanocrystals with a planar layered internal structure (Figure 6.5). One reason for this change in shape is that α-nucleation occurs quickly at multiple locations throughout the droplet to form a polycrystalline particle. The α to β transition is slower and perhaps one β crystal forms initially, which grows at the expense of existing α crystals within the droplet yielding a single monocrystalline particle. For the larger droplet sizes (i.e., >1 μm) shape change and stack formation were not observed on crystallization (Bunjes *et al.*, 2003). This may be because there are a greater number of β-nucleation events within the larger droplets yielding a grained structure, with approximately the same shape as the starting droplet.

The β-form plates can pack together in stacks. The interpolate distance has been measured by small angle X-ray scattering (SAXS) and transmission electron microscopy (TEM) to be in the order of 20 nm (Illing *et al.*, 2004; Unruh *et al.*, 2002). It may be possible in some cases to reduce the tendency of plate-like particles to associate by adding more surfactant to cover the increased interfacial area (Westesen and Siekmann, 1997; Helgason *et al.*, 2008). Presumably, there is an attractive force between fat crystals suspended in water but the effective magnitude of the force is greater between flat plates than between spheres as it acts over a larger area.

6.6 LOCALIZATION OF BLI IN SOLID LIPID NANOPARTICLES

BLI dissolved in liquid oil nanoemulsions will redistribute on crystallization of the fat (Figure 6.6). Even if the BLI is soluble in the liquid oil (i.e., forms a one-phase solution), crystallization of the carrier lipid will likely cause a phase separation. Figure 6.6a shows the

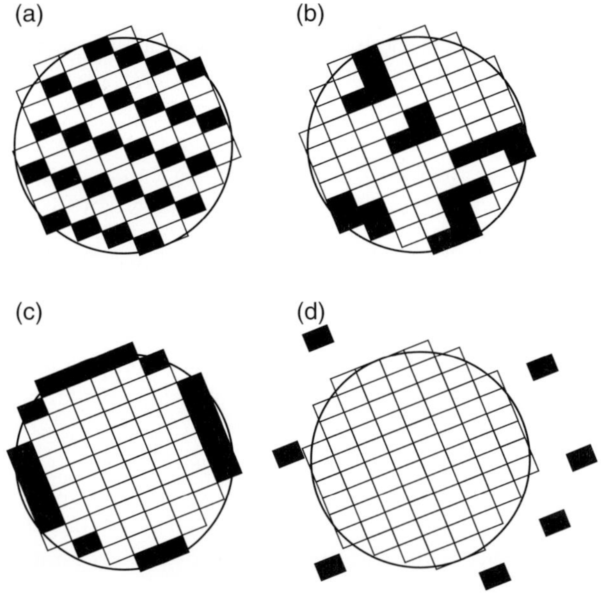

Figure 6.6 Possible effects of droplet crystallization on solute distribution. Solute is shown schematically as filled rectangles and crystalline carrier lipid as open rectangles. (a) The solute may be incorporated into the crystal lattice. On the other hand, if the solute is excluded from the lattice it can accumulate (b) at grain boundaries inside the droplet, (c) at the surface of the droplet or even (d) partition into the aqueous phase. Note the droplets are drawn as spheres although different shapes are possible as discussed previously. (Highly schematic and not to scale.)

unlikely event where phase separation does not follow lipid crystallization, that is, the BLI can form a single solid phase with the carrier lipid. Co-crystallization would presumably be detectable as large shifts in the XRD but unsurprisingly, considering the wide disparity in molecular shapes of solute and solvent, co-crystallization appears to be uncommon. For example, Bunjes and co-workers (2001) found no evidence that ubidecarenone was incorporated into the lattice of SLN although the kinetics of polymorphic transition in the lipid increased in the presence of the drug. It may, however, be possible to form compound crystals if the BLI of interest is itself a TAG (Takeuchi *et al.*, 2002a, 2002b). Presuming co-crystallization is the exception rather than the rule, then some level of phase separation must occur and the question becomes where are the separate phases localized?

Figures 6.6b, 6.6c and 6.6d show a phase separation of the BLI as either a solid precipitate or as a solution with residual liquid lipid. In Figure 6.6b, the BLI is entrapped within the solid droplet; in Figures 6.6c and 6.6d, the BLI is excluded from the crystalline lipid core and either adsorbed at the surface or dispersed in the aqueous phase, respectively. The major thermodynamic difference between the models is that the surface-bound model (Figure 6.6c) has more crystalline lipid–BLI interactions and fewer water–BLI and water–crystalline lipid interactions than the BLI-excluded model (Figure 6.6d). While there may be some exceptions, the surface-bound model is probably energetically preferable in most cases. In conclusion, amongst the models presented for BLI encapsulation in Figure 6.6 the core-trapped (Figure 6.6b) and surface-bound (Figure 6.6c) seem the most probable. The most likely suitable location for entrapment in the solid droplet core (i.e., Figure 6.6b) is the grain boundaries between TAG crystals or simply dissolved in any residual liquid oil.

The imperfection density in the crystalline matrix is equal to the number of nucleation events, so it may be possible to modify the internal structure of the droplet by using surfactants or alterations in cooling rate to change the rate of nucleation. If a droplet crystallizes as a single crystal, then there is no phase volume available in the droplet core so the surface-bound model is the only one available. For example, Ghosh and co-workers (2006, 2007) used headspace measurements to show that the crystallization of eicosane droplets forced an encapsulated BLI out of the droplet core. In SLN from polymorphic TAGs, the α–β transition is associated with the rapid release of encapsulated molecules (i.e., "burst release") (Muhlen *et al.*, 1998; Muller *et al.*, 2000) presumably because the β-crystalline particle is a single crystal. Structuring the SLN as a mixture of different types of crystals may allow the development of more internal structures to entrap BLI in the core of the droplet. For example, TAGs with different fatty acid composition can crystallize into double or triple chain length structures, which are immiscible, and a mixture of them may provide desired crystalline lipid network characteristics (Muller *et al.*, 2000).

Finally, a small fraction of non-crystallizing lipid may provide superior performance to a purely solid SLN. These semi-crystalline particles are sometimes known as nano-structured lipid carriers (NLC). Ideally, the small fraction of liquid oil would provide a reservoir for the BLI, and hence slower release, while the crystalline fat provides some physical protection from chemical degradation (Jenning *et al.*, 2000a, 2000b). The solid fat content of the droplets can be modified by mixing solid fats with compatible or partially compatible liquid oils. Jenning and co-workers (2000a, 2000b) manufactured NLCs by replacing solid lipid with medium chain and showed that liquid lipid phase at low concentrations (8–16%) was evenly distributed within the crystal matrix. In contrast, Jores and co-workers (2004) produced similar NLC particles from similar lipids and observed irregular shapes and uneven distribution of the oil, that is, on the surface or edge of the solid portion. At high liquid to solid oil ratios, liquid oil attached as distinct droplet on platelet solid lipid, a so called "nano-spoon". Similar conclusions were reached by Ahlin and co-workers (2003), who showed that the affinity of BLI for the lipid core increases with the lipophilicity, while amphiphilic molecules tend to be carried in the amphiphilic surface region.

6.7 CONCLUSIONS

There is a need to control and modify the stability and functionality of lipophilic bioactive molecules in foods. One approach is to use microencapsulation to alter the microenviroment of the molecules of interest. There are a range of microencapsulation technologies available and, while different applications require different approaches, in aqueous foods emulsion-based delivery systems remain dominant.

One reason for the continued dominance of emulsion-based delivery systems is that they are so common in foods. Many foods are traditionally emulsions and if they are fortified with a lipophilic ingredient then that ingredient will partition into the lipid phase. The performance of these emulsions as delivery vehicles can be improved, particularly by modifying the structure of the surface and, to a lesser extent, the particle size. More recent research has examined the potential of improving the EBDS by modifying the crystal structure of the dispersed lipid phase.

SLN can be formed using conventional food processing technology and a hot solution of a high-melting fat. Crystallization of the carrier lipid alters the chemical environment of other lipophilic ingredients dissolved in the liquid oil. However, the evidence to date appears to suggest that resultant changes in functional properties are either unhelpful or neutral. Ingredients encapsulated in SLN tend to be more prone to chemical degradation compared to the same ingredients in liquid oil nanoemulsions. Furthermore, SLN seem to be less physically stable than liquid oil nanoemulsions (particularly in the β-polymorph). It may be that the most effective use of SLN will be for ingredients designed to react quickly, perhaps antioxidants, or to promote adsorption.

ACKNOWLEDGEMENT

This work was supported by a grant from the USDA–AFRI program (Award number: 2009–65503–05960, Program code: 93430).

REFERENCES

Acosta E. 2009. Bioavailability of nanoparticles in nutrient and nutraceutical delivery. Curr. Opin Colloid Interface Sci 14(1):3–15.

Ahlin P, Julijana K, Slavko P, Janez S, Marjeta S. 2003. The effect of lipophilicity of spin-labeled compounds on their distribution in solid lipid nanoparticle dispersions studied by electron paramagnetic resonance. J Pharm Sci 92(1):58–66.

Amar I, Aserin A, Garti N. 2003. Solubilization patterns of lutein and lutein esters in food grade nonionic microemulsions. J Agric Food Chem 51(16):4775–4781.

Anton N, Benoit JP, Saulnier P. 2008. Design and production of nanoparticles formulated from nano-emulsion templates – A review. J Control Release 128:185–199.

Arboleya JC, Sutcliffe LH, Wilde PJ, Fairhust SA. 2005. Density and microviscosity studies of palm oil/water emulsions. J Agric Food Chem 53(11):4448–4453.

Awad TS, Sato K. 2002. Acceleration of crystallisation of palm kernel oil in oil-in-water emulsion by hydrophobic emulsifier additives. Colloids Surf B–Biointerfaces 25(1):45–53.

Awad TS, Helgason T, Kristbergsson K, Decker EA, Weiss J, McClements DJ. 2008. Effect of cooling and heating rates on polymorphic transformations and gelation of tripalmitin solid lipid nanoparticle (SLN) suspensions. Food Biophys 3:155–162.

Bonacucina G, Cespi M, Misici-Falzi M, Palmieri GF. 2009. Collodal soft matter as drug delivery system. J Pharm Sci 98(1):1–42.

Boon CS, Xu Z, Yue X, McClements DJ, Weiss J, Decker EA. 2008. Factors affecting lycopene oxidation in oil-in-water emulsions. J Agric Food Chem 56(4):1408–1414.

Bunjes H, Koch MHJ. 2005. Saturated phospholipids promote crystallization but slow down polymorphic transitions in triglyceride nanoparticles. J Control Release 107:229–243.

Bunjes H, Unruh T. 2007. Characterization of lipid nanoparticles by differential scanning calorimetry, x-ray and neutron scattering. Adv Drug Deliv Rev 59:379–402.

Bunjes H, Westesen K. 2001. Influences of colloidal state on physical properties of solid fats. In: Garti N & Sato K (editors). Crystallization Processes in Fats and Lipid Systems. New York: Marcel Dekker, Inc. pp. 457–484.

Bunjes H, Westesen K, Koch MHJ. 1996. Crystallization tendency and polymorphic transitions in triglyceride nanoparticles. Int J Pharm 129:159–173.

Bunjes H, Koch MHJ, Westesen K. 2000. Effect of particle size on colloidal solid triglycerides. Langmuir 16:5234–5241.

Bunjes H, Drechsler M, Koch MHJ, Westesen K. 2001. Incorporation of the model drug Ubidecarenone into solid lipid nanoparticles. Pharm Res 18(3):287–293.

Bunjes H, Koch MHJ, Westesen K. 2002. Effects of surfactants on the crystallization and polymorphism of lipid nanoparticles.ProgColloid PolymSci 121:7–10.

Bunjes H, Koch MHJ, Westesen K. 2003. Influence of emulsifiers on the crystallization of solid lipid nanoparticles. J Pharm Sci 92(7):1509–1520.

Bunjes H, Steiniger F, Richter W. 2007. Visualizing the structure of triglyceride nanoparticles in different crystal modification. Langmuir 23(7):4005–4012.

Chaiyasit W, Silvestre MP, McClements DJ, Decker EA. 2000. Ability of surfactant hydrophobic tail group size to alter lipid oxidation in oil-in-water emulsions. J Agric Food Chem 48(8):3077–3080.

Chaiyasit W, Elias RJ, McClements DJ, Decker EA. 2007. Role of physical structures in bulk oils on lipid oxidation. Crit Rev Food Sci Nutr 47:299–317.

Chee CP, Gallaher JJ, Djordjevic D, *et al.* 2005. Chemical and sensory analysis of strawberry flavoured yogurt supplemented with an algae oil emulsion. J Dairy Res 72(3):311–316.

Chee CP, Djordjevic D, Faraji H, *et al.* 2007. Sensory properties of vanilla and strawberry flavored ice cream supplemented with omega-3 fatty acids. Milchwissenschaft, 62(1):66–69.

Claesson PM, Blomberg E, Poptoshev E. 2004. Surface forces and emulsion stability. In: Friberg SE, Larsson K, & Sjoblom J (editors). Food Emulsions, 4th Edn. New York: Marcel Dekker, Inc. pp. 257–297.

Coupland JN. 2002. Crystallization in emulsions. Curr Opin Colloid Interface Sci 7:445–450.

Dafler JR. 1977. Polymorphism behavior in fully hydrogenated mono acid triglycerides. J Am Oil Chem Soc 54(6):249–254.

Dalgleish DG, Srinivasan M, Singh M. 1995. Surface properties of oil-in-water emulsion droplets containing casein and Tween-60. J Agric Food Chem 43:2351–2355.

Dickinson E. 1992. An Introduction to Food Colloids. New York: Oxford University Press.

Donnelly JL, Decker EA, McClements DJ. 1998. Iron-catalyzed oxidation of Menhaden oil as affected by emulsifiers. J Food Sci 63(6):997–1000.

Garside J. 1985. Industrial crystallization from solution. Chem Eng Sci 40(1):3–26.

Ghosh S, Peterson DG, Coupland JN. 2006. Effects of droplet crystallization and melting on the aroma release properties of a model oil-in-water emulsion. J Agric Food Chem 54 (5):1829–1837.

Ghosh S, Peterson DG, Coupland JN. 2007. Flavor binding by solid and liquid emulsion droplets. In: Dickinson E & Lesser ME (editors). Food Colloids: Self-assembly and material science. London: Royal Society of Chemistry. pp. 423–432.

Gulseren I, Coupland JN. 2008. Surface melting in alkane emulsion droplets as affected by surfactant type. J Am Oil Chem Soc 85(5):413–419.

Haumann BF. 1994. Tools: Hydrogenation, interesterification. INFORM, 5:668–674.

Helgason T, Awad TS, Kristbergsson K, McClements DJ, Weiss J. 2008. Influence of polymorphic transformations on gelation of tripalmitin solid lipid nanoparticle suspensions. J Am Oil Chem Soc 85:501–511.

Helgason T, Awad TS, Kristbergsson K, Decker EA, McClements DJ, Weiss J. 2009. Impact of surfactant properties on oxidative stability of β-carotene encapsulated within solid lipid nanoparticles. J Agric Food Chem 57:8033–8040.

Higami M, Ueno S, Segawa T, Iwanami K, Sato K. 2003. Simultaneous synchrotron radiation X-ray diffraction – DSC analysis of melting and crystallization behavior of trilauroylglycerol in nanoparticles of oil-in-water emulsion. J Am Oil Chem Soc 80(8):731–739.

Himawan C, Starov VM, Stapley AGF. 2006. Thermodynamic and kinetic aspects of fat crystallization. Adv Colloid Interface Sci 122:3–33.

Hu M, McClements DJ, Decker EA. 2003. Impact of whey protein emulsifiers on the oxidative stability of salmon oil-in-water emulsions. J Agric Food Chem 51(5):1435–1439.

Illing A, Unruh T, Koch MHJ. 2004. Investigation on particle self-assembly in solid lipid-based colloidal drug carrier systems. Pharm Res 21(4):592–597.

Jani P, Halbert GW, Langridge J, Florence AT. 1990. Nanoparticle uptake by the rat gastrointestinal mucosa: quantitation and particle size dependency. J Pharm Pharmacol 42:821–826.

Jenning V, Gohla SH. 2001. Encapsulation of retinoids in solid lipid nanoparticles (SLN). J Microencapsulation 18(2):149–158.

Jenning V, Thunemann AF, Gohla SH. 2000a. Characterization of a novel solid lipid nanoparticle carrier system based on binary mixtures of liquid and solid lipids. Int J Pharm 199:167–177.

Jenning V, Mader K, Gohla SH. 2000b. Solid lipid nanoparticles (SLN) based on binary mixtures of liquid and solid lipids: a H-NMR study. Int J Pharm 205:15–21.

Jores K, Mehnert W, Drechsler M, Bunjes H, Johann C, Mader K. 2004. Investigations on the structure of solid lipid nanoparticles (SLN) and oil-loaded solid lipid nanoparticles by photon correlation spectroscopy, field-flow fractionation and transmission electron microscopy. J Control Release 95:217–227.

Kaneko N, Horie T, Ueno S, Yano J, Katsuragi T, Sato K. 1999. Impurity effects on crystallization rates of *n*-hexadecane in oil-in-water emulsions. J Cryst Growth 197:263–270.

Kiokias S, Dimakou C, Oreopoulou V. 2007. Effect of heat treatment and droplet size on the oxidative stability of whey protein emulsions. Food Chem 105(1):94–100.

Klinkesorn U, Sophanodora P, Chinachoti P, Decker EA, McClements DJ. 2005. Encapsulation of emulsified tuna oil in two-layered interfacial membranes prepared using electrostatic layer-by-layer deposition. Food Hydrocolloids 19(6):1044–1053.

Kreuter J. 1991. Peroral administration of nanoparticles. Adv Drug Deliv Rev 7:71–86.

Lee S, Decker EA, Faustman C, Mancini RA. 2005. The effects of antioxidant combinations on color and lipid oxidation in n-3 oil fortified ground beef patties. Meat Sci 70(4):683–689.

Lee S, Faustman C, Djordjevic D, Faraji H, Decker EA. 2006a. Effect of antioxidants on stabilization of meat products fortified with n-3 fatty acids. Meat Sci 72(1):18–24.

Lee S, Hernandez P, Djordjevic D, *et al.* 2006b. Effect of antioxidants and cooking on stability of n-3 fatty acids in fortified meat products. J Food Sci 71(3):C233–238.

Let MB, Jacobsen C, Sorensen ADM, Meyer AS. 2007. Homogenization conditions affect the oxidative stability of fish oil enriched milk emulsions: Lipid oxidation. J Agric Food Chem 55(5):1773–1780.

Losso JN, Khachatryan A, Ogawa M, Godber JS, Shih F. 2005. Random centroid optimization of phosphatidylglycerol stabilized lutein-enriched oil-in-water emulsions at acidic pH. Food Chem 92(4):737–744.

Luykx DMAM, Peters RJB, Van Ruth SM, Bouwmeester H. 2008. A review of analytical methods for the identification and characterization of nano delivery systems in food. J Agric Food Chem 56:8231–8247.

McClements DJ. 2005. Food Emulsions: Principles, Practice, and Techniques, 2nd ed. Boca Raton, FL: CRC Press.

McClements DJ, Decker EA. 2000. Lipid oxidation in oil-in-water emulsions: Impact of molecular environment on chemical reactions in heterogeneous food systems. J Food Sci 65(8):1270–1282.

McClements DJ, Decker E, Weiss J. 2007. Emulsion-based delivery systems for lipophilic bioactive components. J Food Sci 72(8):109–124.

McClements DJ, Decker EA, Park Y, Weiss J. 2009. Structural design principles for delivery of bioactive components in nutraceutical and functional foods. Crit Rev Food Sci Nutr 49:577–606.

Mehnert W, Mader K. 2001. Solid lipid nanoparticles: production, characterization and applications. Adv Drug Deliv Rev 47:165–196.

Mei L, Decker EA, McClements DJ. 1998a. Evidence of iron association with emulsion droplets and its impact on lipid oxidation. J Agric Food Chem 46:5072–5077.

Mei L, McClements DJ, Wu J, Decker EA. 1998b. Iron-catalyzed lipid oxidation in emulsion as affected by surfactant, pH and NaCl. Food Chem 61(3):307–312.

Muhlen A, Schwarz C, Mehnert W. 1998. Solid lipid nanoparticles (SLN) for controlled drug delivery – Drug release and release mechanisms. Eur J Pharm Biopharm 45:149–155.

Muller RH, Mader K, Gohla S. 2000. Solid lipid nanoparticles (SLN) for controlled drug delivery – a review of the state of the art. Eur J Pharm Biopharm 50:161–177.

Nakazawa R, Shima M, Adachi S. 2008. Effect of Oil Droplet Size on the Oxidation of Microencapsulated Methyl Linoleate. J Oleo Sci 57(4):225–232.

Povey MJW. 2001. Crystallization of oil-in-water emulsions. In: Garti N a & Sato K (editors). Crystallization Processes in Fats and Lipid Systems. New York: Marcel Dekker Inc. pp. 251–288.

Reineccius G. (2006). Flavor Chemistry and Technology. Boca Raton, FL: CRC Press.

Ribeiro HS, Rico LG, Badolato GG, Schubert H. 2005. Production of O/W emulsions containing astaxanthin by repeated Premix membrane emulsification. J Food Sci 70(2):E117–123.

Ribeiro HS, Guerrero JMM, Briviba K, Rechkemmer G, Schuchmann HP, Schubert H. 2006. Cellular uptake of carotenoid-loaded oil-in-water emulsions in colon carcinoma cells in vitro. J Agric Food Chem 54(25):9366–9369.

Richelle M, Bortlik K, Liardet S, *et al.* 2002. A food-based formulation provides lycopene with the same bioavailability to humans as that from tomato paste. J Nutr 132:404–408.

Rosenblatt KM, Bunjes H. 2009. Poly(vinyl alcohol) as Emulsifier Stabilizes Solid Triglyceride Drug Carrier Nanoparticles in the alpha-Modification. Mol Pharm 6(1):105–120.

Rothwell JA, Day AJ, Morgan MRA. 2005. Experimental determination of octanol–water partition coefficients of quercetin and related flavors. J Agric Food Chem 53:4355–4360.

Santipanichwong R, Suphantharika M. 2007. Carotenoids as colorants in reduced-fat mayonnaise containing spent brewer's yeast beta-glucan as a fat replacer. Food Hydrocolloids 21(4):565–574.

Sato K. 1999. Solidification and phase transformation behavior of food fats – a review. Fett/Lipid 12:467–474.

Sato K. 2001. Crystallization behavior of fats and lipids – a review. Chem Eng Sci 56:2255–2265.

Semo E, Kesselman E, Danino D, Livney Y. 2007. Casein micelles as a nano-capsular vehicle for neutraceuticals. Food Hydrocolloids 21:936–942.

Sharma R. 2005. Market trends and opportunities for functional dairy beverages. Aust J Dairy Technol 60(2):195–198.

Shinohara Y, Takamizawa T, Ueno S, *et al.* 2008. Microbeam X-ray diffraction analysis of interfacial heterogeneous nucleation of n-hexadecane inside oil-in-water emulsion droplets. Cryst Growth Des 8(9):3123–3126.

Silvestre MPC, Chaiyasit W, Brannan RG, McClements DJ, Decker EA. 2000. Ability of surfactant head group size to alter lipid and antioxidant oxidation in oil-in-water emulsions. J Agric Food Chem 48:2057–2061.

Sirota EB. 1998. Supercooling, nucleation, rotator phases, and surface crystallization of n-Alkane melts. Langmuir 14:3133–3136.

Solans S, Izquierdo P, Nolla J, Azemar N, Garcia-Celma MJ. 2005. Nano-emulsions. Curr Opin Colloid Interface Sci 10:102–110.

Takeuchi M, Ueno S, Floter E, Sato K. 2002a. Binary phase behavior of 1,3-Distearoyl-2-oleoyl-sn-glycerol (SOS) and 1,3-Distearoyl-2-linoleoyl-sn-glycerol (SLS). J Am Oil Chem Soc 79(7):627–632.

Takeuchi M, Ueno S, Sato K. 2002b. Crystallization kinetics of polymorphic forms of a molecular compound constructed by SOS (1,3-distearoyl-2-oleoyl-sn-glycerol) and SSO (1,2-distearoyl-3-oleoyl-rac-glycerol). Food Res Int 35:919–926.

Tyssandier V, Lyan B, Borel P. 2001. Main factors governing the transfer of carotenoids from emulsion lipid droplets to micelles. Biochimica et Biophysica Acta (BBA)/Mol Cell Biol Lipids 1533(3):285–292.

Unruh T, Westesen K, Bosecke P, Linder P, Koch MHJ. 2002. Self-assembly of triglyceride nanocrystals in suspension. Langmuir 18:1796–1800.

Vanapalli SA, Coupland JN. 2004. Orthokinetic stability of food emulsions. In: Friberg SE, Larsson K & Sjoblom J (editors). Food Emulsion, 4th Edn. New York: Marcel Dekker, Inc. pp. 327–351.

Velikov KP, Pelan E. 2008. Colloidal delivery systems for micronutrients and nutraceuticals. Soft Matter 4:1964–1980.

Yannai S. 2004. Dictionary of Food Compounds. Boca Raton, FL: Chapman & Hall/CRC Press.

Wagner LA, Warthesen JJ. 1995. Stability of spray-dried encapsulated carrot carotenes. J Food Sci 60(5):1048–1053.

Weiss J, Coupland JN, McClements DJ. 1996. Solubilization of hydrocarbon emulsion droplets suspended in nonionic surfactant micelle solutions. J Phys Chem 100:1066–1071.

Weiss J, McClements DJ, Kristbergsson K, Helgason T, Awad T. 2008. Solid lipid nanoparticles as delivery systems for bioactive food components. Food Biophys 3:146–154.

Westesen K, Siekmann B. 1997. Investigation of the gel formation of phospholipid stabilized solid lipid nanoparticles. Int J Pharm 151:35–45.

Wu XZ, Sirota EB, Sinha SK, Ocko BM, Deutsh M. 1993. Surface crystallization of liquid normal-alkanes. Phys Rev Lett 70(7):958–961.

7 Fermentation

Mark R. Wilkins and Hasan Atiyeh

7.1 INTRODUCTION

From ancient times man has used microorganisms to produce valuable products. The first instances of the use of microbiology were the production of bread with yeast, followed closely by beer and wine fermentation. Fermented milk products such as yogurt, kefir, and ghee are found throughout the world. Though the mechanisms behind the production of these products have only been understood for the last 100–150 years, their importance in preserving foods and beverages from spoilage and the prevention of food-borne diseases has long been appreciated.

In addition to food products, microorganisms have also played an important role in the production of industrial chemicals and pharmaceuticals. Many products, such as acetone, butanol, glycerol and ethanol, were exclusively produced from microbial fermentations during the two World Wars due to lack of petroleum and other raw materials. As the Second World War ended and the price of petroleum declined, petrochemical methods of producing these and other chemicals emerged as cheaper alternatives to microbial production methods; thus, microbial chemical production methods were abandoned.

In recent years, the price of petroleum has increased again. As emerging economies such as China and India increase their demand for petroleum and put more pressure on petroleum prices and supplies, the need for alternative chemical production methods using renewable resources has been recognized by governments and industries throughout the world. Also, the recognition of carbon dioxide (CO_2) emissions from combustion of fossil fuels as a cause of global climate change has further highlighted the need for production methods that produce less carbon dioxide. As a result of increased petroleum prices and awareness of the role of carbon dioxide in climate change, microbial-based chemical production processes that were previously abandoned have become the subject of intense research and development effort.

The word *fermentation* conventionally is used in several different ways. However, in this chapter it will be used only in the strict definition of anaerobic and/or microaerobic microbial processes. Anaerobic processes are those where oxygen is not present and microaerobic processes are those where oxygen is present at concentrations much lower than those in air (21% by volume). Some discussion of aerobic alternatives to fermentation processes will also be discussed.

Food and Industrial Bioproducts and Bioprocessing, First Edition. Edited by Nurhan Turgut Dunford.

7.2 FERMENTATIVE PATHWAYS

Mono- or disaccharide carbohydrates are the substrate for the majority of industrial fermentations. Carbohydrates are reduced in a series of reactions, known as glycolysis, that result in the production of two moles of pyruvate (Figure 7.1). In glycolysis sugars are degraded to produce ATP, NADH, and pyruvate (Pyr) by the overall reaction:

$$\text{Glucose} + 2\ \text{ADP} + 2\ \text{NAD}^+ + 2\ \text{PO}_3^- \rightarrow 2\ \text{Pyr} + 2\ \text{ATP} + 2\ (\text{NADH} + \text{H}^+)$$

The overall glycolysis reaction is accomplished by a series of ten enzymatic reactions (Figure 7.1). Glycolysis is also known as the Embden–Meyerhoff–Parnas pathway. A few organisms use an alternate pathway to produce pyruvate from glucose known as the Entner–Doudoroff pathway (Shuler and Kargi, 2002). This pathway produces only one mole of ATP per mole of glucose instead of two moles of ATP. The production of only one mole of ATP leaves the cell with less energy to produce cell mass and results in a greater yield of fermentation products. *Zymomonas mobilis*, a bacterium that produces ethanol from glucose and fructose, uses the Entner–Doudoroff pathway.

If oxygen is present, pyruvate is converted to acetyl-CoA, which enters the citric acid cycle. The citric acid cycle generates more ATP, reducing compounds, and chemical

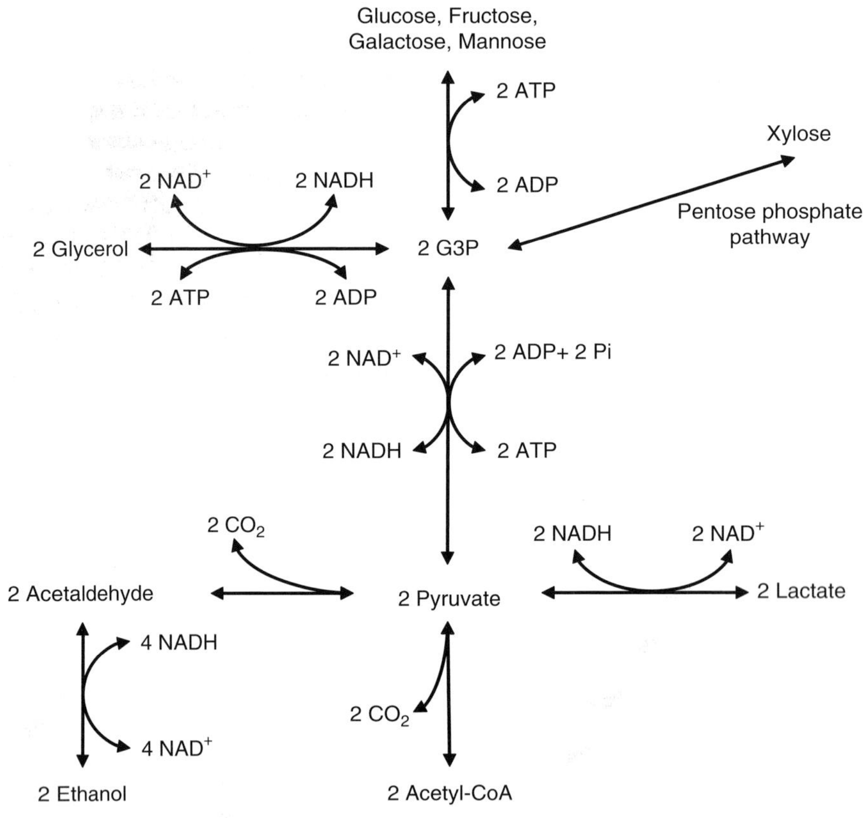

Figure 7.1 Reduction of carbohydrates by glycolysis.

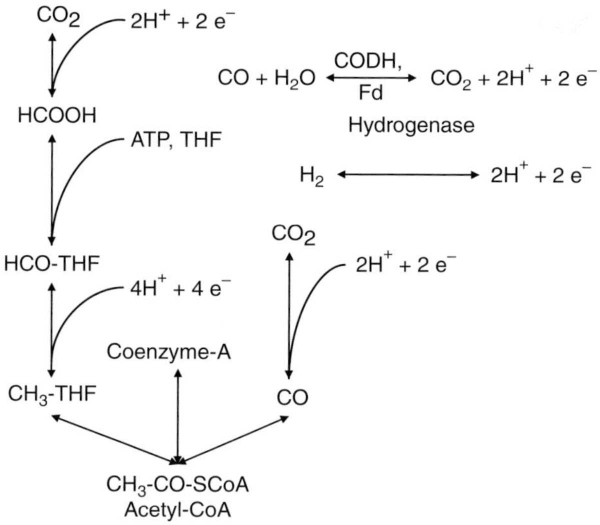

Figure 7.2 Wood–Ljungdahl pathway (CODH: carbon monoxide dehydrogenase; Fd: ferredoxin).

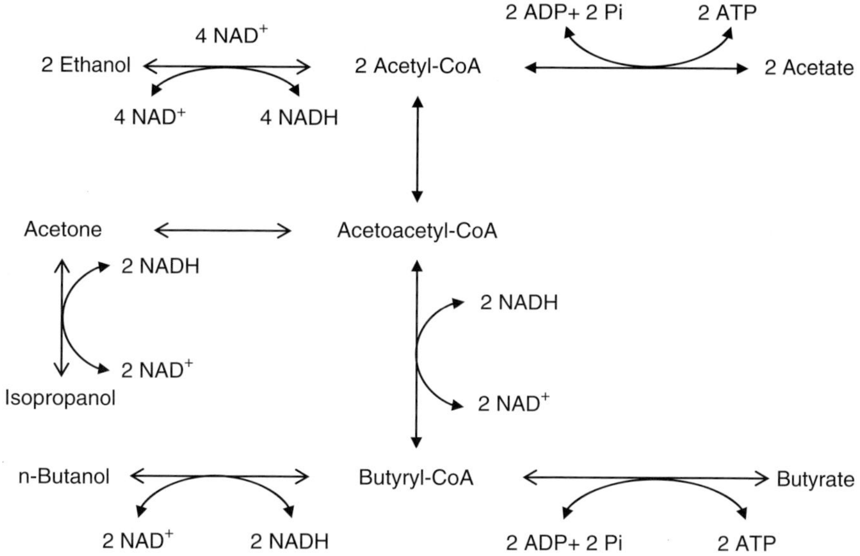

Figure 7.3 Fermentative pathway utilizing acetyl-CoA.

precursors for the formation of the various compounds necessary for cellular metabolism. If oxygen is not present, pyruvate can still be converted to acetyl-CoA or it can be reduced to form various products.

There are also organisms that are capable of utilizing carbon monoxide (CO), carbon dioxide (CO_2), and hydrogen (H_2) to form acids and alcohols instead of, or in addition to, utilizing carbohydrates. These organisms use a metabolic pathway known as the Wood–Ljungdahl pathway to convert carbon monoxide and/or carbon dioxide/hydrogen to acetyl-CoA (Figure 7.2). Acetyl-CoA is then converted to organic acids and/or alcohols (Figure 7.3). The acetyl-CoA pathway involves two branches, known as the methyl branch and the carbonyl branch. The methyl branch involves the reduction of carbon dioxide by six electrons and the consumption of one ATP to form a methyl group (CH_3), which is bound to tetrahydrofolate (THF) (Henstra *et al.*, 2007). The methyl group is transferred from THF to a cobalt ion that is part of a corrinoid–iron–sulfur protein, which is derived from vitamin B12, by methyltransferase (Ragsdale, 2004). The carbonyl branch involves the reduction of carbon dioxide to carbon monoxide with two electrons by carbon monoxide dehydrogenase. Finally, the enzyme acetyl-CoA synthase joins the methyl group, carbon monoxide and coenzyme-A to form acetyl-CoA (Ragsdale, 2004).

In the absence of oxygen, microorganisms convert pyruvate and/or acetyl-CoA to a number of different organic acids or alcohols, which are then excreted from the cell as a waste product (Figures 7.1 and 7.3). The most common products formed are acetone, acetic acid, n-butanol, butyric acid, 2,3 butanediol, ethanol, isopropanol, lactic acid, propionic acid and succinic acid (Shuler and Kargi, 2002). All of these compounds have industrial importance as bulk chemicals, precursors for other products, or fuels. The determination of which of these compounds is formed is dependent on a number of factors, including the species of organism, the redox balance, media pH, nutrient composition and temperature of the reactor.

Under anaerobic conditions, cells divert pyruvate to fermentative pathways, as shown in Figure 7.1, in order to oxidize NADH produced in glycolysis. The most common products produced in these pathways are lactate, ethanol or acetyl-CoA. Acetyl-CoA can then be further utilized to form other products, which are described later (Figure 7.3). Lactate is formed by a reduction of pyruvate with NADH. Ethanol is formed by first reducing pyruvate with NADH to form acetylaldehyde and carbon dioxide, then reducing acetylaldehyde with NADH to form ethanol.

In heterotrophic anaerobic bacteria, pyruvate is combined with Coenzyme-A to produce acetyl-CoA and one carbon dioxide (Figure 7.1). In autotrophic anaerobic bacteria, acetyl-CoA is formed by the Wood–Ljungdahl pathway as previously described (Figure 7.2). A number of industrially important chemicals can be formed from acetyl-CoA by various bacteria (Figure 7.3). Firstly, acetyl-CoA can be (1) converted to acetic acid through a phosphorylation step with phosphate followed by dephosphorylation with ADP resulting in one ATP and one acetic acid, (2) reduced to acetylaldehyde followed by further reduction to ethanol using two NADH, or (3) combined with another acetyl-CoA to form acetoacetyl-CoA or (4) used by the cell to produce cell material. Acetoacetyl-CoA can then be converted to acetone with the emission of one carbon dioxide, or it can be reduced by two NADH to form butyryl-CoA. Some bacteria can reduce acetone with two NADH to form isopropanol. Butyryl-CoA can then be converted either to butyric acid in a manner similar to the conversion of acetyl-CoA to acetic acid, or it can be reduced by two NADH to form n-butanol (hereafter referred to as butanol). Butanol and ethanol can also be produced by cells reducing butyric acid and acetic acid to their respective aldehydes followed by reduction of the aldehydes to alcohols.

7.3 MICROBIAL GROWTH

Microorganisms generally follow a common pattern of growth in batch cultures, as shown in Figure 7.4. As cells become acclimated to the media and other environmental conditions in the reactor, a lag phase occurs. During this time, cells are producing enzymes that they will need to survive in the reactor. Lag time can be reduced or eliminated by selecting an inoculum that has been actively growing in an environment similar to that in the reactor.

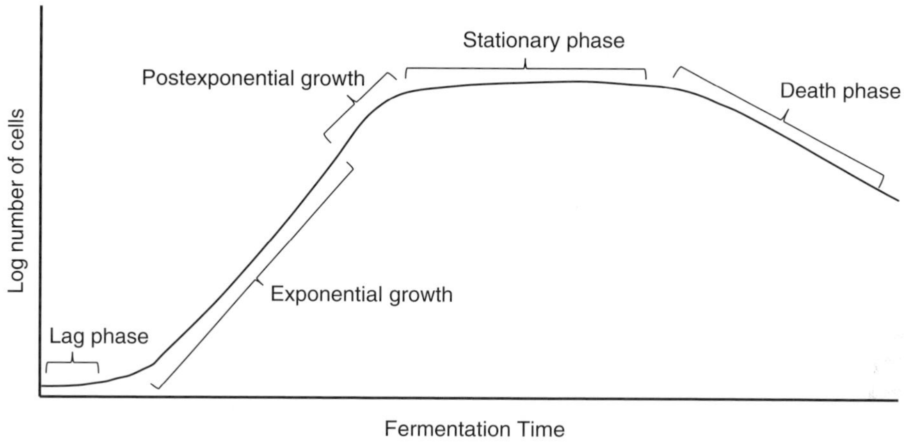

Figure 7.4 A typical growth curve for batch fermentation.

After the lag phase, cells begin multiplying. This phase of growth is known as the exponential phase. Growth rate varies greatly among different organisms and reactor conditions. The specific growth rate (μ_g) can be modeled as:

$$\mu_g = \ln(X/X_0)/t$$

where X = the cell mass concentration at time t, X_0 = the initial cell mass concentration, and t = time. The specific growth rate has units of t^{-1}. A common means of obtaining μ_g is to plot ln (X/X_0) vs. time and calculate the slope of the linear portion of the curve. Another common way to characterize microbial growth rates is by the doubling time (t_d), which equals the time required for actively growing cells to double in concentration:

$$t_d = \ln 2/\mu_g$$

Many valuable fermentation products are produced during the exponential phase. These are known as growth associated products.

After nutrients and substrate begin to diminish, cell growth decelerates and enters a period of no apparent growth. This phase is known as the stationary phase. In the stationary phase, the rate of growth equals the rate of cell death and no net change in cell mass concentration occurs. Fermentation products produced during the stationary phase are known as non-growth associated products.

If additional substrate is not added to the culture, the rate of cell death becomes greater than the rate of growth. This is known as the death phase. Additional substrate and inoculum must be added to restart the culture.

7.4 REACTOR DESIGN

Bioreactors are designed to enhance cell growth and maximize the conversion of substrate to desired product by providing a controlled environment for biological processes. The selection of a specific bioreactor design requires understanding of basic microbiology and biochemistry, process engineering and process economics (Cooney, 1983; Maiorella *et al.*, 1984). The main goal in bioreactor design is to minimize the overall cost of production of the desired product. These include costs associated with raw material conversion to product, cost of product recovery and capital cost.

The performance of a desired bioreactor depends on many parameters such as the ability to (Aiba *et al.*, 1973):

(1) maintain high cell concentrations to provide high product yields;
(2) sustain cell metabolic activities by controlling process parameters such as pH, dissolved oxygen, redox potential, nutrient concentrations and temperature;
(3) achieve and preserve sterile conditions throughout the process to avoid contamination of pure cultures;
(4) provide adequate agitation to achieve homogeneous mixing of substrate and suspended cells and reduce heat and mass transfer limitations.

7.4.1 Types of reactors

The selection of a bioreactor and operation mode starts with the type of biocatalysts (enzymes or whole cells) that will be used to produce the desired product, and if the biocatalysts will be suspended or immobilized with a recycle option. Each of these possibilities should be examined while taking into consideration the reactions stoichiometry and kinetics, mixing needs, gas exchange, heat removal and capital and operating costs (Merchuk and Asenjo, 1995).

For example, different fermentation schemes have been developed for the production of ethanol. Conventional batch, continuous, cell recycle and immobilized cell processes, as well as membrane, extraction and vacuum processes, which selectively remove ethanol from the fermentation medium as it is formed, were compared on identical bases using a consistent model for yeast metabolism (Maiorella *et al.*, 1984). The continuous flow stirred tank reactor (CSTR) with cell recycle, tower and plug flow reactors all showed similar cost savings of about 10% compared to batch fermentation. Cell recycle increases cell density inside the fermentor, which is important in reducing fermentation cost.

Several bioreactors of interest are discussed in the next section, including the stirred tank reactor (STR), bubble column, hollow fiber and monolithic reactors.

7.4.1.1 *Stirred tank reactor*

The stirred tank reactor is one of the most commonly used bioreactors for fermentation (Figure 7.5). The fermentor shape can be rectangular or cylindrical. Currently, cylindrical fermentor design is most commonly used. The height-to-diameter ratio of the cylindrical vessel varies between 2:1 and 6:1. Agitation is facilitated with multiple impellers mounted on a shaft driven by a motor. Agitation is required to facilitate mixing and heat and mass transfer to obtain homogenous suspension of cells and nutrients, pH and temperature. Rushton turbine, hollow blade, three blade pitched and marine are some of the types of impellers that can be used. The number of impellers used depends on the size of the vessel (Doran, 1995). A gas sparger is placed below the impeller for aeration. A STR can be operated in batch, fed-batch or continuous modes.

The temperature inside a large bioreactor is maintained by heating and cooling coils placed in the vessel. However, a heating jacket maintains the operating temperature for small bioreactors.

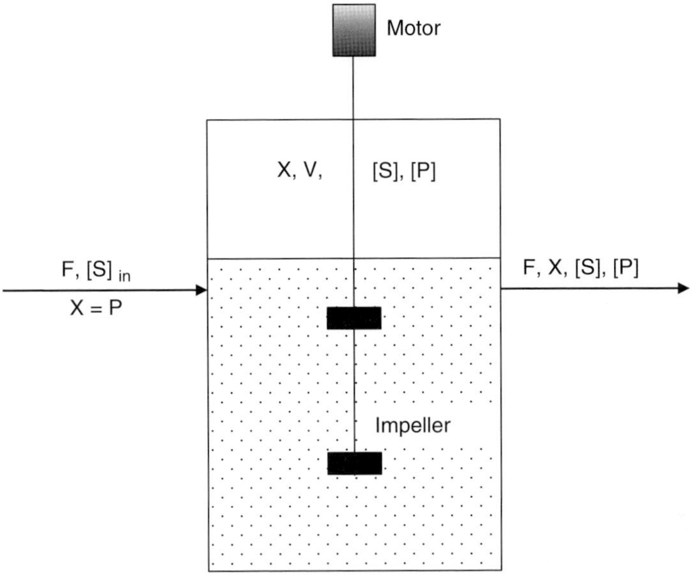

Figure 7.5 Stirred tank reactor.

High agitation and aeration rates can result in foam formation inside the bioreactor, which can block the exhaust gas line and cause loss of cells and media and possible contamination when the foam is washed out from the bioreactor. Mechanical foam breakers and antifoam chemicals are used to control foam formation inside the bioreactor. The working volume of the bioreactor is usually 70–80% of the vessels' physical volume.

The advantages of the STR include the high mass transfer coefficients that can be achieved at high agitation rates and impellers that provide uniform mixing to maintain homogenous composition in the bioreactor. However, high agitation speeds (i.e., high power-to-volume ratios) are required to achieve good mixing and high gas–liquid mass transfer rates. This increases operating costs. In addition, high agitation speeds increase the shear rate, which could damage cells.

7.4.1.2 *Bubble column bioreactors*

Bubble column bioreactors are structurally very simple and have no mechanical agitators. A bubble column bioreactor and its modifications are shown in Figure 7.6. Aeration and mixing are achieved by gas sparging, which requires less energy input compared to a mechanically agitated STR. The sparger is located at the bottom of the column and it introduces gas into the bioreactor. Bubbles rise due to their low density compared to the liquid. The height-to-diameter ratio of the column is greater than two. In the internal and external gas-lift bioreactors (Figures 7.6b and 7.6c), the gas is separated from the liquid and the liquid is recirculated in the bioreactor. Bubble columns and their modifications can be operated counter-currently or co-currently in batch or continuous mode. There are many important design parameters for bubble column reactors, which include the flow pattern, bubble size distribution, gas holdup, gas–liquid interfacial area, mass transfer coefficients, bubble coalescence characteristics, dispersion coefficients and heat transfer characteristics (Shah *et al.*, 1982).

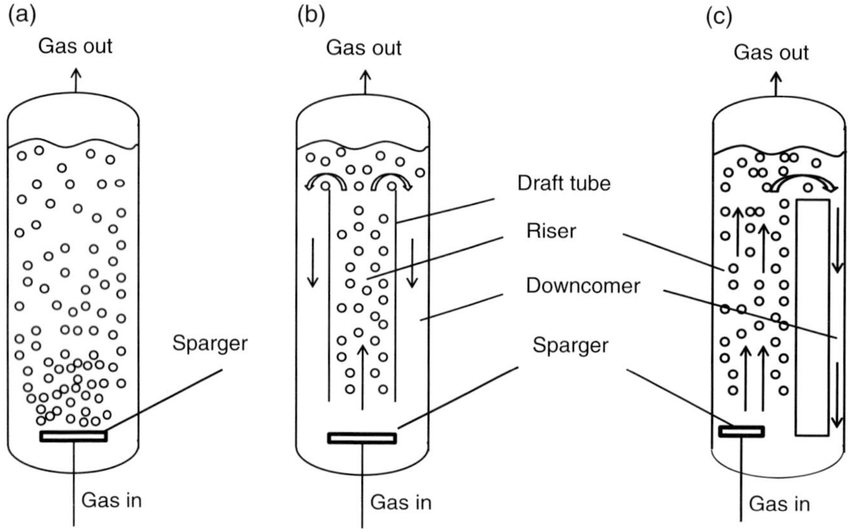

Figure 7.6 Types of bubble column reactors. (a) Bubble column; (b) gas-lift with internal loop; (c) gas-lift with external loop.

Bubble columns have been used in the industrial production of single cell protein, in which low back-mixing is desired to achieve high substrate conversion. The main advantages of bubble column reactors compared to STRs include (Shah *et al.*, 1982):

(1) less maintenance due to simple construction;
(2) high heat transfer rates per unit volume of the reactor;
(3) high liquid residence time for slow reactions to proceed.

On the other hand, bubble coalescence, back mixing and high pressure drop are disadvantages of bubble column reactors. In addition, it is difficult to maintain consistent concentration of substrate, nutrients and oxygen along the height of the column, which could starve the microorganism at points where these levels are low. Multiple feed points to the column can reduce this risk.

7.4.1.3 *Hollow fiber membrane reactor*

A hollow fiber membrane module is the most commonly used membrane bioreactor (Figure 7.7). The hollow fiber membrane module contains hundreds to thousands of hollow fibers in an assembly similar to a shell and tube heat exchanger. The feed can be applied to either the fiber (lumen) side or to the shell side. However, the feed is usually applied to the shell side because the flow path from the lumen side to shell side is too narrow, which increases the risk for membrane clogging. The permeate stream from the membrane module usually contains the product and the retentate contains the concentrated feed stream.

The driving force in membrane bioreactors with conventional biomass separation is the transmembrane pressure, which also affects the flux rate through the membrane. Microfiltration membranes with pore sizes between 0.1 and 5 μm can be used to confine cells in the reactor (Salmon and Robertson, 1995). The selection of the pore size depends on

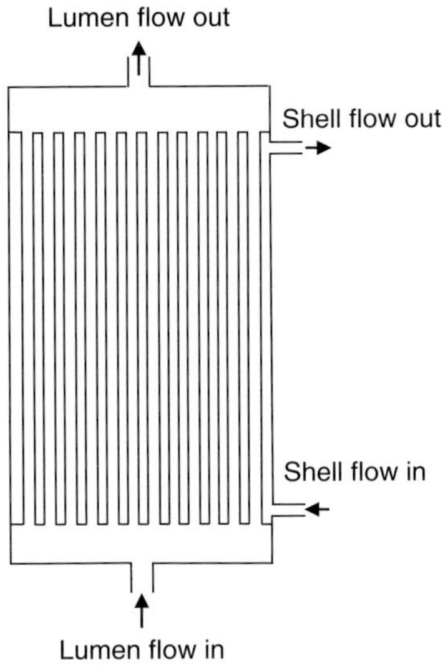

Figure 7.7 Hollow fiber membrane reactor.

the size of the cells. Ultrafiltration membranes typically have pore sizes of 20–1000 Å. These membranes can retain macromolecules with molecular weights between 10^3 and 10^6 daltons. Nanofiltration and reverse osmosis membranes, which have lower pore sizes, can be used to retain smaller molecules.

Membrane bioreactors have applications in fermentation, pharmaceutical production, protein extraction and wastewater treatment (Inloes *et al.*, 1983; Gabelman and Hwang, 1999). Membranes can be used in the recycle loop with other types of reactor, such as CSTR, in which the membrane module is used to separate the cells from the product stream and then recycle the cells back into the reactor. In another arrangement, cells or enzymes can be entrapped on the surface of the membrane or incorporated into the membrane's porous structure. The substrate is fed through the lumen and product leaves the other side of the lumen through the shell side of the module.

The advantages of using hollow fiber membrane reactors include high membrane surface area that allows high cell density, improvement in mass transfer, reduction in product inhibition and increased productivity. However, there are limitations for the use of membrane bioreactors; these include low membrane permeability, selectivity and durability and membrane fouling. These factors generally make membrane bioreactors more expensive than conventional bioreactors.

7.4.1.4 Monolithic reactors

Monolithic reactors are used in multiphase catalytic reactions in which the monolithic support consists of a large number of narrow parallel channels of well defined geometry separated by thin walls that are coated with catalyst. Monolithic reactors have many applications in the chemical and environmental industries, such as hydrogenation and oxidation of organic compounds in wastewater. There are several applications of monolithic reactors in biotechnology, such as enzymatic oxidation of glucose (Kawakami *et al.*, 1989), gluconic acid formation by *Gluconobacter suboxydans* (Shiraishi *et al.*, 1989) and hydrogen production by *Clostridium butyricum* (Fritsch *et al.*, 2008).

The main advantages of monolithic reactors in catalytic gas–liquid reactions include very low pressure drop, high mass transfer, high surface-to-volume ratio, good contact between gas and liquid and ease of reactor scale up. However, there are some drawbacks for monolithic reactors, such as the high manufacturing cost of the monolithic catalysts, short residence time, difficulty in uniformly distributing fluid across the monolithic reactor cross-section and the possibility of clogging monoliths with biofilm formation (Ebrahimi *et al.*, 2005; Cybulski *et al.*, 2006).

7.5 FERMENTATION SCHEMES

Fermentation processes can be conducted in batch, fed-batch or continuous modes. However, many fermentation processes are operated in batch or fed-batch modes due to the advantages of these schemes compared to continuous operation. These advantages include easier set-up and operation, less complication in process control, less susceptibility to microbial contamination, and high final product titer and yields. On the other hand, continuous bioreactors have advantages such as consistent quality of product, high productivity and flexibility in system investigation and analysis.

7.5.1 Batch fermentation

In batch operation, the medium containing the substrate and nutrients is added at time zero. Then, the medium is sterilized and cooled before addition of the desired microorganism. Chemical agents are added to control pH, foam formation and contaminating bacteria, resulting in a minimal change in the net volume of the medium.

Batch fermentation has the disadvantage of low productivity due to the long turn-around times associated with cleaning and preparation of fermentors and the initial lag phase during growth of the desired microorganism. Cell recycle from one batch to the second batch has been successfully used in ethanol fermentation, which significantly reduced the fermentation cycle time. In addition, the use of high performing strains of microorganisms can increase product yield and/or reduce total fermentation cycle time. This substantially reduces the size of fermentor required and lowers operating and capital costs.

7.5.2 Fed-batch fermentation

The fed-batch mode starts the same as batch operation. However, concentrated medium is at some time continuously or intermittently added into the fermentor until the desired working volume is reached. The volume of the medium inside the fermentor changes in the fed-batch mode due to the frequent additions of substrate and nutrients. This is done to keep the substrate concentration low to minimize substrate inhibition to microorganisms. Higher product concentration and yield can be obtained in fed-batch compared to batch operation. Both batch and fed-batch fermentations are run until completion and no products are taken out from the fermentor until the end of fermentation.

7.5.3 Continuous fermentation

In the continuous fermentation scheme, fermentation is done at a relatively low substrate concentration and high biomass concentration. The addition of high substrate concentration in the feed stream can inhibit cells' growth and fermentation capability. Continuous fermentation usually involves two stages: start up and steady state. During the start-up period, the fermentor is operated in batch mode until cells grow to the desired cell concentration. Once the active culture is established, the substrate and nutrients are continuously fed into the fermentor and products are removed at the same rate. The volume inside the fermentor is kept constant at steady state operation. Cells can be either suspended or immobilized in the fermentor. A recycle system, which is usually a membrane module, is used to separate the cells in the product stream and recycle them back into the fermentor to keep its cell mass concentration high. To ensure consistent performance, the feed rate should be properly controlled, microorganism population in cell number and viability should be maintained and contamination should be avoided.

The main advantage of the continuous scheme is its high productivity. However, the main problems of the continuous scheme are the high possibility of contamination and wash out of cells from the bioreactor at high flow rate. In addition, the final product concentration is generally lower than in batch or fed-batch fermentations, which increases the cost of product recovery.

7.6 FERMENTATION PRODUCTS

7.6.1 Acetone–Butanol–Ethanol (ABE) fermentation

One of the first processes to produce bulk chemicals through microbial fermentation was the acetone–butanol–ethanol fermentation, or the ABE fermentation. In the early twentieth century, butanol was desired for the production of synthetic rubber (Jones and Woods, 1986). This lead to several researchers pursuing the isolation of bacteria that could produce butanol from starch, including Chaim Weizmann, who isolated the bacterium *Clostridium acetobutylicum* that could produce a mixture of acetone, butanol, and ethanol from starch under anaerobic conditions (Jones and Woods, 1986). Upon the outbreak of the First World War, the British government was particularly interested in acetone, since it was needed for the production of cordite, a component of smokeless gunpowder. As a result several fermentation facilities were built in Britain, Canada, and the United States during the war to produce acetone by fermentation using *C. acetobutylicum* (Jones and Woods, 1986). Production of acetone and butanol by fermentation was mostly discontinued after the Second World War, when petrochemical butanol production and the high cost of substrate made butanol fermentation economically uncompetitive (Jones and Woods, 1986).

As stated earlier, interest in the ABE fermentation has recently increased due to the desire to produce chemicals from renewable resources because of high petroleum costs and greenhouse gas emissions from fossil fuel combustion. Research into the ABE fermentation has generally focused on two species, *C. acetobutylicum* and *Clostridium beijerinkii*. Both species are heterotrophs, which are organisms that derive energy from organic molecules. They can utilize starch and simple sugars for ABE production. Native strains of these species are not able to utilize cellulose due to their lack of cellulose degrading enzymes. In addition to heterotrophs, some autotrophic bacteria have also been observed to produce butanol. *Clostridium carboxidivorans* and *Butyribacterium methylotrophicum* both have been observed to produce a mixture of butanol, ethanol, acetic acid, and butyric acid from carbon monoxide (Datar *et al.*, 2004; Worden *et al.*, 1991).

All three of the primary products of ABE fermentation have industrial value. Acetone is an effective solvent and is the principal ingredient in nail polish remover. Acetone also is used as a building block in plastics manufacture, particularly production of methyl methacrylate, which is polymerized to form PMMA, also known as plexiglass. Butanol, too, is used as a solvent. It is also used to produce butyl esters, some of which are flavorings used in the food industry, and as a precursor for plastics. Recently, butanol has been proposed for use as a transportation fuel. Ethanol is found in numerous beverages and is used as a transportation fuel.

ABE fermentations are characterized by the presence of two distinct phases. The first phase is called acidogenesis. During this phase acetate and butyrate are formed and cell growth is observed. This phase is also characterized by a media pH greater than 5. As the pH decreases to below 5 with the production of more acid, cell growth ceases and the cells enter the second phase, known as solventogenesis. During solventogenesis acetone, butanol and ethanol are produced. Often during solventogenesis acids produced during acidogenesis are taken up by cells and reduced to form alcohols. This action increases extracellular pH and reduces the pH gradient across the cell membrane.

7.6.2 Glycerol

Glycerol is produced primarily from petrochemical catalysis, although some is produced as a by-product of soap manufacturing. An increasing supply of glycerol is also produced by the biodiesel industry as a result of the hydrolysis of fatty acids from glycerol prior to methyl esterification of the fatty acids to produce biodiesel. Glycerol is an excellent substrate for the production of succinate and propionate, as described below. It can be used also to produce ethanol using recombinant *Escherichia coli* (Dharmadi *et al.*, 2006). Glycerol is a valuable chemical in its own right for food and pharmaceutical production, as well as the production of many other consumer products. One example is the production of propylene glycol, used in antifreeze, by the catalytic upgrading of glycerol (Dasari *et al.*, 2005).

Production of glycerol by fermentation has not been pursued commercially since before the Second World War. During the First World War, Germany produced glycerol with yeast through a process called sulfite steering (Wang *et al.*, 2001). In this process sulfite ions were added to the media to form complexes with acetaldehyde. These complexes cannot be reduced by NADH to ethanol, so the yeast cell must reduce dihydroxyacetone phosphate (DHAP) to glycerol in order to maintain its redox balance. This process results in an approximately 25% mass yield of glycerol from glucose. Due to low yield and the low cost of petroleum, this process was abandoned.

Another process that also was used for microbial production of glycerol was the neutral pH process (Wang *et al.*, 2001). This process involved fermenting glucose with yeast at pH 7. At neutral pH acetaldehyde dehydrogenase activity is increased; thus, acetaldehyde is oxidized to acetate and NADH is formed (Wang *et al.*, 2001). The cell then reduces DHAP to glycerol to maintain its redox balance. This process also suffered from low yields and the production of other side products, such as acetate and ethanol, and was abandoned. (Wang *et al.*, 2001).

The microbial glycerol production process that has generated the highest yields to date is the aerobic conversion of glucose by osmotolerant yeast (Wang *et al.*, 2001; Overkamp *et al.*, 2002). Glycerol is a protectant against high osmotic pressures; thus, it is produced by osmotolerant yeast under osmotic stress (Blomberg and Adler, 1992). Mass yields of glycerol from glucose of up to 64% have been reported using osmotolerant yeast (Zhuge *et al.*, 2001).

7.6.3 Propionate

Propionate is a common industrial chemical that is used in the food and feed industry as a preservative. It is also used for the production of various plastics, pharmaceuticals, and perfumes (Himmi *et al.*, 2000). Propionic acid produced from fermentation is primarily produced by bacteria from the genus *Propionibacterium*. Propionic acid can be produced using either glycerol or glucose as the carbon source. The precursor of propionic acid is phosphenolpyruvate (PEP). Instead of PEP being dephosphorylated to produce pyruvate, propionic acid bacteria dephosphorylate PEP while adding a carboxyl group from carbon dioxide to form oxaloacetate. Oxaloacetate is then converted through four enzymatic reactions to form propionate. Three of these reactions are actually part of the citric acid cycle; however, the reactions are carried out in the reverse direction. The final reaction combining water and succinate to form propionate and carbon dioxide is not part of the citric acid cycle.

Propionic acid fermentations can achieve high yields and productivities. Himmi and co-workers (2000) used *Propionibacterium acidipropionici* and achieved a propionate yield of 0.79 mol/mol with a productivity of 0.42 g/l/h and a final concentration of 12 g/l when using 20 g/l glycerol. They also found that with *P. acidipropionici*, using glycerol as the carbon source resulted in more propionate and less acetate production than when they used glucose. One of the keys for successful propionate fermentation is to minimize acetate production, as acetate interferes with the distillation process used to purify propionate (da Silva *et al.*, 2009).

7.6.4 Succinate

Succinate is an important precursor for plastics, pharmaceuticals, detergents, and other consumer products (da Silva *et al.*, 2009). Several biodegradable plastics can be produced with succinate, and the production of these plastics with renewable resources is desirable. Also, one mole of carbon dioxide is fixed for every mole of succinate produced; thus, fermentative production of succinate can be used to reduce carbon dioxide emissions (Song and Lee, 2006). Succinate production can be accomplished using several metabolic pathways, but the most prevalent in bacteria is PEP carboxylation. This is the same pathway previously described for propionate production, with the exception that succinate is not converted to propionate.

Most of the bacteria that have been isolated for succinate fermentation are rumen bacteria, such as *Anaerobiospirillum succiniciproducens*, *Actinobacillus succinogenes* and *Mannheimia succiniciproducens* (Guettler *et al.*, 1999; Lee *et al.*, 2000; Beauprez *et al.*, 2010). Strains of *E. coli* have also been genetically modified to produce succinate (Hong and Lee, 2000). Recent developments in fermentative succinate production have involved the use of lignocellulosic substrates, such as corn stover (Zheng *et al.*, 2010) and various woods (Hodge *et al.*, 2009; Lee *et al.*, 2003), as feedstocks for succinate production.

7.6.5 1,3 Propanediol

1,3 propanediol (PDO) is an important precursor for plastic manufacturing as well as an effective solvent. Although PDO has been known as a product of glycerol fermentation for many years (Freund, 1881), its commercial relevance has increased since Dupont began producing PDO through fermentation of glucose for the manufacture of polytrimethylene terephthalate under the brand name Sorona® (da Silva *et al.*, 2009). PDO is produced from glycerol by a reductive pathway consisting of a dehydration of glycerol followed by a reduction by NADH (Yang *et al.*, 2007). To generate NADH for the reductive pathway, an oxidative pathway oxidizes glycerol to eventually form pyruvate; thus, not all glycerol can go to the formation of PDO.

Numerous bacteria naturally form PDO from glycerol, including *Klebsiella oxytoca* (Yang *et al.*, 2007), *Klebsiella pneumoniae* (Lin *et al.*, 2005), *Citrobacter freundii* (Pflugmacher and Gottschalk, 1994; Boenigk *et al.*, 1993), *Enterobacter agglomerans* (Barbirato *et al.*, 1997), and *C. butyricum* (Boenigk *et al.*, 1993; Biebl, 1991). These bacteria also produce numerous other products in conjunction with PDO that affect their productivity and may inhibit growth due to the need to produce NADH. Efforts to improve strains through adaptation or genetic manipulation strategies are ongoing to increase PDO yields, reduce the effect of inhibitors, and utilize sugars for PDO production.

7.6.6 Butanediol

2,3 Butanediol (BDO) is another industrially important chemical that has a long history of production from fermentation. The first published report of microbial BDO production was by Harden and Walpole (1906). Like butanol production, microbial BDO production was also driven by necessity during the world wars, especially for the manufacture of synthetic rubber (Syu, 2001). Current uses include food flavorings, solvent, antifreeze, and as an ingredient for plastic manufacturing (Syu, 2001).

The main pathway used for the production of BDO from two pyruvate moelcules is combination of these molecules to form carbon dioxide and α-acetolactate, followed by decarboxylation of α-acetolactate to form carbon dioxide and acetoin, and then reduction of acetoin by NADH to form BDO and NAD^+ (Juni and Heym, 1956). The reduction of acetoin is reversible (Syu, 2001). If acetoin is oxidized to form diacetyl instead of being reduced, BDO can be formed by combining two diacetyl molecules to form acetate and acetylacetoin, which is then reduced by NADH or NADPH to acetylbutanediol followed by another reduction to form acetate and BDO (Juni and Heym, 1956). The sum of these reactions forms a cycle known as the BDO cycle (Juni and Heym, 1956).

BDO is a chiral molecule with two chiral centers. Thus, it can be present in levarotatory, dextrorotatory, or meso (optically inactive) forms (Syu, 2001). The stereoisomers depend on the stereoisomers of acetoin present, the specificity of BDO dehydrogenase, and the reducing agent specificity for the acetylacetoin reductase enzyme, which catalyzes the reduction of acetylacetoin to acetylbutanediol (Voloch *et al.*, 1983; Ui *et al.*, 1998). Voloch and co-workers (1983) proposed that there are two types of BDO dehydrogenase in *Klebsiella pneumoniae*, a D form and an L form. The D form catalyzes reduction of D(–) acetoin to meso-BDO and the L form catalyzes reduction of L(+)acetoin to levarotatory BDO. They also proposed that an enzyme called acetoin racemase catalyzes the isomerization of D(–)acetoin to L(+)acetoin, and vice versa (Voloch *et al.*, 1983). Other BDO producers have other BDO dehydrogenases that catalyze the production of dextrorotatory BDO. In addition, when the acetylacetoin reduction pathway is used, the reducing agent specificity of acetylacetoin reductase determines the stereochemistry of BDO. If NADH is the reducing agent, dextrorotatory BDO is formed. If NADPH is the reducing agent, meso BDO is formed (Ui *et al.*, 1998).

7.7 SEPARATION

The broth at the end of the fermentation process contains the product (or mixture of products), unconverted substrate, cells and other soluble and insoluble components. To recover the desired product from the broth, a separation scheme is required. The separation scheme used depends on the location of the product (intracellular or extracellular), size, solubility, charge and concentration of the product in the broth, the product value and purity of the product. The higher the purity of the product needed, the higher the separation cost. Different separation methods are used to recover acids, alcohols and diols from the fermentation broth, which is discussed in the next section.

7.7.1 Separation of acids

Organic acids, such as acetic and lactic acids, are recovered from the fermentation broth by first removing cells and other insoluble components followed by a precipitation step. For example, for citric acid, calcium hydroxide is used to produce calcium or sodium citrate. The precipitate is filtered and transferred to a stirred tank to which sulfuric acid is added to form calcium sulfate precipitate, which is then removed by filtration. The acetic acid is further purified by evaporation followed by crystallization, centrifugation and drying (Bailey and Ollis, 1986; Rogers *et al.*, 2006). The traditional separation method for succinic acid is similar to acetic acid. Zeikus and co-workers (1999) reviewed succinic acid production and highlighted a method called simultaneous acidification and purification by electrodialysis and crystallization as an alternative to the traditional separation method. This method was claimed to be cost effective and easy to scale up for commercial use. Other researchers used adsorbents to separate succinic acid from the fermentation broth (Davison *et al.*, 2004). The preferred choice for acetic acid recovery from the fermentation broth containing between 1.2 and 3.5% acid has been solvent extraction (Rogers *et al.*, 2006). Solvent extraction is a method to separate a selected compound from a mixture based on its relative solubility when two different immiscible solvents are used, which form two phases.

7.7.2 Separation of alcohols

Separation of alcohols, such as ethanol and butanol, from the fermentation broth is traditionally done by distillation. The higher the alcohol concentration in the fermentation broth, the lower the energy required for distillation. For ethanol fermentation, the broth usually contains 10–15% (w/w) ethanol. After distillation, ethanol concentration in the distillate is about 90% (w/w). The distillation process will not yield more than 93% (w/w) ethanol, which is the azeotropic mixture of ethanol–water. Azeotropic mixtures cannot be separated by distillation because the compositions of ethanol in the vapor and liquid phases are the same. Therefore, azeotropic distillation with benzene or dehydration with molecular sieves is usually used to remove the remaining water and produce fuel grade ethanol (99.9 wt-%).

Distillation is based on the differences between the boiling points of the desired alcohol and water. The distillation column has two distinct sections, the rectifying section above the feed stream and the stripping section below the feed stream. In the rectifying section the alcohol increases in purity as it travels as a vapor to the top of the column, where it is condensed. Part of the condensate will be taken as the purified product stream and the other part is returned back to the column as a flux to enhance the rectifying process. In the stripping section below the feed stream, ethanol is removed from water as the feed moves towards the bottom of the column. Vapor flows up the column from tray to tray and the liquid flows down the column. The purpose of the tray is to provide mixing and separation of vapor and liquid.

In addition, unfermented sugars can negatively affect the efficiency of the distillation process, as these sugars adhere to the surfaces of the trays in the distillation column. There are several design factors that affect the size of distillation columns; these include the amount of vapor generated, the required purity of the product, type of contacting device (tray or packing), the amount of undissolved solids in the feed and the energy requirement.

For acetone–butanol–ethanol (ABE) fermentation, the broth contains about 25–35 g/l of mixed solvents. Butanol concentration is usually less than 20 g/l, which makes its recovery by distillation expensive. This low butanol concentration in the fermentation broth is related to the inability of *Clostridium* species to produce more butanol due to solvent toxicity (Lee *et al.*, 2008). Metabolic engineering and advanced fermentation techniques are ongoing to enhance the organisms' abilities to produce and tolerate higher concentrations of butanol and increase productivity. Several integrated fermentation and recovery processes, such as fed-batch fermentation with pervaporation and continuous fermentation with gas stripping, have been reviewed elsewhere (Lee *et al.*, 2008).

7.7.3 Separation of diols and triols

The recovery of glycerol (1,2,3-propanetriol) from the fermentation broth is difficult because glycerol is miscible in water and has a high boiling point. Glycerol is separated via vacuum distillation after the removal of ethanol and precipitation of sulfite, sulfate and phosphate salts from the fermentation broth (Rehm, 1988). Even so, only about 50% of the glycerol in the broth is recovered by vacuum distillation. Other methods for glycerol recovery using ion exchange resins and extraction with different solvents have been reported by Rehm (1988) and Wang and co-workers (2001).

The recovery of 1,3 propanediol (1,3-PD) from the fermentation broth is done by vacuum distillation and/or preparative liquid chromatography after removing the cells and other impurities (Xiu and Zeng, 2008). Other methods for recovery of 1,3-PD, such as extraction and membrane separation, have been investigated (Xiu and Zeng, 2008). On the other hand, Syu (2001) summarized several separation techniques for 2,3-butanediol (2,3-BD), which include pervaporation, membrane distillation and extraction.

7.8 FUTURE APPLICATION AREAS AND EMERGING DEVELOPMENTS

Emerging areas in fermentation technologies include metabolic engineering, genetic engineering and synthetic biology, with focus on advanced biofuels and bio-based products (Ragauskas *et al.*, 2006; Atsumi and Liao, 2008; Sillers *et al.*, 2008; Park *et al.*, 2008; Peralta-Yahya and Keasling, 2010). These areas are being used to develop new organisms that can tolerate the inhibitory effects of substrate and product, can utilize various substrates and produced desired products with high titer and productivity.

Another recent trend in fermentation is the use of disposable bioreactors instead of stainless steel reactors for process development, especially in pharmaceutical manufacturing (Hanson *et al.*, 2009). Disposable bioreactors are typically plastic devices, such as microtiter plates, T-flasks, shake flasks and wave reactors, with different sizes. Disposable bioreactors can be used as a seed fermentor or as a production fermentor for products on a small scale (Mikola *et al.*, 2007). The main advantages of disposable reactors include more flexibility in operation and use for different products, elimination of cross-contamination, less time needed to set up because the reactor is ready to use, low cost and less labor needed. However, disposable reactor sizes do not exceed 2,000 L because of physical limitations, stability issues and heat and mass transfer limitations, as these reactors do not have impellers for mixing. In addition, disposable plastic reactors may leach chemical components into the media that could negatively impact the quality of the final product (Hanson *et al.*, 2009).

Another emerging application in fermentation is the use of high throughput miniature (mL scale) bioreactors (HT MBR) for assessment of microbial systems for strain screening, nutrient optimization and productivity screening (Isett *et al.*, 2007). The advantages of HT MBR compared to conventional shake flasks are lower labor and material costs and higher throughput because of the ability to run many cell cultivations in parallel. However, HT MBRs do not contain online sensors and do not provide enough samples for offline analysis due to their small size. Several miniature bioreactors, such as microwell plates (μL scale), miniature stirred bioreactors and miniature bubble column bioreactors, have been reviewed by Betts and Baganz (2006).

The use of disposable bioreactors and miniature bioreactors resulted in new developments in online sensor technology to enable direct measurements of pH, dissolved oxygen, optical sensing for cell concentration, fluorometry and infrared spectroscopy (Glindkamp *et al.*, 2009). Fluorescence sensors are used to measure several analytes, such as proteins, vitamins, glucose, ethanol, ATP, NAD(P)H and pyruvate, by scanning the excitation and emission wavelengths for these compounds. Infrared spectroscopy (IR) can be used to measure glucose, fructose and ethanol concentration in the fermentation medium. IR can also be used to monitor gas phase composition, in particular carbon dioxide. Online sensors will help to improve control of fermentation processes. These sensors should be low cost and reliable, which may be challenging if the sensors are disposable.

REFERENCES

Aiba, S, Humphrey A, Millis N. 1973. Biochemical Engineering. 2nd edn. New York: Academic Press.

Atsumi S, Liao JC. 2008. Directed evolution of *Methanococcus jannaschii* citramalate synthase for biosynthesis of 1-propanol and 1-butanol by *Escherichia coli*. Appl Environ Microb 74(24):7802–7808.

Bailey, JE, Ollis D. 1986. Biochemical Engineering Fundamentals, 2nd edn. New York: McGraw-Hill.

Barbirato F, Larguier A, Conte T, Astruc S, Bories A. 1997. Sensitivity to pH, product inhibition, and inhibition by NAD(+) of 1,3-propanediol dehydrogenase purified from *Enterobacter agglomerans* CNCM 1210. Arch Microbiol 168(2):160–163.

Beauprez JJ, De Mey M, Soetaert WK. 2010. Microbial succinic acid production: Natural versus metabolic engineered producers. Process Biochem 45(7):1103–1114.

Betts JI, Baganz F. 2006. Miniature bioreactors: current practices and future opportunities. Microbial Cell Factories 5:21.

Biebl H. 1991. Glycerol fermentation of 1,3-propanediol by *Clostridium butyricum*. Measurement of product inhibition by use of a pH-auxostat. Appl Microbiol Biotechnol 35(6):701–705.

Blomberg A, Adler L. 1992. Physiology of osmotolerance in fungi. In: Rose, A. (editor). Advances in Microbial Physiology. London: Academic Press. pp. 145–212.

Boenigk R, Bowien S, Gottschalk G. 1993. Fermentation of Glycerol to 1,3-Propanediol in Continuous Cultures of *Citrobacter-Freundii*. Appl Microbiol Biotechnol 38(4):453–457.

Cooney CL. 1983. Bioreactors – design and operation. Science 219:728–733.

Cybulski A, Albers R, Moulijn JA. 2006. Monolithic catalysts for three-phase processes. In: Cybulski A, Moulijn JA (editors). Structure catalysis and reactors, 2nd edn. Boca Raton, FL: Taylor & Francis. pp. 355–392.

Dasari MA, Kiatsimkul PP, Sutterlin WR, Suppes GJ. 2005. Low-pressure hydrogenolysis of glycerol to propylene glycol. App Catal A: General 281(1–2):225–231.

Datar RP, Shenkman RM, Cateni BG, Huhnke RL, Lewis RS. 2004. Fermentation of biomass-generated producer gas to ethanol. Biotechnol Eng 86(5):587–594.

Davison BH, Nghiem NP, Richardson GL. 2004. Succinic acid adsorption from fermentation broth and regeneration. Appl Biochem Biotechnol 114(1–3):653–669.

Dharmadi Y, Murarka A, Gonzalez R. 2006. Anaerobic fermentation of glycerol by *Escherichia coli*: A new platform for metabolic engineering. Biotechnol Bioeng 94(5):821–829.

Doran, PM. 1995. Bioprocess Engineering Principles, 1st edn. London: Elsevier Academic Press.

Ebrahimi S, Picioreanu C, Xavier JB, *et al.* 2005. Biofilm growth pattern in honeycomb monolith packings: Effect of shear rate and substrate transport limitations. Catal Today 105(3–4):448–454.

Freund A. 1881. Über die bldung und darstellung von trimethylenalkohol aus glycerin. Monatsh Chem 2(1):636–641.

Fritsch M, Hartmeier W, Chang J-S. 2008. Enhancing hydrogen production of *Clostridium butyricum* using a column reactor with square-structured ceramic fittings. Int J Hydrogen Energ 33(22):6549–6557.

Gabelman A, Hwang S-T. 1999. Hollow fiber membrane contactors. J Membrane Sci 159(1–2):61–106.

Glindkamp A, Riechers D, Rehbock C, Hitzmann B, Scheper T, Reardon KF. 2009. Sensors in disposable bioreactors status and trends. Advances in Biochemical Engineering/ Biotechnology 115:145–169.

Guettler MV, Rumler D, Jain MK. 1999. *Actinobacillus succinogenes* sp. nov., a novel succinic-acid-producing strain from the bovine rumen. Int J Syst Bacteriol 49:207–216.

Hanson MA, Brorson KA, Moreira AR, Rao G. 2009. Comparisons of optically monitored small-scale stirred tank vessels to optically controlled disposable bag bioreactors. Microb Cell Fact 8:44.

Harden A, Walpole GS. 1906. 2,3-Butylene glycol fermentation by *Aerobacter aerogenes*. Proc Royal Society B 77:399–405.

Henstra AM, Sipma J, Rinzema A, Stams AJM. 2007. Microbiology of synthesis gas fermentation for biofuel production. Curr Opin Biotechnol 18(3):200–206.

Himmi EH, Bories A, Boussaid A, Hassani L. 2000. Propionic acid fermentation of glycerol and glucose by *Propionibacterium acidipropionici* and *Propionibacterium freudenreichii* ssp shermanii. Appl Microbiol Biotechnol 53(4):435–440.

Hodge DB, Andersson C, Berglund KA, Rova U. 2009. Detoxification requirements for bioconversion of softwood dilute acid hydrolyzates to succinic acid. Enzyme Microb Tech 44(5):309–316.

Hong S-H, Lee S-Y. 2000. Metabolic flux distribution in a metabolically engineered *Escherichia coli* strain producing succinic acid. J Microbiol Biotechnol 10(4):496–501.

Inloes DS, Smith WJ, Taylor DP, Cohen SN, Michaels AS, Robertson CR. 1983. Hollow-fiber membrane bioreactors using immobilized *E. coli* for protein synthesis. Biotechnol Bioeng 25(11):2653–2681.

Isett K, George H, Herber W, Amanullah A. 2007. Twenty-four-well plate miniature bioreactor high-throughput system: assessment for microbial cultivations. Biotechnol Bioeng 98(5):1017–1028.

Jones DT, Woods DR. 1986. Acetone-butanol fermentation revisited. Microbiol Reviews 50(4): 484–524.

Juni E, Heym GA. 1956. A cyclic pathway for the bacterial dissimilation of 2,3-butanediol, acetylmethylcarbinol, and diacetyl. 1. General aspects of the 2,3-butanediol cycle. J Bacteriol 71(4):425–432.

Kawakami K, Kawasaki K, Shiraishi F, Kusunoki K. 1989. Performance of a honeycomb monolith bioreactor in a gas-liquid solid three-phase system. Ind Eng Chem Res 28:394–400.

Lee PC, Lee WG, Lee SY, Chang HN, Chang YK. 2000. Fermentative production of succinic acid from glucose and corn steep liquor by *Anaerobiospirillum succiniciproducens*. Biotechnol Bioprocess Eng 5:379–381.

Lee SY, Lee PC, Hong SH, Chang HN, Park SC. 2003. Biological conversion of wood hydrolysate to succinic acid by *Anaerobiospirillum succiniciproducens*. Biotechnol Lett 25(2):111–114.

Lee SY, Park JH, Jang SH, Nielsen LK, Kim J, Jung KS. 2008. Fermentative butanol production by clostridia. Biotechnol Bioeng 101(2):209–228.

Lin RH, Liu HJ, Hao J, Cheng K, Liu DH. 2005. Enhancement of 1,3-propanediol production by *Klebsiella pneumoniae* with fumarate addition. Biotechnol Lett 27(22):1755–1759.

Maiorella BL, Blanch HW, Wilke CR. 1984. Economic evaluation of alternative ethanol fermentation processes. Biotechnol Bioeng 26:1003–1025.

Merchuk JC, Asenjo JA. 1995. The Monod equation and mass transfer. Biotechnol Bioeng 45(1):91–94.

Mikola M, Seto J, Amanullah A. 2007. Evaluation of a novel Wave Bioreactor cellbag for aerobic yeast cultivation. Bioprocess Biosyst Eng 30:231–241.

Overkamp KM, Bakker BM, Kotter P, Luttik MAH, van Dijken JP, Pronk JT. 2002. Metabolic engineering of glycerol production in *Saccharomyces cerevisiae*. Appl Environ Microb 68(6):2814–2821.

Park JH, Lee SY, Kim TY, Kim HU. 2008. Application of systems biology for bioprocess development. Trends Biotechnol 26(8):404–412.

Peralta-Yahya PP, Keasling JD. 2010. Advanced biofuel production in microbes. Biotechnol J 5:147–162.

Pflugmacher U, Gottschalk G. 1994. Development of an immobilized cell reactor for the production of 1,3-propanediol by *Citrobacter freundii*. Appl Microbiol Biotechnol 41(3):313–316.

Ragauskas AJ, Williams CK, Davison BH, *et al*. 2006. The path forward for biofuels and biomaterials. Science 311(5760):484–489.

Ragsdale SW. 2004. Life with carbon monoxide. Critical Reviews in Biochemistry and Molecular Biology 39(3):165–195.

Rehm, H–J. 1988. Microbial production of glycerol and other polyols. In: Rehm H-J (editor). Biotechnology, Vol. 6B, Special Microbial Processes. Weinheim, Germany, VCH. pp. 51–69.

Rogers P, Chen J-S, Zidwick MJ. 2006. Organic acid and solvent production part i: acetic, lactic, gluconic, succinic and polyhydroxyalkanoic acids. In: Dworkin, M., Falkow, S., Rosenberg, E., Schleifer, K.-H., Stackebrandt, E. (editors). The Prokaryotes. New York: Springer. pp. 511–755.

Salmon PM, Robertson CR. 1995. Membrane reactors. In: Asenjo JA, Merchuk JC (editors). Bioreactor System Design. New York: Marcel Dekker, Inc. pp. 305–338.

Shah YT, Kelkar BG, Godbole SP, Deckwer W-D. 1982. Design parameters column reactors estimations for bubble. AIChE J 28(3):353–379.

Shiraishi F, Kawakami K, Kono S, Tamura A, Tsuruta S, Kusunoki K. 1989. Characterization of production of free gluconic acid by *Gluconobacter suboxydans* adsorbed on ceramic honeycomb monolith. Biotechnol Bioeng 33(11):1413–1418.

Shuler ML, Kargi F. 2002. Bioprocess Engineering: Basic Concepts, 2nd edn. Upper Saddle River, NJ: Prentice Hall.

Sillers R, Chow A, Tracy B, Papoutsakis ET. 2008. Metabolic engineering of the non-sporulating, non-solventogenic *Clostridium acetobutylicum* strain M5 to produce butanol without acetone demonstrate the robustness of the acid-formation pathways and the importance of the electron balance. Metab Eng 10(6):321–332.

da Silva GP, Mack M, Contiero J. 2009. Glycerol: A promising and abundant carbon source for industrial microbiology. Biotechnol Adv 27(1):30–39.

Song H & Lee SY. 2006. Production of succinic acid by bacterial fermentation. Enzyme Microb Tech 39(3):352–361.

Syu MJ. 2001. Biological production of 2,3-butanediol. Appl Microbiol Biotechnol 55(1):10–18.

Ui S, Hosaka T, Watanabe K, Mimura A. 1998. Discovery of a new mechanism of 2,3-butanediol stereoisomer formation in *Bacillus cereus* YUF-4. J Ferment Bioeng 85(1):79–83.

Voloch M, Ladisch MR, Rodwell VW, Tsao GT. 1983. Reduction of acetoin to 2,3-butanediol in *Klebsiella-Pneumoniae* – a new model. Biotechnol Bioeng 25(1):173–183.

Wang ZX, Zhuge J, Fang HY, Prior BA. 2001. Glycerol production by microbial fermentation: A review. Biotechnol Adv 19(3):201–223.

Worden RM, Grethlein AJ, Jain MK, Datta R. 1991. Production of butanol and ethanol from synthesis gas via fermentation. Fuel 70(5):615–609.

Xiu Z-L, Zeng A-P. 2008. Present state and perspective of downstream processing of biologically produced 1,3-propanediol and 2,3-butanediol. Appl Microbiol Biotechnol 78:917–296.

Yang G, Tian JS, Li JL. 2007. Fermentation of 1,3-propanediol by a lactate deficient mutant of *Klebsiella oxytoca* under microaerobic conditions. Appl Microbiol Biotechnol 73(5):1017–1024.

Zeikus JG, Jain MK, Elankovan P. 1999. Biotechnology of succinic acid production and markets for derived industrial products. Appl Microbiol Biotechnol 51(5):545–552.

Zheng P, Fang L, Xu Y, Dong J-J, Ni Y, Sun Z-H. 2010. Succinic acid production from corn stover by simultaneous saccharification and fermentation using *Actinobacillus succinogenes*. Bioresource Technol 101(20):7889–7894.

Zhuge J, Fang HY, Wang ZX, Chen DZ, Jin HR, Gu HL. 2001. Glycerol production by a novel osmotolerant yeast *Candida glycerinogenes*. Appl Microbiol Biot 55(6):686–692.

8 Fungal Cell Factories

Sue A. Karagiosis and Scott E. Baker

8.1 FUNGI AND FUNGAL BIOTECHNOLOGY

Human history and fungi have been profoundly intertwined for centuries. This large and diverse group of eukaryotic, heterotrophic microbes derives its energy through extracellular digestion of organic matter. Some of these microbes exist as the bane of mankind, notably pathogens of crops, plants or people. Alternatively, a multitude of beneficial and valuable resources have been obtained by harnessing the metabolic activities of fungi. Fungi are a major food source and essential for the creation of bread, veined cheeses, alcoholic beverages, and traditionally fermented foods such as tempeh, *miso* and soy sauce to just name a few. Commercial manufacture of antibiotics, many organic acids and enzymes are driven by fungal biotechnology. Additionally, these organisms are a treasure trove of yet undiscovered bioactive secondary metabolites with possible drug discovery and pharmaceutical applications.

Applications of fungal biotechnology allow for the industrial exploitation of a fungal culture's innate biochemical activities. Molds secrete an impressive diversity of enzymes and metabolites, some of which are important commodities for the food, agricultural, enzyme and pharmaceutical industries. Fungi are commercially cultured by various means to obtain high process productivity and product yields. Highlighted here are a few of the key processes, as these techniques have been reviewed extensively elsewhere (Murphy and Horgan, 2005; Ruijter *et al.*, 2002; Sahasrabudhe and Sankpal, 2001; Bentley and Bennett, 2008; Fazenda *et al.*, 2008).

One of the oldest methods, **solid-state fermentation** or the **koji process**, uses solid substrates, such as steam sterilized fibers or wheat bran, to cultivate the mold inoculum. This technique has been central to traditional food and beverage fermentations throughout Asia for thousands of years. Today solid-state fermentation is the cultivation method of choice for the commercial production of many enzymes and is gaining popularity in areas including bioremediation, detoxification of agro-industrial waste and biotransformation of crop residues for nutritional enrichment (Pandey *et al.*, 2000; Singhania *et al.*, 2009).

Surface fermentation, used in the early production of both penicillin and citric acid, is carried out in shallow, nutrient medium-filled stainless steel trays instead of on a solid matrix. These trays are stacked within an aseptic fermentation chamber, and the fungal mycelium develops as a mat on the surface of the liquid.

Food and Industrial Bioproducts and Bioprocessing, First Edition. Edited by Nurhan Turgut Dunford.

In **submerged fermentation**, the microorganism is grown in nutrient broth in large, aerated vessels, such as stirred tank reactors or tower fermentors, under rigorously controlled environmental conditions. Submerged fermentation has several advantages over other fermentation methods, as it requires less space, is less labor intensive and higher production rates are obtained. The majority of citric acid is now produced by submerged fermentation (Sahasrabudhe and Sankpal, 2001; Bentley and Bennett, 2008). However, facilities that use surface culturing techniques to produce citric acid are still operated today, as they are simple and economical to maintain.

In addition to optimizing fermentation parameters, fungal strain improvement and selection, either by classical mutagenesis or more sophisticated targeted molecular genetic approaches, are important research avenues to increase productivity and yield of industrial fermentations (Ruijter *et al.*, 2002).

8.2 HISTORICAL PERSPECTIVE

The first known utilization of filamentous fungi for beneficial purposes is the traditional food and beverage fermentations used in many Asian cultures.

8.2.1 Koji

Fungal fermentation has been employed for thousands of years in China, Japan, Korea, Indonesia and other parts of Asia to enhance the palatability of rice, soybeans and other plant-based foods. In a process dating to the Chou dynasty (1121–220 BC), the mold inoculum was isolated and identified by its characteristic yellow-green hue (Bentley, 2006). Depending on what is being made, *Aspergillus oryzae* or other *Aspergillus* species, including *A. sojae* and *A. awamori*, is cultivated on steamed cereal grains, commonly rice or barley, with or without soy, to produce koji with distinctive characteristics and flavors. Koji, or "bloom of the mold", is the Japanese name for mold-fermented grains or soybeans. The resulting koji "starters" make a variety of fermented foods and beverages, including *miso* (soybean paste), *shoyu* (soy sauce), *sake* (rice wine), *shochu* (spirits) and *yonezu* (rice vinegar) (Reddy *et al.*, 1986).

Filamentous fungi secrete enzymes including proteases, amylases, lipases and pectinases as they invade and degrade their substrate. These enzymes in turn hydrolyze proteins to peptides and amino acids, and degrade starches to simple sugars. Fermentation enhances flavor, preservation and nutritional value as well as providing substrates for additional yeast or lactic bacterial fermentations. These fermentations, now collectively known as koji processes, are a very early example of solid-state fermentation (Machida *et al.*, 2008; Murooka and Yamshita, 2008). Solid-state fermentation is the basis for many modern commercial processes to produce a variety of enzymes, organic acids and secondary metabolites.

Japanese-born scientist Jokichi Takamine developed the commercial use of mold-derived enzymes and introduced these fungal products to industry. Takamine, the "Japanese father of American biotechnology", brought a modification of the koji process to the United States in the late nineteenth century (Bennett, 1985; Machida *et al.*, 2008). He believed koji-derived amylolytic enzymes could revolutionize the American distillery industry. His method involved growing *A. oryzae* on steamed bran and using aqueous alcohol to extract diastase, now known as amylase. While working in Peoria, IL, he originally marketed his diastase to the beer and whiskey trade as a more enzymatically-active alternative to malting enzymes. Takamine's efforts in this endeavor were thwarted due to personal ill health as well as labor unrest and intimations of arson. However, he ultimately achieved business success for his commercialized enzyme "Taka-diastase" which was marketed as a dyspepsia panacea by Parke, Davis and Company of Detroit, MI. He was granted a US patent in 1894 for his method of producing koji diastatic enzymes; this is thought to be the first patent for a microbial enzyme (Bennett, 1988, 2001; Machida *et al.*, 2008).

A variety of fungal enzymes, including cellulases, amyloglucosidases, lipases and proteases, are now commercially available, and their applications extend well beyond the traditional use in food fermentation processes. *Aspergilli*, including *A. niger* and *A. oryzae*, enjoy an extended history of safe use and play a key role in the industrial manufacture of many of these enzymes (Baker and Bennett, 2008). Some of Takamine's methods for culturing filamentous fungi are still used in today's enzyme industry. Solid-state fermentation produces higher yields of certain hydrolytic fungal enzymes than liquid fermentation, and this process is utilized for the commercial production of glucoamylase (GlaB) and aspartic protease (PepA) of *A. oryzae* and an acid-stable α-amylase of *A. kawachii* (Abe and Gomi, 2008).

8.2.2 Penicillin

The discovery of the fungal metabolite penicillin and its subsequent development into a "wonder drug" was central to the innovation of modern fungal biotechnology (Wainwright, 1990; Bentley and Bennett, 2008). Bacteriologist Alexander Fleming isolated penicillin from *Penicillium notatum* while at St. Mary's Hospital, London, in 1928. Fleming noted his discovery might have therapeutic value if it could be produced in sufficient quantity (Fleming, 1929).

The industrial microbial production of penicillin became a reality after the technological breakthrough of submerged (also referred to as "deep") fermentation methods at United States Department of Agriculture's Northern Regional Research Laboratories (NRRL) in Peoria, IL. With the world at war in 1941, Oxford researchers led by Drs Florey, Ernst Boris Chain and Norman Heatley collaborated with NRRL to produce the quantities of penicillin needed for clinical trials. In 1945 Fleming, Florey and Chain were awarded the Nobel Prize in Medicine or Physiology *"for the discovery of penicillin and its curative effect in various infectious diseases."*

Orville May, Percy Wells and Andrew Moyer of the NRRL Fermentation Division had previously patented a novel submerged fermentation process in 1935, and these methods were applied to penicillin production. Submerged culture conditions in large, aerated vessels generated greater penicillin yields than the previous surface growth fermentation. In addition to the development of submerged fermentation methods, penicillin yield was greatly improved by substituting corn steep liquor in the growth medium and utilizing higher yielding penicillin strains. After a worldwide search for higher yielding isolates, *P. chrysogenum* (NRRL, 1951) cultured from a moldy cantaloupe found in a Peoria market generated the greatest penicillin yield when grown in the submerged culture conditions.

The success of penicillin stimulated screens for other antibacterial secondary metabolites. This golden age of antibiotics of the 1940s and 1950s promoted research on fungal physiology, industrial strain development and large-scale fermentation methods. In addition to anti-infectives, many other important fungal-derived pharmaceuticals were isolated during this systematic search for beneficial metabolites, including cholesterol-lowering mevalonins (the "statins") and immunosuppressant cyclosporins (Monghan and Tkacz, 1990; Von Wartburg and Trabor, 1986). Understanding of penicillin synthesis at the molecular level was recently advanced by the sequencing of *P. chrysogenum*'s 32.19 Mb genome (van den Berg *et al.*, 2008).

8.2.3 Citric acid

Citric acid is another filamentous fungal product that was key to the development of modern industrial microbiology (Bentley and Bennett, 2008). This organic acid was originally isolated from citrus fruit and is used as a natural preservative and flavoring to impart a sour or acidic taste to foods and beverages. In 1893, Carl Wehmer discovered that *Penicillium* species produce citric acid from sugars, but sufficient quantities could not be produced for industrial use. American chemist James Currie at the United States Department of Agriculture demonstrated in 1917 the abundant production of citric acid by *A. niger* when cultured under the appropriate conditions (Currie, 1917). Commercial microbial production of citric acid became a possibility in 1919 when Currie took his fermentation method to Pfizer in Brooklyn, NY. Concomitantly, Europe began commercial production of citric acid at Citrique Belge, now DSM, by surface fermentation of *A. niger* on beet molasses.

Currie successfully pioneered the commercial production of citric acid by *A. niger* and eliminated the dependence of citric acid producers on citrus fruit import from Europe during World War I. The original Pfizer process used surface culture methods, but after the concurrent development of deep-tank fermentation for large-scale penicillin production, the citric acid process was adapted for large tanks and submerged culture. Techniques including manipulating culture conditions, perfecting deep-tank fermentation, and improving product recovery were innovated at Pfizer in the context of improving citric acid yields and laid the foundation for modern fermentation technologies (Rodenger, 1999). Today, highly productive *A. niger* strains developed through random mutagenesis generate almost all of the entire world supply of citric acid.

8.3 INDUSTRY

Filamentous fungal fermentation is a cornerstone of bioprocess manufacturing and generates a wide array of products used in the food, enzyme and pharmaceutical industries. Some of the key products for these applications are reviewed here.

8.3.1 Organic acids

Fungi, especially *Aspergilli*, are recognized for their prolific ability to yield a repertoire of organic acids. Conditions that produce an almost quantitative conversion of carbon substrate into acid are exploited in the large-scale manufacture of organic acids. Moreover, many organic acid-producing organisms tolerate low pH, thereby reducing production costs by eliminating the need for neutralization and minimizing contamination. Several fungal-derived acids have significant commercial value and have found wide ranging applications in the food, feed, pharmaceutical and polymer industries. Citric, gluconic, itaconic and lactic acids are manufactured by fungal fermentations and are of the most commercial importance. Oxalic, fumaric and malic acids can also be made via industrial bioprocesses, but the market demand is currently small because competing chemical conversion routes are more economical (Magnuson and Lasure, 2004). The commercial success of fungal fermentation is ultimately based on the rapid and economical conversion of sugar to acid and the process of product recovery (Magnuson and Lasure, 2004). Thus, only a small number of fungal-produced acids are manufactured on an industrial scale, despite the fact that fungi make a diversity of organic acids.

Citric acid is an excellent example of an industrially produced fungal primary metabolite and is a true bulk commodity. Almost the entire world supply of citric acid is made by *A. niger* in submerged fermentation processes, with an estimated global production of 1.7 million tons in 2007 (Berovic and Legisa, 2007). Citric acid is the most ubiquitous acidulant in foods, beverages and pharmaceuticals. The US Food and Drug Administration classifies citric acid and many *A. niger*-derived enzymes as GRAS (Generally Recognized as Safe), a highly desired category, because this organism has a long history of safe, non-pathogenic and non-toxigenic use (Schuster *et al.*, 2002). The properties of citric acid that form the basis for its broad range of applications include acidity and buffering capacity, taste and flavor, and chelation of metal ions. Industrial uses of citric acid include detergent manufacturing, electroplating and leather tanning. Citric acid is also used as a preservative for stored blood and a buffer and antioxidant in the pharmaceutical and cosmetic industries (Schuster *et al.*, 2002).

The manufacture of *A. niger*-derived citric acid conducted today is much more efficient and economical than the original Pfizer method. The majority of citric acid is currently generated by submerged fermentation methods, though the more labor intensive, original method of surface fermentation is still in use (Sahasrabudhe and Sankpal, 2001). Modern fermentation technologies can yield 95 kg citric acid per 100 kg of supplied sugar with the best producer strains and optimized production conditions (Karaffa and Kubicek, 2003). High yields of citric acid accumulation require a unique combination of several unusual nutrient conditions, including 15–20% w/v readily metabolized sugars, suboptimal amounts of phosphate and nitrogen to limit fungal growth and low levels of heavy metal ions including manganese. Maintaining a distinct fungal morphology in submerged culture is a good indicator of high levels of citric acid outflow (Karaffa and Kubicek, 2003). Germinating spores form short, stubby, forked, bulbous hyphae that aggregate into smooth, round pellets (Figure 8.1). Individual pellets are 200–500 microns in diameter. This tight-pelleted morphology is dependent on appropriate nutrient composition, especially maintaining low manganese levels (Kisser *et al.*, 1980).

Fungal fermentation also produces gluconic and itaconic acids, which are of lesser economic importance than citric acid. Gluconic acid, produced by *A. niger*, is used primarily as a food additive as well as for cleaning and finishing of metal surfaces and some pharmaceutical applications (Ramachandran *et al.*, 2006). The estimated annual global production of gluconic acid is 60 000 tons. Itaconic acid is produced by *A. terreus* in submerged fermentation processes similar to that for citric acid accumulation (Willke

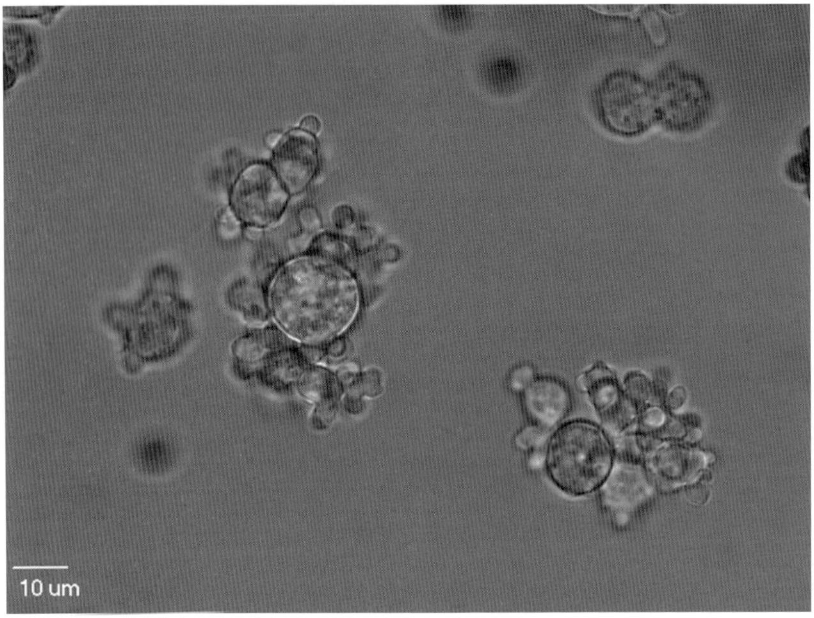

Figure 8.1 Light micrograph showing the *A. niger* pelleted morphology important for high citric acid production (Image: Sue A. Karagiosis, Pacific Northwest National Laboratory).

and Vorlop, 2001; Okabe *et al.*, 2009). This organic acid has an annual global production of 70 000 tons. Unlike citric and gluconic acid, itaconic acid is used exclusively in non-food applications, such as making synthetic fibers, coatings, adhesives, thickeners and binders.

The zygomycete fungus *Rhizopus oryzae* is a significant industrial source of enantiomerically pure L(+)-lactic acid. Lactic acid production has an annual global market of over 275 000 tons, with forecasted increasing demands due to expanding use in cosmetics and biodegradable packaging. Lactic acid is used to generate poly lactic acid and ethyl lactate, a biodegradable polymer and solvent, respectively. Additionally, this mild organic acid is utilized as a food preservative and flavor enhancer. Unlike *Lactobacillus* species, which produce mixed isomers, *Rhizopus* generates the preferred L(+)-lactic acid. The principal limitation of the *R. oryzae* fermentation process is the diversion of carbon away from lactic acid production to ethanol and fumaric acid fermentation. Fungal fermentation of glucose yields lactic acid at 60–80% efficiency (Skory, 2004), and productivity rates and yields are lower than those obtained for commercial *Lactobacillus* fermentations.

A clearer understanding of *R. oryzae*'s fermentative pathways and lactic acid biosynthesis may provide novel insight into more efficient and cost-effective methods of lactic acid synthesis. Targeted molecular genetic approaches, such as increasing lactate dehydrogenase activity to redirect pyruvate away from ethanol fermentation and increase the overall yield of lactic acid, are promising (Skory 2000, 2004). However, the current paucity of molecular tools for genetic manipulation of *Rhizopus* and an incomplete understanding of the recombination and replication mechanisms that affect the fate of introduced DNA have challenged this process (Skory, 2005).

8.3.2 Enzymes

Saprophytic fungi are efficient producers of a myriad of extracellular enzymes. More than 2000 fungal enzymes have been isolated and characterized, but only a small number are commercially exploited (Saxena *et al.*, 2005). In the food industry, fruit juices and wines are clarified using *Aspergillus* and *Rhizopus* pectinases; milk is coagulated with *Mucor* rennet for making cheese; and high fructose corn syrup is converted via an enzymatic route using *Aspergillus* and *Mucor* species-derived glucose isomerase. Industrial applications for fungal amylases and phytases include the removal of sizing from fabrics and as a nutritive supplement in animal feeds for liberation of phosphate from plant material, respectively. Microbial cellulases are used in the biorefinery to hydrolyze plant biomass to mixed sugars that are converted to chemical intermediates and biofuels. Genome sequence analyses indicate that filamentous fungi have undiscovered and unexploited enzymatic capacities. Thus, the number of commercial applications that employ fungal-derived enzymes is likely to expand.

Research efforts are aimed at developing biofuels and bioproducts using lignocellulosic biomass from agricultural crop residue and grasses as a sustainable source of mixed sugars for fermentation. Fungal-derived cellulolytic enzymes degrade plant cell wall material, a key step in the conversion of cellulosic material into glucose, which is then fermented to bioethanol by yeast. The mesophilic ascomycete fungus *Trichoderma reesei* is an attractive candidate for inexpensive cellulase and hemicellulase production because of its capacity to secrete abundant amounts of extracellular enzyme, the availability of genetic tools and its simple fermentation. This organism is one of the most prolific generators of cellulolytic enzymes and presents a paradigm for the efficient degradation of plant cell wall polysaccharides. *T. reesei*, the anamorph of *Hypocrea jecorina*, was originally isolated from rotting cotton fabric in the Solomon Islands during World War II (Reese, 1976; Reese and Levinson, 1952; Reese *et al.*, 1950). Mutant lines derived from this original isolate, selected for their potent secretory system and high enzyme expression levels, are the industrial workhorse organisms for cellulase and hemicellulase production (Mandels *et al.*, 1971; Montenecourt and Eveleigh, 1977). Derivatives of these strains are unofficially reported to produce over 100 g/l of secreted protein. In addition to the conversion of plant biomass into sugars and bioethanol, these plant cell wall degrading enzymes have wide ranging commercial applications in the food, pulp and paper and textile industries (Kubicek *et al.*, 2009; Schuster and Schmoll, 2010).

Presently, *T. reesei* strain development to boost cellulase production is a key focus of industrial research. In addition to the classical mutagenesis approaches, a wide array of genetic engineering tools, DNA-mediated transformation protocols and gene knock-out techniques (Nevalainen *et al.*, 1994) has been developed to improve *T. reesei*'s enzyme production capacity. The recent sequencing of the 34 million base pairs of the *T. reesei* genome (Martinez *et al.*, 2008) opens other molecular genetic avenues, such as manipulating inducer forming pathways, signaling cascades or transcriptional activation to enhance cellulase expression. The recent discovery that *T. reesei* has a mating type-dependent sexual cycle will also yield fast and efficient strain improvement and boost research toward economically feasible biofuel production (Seidl *et al.*, 2009).

8.3.3 Lovastatin

Filamentous fungi generate a chemically diverse array of bioactive, low molecular weight compounds, termed secondary metabolites. Secondary metabolites are dispensable for

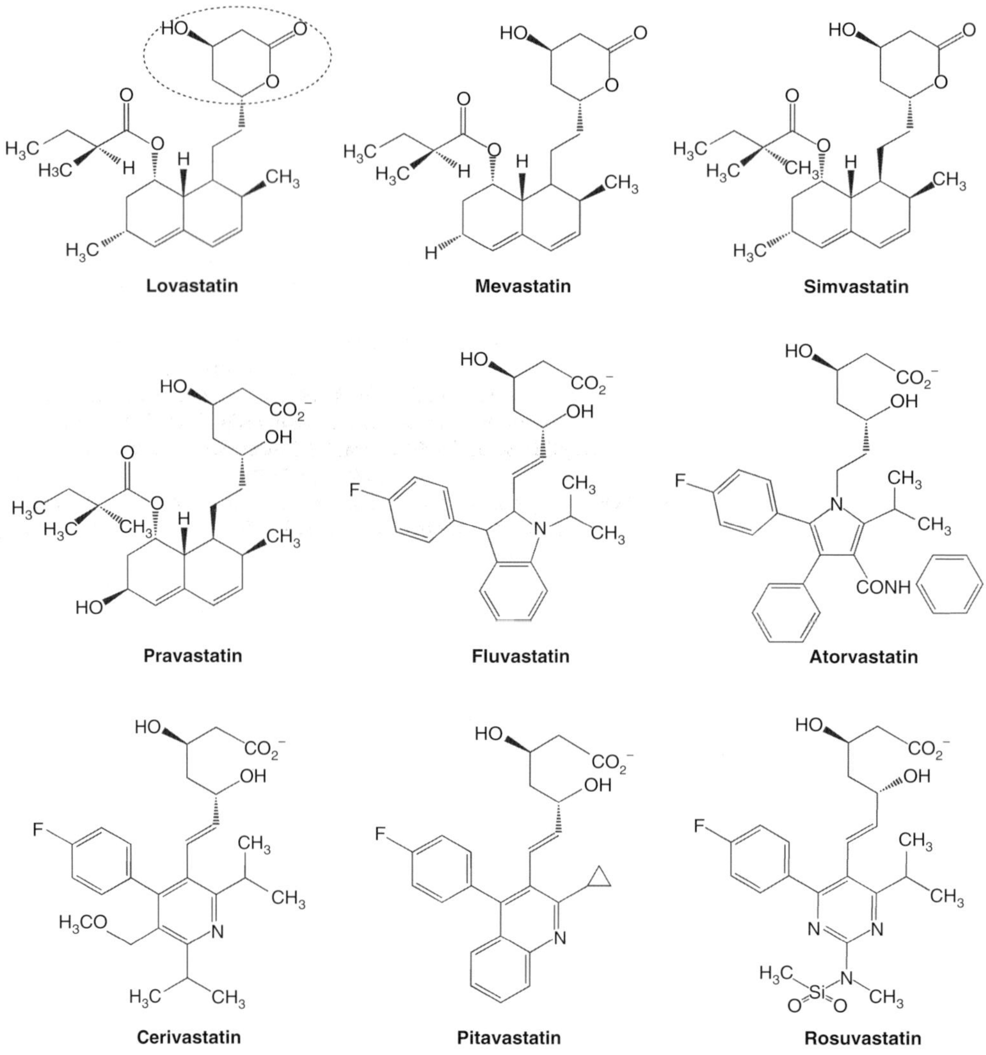

Figure 8.2 Chemical structures of statins. HMG-CoA analog is highlighted with a dashed circle. Lovastatin and mevastatin are natural statins of fungal origin. Simvastatin and pravastatin are chemically modified derivatives of lovastatin and mevastatin, respectively. Fluvastatin, atorvastatin, cerivastatin (withdrawn from clinical use in 2001), pitavastatin and rosuvastatin are fully synthetic compounds (Schachter, 2005).

growth and development but are thought to confer a selective advantage to the organism that produces them. Well-known fungal secondary metabolites with pharmaceutical applications include penicillin, cyclosporins, ergot alkaloids and statins.

Aspergillus-derived statins are an excellent example of fungal secondary metabolites that have made a great impact to human health as a drug. Statins are widely used for reducing plasma concentrations of total and LDL cholesterol in the treatment of hypercholesterolemia and reducing the risk of heart attacks. Statins are hydroxymethylglutaryl-coenzyme

A (HMG-CoA) reductase inhibitors that block the HMG-CoA reductase (HMGR) conversion of HMG-CoA into mevalonic acid, an early and rate-limiting step in *de novo* cholesterol biosynthesis. The lactone ring of the statin molecule is reversibly converted to an open acid that resembles the HGM portion of HMG-CoA (Figure 8.2), the substrate of the reaction catalyzed by the HMGR enzyme (Endo, 2004). The mechanism by which statins lower cholesterol is an important factor in its success. Previous attempts to reduce cholesterol biosynthesis, such as the Triparanol compound that blocks a late step in the pathway, were ultimately unsuccessful due to accumulation of potentially toxic precursors and adverse effects.

The first identified HMGR inhibitor mevastatin, also known as compactin and ML-236B, was isolated from *Penicillium citrinum* in 1971 by Japanese microbiologist Endo Akira at Sankyo (Endo, 1979). However, the chemically distinct lovastatin, isolated from the soil fungus *Aspergillus terreus* in 1978 by Alberts and colleagues at Merck Research Laboratories, became the first statin developed for human drug use (Alberts *et al.*, 1980). Lovastatin proved to be slightly more effective in lowering total serum cholesterol than mevastatin (Alberts *et al.*, 1980; Endo, 1980), and Merck patented lovastatin in 1980 (Monaghan *et al.*, 1980; Tobert, 2003). The US Food and Drug Administration approved lovastatin in 1987, and it became the first prescribed statin used in humans for lowering cholesterol. Lovastatin was a blockbuster for Merck, with over $1 billion of sales at its peak (Tobert, 2003).

The discovery of these early statins paved the way for the worldwide development of other drugs based on the statin chemical structure (Figure 8.2). Sankyo and Merck directed their later efforts at manufacturing synthetic analogs. Many different statins are currently available for therapeutic use, but lovastatin and mevastatin remain the only fermentation-derived statins. The lovastatin biosynthetic pathway in *A. terreus* is well understood. This pathway was the first example of a polyketide synthetic pathway in which two fungal type I polyketide synthases work in combination to produce a product (Hendrickson *et al.*, 1999; Kennedy *et al.*, 1999). Since then several statins, including simvastatin, pravastatin, fluvastatin and atorvastatin, have been approved in many countries and are currently used by millions.

8.4 GENOMICS AND THE FUTURE

8.4.1 Citric acid and *Aspergillus niger*

Current *A. niger* industrial strain development combines information garnered from the reported genome sequences and targeted reverse genetic engineering to boost enzyme and acid production. Pel and colleagues reported the genome sequence of *A. niger* CBS 513.88, a predecessor to the now widely used workhorse strain for industrial enzyme and organic acids production, including citric acid (Pel *et al.*, 2007). The authors integrated the annotated genome data into a model of central carbon metabolism to better understand the metabolic processes behind the organism's high productivity. Genetic redundancy may be one factor that influences citric acid overflow; the model highlights the involvement of multiple copies of citrate synthases, aconitases and malate dehydrogenase-like sequences.

Molecular, genetic and cell biology analysis will aid in illuminating the functions and features of these annotated gene products in enzyme or acid synthesis. Gene targeting methods, including replacing the coding sequence with selectable markers or fusing it to fluorescent labels, can be used to evaluate the protein's cellular role. Fast and efficient high throughput gene deletions are facilitated by the recent development of the *A. niger kusA* (*ku*70) non-homologous end joining-deficient mutant that suppresses double stranded break repair (Meyer *et al.*, 2007; Carvalho *et al.*, 2010). Polarized growth is thought to be an important feature of fermentative output, and imaging analysis has demonstrated that productive citric acid overflow is dependent on cell morphology (Cox and Thomas, 1992; Papagianni, 2004). Microscopy imaging of *A. niger* during fermentative growth will allow investigation of recognized key polarization components, including the Spitzenkörper, F-actin, GTPases and other signaling molecules for metabolite production and secretion (Harris *et al.*, 2005). A detailed understanding of individual genes combined with high throughput characterization of the metabolome, proteome and transcriptome will improve targeted strain development of this industrial workhorse.

Systems biology approaches to understanding metabolism and accelerating strain improvement are gaining momentum across the genus *Aspergillus* (Andersen and Nielsen, 2009). In 2008, a comprehensive metabolic reconstruction for *Aspergillus niger* based on published literature (aka the "bibliome"), genome sequence, metabolomic and transcriptomic data was published; it lays a solid foundation for a systems biology approach to strain improvement (Andersen *et al.*, 2008). This model has been used to predict essential genes (Thykaer *et al.*, 2009) as well as generate a deeper understanding of the biochemical pathways that feed production of various metabolites (Andersen *et al.*, 2009; Panagiotou *et al.*, 2008; Sorensen *et al.*, 2009).

8.4.2 Cellulase production

Enzyme production costs present a considerable barrier to economic lignocellulose fuel ethanol. A clearer understanding of the molecular mechanisms of cellulase synthesis will accelerate industrial strain development for the generation of more economical enzymes. Le Crom and colleagues performed massively parallel sequencing to identify mutations in the genomes of two cellulase hyper-producing mutants NG14 and RUT C30 (Le Crom *et al.*, 2009). They detected a surprisingly high number of mutagenic events: 223 single nucleotide variants, 15 small deletions or insertions, and 18 larger deletions, leading to the loss of more than 100 kb of genomic DNA. The authors confirmed previously reported mutations and uncovered novel mutations in 43 genes that are involved in nuclear transport, mRNA stability, transcription, secretion/vacuolar targeting, and metabolism. This heterogeneity of functional categories suggests that multiple changes may be necessary to improve cellulase production (Le Crom *et al.*, 2009). The results of Martinez and colleagues and Le Crom and colleagues build a genomic foundation to test a large number of hypotheses about the mechanisms underlying *T. reesei* protein secretion, carbon catabolite repression and cellulose induction of cellulase (Le Crom *et al.*, 2009; Martinez *et al.*, 2008). Their findings also highlighted previously unexplored targets for directed strain improvement, including nucleocytoplasmic transport, vacuolar protein trafficking and mRNA turnover.

8.4.3 Bioactive secondary metabolites

Fungi are a particularly rich source of bioactive secondary metabolites with potent physiological activities and possible pharmaceutical applications; these microbial compounds continue to be a key source of molecules for drug discovery. More than half of the isolated fungal compounds have antibacterial, antifungal or antitumor activity (Palaez, 2005), and 60% of anticancer agents and 70% of anti-infectives presently in clinical use are natural products or natural product-based compounds (Yin *et al.*, 2007). Secondary metabolites are synthesized by a few common biosynthetic pathways, often in conjunction with a specific stage of morphological differentiation. The polyketide synthases (PKSs) and non-ribosomal peptide synthetases (NRPSs) generate many of these biologically active compounds. The conserved features within PKS and NRPS gene sequences are the foundation for genomic-guided discovery of nature products. Bioinformatic analysis of these biosynthetic genes in the burgeoning number of reported fungal genomes indicates that these organisms encode the information for the production of a multitude of yet undiscovered compounds.

A major obstacle for mining novel fungal metabolites is uncovering the conditions in which they are synthesized. Because these compounds are not required for the organism's growth or development, in the absence of a particular trigger these gene loci remain silent. Genome mining has proven to be a fruitful approach for the discovery of novel microbial metabolites. Bergmann and colleagues showed that an *A. nidulans* cryptic gene cluster could be chemically switched on by the ectopic expression of the pathway specific regulator by an inducible promoter (Bergmann *et al.*, 2007). The authors isolated and identified two novel PKS-NRPS hybrid metabolites: the cytotoxic aspyridones A and B (Bergmann *et al.*, 2007). Chiang and colleagues induced an otherwise silent pathway by a similar strategy to generate the novel polyketide asperfuranone (Chiang *et al.*, 2009).

Some genes encoding specific fungal secondary metabolites are clustered near telomeres or in heterochromatin regions, and their activation is controlled by regulation at the chromatin level by processes that include DNA methylation and histone deacetylation. This productive area of research was initiated with the discovery of *laeA*, a gene whose deletion led to decreased expression of known secondary metabolites in a number of *Aspergilli* (Bok and Keller, 2004). Deletion of *hdaA*, encoding an *A. nidulans* histone deacetylase, relieved chromatin-level suppression of two secondary metabolites located in the telomere region (Shwab *et al.*, 2005). Bok and colleagues showed that the loss-of-function *A. nidulans* CclA, involved in histone H3 lysine 4 methylation, activated the expression of cryptic secondary metabolite clusters and generated monodictyphenone and anti-osteoporosis polyketides F9775A and F9775B (Bok *et al.*, 2009). Orthologues of *laeA* have been identified in additional ascomycetes outside of the genus *Aspergillus*, including *Penicillium* and *Fusarium*, and characterized (Wiemann *et al.*, 2010; Xing *et al.*, 2010; Kosalkova *et al.*, 2009).

8.5 CONCLUSIONS

For centuries, fungi have been utilized by humans for their ability to convert complex and simple substrates into more usable, palatable and/or valuable forms. Optimization of the fungal strains and bioprocesses responsible for these amazing transformations takes time.

However, we are poised to enter a new era of fungal bioprocess development. The rapid pace of genomics in both technology and in the number of organisms for which there is genome sequence available has opened the door to novel methods for rapid strain development, bioprocess improvement and small molecule discovery. With their ability to produce large quantities of enzymes and small molecules, the Kingdom Fungi is an attractive branch of the tree of life on which to use these new genomics-based methods for new medicines, food processes and biofuels.

REFERENCES

Abe K & Gomi K. 2008. Food products fermented by *Aspergillus oryzae*. In: Goldman, G.H. & Osmani, S.A. (editors). The Aspergilli Genomics, Medical Aspects, Biotechnology, and Research Methods. Boca Raton, FL: CRC Press. pp. 429–439.

Monaghan RL, Alberts AW, Hoffman CH, Alberts-Schonberg GA. 1980. Hypocholesteremic fermentation products and process of preparation. US Patent 4,231,938.

Alberts AW, Chen J, Kuron G, *et al.* 1980. Mevinolin: a highly potent competitive inhibitor of hydroxymethylglutaryl-coenzyme A reductase and cholesterol-lowering agent. Proc Natl Acad Sci USA – Biological Sciences 77(7):3957–3961.

Andersen MR & Nielsen J. 2009. Current status of systems biology in Aspergilli. Fungal Genet Biol 46(Suppl 1):S180–190.

Andersen MR, Nielsen ML & Nielsen J. 2008. Metabolic model integration of the bibliome, genome, metabolome and reactome of Aspergillus niger. Mol Syst Biol 4:178.

Andersen MR, Lehmann L & Nielsen J. 2009. Systemic analysis of the response of Aspergillus niger to ambient pH. Genome Biol 10(5):R47.

Baker SE & Bennett JW. 2008. An overview of the genus *Aspergillus*. In: Goldman, G.H. & Osmani, S. (editors). The Aspergilli: Genomics, Medicine, Biotechnology and Research Methods. Boca Raton, FL: CRC Press, Taylor & Francis. pp. 3–13.

Bennett JW. 1985. Taxonomy of fungi and biology of the *Aspergilli*. In: Demain, A.L. & Solomon, N.A. (editors). Biology of Industrial Microorganisms. Menlo Park, CA: Benjamin-Cummings Pub. Co. pp. 359–406.

Bennett JW. 1988. Documents from the Dawn of Industrial Biotechnology. Elkhart, IN: Miles Laboratory.

Bennett JW. 2001. Adrenalin and cherry trees. Modern Drug Discoveries 4(12):47–51.

Bentley R. 2006. From miso, sake and shoyu to cosmetics: a century of science for kojic acid. Natural Product Reports 23(6):1046–1062.

Bentley R & Bennett JW. 2008. A ferment of fermentations: reflections on the production of commodity chemicals using microorganisms. Adv Appl Microbiol 63:1–32.

van den Berg MA, Albang R, Albermann K, *et al.* 2008. Genome sequencing and analysis of the filamentous fungus *Penicillium chrysogenum*. Nature Biotechnology 26(10):1161–1168.

Bergmann S, Schumann J, Scherlach K, Lange C, Brakhage AA & Hertweck C. 2007. Genomics-driven discovery of PKS-NRPS hybrid metabolites from *Aspergillus nidulans*. Nature Chemical Biology 3(4):213–217.

Berovic M & Legisa M. 2007. Citric acid production. Biotechnol Annu Rev 13:303–343.

Bok JW, Chiang YM, Szewczyk E, *et al.* 2009. Chromatin-level regulation of biosynthetic gene clusters. Nature Chemical Biology 5(7):462–464.

Bok JW & Keller NP. 2004. LaeA, a regulator of secondary metabolism in Aspergillus spp. Eukaryotic Cell 3(2):527–535.

Carvalho N, Arentshorst M, Kwon MJ, Meyer V & Ram AFJ. 2010. Expanding the ku70 toolbox for filamentous fungi: establishment of complementation vectors and recipient strains for advanced gene analyses. Applied Microbiology and Biotechnology 87(4):1463–1473.

Chiang YM, Szewczyk E, Davidson AD, Keller N, Oakley BR & Wang CCC. 2009. A gene cluster containing two fungal polyketide synthases encodes the biosynthetic pathway for a polyketide, asperfuranone, in *Aspergillus nidulans*. Journal of the American Chemical Society 131(8):2965–2970.

Cox PW & Thomas CR. 1992. Classification and measurement of fungal pellets by automated image-analysis. Biotechnology and Bioengineering 39(9):945–952.

Currie JN. 1917. The citric acid fermentation of *Aspergillus niger*. Journal of Biological Chemistry 31: 15–37.

Endo A. 1979. Monacolin-K, a new hypocholesterolemic agent produced by a *Monascus* species. Journal of Antibiotics 32(8):852–854.

Endo A. 1980. Monacolin-K, a new hypocholesterolemic agent that specifically inhibits 3-hydroxy-3-methylglutaryl coenzyme A reductase. Journal of Antibiotics 33(3):334–336.

Endo A. 2004. The origin of the statins. Atherosclerosis Supplements 5(3):125–130.

Fazenda ML, Seviour R, McNeil B & Harvey LM. 2008. Submerged culture fermentation of "higher fungi": the macrofungi. Adv Appl Microbiol 63:33–103.

Fleming A. 1929. On the antibacterial action of cultures of a *Penicillium*, with special reference to their use in the isolation of *B. influenzae*. British Journal of Experimental Pathology 10:226–236.

Harris SD, Read ND, Roberson RW, *et al.* 2005. Polarisome meets Spitzenkorper: Microscopy, genetics, and genomics converge. Eukaryotic Cell 4(2):225–229.

Hendrickson L, Davis CR, Roach C, *et al.* 1999. Lovastatin biosynthesis in *Aspergillus terreus*: characterization of blocked mutants, enzyme activities and a multifunctional polyketide synthase gene. Chemistry & Biology 6(7):429–439.

Karaffa L & Kubicek CP. 2003. *Aspergillus niger* citric acid accumulation: do we understand this well working black box? Applied Microbiology and Biotechnology 61(3):189–196.

Kennedy J, Auclair K, Kendrew SG, Park C, Vederas JC & Hutchinson CR. 1999. Modulation of polyketide synthase activity by accessory proteins during lovastatin biosynthesis. Science 284(5418): 1368–1372.

Kisser M, Kubicek CP & Rohr M. 1980. Influence of manganese on morphology and cell wall composition of *Aspergillus niger* during citric acid fermentation. Archives of Microbiology 128(1):26–33.

Kosalkova K, Garcia-Estrada C, Ullan RV, *et al.* 2009. The global regulator LaeA controls penicillin biosynthesis, pigmentation and sporulation, but not roquefortine C synthesis in *Penicillium chrysogenum*. Biochimie 91(2):214–225.

Kubicek CP, Mikus M, Schuster A, Schmoll M & Seiboth B. 2009. Metabolic engineering strategies for the improvement of cellulase production by *Hypocrea jecorina*. Biotechnology for Biofuels 2(19).

Le Crom S, Schackwitz W, Pennacchio L, *et al.* 2009. Tracking the roots of cellulase hyperproduction by the fungus *Trichoderma reesei* using massively parallel DNA sequencing. Proc Natl Acad Sci USA 106(38):16151–16156.

Machida M, Yamada O & Gomi K. 2008. Genomics of *Aspergillus oryzae*: learning from the history of Koji mold and exploration of its future. DNA Res 15(4):173–183.

Magnuson JK & Lasure LL. 2004. Organic acid production by filamentous fungi. In: Tkacz, J.S. & Lange, L. (editors). Advances in fungal biotechnology for industry, agriculture and medicine. Norwell, MA: Kluwer Academic Publishers. pp. 307–340.

Mandels M, Weber J & Parizek R. 1971. Enhanced cellulase production by a mutant of *Trichoderma viride*. Appl Microbiol 21(1):152–154.

Martinez D, Berka RM, Henrissat B, *et al.* 2008. Genome sequencing and analysis of the biomass-degrading fungus *Trichoderma reesei* (syn. *Hypocrea jecorina*). Nature Biotechnology 26(5):553–560.

Meyer V, Arentshorst M, El-Ghezal A, *et al.* 2007. Highly efficient gene targeting in the *Aspergillus niger* kusA mutant. Journal of Biotechnology 128(4):770–775.

Monghan RL & Tkacz JS. 1990. Bioactive microbial products: focus upon mechansims of action. Annual Review of Microbiology 44:271–301.

Montenecourt BS & Eveleigh DE. 1977. Preparation of mutants of Trichoderma reesei with enhanced cellulase production. Appl Environ Microbiol 34(6):777–782.

Murooka Y & Yamshita M. 2008. Traditional healthful fermented products of Japan. J Ind Microbiol Biotechnol 35(8):791–798.

Murphy RA & Horgan KA. 2005. Antibiotics, enzymes and chemical commodities from fungi. In: Kavanagh, K. (editor). Fungi: biology and applications. Chichester: John Wiley & Sons Ltd, pp. 113–144.

Nevalainen H, Suominen P & Taimisto K. 1994. On the safety of *Trichoderma reesei*. Journal of Biotechnology 37(3):193–200.

Okabe M, Lies D, Kanamasa S & Park EY. 2009. Biotechnological production of itaconic acid and its biosynthesis in *Aspergillus terreus*. Appl Microbiol Biotechnol 84(4):597–606.

Palaez F. 2005. Biological activities of fungal metabolites. In: An, Z. (editor). Handbook of Industrial Mycology. New York, NY: Marcel Dekker, Inc. pp. 54–104.

Panagiotou G, Andersen MR, Grotkjaer T, *et al.* 2008. Systems analysis unfolds the relationship between the phosphoketolase pathway and growth in *Aspergillus nidulans*. PLoS One 3(12):e3847.

Pandey A, Soccol CR & Mitchell D. 2000. New developments in solid state fermentation: I – bioprocesses and products. Process Biochemistry 35(10):1153–1169.

Papagianni M. 2004. Fungal morphology and metabolite production in submerged mycelial processes. Biotechnology Advances 22(3):189–259.

Pel HJ, de Winde JH, Archer DB, *et al.* 2007. Genome sequencing and analysis of the versatile cell factory *Aspergillus niger* CBS 513.88. Nature Biotechnology 25(2):221–231.

Ramachandran S, Fontanille P, Pandey A & Larroche C. 2006. Gluconic acid: Properties, applications and microbial production. Food Technology and Biotechnology 44(2):185–195.

Reddy NR, Pierson MD & Salunkhe DK. 1986. Legume based fermented foods. Boca Raton, FL: CRC Press.

Reese ET. 1976. History of the cellulase program at the U.S. army Natick Development Center. Biotechnol Bioeng Symp (6):9–20.

Reese ET & Levinson HS. 1952. A comparative study of the breakdown of cellulose by microorganisms. Physiologica Plantarum 5:345–366.

Reese ET, Levinson HS, Downing M & White WL. 1950. Quartermaster culture collection. Farlowia 4(1):45–86.

Rodenger JL. 1999. The legend of Pfizer. Ft. Lauderdale, FL: Write Stuff Syndicate, Inc.

Ruijter GJG, Kubicek CP & Visser J. 2002. Production of organic acids by fungi. In: Esser, K. & Osiewacz, H.D. (editors). The mycota: industrial applications. New York: Springer-Verlag. pp. 213–230.

Sahasrabudhe NA & Sankpal NV. 2001. Production of organic acids and metabolites of fungi for food industry. In: Khachatourians, G.G. & Arora, D.K. (editors). Applied mycology and biotechnology Amsterdam: Elsevier Science BV. pp. 387–426.

Saxena RK, Malhotra B & Batra A. 2005. Commercial importance of some fungal enzymes. In: Arora, D.K. (editor). Handbook of Fungal Biotechnology. New York, NY: Marcel Dekker, Inc. pp. 498–518.

Schachter, M. (2005). Chemical, pharmacokinetic and pharmacodynamic properties of statins: an update. Fundam Clin Pharmacol 19(1):117–125.

Schuster A & Schmoll M. 2010. Biology and biotechnology of Trichoderma. Appl Microbiol Biotechnol 87(3):787–799.

Schuster E, Dunn-Coleman N, Frisvad JC & van Dijck PWM. 2002. On the safety of *Aspergillus niger* – a review. Applied Microbiology and Biotechnology 59(4–5):426–435.

Seidl V, Seibel C, Kubicek CP & Schmoll M. 2009. Sexual development in the industrial workhorse *Trichoderma reesei*. Proc Natl Acad Sci USA 106(33):13909–13914.

Shwab E, Bok J, Tribus M, Galehr J, Graessle S & Keller N. 2005. A role for histone deacetylases in regulation of *Aspergillus nidulans* secondary metabolism. Phytopathology 95(6):S96–S96.

Singhania RR, Patel AK, Soccol CR & Pandey A. 2009. Recent advances in solid-state fermentation. Biochemical Engineering Journal 44(1):13–18.

Skory CD. 2000. Isolation and expression of lactate dehydrogenase genes from *Rhizopus oryzae*. Applied and Environmental Microbiology 66(6):2343–2348.

Skory CD. 2004. Lactic acid production by *Rhizopus oryzae* transformants with modified lactate dehydrogenase activity. Applied Microbiology and Biotechnology 64(2):237–242.

Skory CD. 2005. Inhibition of Non-Homologous End Joining and integration of DNA upon transformation of *Rhizopus oryzae*. Molecular Genetics and Genomics 274(4):373–383.

Sorensen LM, Lametsch R, Andersen MR, Nielsen PV & Frisvad JC. 2009. Proteome analysis of *Aspergillus niger*: lactate added in starch-containing medium can increase production of the mycotoxin fumonisin B2 by modifying acetyl-CoA metabolism. BMC Microbiol 9:255.

Thykaer J, Andersen MR & Baker SE. 2009. Essential pathway identification: from in silico analysis to potential antifungal targets in Aspergillus fumigatus. Med Mycol 47 (Suppl 1):S80–87.

Tobert JA. 2003. Lovastatin and beyond: The history of the HMG-CoA reductase inhibitors. Nature Reviews Drug Discovery 2(7):517–526.

Von Wartburg A & Trabor R. 1986. Chemistry of the natural cyclosporin metabolites. Progress in Allergy 38:28–45.

Wainwright M. 1990. Miracle Cure – The Story of Antibiotics. Oxford, UK: Blackwell Publishing Ltd.

Wiemann P, Brown DW, Kleigrewe K, *et al.* 2010. FfVel1 and FfLae1, components of a velvet-like complex in Fusarium fujikuroi, affect differentiation, secondary metabolism and virulence. Mol Microbiol 77(4):972–994.

Willke T & Vorlop KD. 2001. Biotechnological production of itaconic acid. Appl Microbiol Biotechnol 56(3–4):289–295.

Xing W, Deng C & Hu CH. 2010. Molecular cloning and characterization of the global regulator LaeA in Penicillium citrinum. Biotechnol Lett 32(11):1733–1737.

Yin J, Straight PD, Hrvatin S, *et al.* 2007. Genome-wide high-throughput mining of natural-product biosynthetic gene clusters by phage display. Chemistry & Biology 14(3):303–312.

9 Microalgae: A Renewable Source of Bioproducts

Susan I. Blackburn and John K. Volkman

9.1 INTRODUCTION

Microalgae are single-celled plants that are found in aquatic environments globally. With a vast biodiversity they are an essential part of the aquatic food web and produce a range of bioproducts, including bioactive compounds, lipids of nutritional and energy importance, pigments, proteins, polysaccharides, vitamins, and antioxidants. This, combined with their capacity to be cultured, means they are a renewable resource of considerable economic potential. Used historically as human foods and in recent decades developed for aquaculture feeds, microalgae are still a relatively untapped source for bioproducts. The intense interest in biofuels for renewable energy has given a new focus on microalgae not only for biofuels but also for the relatively untapped potential for algae for a range of products as part of a biorefinery concept. New, more efficient growth technologies and genetic engineering pave the way for microalgae to be a significant future source of renewable bioproducts.

9.2 MICROALGAE AND THEIR GLOBAL IMPORTANCE

Microalgae are found in all aquatic environments – oceans, lakes, rivers, even small ponds – as well as some extreme terrestrial environments, such as deserts and polar environments. With sizes (length or breadth) ranging from 10^{-6} of a metre (1 μm) to less than a millimetre they are mostly invisible to the naked eye, except when they form natural proliferations called algal blooms. Microalgae have many varied and often beautiful morphologies that may only be revealed at high magnification. Examples of microalgae viewed with the scanning electron microscope are shown in Figure 9.1.

Microalgae are the world's fastest-growing plants. Typically they reproduce vegetatively, dividing by binary fission, that is, one cell divides to form two cells, two to four cells, and so on. Doubling times range from more than once per day for some species (e.g. the cyanobacterium (blue green alga) *Synechococcus*), to doubling times of two or more days for some dinoflagellates. Some microalgae can also reproduce sexually with consequent genetic exchange, thus maintaining genetic variation within a population. Some species form resting stages that can survive long periods in aquatic environments. These are often associated with sexual reproduction, thus combining the survival strategies of resistance to

Food and Industrial Bioproducts and Bioprocessing, First Edition. Edited by Nurhan Turgut Dunford.

Figure 9.1 Microalgal morphological diversity viewed with the scanning electron microscope: (a) the diatom *Bacteriastrum*; (b) naviculoid diatom; (c) the coccolithophorid (Prymnesiophyceae) *Emiliania huxleyi*; (d) the dinoflagellate *Gymnodinium catenatum* (CSIRO images).

adverse environmental conditions with genetic variability (Blackburn and Parker, 2005). Most microalgae are phototrophic (autotrophic), that is they use the sun's energy through photosynthesis, capturing it as chemical energy in biological molecules, for growth, using carbon dioxide (CO_2) and producing oxygen (O_2) as part of that process. Indeed, microalgae are responsible for over half of global primary productivity, more than all the land plants together, and with their capacity to use carbon dioxide they play a key role in global biogeochemical cycles.

Some microalgae are heterotrophic, that is they use organic carbon compounds, for example, glucose, acetate, lactate and glutamate, and do not require light as part of the process. Other microalgae have both autotrophic and heterotrophic capacity, depending on the light and nutritional status of the environment they inhabit. This dual capacity is known as mixotrophy (Lee, 2004).

Globally there is a vast biodiversity of microalgae. The numbers of different phytoplankton species in the sea can only be guessed at. Twenty years ago, Sournia and co-workers (1991) estimated that there were 474–504 genera and 3444–4375 species. Hallegraeff (2003) put the figure at about 5000 species. Much larger estimates are commonplace. However, with the use of modern molecular biology tools it is now appreciated that some microalgae are not culturable using standard methods and that there is an incredible diversity at each taxonomic level including infra-specific diversity, down to the level of within and between populations (Norton *et al*., 1996; Simon *et al*., 2009). Even within the known species there

Plate 9.1
(a) Microalgal cultures maintained in constant environment room;
(b) Microalgal cultures grown in disposable plastic bags supported vertically in metal frames;
(c) Raceway pond (Seambiotic);
(d) Green Wall Panel photobioreactor from Fotosintetica & Microbiologica Srl., Italy. Used with permission.

Food and Industrial Bioproducts and Bioprocessing, First Edition. Edited by Nurhan Turgut Dunford.

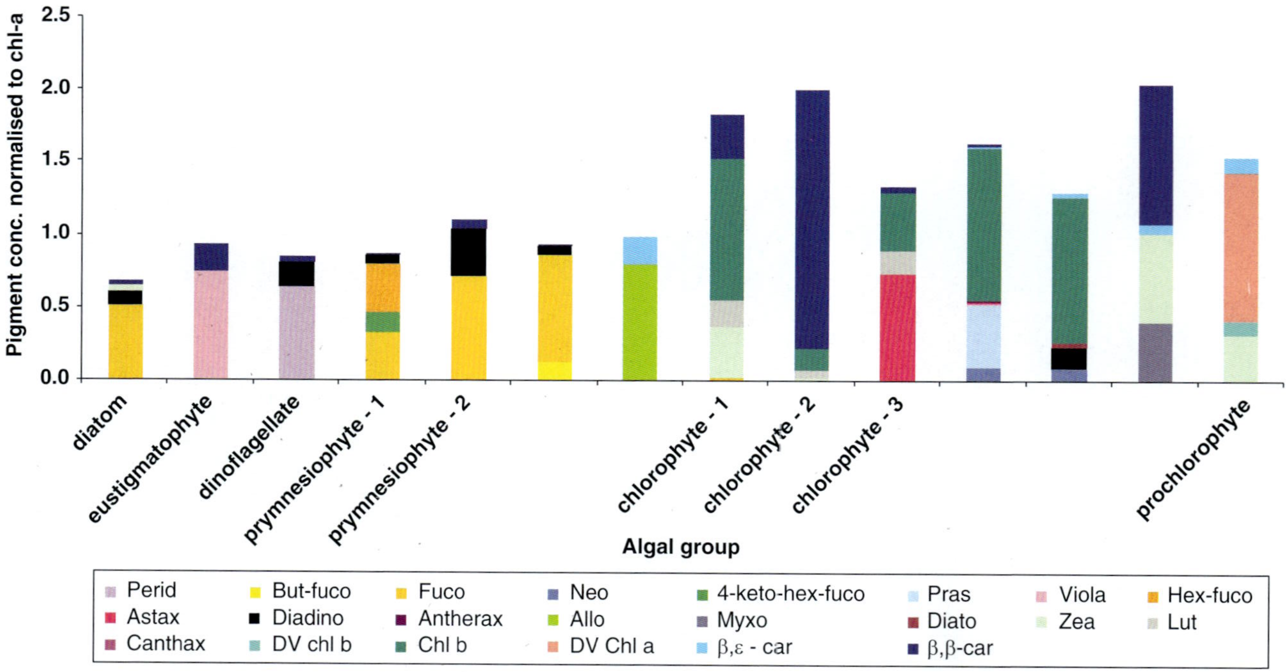

Plate 9.2 Pigment profiles of different microalgal classes from strain representatives held in the Australian National Algae Culture Collection (data compiled by Lesley Clementson). Within one class there can be variation between genera, species and even strains. P gment key: But-fuco=19′Butanoyloxyfucoxanthin, Fuco=Fucoxanthin; Neo=9′-cis-Neoxanthin; 4-keto-hex-fuco=4-keto-19′-Hexanoyloxyfucoxanthin; Pras=Prasinoxanthin; Viola=Violaxanthin; Hex-fuco=19′-Hexanoyloxyfucoxanthin; Astax=Astaxanthin; Diadino=Diadinoxanthin; Antherax=Antheraxanthin; Allo=Alloxanthin; Myxo=Myxoxanthophyll; Diato=Diatoxanthin; Zea=Zeaxanthin; Lut=Lutein; Canthax=Canthaxanthin; DV chl b=Divinyl chlorophyll b; Chl b=Monovinyl chlorophyll b; DV chl a=Divinyl chlorophyll a; β,ε-car=β,ε-carotene; β,β-car=β,β-carotene.

can be considerable biochemical diversity such that different strains of a single species can have different properties. An understanding of microalgal biodiversity and the associated effect on cellular properties is of great importance when selecting microalgae for bioapplications. Selection at the species level may be a guide to potential, but there will be variability between strains so testing and selection of individual strains will be essential.

9.3 CULTURED MICROALGAE

Microalgae can be cultivated (i.e. cultured) in the laboratory or in industrial or aquaculture situations using aqueous media containing nutrients required for growth. Cultured microalgae, in essence, behave rather like natural algal blooms that form when there are good growth conditions (e.g. light, temperature, nutrient supply), but with the advantage that conditions and growth systems can be controlled and optimised. Typical growth and proliferation of microalgae in a batch culture is characterised by a lag phase, where there is no increase in cell numbers, followed by an exponential or logarithmic phase of growth until some growth factor becomes limiting (e.g. light or one or more nutrients) at which stage the growth pattern changes to a phase of declining relative growth and then stationary phase, followed by a death phase (Figure 9.2). The duration of a culture may be extended by using semi-continuous or continuous culturing. In such cultures, fresh medium (nutrients and water) are supplied, either intermittently (semi-continuous) or continuously, along with removal of a matching amount of the culture. This prevents the culture becoming nutrient limited, thus avoiding the stationary and death phases. More details on the principles of continuous flow culture can be found elsewhere (Lee and Shen, 2004).

Autotrophic and heterotrophic microalgae have different growth requirements. Autotrophic microalgae have an absolute requirement for light for photosynthesis and may

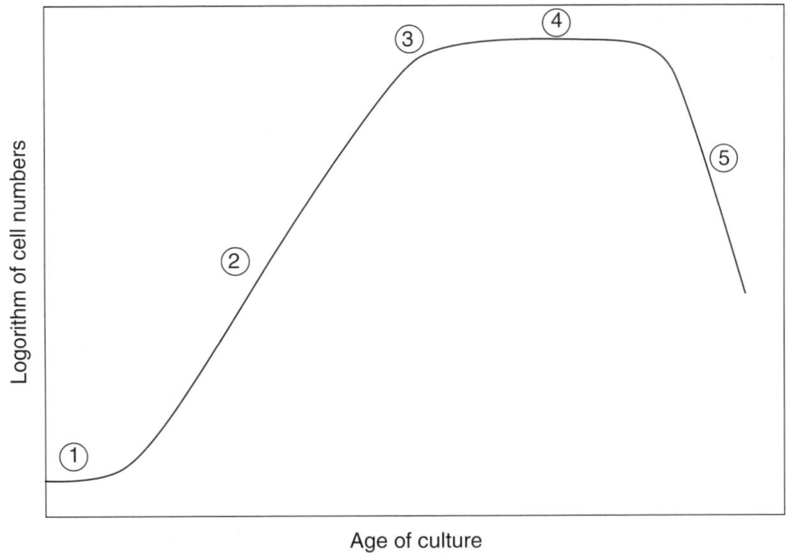

Figure 9.2 The characteristic pattern of growth shown by a unicellular microalga in a culture of limited volume: (1) lag phase, (2) exponential phase, (3) phase of declining growth, (4) stationary phase, and (5) death phase. (Adapted from Fogg, copyright 1965, by the Board of Regents of the University of Wisconsin System. Reprinted by permission of The University of Wisconsin Press.)

need additional carbon dioxide supplied for optimal growth and photosynthesis, as well as an adequate supply of inorganic nutrients. Heterotrophic microalgae require an organic carbon source such as glucose and can be grown in fermentors that have been developed for cultivation of other microorganisms, such as bacteria. Because the organic carbon also supports growth of bacteria, it is important that microalgae grown in fermentors are axenic (i.e. bacteria free). Otherwise the microalgae are soon overwhelmed by bacteria that multiply rapidly under such conditions. For further details on heterotrophic growth see Lee (2004) and references therein. Whether autotrophic or heterotrophic, controlled microalgal production is a renewable resource and offers the potential to produce a range of bioproducts.

9.4 ALGAL CULTURE COLLECTIONS

Of the vast diversity of species of naturally occurring microalgae only several thousand have been isolated from nature and cultured. Many of these cultured microalgal strains are maintained in algal culture collections in various parts of the world. Information on the major algal culture collections is available from the World Federation of Culture Collections (http://www.wfcc.nig.ac.jp/wfcc.html). Some of these collections are listed in Table 9.1. The physiological, biochemical and genetic characterisation of cultured microalgae, including type species held long term in algal culture collections, highlights the value of collections as repositories of global microalgal biodiversity. Many of the culture collections make their strains available to researchers, educational institutions and industry, often for a modest charge, thus providing a reliable base for all microalgal applications, including bioproducts. Note that some countries may impose strict import restrictions on microalgae from outside sources, making the isolation, identification and maintenance of native microalgae an essential part of the development of algal industries.

Culture collections usually maintain microalgal strains in liquid or solid agar (nutrient media in water) contained in small scale culture vessels, such as glass tubes and flasks or sterile disposable plastic tissue culture flasks or Petri dishes. For autotrophic (photosynthetic) microalgae cultures are usually maintained in constant-environment rooms or cabinets with artificial lighting, usually with a photoperiod and dark period, simulating night and day (Plate 9.1a). Jeffrey and LeRoi (1997) recommended the use of a light–dark cycle for algal culture, although some algal production systems successfully use continuous light. The choice of growth medium is very species-dependent and will depend on whether the microalga comes from a fresh, brackish or seawater environment. Media contain

Table 9.1 Selected major algal culture collections.

Algal Culture Collection	Web site
Australian National Algae Culture Collection – ANACC	www.csiro.au/ANACC
Culture Collection of Algae and Protozoa – CCAP	www.ccap.ac.uk
The Provasoli–Guillard National Centre for Culture of Marine Phytoplankton – CCMP	ccmp.bigelow.org
Microbial Culture Collection at the National Institute of Environmental Studies – NIES	mcc.nies.go.jp
Culture Collection of Algae at Goettingen – SAG	www.epsag.uni-goettingen.de
The Culture Collection of Algae – UTEX	web.biosci.utexas.edu/utex/

macronutrients, such as nitrogen and phosphorus, as well as micronutrients, such as iron, zinc, cobalt, cadmium and molybdenum, and vitamins, such as vitamins B_1 and B_{12} as well as biotin (vitamin H). Some media, such as f medium (Guillard and Ryther, 1962), have been used for over 40 years for diverse microalgae. Several media have been developed that contain low levels of selenium as selenite, as this has been found to be essential for some oceanic species (K medium; Keller *et al.*, 1987) or assists with long term maintenance of fastidious microalgae such as dinoflagellates (GSe medium; Blackburn *et al.*, 2001). Details of methods for growing microalgae, including information on the different media available, can be found in Andersen (2005) as well as on the web sites of some algal culture collections (Table 9.1).

There is an increasing move to cryopreserve strains in liquid nitrogen, either as an alternative to culturing in 'fresh' media or as the sole form of preservation for some classes of microalgae that have proven robust enough for long term preservation by this method (Andersen and Sexton, 2009). Cryopreservation is a reliable form of maintenance for chlorophytes and cyanobacteria (Mori *et al.*, 2002), but further testing of genetic stability is needed for other classes.

9.5 MICROALGAL PRODUCTION SYSTEMS

To grow microalgae for bioproducts and other industrial applications it is necessary to scale up production of microalgae using large scale production systems. There are many different types of algal growth technologies ranging from very simple to complicated engineering works. The main types of growth systems are described in the following sections.

9.5.1 Plastic bags and tanks

With the development of the aquaculture industry and the need by hatcheries to rear larval and juvenile aquaculture animals came the need for production of microalgae as live feeds. While some hatcheries grow microalgae in multiples of relatively small scale culture containers, for example 20 litre carboys, it is more common for microalgal cultures to be grown in disposable plastic bags (200–1000 litres is typical) that are either supported vertically (Plate 9.1b) in metal frames or lie horizontally.

9.5.2 Open ponds

Open ponds include naturally occurring ponds as well as low technology constructions that are low cost and relatively easy to build and operate. Different types of ponds varying in size, shape, construction material, method of mixing, and inclination have been designed. Overall, other than natural ponds or lakes, there are three major types constructed: (1) inclined systems where mixing is achieved through pumping and gravity flow, (2) circular ponds with agitation provided by a rotating arm, and (3) raceway ponds constructed as an endless loop, in which the culture is circulated by paddle wheels (Tredici, 2004). For commercial algal production, natural ponds and lakes, circular ponds with agitation, and raceways are successfully used. Examples of commercial production from natural ponds or lakes include the harvesting of natural blooms of the cyanobacterium (blue–green alga) *Aphanizomenon flos-aquae* var. flos aquae (Carmichael *et al.*, 2000). In Australia, artificial hypersaline 'lakes' are used by Cognis Australia to grow *Dunaliella salina* for β-carotene production. Inclined systems have not received much attention, although they have promise because of highly turbulent flow and very thin culture layers, thus promoting high cell densities and a higher surface area to volume ratio compared with other open systems (Richmond, 1999; Tredici, 2004). Although circular ponds require high energy input for mixing, they are widely used in Asia (Japan, Korea and Indonesia) for *Chlorella* mass production (Lee, 2001).

Raceway ponds are popular algal pond systems and are the preferred production system for many of the commercial operations, including Cyanotech Corporation in Hawaii and Earthrise in California, both of which produce *Arthrospira* (*Spirulina*), as well as Seambiotic in Israel (Plate 9.1c). Excessive evaporation is a problem in these regions and a minimum depth of at least 15 cm is required to avoid a reduction in flow which can result in light limitation (Tredici, 1999).

9.5.3 Photobioreactors

Photobioreactors are bioreactors where phototrophic microalgae and other microorganisms and plant cells are grown under conditions that aim to ensure that algal growth can proceed without light limitation. They are usually considered to be closed systems, that is, in contrast to the low technology systems considered above, they do not allow the direct exchange of gases such as carbon dioxide and oxygen or contaminants (other microorganisms, atmospheric particles etc.) between the culture and the atmosphere (Tredici, 1999).

The status of photobioreactor technologies has been reviewed a number of times (Tredici, 1999; 2004; Pulz, 2001). While being more expensive to operate than low technology systems such as ponds and raceways, photobioreactors ensure high productivities along with quality control with low contamination rates. Some advantages of these systems and technical issues have been discussed by Grobbelaar (2009). The dense cultures obtainable using photobioreactors are easier to harvest than is the case for dilute open pond cultures. Indeed, it is feasible to grow a dense culture that is already a slurry and, therefore, requires little effort to harvest. Closed systems allow efficient utilisation of carbon dioxide and other gases, such as flue gases, for microalgal production. Moreover, photobioreactors can use both natural sunlight as well as artificial illumination. There are many different designs that have been developed (Plate 9.1d), but most are not amenable to scale up to the levels required for commercial production of microalgae.

The increased productivities achievable in photobioreactors versus open ponds were demonstrated by Tredici and Materassi (1992). However, comparison of open pond systems and various photobioreactors for a number of species, including *Chlorella pyrenoidosa*, *Tetraselmis chui* and *Spirulina* spp., made by Lee (2001) showed that productivities from photobioreactors need not surpass those of open pond systems. More recent estimates of productivities (Chisti, 2007; Wiffjels and Barbosa, 2010) suggest that photobioreactors will be essential to achieve the high production necessary for commercial production of bioenergy and other bioproducts.

Geographic location can play an important part in deciding whether microalgae culture will be effective. A temperature of 15 °C or above is considered essential for sustained algal production; outdoor productivities are affected by environmental variables such as low seasonal and night-time temperatures as well as variable irradiance (Von Hamelen and Oonk, 2006). Chini Zittelli and co-workers (2003) demonstrated that a combination of natural and artificial illumination gave optimum productivities of *Nannochloropsis* between December and May in the continental Mediterranean climate of Italy.

9.5.4 Hybrid or combination growth systems

The choice of growth systems is not necessarily restricted to a single technology. Various multiphase cultivation strategies have been considered. Richmond (1987) suggested combining a tubular reactor connected to an open raceway to maximise biomass production by optimising environmental variables. With this system the raceway cultures could be used in the hot part of the day and photobioreactors used for high productivities during the periods of the day with lower environmental stress. Multiphase cultivation strategies can be devised that ensure maximum production of biomass in one stage and maximum induction and accumulation of desired products in the other. This concept has been successfully applied to outdoor cultures of *Dunaliella* for β-carotene production, as well as to photobioreactor cultures of *Haematococcus* for astaxanthin production (Hu, 2004).

The approach for the next 5–10 years suggested by the International Network on Biofixation of CO_2 and Greenhouse Gas Abatement with Microalgae (Von Harmelen and Oonk, 2006) and Benemann (personal communication, 2010) is to produce microalgae in conjunction with wastewater treatment. Benemann (personal communication, 2010), with a background of decades in microalgal production including the US DOE Aquatic species program (Sheehan *et al.*, 1998), expressed the opinion that considerable R&D is still required for economic algal biofuels and other products, with the near-term opportunity being in wastewater treatment, in particular municipal wastewaters. By adding carbon dioxide to such wastewater treatment ponds, it would be possible to grow enough algae to remove all nitrogen and phosphorus nutrients, thus achieving an improved level of treatment of the wastewaters, while generating algal biomass, potentially with a high oil content. The carbon dioxide required for the process could be provided by the organic matter in the wastewater itself. In such scenarios algae wastewater treatment combined with biofuels production, especially if municipal, agricultural and industrial wastes are used, has the potential to reduce greenhouse gas emissions by perhaps 1%. Such systems could be further developed to recycle available nutrients and waters, greatly increasing biofuel outputs.

Cost of production will obviously vary with the species and technology used, whether algal growth is indoors using artificial illumination or outside using solar radiation, or a combination of both. Zhang and co-workers (2001) made estimates of production costs for the eustigmatophyte *Nannochloropsis* sp. in a flat-plate glass reactor. They determined that a 2000 litre reactor was sufficient for industrial scale production of microalgae fed to rotifers which in turn were fed to eight million Denise fish fingerlings annually.

9.5.5 Fermentation systems

There has been extensive research on fermentation technologies for growing microorganisms such as bacteria, yeast and fungi for industrial and medical applications. Fermentation technology is, therefore, much more mature and established at an industrial scale compared with phototrophic production systems. However, the number of microalgae produced heterotrophically remains very small. Technology development has, in general, not focused on the particular requirements of microalgae and only a handful of microalgae have been shown to grow effectively under heterotrophic conditions. These include species that can produce bioproducts of interest, such as the chlorophytes *Chlorella*, *Dunaliella* and *Haematococcus* species (Lee, 2004). Harel and Place (2004) considered that heterotrophic microalgal production has high potential for aquaculture feeds and examined the industrial potential including production issues.

9.6 HISTORICAL NATURAL FOODS

Utilization of microalgae for human health and as a foodstuff by indigenous populations has a long history. Indeed, edible blue–green algae, including *Nostoc, Arthrospira* (*Spirulina*) and *Aphanizomenon* species, have been used for food for thousands of years (Jensen *et al.*, 2001).

The first commercial large scale microalgal cultivation started in the early 1960s in Japan with the culture of *Chlorella* by Nihon Chlorella. As a dry product, *Chlorella* is composed of about 45% protein, 20% fat, 20% carbohydrate, 5% fiber, and 10% minerals and vitamins (Belasco, 1997). In the 1960s health foods became available in the form of tablets, granules and drinks as well as food additives in Japan (Yamaguchi, 1997). However, *Chlorella* has proved difficult to produce on a large scale and digestibility problems have also arisen in nutritional studies due to the tough cell wall of these species. An interesting side-line to this story is the use of so-called 'Japanese *Chlorella*' as a health food and for aquaculture. While superficially similar to genuine *Chlorella*, detailed studies and biochemical analyses showed that this was actually a eustigmatophyte of the genus *Nannochloropsis*, which is notable for having a high content of the polyunsaturated fatty acid 20:5n-3 (Volkman *et al.*, 1993).

In the early 1970s a harvesting and culturing facility for *Arthrospira* was established in Mexico by Sosa Texcoco SA (Spolaore *et al.*, 2006). *Arthrospira* is used in human nutrition because of its high protein content and its excellent nutritive value. A number of possible health-promoting effects have also been claimed, including the alleviation of hyperlipidemia, suppression of hypertension, protection against renal failure, growth promotion of intestinal *Lactobacillus*, and suppression of elevated serum glucose level (an overview has been given by Spolaore *et al.*, 2006).

Arthrospira (*Spirulina*) production for nutraceuticals is a commercial success story. Companies such as Cyanotech Corporation and Earthrise in the United States have established a market for nutraceuticals from *Spirulina*, which is sold as a 'superfood', and when grown using naturally occurring deep-sea nutrients these products can be accredited with organic status. However, most *Arthrospira* (*Spirulina*) production occurs in China and India.

The cyanobacterium *Aphanizomenon flos-aquae* contains approximately 60% protein by dry weight and is known to contain an array of digestible nutrients, including essential fatty acids such as linoleic acid (18:2n-6) and linolenic acid (18:3n-3). *A. flos-aquae* has been harvested commercially in Oregon, USA, from naturally occurring algal blooms as a source of nutraceuticals (Carmichael *et al.*, 2000).

9.7 LIVE FEEDSTOCKS FOR AQUACULTURE

Microalgae are extensively utilized in aquaculture as live feeds for all growth stages of bivalve molluscs (e.g. oysters, scallops, clams and mussels), for the larval and early juvenile stages of abalone, crustaceans and some fish species, and as booster feeds for zooplankton such as rotifers, copepods and *Artemia* used as live feeds. The microalgae must be of an appropriate size for ingestion, e.g. from 1 to 15 μm for filter feeders; 10 to 100 μm for grazers (Webb and Chu, 1983; Jeffrey and co-workers, 1992) and can be readily digested (Brown *et al.*, 1997; Volkman and Brown, 2006). The algal species must have rapid growth rates, be amenable to mass culture, and also be robust and stable in culture to any fluctuations in temperature, light and nutrients, as often occurs in aquaculture hatcheries. As well, the aquaculture feed species chosen must be toxin free to avoid any possibility of toxins being transferred to the animals being grown.

The nutritional quality of the microalgae used as live feeds is a vital factor; different aquaculture animals or even stages of growth of a particular aquaculture species will have different requirements (Volkman and Brown, 2006). The best nutritional composition for a particular situation is often achieved by using a mixture of several microalgal species, even though this involves the extra cost of maintaining multiple species in culture. Strains fulfilling these attributes include the prymnesiophyte *Pavlova lutheri* and the eustigmatophyte *Nannochloropsis oculata*, the diatoms *Chaetoceros calcitrans, Chaetoceros muelleri, Skeletonema costatum* and *Thalassiosira pseudonana*, and the prasinophyte *Tetraselmis suecica*. Within a local area there is often interest in using local microalgae, particularly in low technology 'green water' situations, where fish larvae or crustaceans are reared together with phytoplankton and zooplankton.

Microalgal production may be up to half the cost of the operation of a hatchery. Therefore, the aquaculture industry generally uses low technology production systems such as plastic bags (Plate 9.1b) and ponds. There have been numerous attempts to use dried algae (de Pauw *et al.*, 1984; Borowitzka, 1997; Muller-Feuga, 2000) as aquaculture feedstocks in order to eliminate the difficulty and expense of in-house live algal production, but results have been mixed and progress in this area remains slow. Algal pastes have also been trialled in aquaculture (McCausland *et al.*, 1999) and some pastes, such as those from *Tetraselmis* spp., are able to be stored for long periods. However, most pastes cannot be stored for more than a few weeks before they lose their nutritional value (Borowitzka, 1997). As well as live microalgal feeds, components of the microalgae are used as feed additives in aquaculture. In particular, pigments such as β-carotene and astaxanthin are used to enhance the colour of fish and crustaceans (Section 9.8.5).

Microalgae, either in dried form or as a slurry, have been investigated as a possible feedstock for cattle and other farmed animals with encouraging results. Chowdury and co-workers (1995) working at the Bangladesh Livestock Research Institute grew a mixed algal culture (*Chlorella* and *Scenedesmus*) in shallow ponds and achieved a daily yield of algal suspension of 95 tonnes, equivalent to 247 kg dry material per hectare. When fed to cattle as a supplement to a straw diet, the algal suspension seemed to have created a more favourable environment in the rumen which enhanced microbial growth and thus increased nutrient availability. A large number of nutritional and toxicological evaluations have demonstrated the suitability of algal biomass as a feed supplement or substitute for conventional protein sources (soybean meal, fish meal, rice bran etc.). There is considerable interest today in the commercial use of microalgae as an animal feed for the poultry industry (Becker, 2006).

9.8 BIOPRODUCTS

9.8.1 Bioactive compounds

It has long been known that many marine organisms produce a variety of chemical signals, toxins, anti-fouling agents, and biochemical means of protection from harmful environments (Volkman, 1999; Skulberg, 2004). An understanding of how these agents work has wide potential for commercial and biomedical applications. For example, many marine plants use chemical means to protect themselves from grazers and fouling, and this knowledge has been used to identify natural anti-foulants (Steinberg *et al.*, 1998) and natural sunscreens (Dunlap and Shick, 1998). It is now recognised that some of these compounds are derived from microalgae and cyanobacteria ingested by these animals or living symbiotically with them (Skulberg, 2004). Indeed, microalgae with their rich and diverse biochemistries represent potential sources for new bioactive compounds that are still largely untapped.

9.8.2 Lipids

Microalgae contain a surprisingly diverse range of lipids, many of which are restricted in distribution to a few species or genera. Thus, they can be useful as chemotaxonomic markers (Volkman *et al.*, 1998). More than 30 years ago the US National Renewable Energy Laboratory (NREL) initiated the Aquatic Species Program (ASP) as a result of the oil crisis in the 1970s. Between 1978 and 1996 the ASP studied the feasibility of using algae grown in ponds and using carbon dioxide from power plants to produce biodiesel. This study provided considerable information about the lipid composition of microalgae and the effects of different growth techniques and environmental conditions. Although the program ended in 1995 it set the scene for renewed interest in algae as a source of biofuels and other chemicals (Sheehan *et al.*, 1998). An unfortunate loss is that many of the strains that were isolated as part of the ASP were not deposited in the main algal culture collections in the United States and are thus no longer available.

One of the most important findings from the Aquatic Species Program is that although nutrient stress causes lipid to increase in many strains (when expressed as a percentage of the total biomass), this increase is generally accompanied by a decrease in total cell and lipid productivity. One way to overcome this is to use a two-phase production system. Rodolfi and co-workers (2009) demonstrated that with careful strain selection and a two-phase cultivation system (a nutrient-sufficient phase to produce the inoculum followed by a nitrogen-deprived phase to boost lipid synthesis) an oil production potential of more than 90 kg per hectare per day could be attained with the final lipid content reaching more than 60% of the cell biomass. This is the first report of an increase in both lipid content and real lipid productivity achieved through nutrient deprivation in an outdoor culture.

There are now efforts to enhance oil production through genetic engineering. Wiffjels and Barbosa (2010) consider that knowledge of the biosynthesis of triacylglycerols and their accumulation in lipid bodies would open the way to inducing lipid accumulation in oil bodies without having to apply a stress factor. Radokovits and co-workers (2010) reviewed the potential of genetic engineering to improve microalgae as a biofuel platform for the production of biohydrogen, starch-derived alcohols, diesel fuel, and alkanes. While in its infancy at present, advances in the development of genetic manipulation tools pave the way for significant future progress.

Already there are a number of biodiscovery programmes searching for new hyper-lipid producing strains (Mutanda *et al.*, 2011). This, in turn, has led to efforts to develop methods for rapid screening techniques for naturally high lipid producers (Dean *et al.*, 2010).

Both the lipid content and fatty acid profile is first and foremost dependent on the class of microalgae, with members of the green algae (Chlorophyceae) containing a high proportion of 16:0 and C_{18} unsaturated fatty acids with relatively low levels (or absence) of longer chain fatty acids. Of the green microalgal C_{18} fatty acids, usually either 18:2(n-6) or 18:3(n-3) can dominate. A few species from the genera *Polytoma*, *Botryococcus* and *Ankistrodesmus* have 18:1 as the major C_{18} fatty acid. Other microalgal classes that have

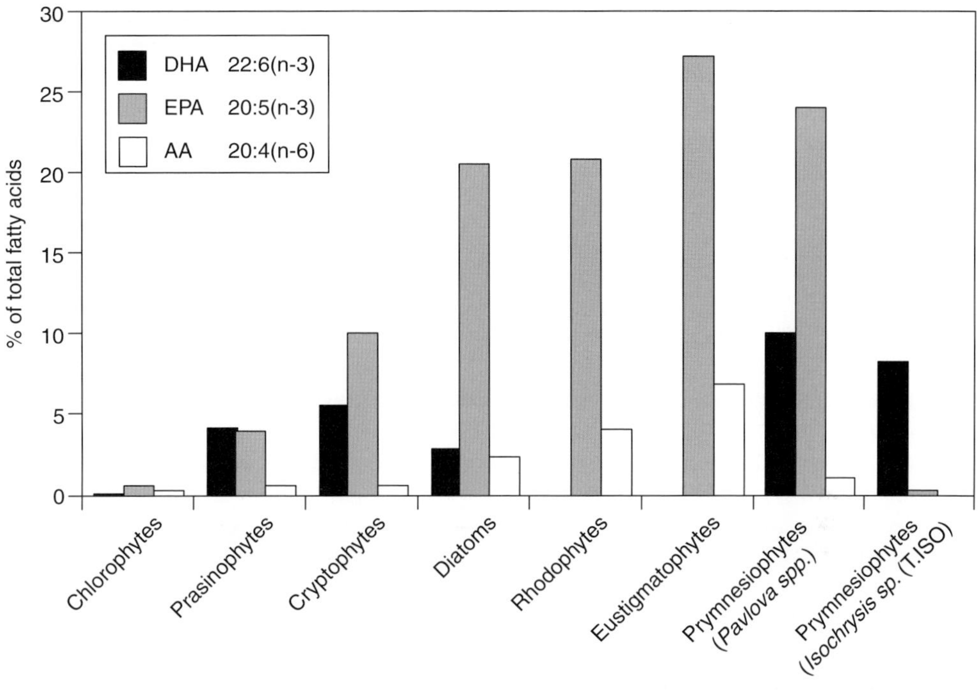

Figure 9.3 Average percentage compositions of the long chain polyunsaturated fatty acids (PUFAs) docosahexaenoic acid (DHA; 22:6n-3), eicosapentaenoic acid (EPA; 20:5n-3) and arachidonic acid (AA; 20:4n-6) of different microalgal classes. Data compiled from over 40 species held in the Australian National Algae Culture Collection.

high lipid contents include the green algal classes Euglenophyceae (e.g. *Euglena gracilis*), Prasinophyceae (e.g. *Tetraselmis chui* and *T. suecica*) and the Eustigmatophyceae (e.g. *Nannochloropsis* spp.,), as well as members of the Prymnesiophyceae (e.g. *Pleurochrysis carterae* and *Isochrysis galbana*) and Bacillariophyceae (diatoms) such as *Cyclotella cryptica*.

Polyunsaturated fatty acids (PUFA) are essential dietary requirements for marine animals and thus information on PUFA contents has been of considerable value to the aquaculture industry. This work revealed significant compositional differences between algal classes (Figure 9.3) and illustrated the ability to manipulate relative proportions of desired fatty acids by changes to the growth conditions (Dunstan *et al.*, 1993). Unusual fatty acids have been identified, including C_{28} polyunsaturated fatty acids in dinoflagellates (Mansour *et al.*, 1999), indicating that there is still much to be learned about the biosynthetic pathways for fatty acid synthesis in marine microalgae.

Commercial production of microalgae for their high fatty acid content has now been achieved by a number of companies. For example, early work by Martek Biosciences identified a dinoflagellate *Crypthecodinium cohnii* that produces a high content of docosahexaenoic acid (22:6n-3, DHA), which plays a key role in infant development and adult health. While DHA is usually thought of as a fish oil constituent, it is also the most abundant omega-3 fatty acid in the human brain and retina. Martek patented a method of deriving DHA-rich oil from the alga which contains 40–50% DHA, but no eicosapentaenoic acid (20:5n-3; EPA). This company also patented a process for producing arachidonic acid (20:4n-6, ARA) from the fungus *Mortierella alpina*. Martek's products *life's DHA*™ and *life's ARA*™ are found in about 99% of infant milk formulas produced in the United States, as well as a range of supplements and functional foods. OmegaTech, also owned by Martek, uses an algal-like protist, *Schizochytrium* (a thraustochytrid), to produce an oil (formerly known as DHA Gold) that is used as a dietary supplement in food and beverages, health foods, animal feeds and mariculture products.

There has been a recent resurgence of interest in the fatty acid content and composition of microalgae as possible source material for the production of biodiesel (Chisti, 2007; Stephens *et al.*, 2010). Biodiesel consists of the fatty acid methyl esters (FAME) produced from natural plant oils, waste cooking oil or animal fats. Microalgae are being reconsidered as a feedstock because of the high productivity per unit area and because microalgae can be grown under conditions that are not favourable for agricultural crop production.

Much of our information on novel lipids in microalgae has come from studies by organic geochemists and chemical oceanographers trying to understand the origins of the complex distributions of lipids found in marine sediments. More recently, the emphasis has switched to understanding the distribution of biochemical pathways amongst the various algal groups and to relating these to the evolution of biochemical pathways. For example, while hydrocarbons are not usually abundant in microalgae, an exception is found in some diatoms. The genera *Haslea* and *Pleurosigma* produce C_{25} unsaturated highly branched isoprenoid (HBI) alkenes while C_{25} and C_{30} alkenes are found in *Rhizosolenia*. It is now clear from detailed studies of diatoms in culture and from examination of well-dated sediment cores that the HBI biosynthetic pathways evolved separately in two different diatom clusters, with the oldest being about 92 million years ago (Sinninghe Damsté *et al.*, 2004).

Many unusual compounds have been identified that have little similarity to the lipids produced by higher plants. These include unusual C_{36}–C_{40} unsaturated methyl and ethyl ketones in some haptophyte microalgae, where the double bonds have *trans* geometry rather than the more usual *cis*. Other compounds include saturated and unsaturated C_{30} and C_{32} alkyl 1,15-diols having a mid-chain hydroxyl group at C-15 that occur in eustigmatophyte microalgae (Volkman, 2006). It has now been shown that these compounds are precursors for aliphatic biopolymers (algaenans) synthesised by these algae. A structurally similar group of saturated and monounsaturated C_{28}–C_{30} diols, but with the hydroxyl group at C-14, has now been identified in diatoms of the genus *Proboscia* (Sinninghe Damsté *et al.*, 2003). It is not known whether the structural similarity is accidental and arises from very different biosynthetic pathways.

Microalgae are an excellent source of sterols, which can comprise up to 10% of the dry weight of an alga. Extensive information on the diverse sterol compositions of marine microalgae has been established over the past decades (Volkman, 1986, 2005; Barrett *et al.*, 1995). Some microalgae contain simple distributions of sterols whereas others can contain 20 or more sterols differing in number of carbon atoms, presence or absence of methyl groups at C-4, patterns of side chain alkylation or positions of double bonds. Some microalgae can contain high contents of cholesterol, more commonly thought of as an animal sterol, or of sitosterol, more commonly associated with higher plants (Volkman, 1986).

Microalgae are a potential source of the phytosterols and stanols added to margarine and have the advantage that the algal oil also contains vitamins and carotenoids.

Phytosterols are absorbed only in trace amounts by the body but inhibit the absorption of intestinal cholesterol, including recirculating endogenous biliary cholesterol, a key step in cholesterol elimination.

9.8.3 Proteins and carbohydrates

Microalgae are good sources of protein (typically 25–40% of the dry weight) and are rich in the essential amino acids (Volkman and Brown, 2006). The amino acid patterns found in most species compare favourably with those of other food proteins. Most microalgae have rather similar amino acid profiles; aspartate and glutamate are usually the most abundant amino acids (8–12% of total amino acids) whereas levels of cystine, methionine, tryptophan and histidine can be quite low (Volkman and Brown, 2006). Interestingly, the arginine content in species of the green alga *Tetraselmis*, which is often used as an aquaculture feed, is twice that of other species (Brown, 1991). The amino acid composition of microalgae does not seem to be greatly affected by changes in light or nutrient conditions.

The carbohydrate content of microalgae can be quite variable, with values as low as 5% and as high as 30% of dry weight in some species (Volkman and Brown, 2006). Sugars can occur in a variety of chemical forms, but polysaccharides typically constitute 80–95% of the total carbohydrate (Brown, 1991). Unlike the amino acids, the sugar composition of algal polysaccharides varies greatly between species and with changing culture conditions. Glucose is usually the dominant sugar (20–90% of sugars) reflecting the fact that glucan is the major energy reserve in microalgae (Handa and Yanagi, 1969). Note that some species excrete polysaccharides (exopolysaccharides, EPS) into the culture medium, especially when nitrogen is limited as in the stationary growth phase. Microalgal exopolysaccharides have a range of potential applications, such as adhesives as well as gelling, flocculant and thickening agents for the glue, oil, paper, paint, textile, cosmetic, food, and beverage industries. Mancuso-Nichols and co-workers (2009) have screened nearly 800 cultured microalgal strains across 18 algal taxonomic groups for their potential as a source of adhesives.

The red microalga *Porphyridum* is seen as a potential commercial source of polysaccharides for the cosmetic, pharmaceutical and health food markets as gelling agents, thickeners, stabilizers, and emulsifiers. The carbohydrate content of this marine species can be very high (up to 57%) (Singh *et al.*, 2000) and it is present as a gel-like acidic heteropolymer envelope (Arad *et al.*, 2010), the thickness of which varies with phase of growth and environmental conditions. Such sulfated polysaccharides form thermally reversible gels similar to agar and carrageenan more usually extracted from marine macroalgae.

9.8.4 Vitamins and antioxidants

Some microalgae can be excellent sources of vitamins and antioxidants, including ascorbic acid (vitamin C), α-tocopherol (vitamin A) and riboflavin (vitamin B_2) (Volkman and Brown, 2006). Ascorbic acid is usually the major vitamin present and contents from 1 to 16 $mg\,g^{-1}$ have been noted (Brown and Miller, 1992). A full list of vitamins and their concentration ranges can be found in Volkman and Brown (2006). The importance of growth conditions for production of α-tocopherol, paramylon and tyrosine from the green alga *Euglena gracilis* was examined by Rodríguez-Zavala and co-workers (2010). By combining different carbon and nitrogen sources as well as inducing a tolerable stress to the cell by adding ethanol, it was possible to increase production of biomass and α-tocopherol as well as paramylon and some amino acids.

9.8.5 Pigments

Microalgae are excellent sources of natural carotenoid and chlorophyll pigments. Over 600 carotenoids are now known and many of these are produced by microalgae. Sophisticated high performance liquid chromatography (HPLC) methods using reverse-phase C_{18} columns to separate both carotenoids and chlorophylls in a single analysis have been developed (Wright *et al.*, 1991; Van Heukelem and Thomas, 2001), which has enabled extensive databases of the characteristic pigment compositions of the different algal classes to become available (Wright and Jeffrey, 1987, 2005; Jeffrey *et al.*, 1997).

A distinction is made between primary and secondary carotenoids. Primary xanthophylls (i.e. oxygenated carotenoids) are structural and functional components of the photosynthetic apparatus of the cell, and hence essential for cellular survival. Secondary xanthophylls are those that microalgae produce in large quantities after exposure to specific environmental stimuli (carotenogenesis), such as high light, oxygenated species and so on (Jin *et al.*, 2003). Plate 9.2 gives examples of pigment profiles of different microalgal classes. Within one class there can be compositional variations between genera, species and even strains; this needs to be considered when selecting strains for pigment production. However, as a preliminary screening device the algal class is generally a good guide to which pigments will predominate.

Aquaculture-reared salmonids require pigment added to their diet to attain commercially desirable pink-coloured flesh. Synthetic pigments have traditionally been used but natural sources of astaxanthin are now available. For example, the green microalga *Haematococcus pluvialis* has a particularly high content of astaxanthin (1–3% of the dry weight) and is now grown commercially as a carotenoid source for use in aquaculture, in poultry feed and as an antioxidant (Lorenz and Cysewski, 2000). There is a striking difference in pigment composition in *Haematococcus* between the different life stages of vegetative cells and the resting stage: vegetative cells can contain significant amounts of lutein and are low in astaxanthin, while the aplanospores contain high levels of astaxanthin. Table 9.2 shows a full pigment composition and Jeffrey and Egeland (2009) provide more information on regulation of pigment synthesis.

Table 9.2 Pigment composition of vegetative cells and aplanospores of *Haematococcus* (Harker and Young, 1995).

Pigment	Vegtative cells	Aplanospores
	(%)	(%)
Neoxanthin	13.0	ND
Violaxanthin	14.5	ND
Lutein-5,6-epoxide	5.2	ND
Lutein	53.8	1.0
β,β-Carotene	13.5	3.0
Canthaxanthin	ND	2.0
Adonirubin	ND	2.0
Echinenone	ND	3.0
Astaxanthin (unesterified)	ND	1.0
Astaxanthin (monoester)	ND	49.0
Astaxanthin (diester)	ND	39.0

ND = Not Detected

Cognis Australia is now a major supplier of β-carotene isolated from the halotolerant green alga *Dunaliella salina* grown in 400 hectare hypersaline 'lakes' at Hutt lagoon in Western Australia and at Whyalla in South Australia.. At Whyalla, the company processes up to one million litres of brine per hour. This pigment has a multitude of uses in food products and as a source of pigmentation in farmed prawns (Borowitzka, 1999). Cognis promotes its product Betatene® as GRAS (Generally Recognized As Safe) for use as a nutrient for food and beverage applications and as a natural colorant in water dispersible powders and oils. Betatene® actually contains a mixture of carotenoids, including β-carotene, α-carotene, lutein, zeaxanthin and cryptoxanthin. It also contains both the *cis* and *trans* isomers of β-carotene, compared to synthetic β-carotene which is almost entirely the *trans* isomer.

Commercial production of carotenoids from microalgae is diversifying, with recent interest in the fast-growing *Chlorella protothecoides* (chlorophyte) as a promising organism for commercial production of lutein by heterotrophic fermentation (Shi *et al.*, 2002). Wei *et al.* (2008) showed that the addition of reactive oxygen species (ROS) could increase yields of carotenoid up to 31.4 mg/l with biomass yields up to 15.9 g/l.

While there is an already established market for some of the microalgal carotenoids, such as β-carotene, there is potential for other pigments in new applications or as an alternative source for established applications. Fucoxanthin is receiving interest for a possible role in combating obesity. While there are macroalgal sources (Miyashita *et al.*, 2011), microalgae offer an attractive alternative with potentially novel applications, with early research demonstrating interesting anti-proliferative activity of well differentiated pathologic cells (Moreau *et al.*, 2006). Lutein, a carotenoid pigment from green algae, is receiving much attention for prevention of age-related macular degeneration of the eyes. Research has focused on screening for microalgae that produce high contents of lutein as well as identifying conditions that optimise lutein production (Shi *et al.*, 2002; Del Campo *et al.*, 2007). Different procedures for recovery of the lutein from microalgal biomass are also being investigated (Cerón *et al.*, 2008).

9.9 PHARMACEUTICALS

Despite interest in microalgae as a source of pharmaceuticals since the 1990s, a commercial product has yet to be achieved. Borowitzka (1995) reviewed efforts to screen for antimicrobials, including antivirals, toxins and other pharmacologically active compounds, as well as the hurdles that are needed to get a new pharmaceutical into the market place. Microalgal toxins that are so significant in environmental issues, aquaculture and human health when produced by harmful algal blooms (HABs) (Hallegraeff, 2003) have a role as toxin standards. As part of biodiscovery for anticancer compounds, the US National Cancer Institute played a major role in the search for bioactive compounds from algae having anticancer properties (Apt and Behrens, 1999). Interestingly, there have been more drugs leads from the cyanobacteria than eukaryotic microalgae, but Gerwick and co-workers (1994) suggest that microalgae deserve more attention. The Natural Products Reports (e.g. Pietra, 1997; Faulkner, 2002) give on-going information on the secondary metabolites of microalgae and other microorganisms and their activity. Overall, the potential of microalgae for pharmaceuticals remains largely untapped.

9.10 MICROALGAE IN COSMETICS AND SKIN CARE

Mycosporine-like amino acids (MAAs) have been identified in a wide range of marine animals and are thought to act as natural sunscreens. These water-soluble compounds contain either a cyclohexenone or cyclohexenimine ring conjugated with the nitrogen substituent of an amino acid. MAAs are thought to absorb the energy of ultraviolet radiation before it reaches cellular targets and dissipates this energy as heat. In marine symbioses such as corals that contain zooxanthellae the MAAs are assumed to be synthesised by endosymbiotic algae and translocated to their invertebrate host (Banaszak and Trench, 1995), although this may not be the case in all invertebrate–algal symbioses (Mason *et al.*, 1998). MAAs are now known to be found in cyanobacteria, fungi, macroalgae and microalgae (Balskus and Walsh, 2010, and references therein). The MAA gene cluster from a cyanobacterium *Anabaena variabilis* has now been sequenced (Balskus and Walsh, 2010). Sunscreen Technologies in Australia developed a lotion that contains a synthetic cousin of MAAs based on research carried out at the Australian Institute of Marine Science. The possibility that natural sources of MAAs might be the basis of natural sunscreen products is still being pursued.

Some microalgal species are established in the skin care market, the main ones being *Arthrospira* and *Chlorella* (Spolaore *et al.*, 2006). Some cosmetic companies have even invested in their own microalgal production system (LVMH in Paris, France and Daniel Jouvance in Carnac, France). Microalgae extracts are mainly used in face and skin care products (e.g. anti-aging cream, refreshing or regenerative care products, emollient and as an anti-irritant in peelers). Microalgae are also represented in sun protection and hair care products. Two examples of commercially available products and their properties claimed by their companies are: a protein-rich extract from *Arthrospira* that repairs the signs of early skin aging, exerts a tightening effect and prevents stria formation (Protulines, Exsymol SAM, Monaco); and an extract from *Chlorella vulgaris* that stimulates collagen synthesis in skin, thereby supporting tissue regeneration and wrinkle reduction (Dermochlorella, Codif, St Malo, France).

Recently, two new products have been launched by Pentapharm (Basel, Switzerland): an ingredient from *Nannochloropsis oculata* with skin-tightening properties (short and long term effects) (Pepha-Tight) and an ingredient from *D. salina*, which shows the ability to stimulate cell proliferation and turnover (Pepha-Ctive)(Stolz and Obermayer, 2003).

9.11 MICROALGAE BIOPRODUCTS: FUTURE POTENTIAL

Bioproducts from microalgae that are currently commercially viable are for high value products, not commodity products. This is a quandary for those with intense interest in developing microalgae for biofuels and other energy applications where low value microalgal production is needed (US DOE, 2010). Previous estimates of the economics of microalgal biotechnological applications (Borowitzka, 1992; Richmond, 2000) in essence still hold true: that current economics and markets are for high value compounds and algal biomass for aquaculture feeds and human consumption. Further development in these areas will pave the way for new products, both high and low value.

The twenty-first century has the potential to be the time when 'microalgae come of age'. Their historical use as a food and medicine by humans, and their essential place in the aquatic food web, along with production of a wide range of bioactive compounds of considerable interest and value to mankind, mean they are increasingly attractive in a world with pressures for renewable resources. With further exploration of the largely untapped global microalgal biodiversity and genetic and technological improvements to production systems, microalgal bioproducts will be a significant feature of a more sustainable world.

REFERENCES

Andersen RA. 2005. Algal Culturing Techniques. Boston, MA: Elsevier/Academic Press.

Andersen RA, Sexton JP. 2009. The importance of algae and culture collections, with comments on marine algal preservation. Phycologia 48(4) Supplement:152–153.

Apt KE, Behrens PW. 1999. Commercial developments in microalgal biotechnology. Journal of Phycology 35:215–226.

Arad S, Lew-Ontman O. 2010. Red microalgal cell-wall polysaccharides: biotechnological aspects. Current Opinion in Biotechnology 21:358–364.

Balskus EP, Walsh CT. 2010. The genetic and molecular basis for sunscreen biosynthesis in cyanobacteria. Science 329:1653–1656.

Banaszak AT, Trench RK. 1995. Effects of ultraviolet (UV) radiation on marine microalgal-invertebrate symbioses. II. The synthesis of mycosporine-like amino acids in response to exposure to UV in *Anthopleura elegantissima* and *Cassiopeia xamachana*. Journal of Experimental Marine Biology and Ecology 194:233–250.

Barrett, SM, Volkman JK, Dunstan GA, LeRoi JM. 1995. Sterols of 14 species of marine diatoms (Bacillariophyta). Journal of Phycology 31:360–369.

Becker EW. 2006. Micro-algae as a source of protein. Biotechnology Advances 25:207–210.

Belasco W. 1997. Algae Burgers for a Hungry World? The Rise and Fall of *Chlorella* Cuisine. Technology and Culture 38:608–634.

Blackburn SI, Bolch CJ, Haskard KA, Hallegraeff GM. 2001. Reproductive compatibility among four global populations of the toxic dinoflagellate *Gymnodinium catenatum* Graham (Dinophyceae). Phycologia 40:78–87.

Blackburn SI, Parker NS. 2005. Microalgal life cycles: encystment and excystment. In: Andersen RA (editor). Algal culturing techniques. Boston, MA: Elsevier/Academic Press. pp. 399–418.

Borowitzka MA. 1992. Algal biotechnology products and processes – matching science and economics. Journal of Applied Phycology 4:267–279.

Borowitzka MA. 1995. Microalgae as sources of pharmaceuticals and other biologically-active compounds. Journal of Applied Phycology 7:3–15.

Borowitzka MA. 1997. Microalgae for aquaculture: Opportunities and constraints. Journal of Applied Phycology 9:393–401.

Borowitzka MA. 1999. Commercial production of microalgae: Ponds, tanks, tubes and fermenters. Journal of Biotechnology 70:313–321.

Brown MR. 1991. The amino acid and sugar content of 16 species of microalgae used in mariculture. Journal of Experimental Marine Biology and Ecology 145:79–99.

Brown MR, Miller KA. 1992. The ascorbic acid content of eleven species of microalgae used in mariculture. Journal of Applied Phycology 4:205–215.

Brown MR, Jeffrey SW, Volkman JK, Dunstan GA. 1997. Nutritional properties of microalgae for mariculture. Aquaculture 151:315–331.

Carmichael WW, Drapeau C, Anderson DM. 2000. Harvesting of *Aphanizomenon flosaquae* Ralfs ex Born & Flah var *flosaquae* (Cyanobacteria) from Klamath Lake for human dietary use. Journal of Applied Phycology 12:585–595.

Cerón MC, Campos I, Sanchez JF, *et al.* 2008. Recovery of lutein from microalgae biomass: development of a process for *Scenedesmus almeriensis* biomass. Journal of Agricultural Food Chemistry 56: 11761–11766.

Chini Zittelli G, Rodolfi L, Tredici MR. 2003. Mass cultivation of *Nannochloropsis* sp. in annular reactors. Journal of Applied Phycology 15:107–114.

Chisti Y. 2007. Biodiesel from microalgae. Biotechnology Advances 25:294–306.

Chowdury A, Huque, KS, Khatun M. 1995. Algae in animal production. In Agricultural Science for Biodiversity and Sustainability in Developing Countries, Tune, Denmark, 3–7 April 1995. Jordbrugsforlaget, Copenhagen.

Dean AP, Sigee DC, Estrada E, Pittman JK. 2001. Using FTIR spectroscopy for rapid determination of lipid accumulation in response to nitrogen limitation in freshwater microalgae. Bioresource Technology 101:4499–4507.

Del Campo JA, Garcia-Gonzalez M, Guerrero MG. 2007. Outdoor cultivation of microalgae for carotenoid production: current state and perspectives. Applied Microbiology and Biotechnology 74:1163–1174.

Dunlap WC, Shick JM. 1998. Ultraviolet radiation-absorbing mycosporine-like amino acids in coral reef organisms: a biochemical and environmental perspective. Journal of Phycology 34:418–430.

Dunstan GA, Volkman, JK, Barrett SM, Garland CD. 1993. Changes in the lipid composition and maximisation of the polyunsaturated fatty acid content of three microalgae grown in mass culture. Journal of Applied Phycology 5:71–83.

Faulkner DJ. 2002. Marine Natural Products. Marine natural products 19:1–48.

Fogg GE. 1965. Algal Cultures and Phytoplankton Ecology. London: Athelone Press.

Gerwick WH, Roberts MA, Proteau PJ, Chen J-L. 1994. Screening cultured marine microalgae for anticancer-type activity. Journal of Applied Phycology 6:143–149.

Grobbelaar JU. 2009. Factors governing algal growth in photobioreactors: the 'open' versus 'closed' debate. Journal of Applied Phycology 21:489–492.

Guillard RRL, Ryther JH. 1962. Studies on *Cyclotella nana* Hustedt and *Detonula confervacea* Cleve. Canadian Journal of Botany 8:229–239.

Hallegraeff GM. 2003. Harmful algal blooms: a global overview. In: Hallegraeff GM, Anderson DM, Cembella AD (editors). Manual on Harmful Marine Microalgae. Paris: UNESCO Publishing. pp. 25–49.

Handa N, Yanagi K. 1969. Studies of water extractable carbohydrates of the particulate matter from the northwest Pacific Ocean. Marine Biology 4:197–207.

Harel M, Place AR. 2004. Heterotrophic production of marine algae for aquaculture. In: Richmond A (editor). Handbook of Microalgal Culture: Biotechnology and Applied Phycology. Oxford, UK: Blackwell Publishing Ltd., pp. 513–524.

Harker M, Young AJ. 1995. Inhibition of astaxanthin synthesis in the green-alga *Haematococcus pluvialis*. European Journal of Phycology 30:179–187.

Hu Q. 2004. Environmental effects on cell composition. In: Richmond A (editor). Handbook of Microalgal Culture: Biotechnology and Applied Phycology. Oxford, UK: Blackwell Publishing Ltd., pp. 83–94.

Jeffrey SW, Egeland ES. 2009. Pigments of green and red forms of *Dunaliella*, and related chlorophytes. In: Ben-Amotz A, Jurgen EW, Suuba Rao DV, editors. The Alga *Dunaliella*. Biodiversity, Physiology, Genomics and Biotechnoloty, Enfiels: Science Publishers. p 111–145.

Jeffery SW, LeRoi J-M. 1997. Simple procedures for growing SCOR reference cultures. In: Jeffrey SW, Mantoura RFC, Wright SW, editors. Phytoplankton Pigments in Oceanography: Guidelines to Modern Methods, Monographs on Oceanographic Methodology. Paris: UNESCO Publishing. pp. 181–205.

Jeffrey, SW, LeRoi, J-M, Brown MR. 1992. Characteristics of microalgal species for Australian mariculture. In: Allan GL, Dall W (editors). Proceedings of the National Aquaculture Workshops, Pt Stephens, NSW Australia, April 1991, pp. 164–173.

Jeffrey SW, Mantoura RFC, Wright SW. 1997. Phytoplankton pigments. In: Jeffrey SW, Mantoura RFC, Wright SW (editors). Oceanography: Guidelines to Modern Methods, Monographs on Oceanographic Methodology. Paris: UNESCO Publishing, 661 pp.

Jensen GS, Ginsberg DI, Drapeau MS. 2001. Blue–green algae as an immuno-enhancer and biomodulator. Journal of the American Nutraceutical Association 3:24–30.

Jin, ES, Polle JEW, Lee HK, Hyun SM, Chang M. 2003. Xanthophylls in microalgae: from biosynthesis to biotechnological mass production and application. Journal of Microbiology and Biotechnology 13:165–174.

Keller MD, Selvin RC, Claus W, Guillard RRL. 1987. Media for the culture of oceanic ultraplankton. Journal of Phycology 23:633–634.

Lee Y-K. 2001. Microalgal mass culture systems and methods: their limitation and potential. Journal of Applied Phycology 13:307–315.

Lee Y-K. 2004. Algal nutrition. Heterotrophic carbon nutrition. In: Richmond A (editor). Handbook of Microalgal Culture: Biotechnology and Applied Phycology. Oxford, UK: Blackwell Publishing Ltd., pp. 116–124.

Lee Y-K, Shen H. 2004. Basic culturing techniques. In Richmond A (editor). Handbook of Microalgal Culture: Biotechnology and Applied Phycology. Oxford, UK: Blackwell Publishing Ltd., pp. 40–56.

Lorenz RT, Cysewski GR. 2000. Commercial potential for *Haematococcus* microalgae as a natural source of astaxanthin. Trends in Biotechnology 18:160–167.

McCausland MA, Brown MR, Barrett SM, Diemar JA, Heasman MP. 1999. Evaluation of live microalgae and microalgal pastes as supplementary food for juvenile Pacific oysters (*Crassostrea gigas*). Aquaculture 174:323–342.

Mancuso-Nichols C, Nairn K, Glattauer V, Blackburn S, Ramshaw J, Graham L. 2009. Screening of microalgal cultures in search of microbial polysaccharides with potential as adhesives. Journal of Adhesion 85:97–125.

Mansour MP, Volkman JK, Holdsworth DG, Jackson AE, Blackburn SI. 1999. Very-long-chain (C_{28}) highly unsaturated fatty acids in marine dinoflagellates. Phytochemistry 50:541–548.

Mason DS, Schafer F, Shick JM, Dunlap WC. 1998. Ultraviolet radiation-absorbing mycosporine-like amino acids (MAAs) are acquired from their diet by medaka fish (*Oryzias latipes*) but not by SKH-1 hairless mice. Comparative Biochemistry and Physiology 120:587–598.

Miyashita K, Nishikawa S, Beppu F, Tsukui T, Abe M and Hosokawa M. 2011. The allenic carotenoid fucoxanthin, a novel marine nutraceutical from brown seaweeds. Journal of the Science of Food and Agriculture 91:1166–1174.

Moreau D, Tomasoni D, Jacquot C. 2006 Cultivated microalgae and the carotenoid fucoxanthin from *Odontella aurita* as potent anti-proliferative agents in bronchopulmonary and epithelial cell lines. Environmental Toxicology and Pharmacology 22:97–103.

Mori F, Erata M, Makoto MM. 2002. Cryopreservation of cyanobacteria and green algae in the NIES-Collection. Microbiological Culture Collections June 2002:45–55.

Muller-Feuga A. 2000. The role of microalgae in aquaculture: situation and trends. Journal of Applied Phycology 12:527–534.

Mutanda T, Ramesh D, Karthikeyan S, Kumari S, Anandraj A, Bux F. 2011. Bioprospecting for hyper-lipid producing microalgal strains for sustainable biofuel production. Bioresource Technology 102:57–70.

Norton TA, Melkonian M, Andersen RA. 1996. Algal biodiversity. Phycologia 35:308–326.

de Pauw, N, Morales J, Persoone G. 1984. Mass culture of microalgae in aquaculture systems: progress and constraints. Hydrobiologia 116/117:121–134.

Pietra F. 1997 Secondary metabolites from marine microorganisms: bacteria, protozoa, algae and fungi. Achievements and prospects. Natural Products Reports 1997:453–464.

Pulz O. 2001. Photobioreactors: production systems for phototrophic microorganisms. Applied Microbiology and Biotechnology 57:287–293.

Radokovits R, Jinkerson RE, Darzins A, Posewitz MC. 2010. Genetic engineering of algae for enhanced biofuel production. Eukaryotic Cell April 2010:486–501.

Richmond A. 1987. The challenge confronting industrial microagriculture: high photosynthetic efficiency in large-scale reactors. Hydrobiologia 151/152:117–21.

Richmond A. 1999. Physiological principles and modes of cultivation in mass production of photoautotrophic microalgae. In: Cohen Z (editor). Chemicals from Microalgae. London, UK: Taylor & Francis. pp. 358–386.

Richmond A. 2000. Microalgal biotechnology at the turn of the millennium: A personal view. Journal of Applied Phycology 12:441–451.

Rodolfi L, Chini Zitelli G, Bassi N, *et al.* 2009. Microalgae for oil: strain selection, induction of lipid synthesis and outdoor cultivation in a low-cost photobioreactor. Biotechnology and Bioengineering 102:100–112.

Rodríguez-Zavala1 JS, Ortiz-Cruz1 MA, Mendoza-Hernandez G, Moreno-Sanchez R. 2010. Increased synthesis of α-tocopherol, paramylon and tyrosine by *Euglena gracilis* under conditions of high biomass production. Journal of Applied Microbiology 109:2160–2172.

Sheehan J, Dunahay T, Benemann J, Roessler P. 1998. A look back at the U.S. Department of Energy's aquatic species program – biodiesel from algae. Report NREL/TP-580-24190. Golden, CO: National Renewable Energy Laboratory.

Shi XM, Jiang Y, Chen F. 2002. High yield production of lutein by the green microalga *Chlorella protothecoides* in heterotrophic fed-batch culture. Biotechnology Progress 18:723–727.

Simon N, Cras A-L, Foulon E, Lemée R. 2009. Diversity and evolution of marine phytoplankton. Comptes Rendus Biologies 332:159–170.

Singh SS, Arad S, Richmond A. 2000. Extracellular polysaccharide production in indoor mass cultures of *Porphyridium* sp. in flat plate glass reactors. Journal of Applied Phycology 12:269–275.
Sinninghe Damsté, JS, Rampen S, Rijpstra WIC, Abbas B, Muyzer G, Schouten S. 2003. A diatomaceous origin for long-chain diols and mid-chain hydroxy methyl alkanoates widely occurring in Quaternary marine sediments: Indicators for high-nutrient conditions. Geochimica et Cosmochimica Acta 67:1339–1348.
Sinninghe Damsté JS, Muyzer G, Abbas B, *et al.* 2004. The rise of the rhizosolenid diatoms. Science 304:584–587.
Skulberg OM. 2004. Bioactive chemicals in microalgae. In: Richmond A (editor). Handbook of Microalgal Culture. Oxford, UK: Blackwell Publishing Ltd., pp. 485–512.
Sournia A, Chrétiennot-Dinet M-J, Ricard M. 1991. Marine phytoplankton: how many species in the world ocean? Journal of Plankton Research 13:1093–1099.
Spolaore P, Joannis-Cassan C, Duran E, Isambert A. 2006. Commercial applications of microalgae. Journal of Bioscience and Bioengineering 101:87–96.
Steinberg PD, de Nys R, Kjelleberg S. 1998. Chemical inhibition of epibiota by Australian seaweeds. Biofouling 12:227–244.
Stephens E, Ross IL, King Z, *et al.* 2010. An economic and technical evaluation of microalgal biofuels. Nature Biotechnology 28:126–128.
Stolz P, Obermayer B. 2003. Manufacturing microalgae for skin care. Cosmetics Toiletries 120:99–106.
Tredici MR. 1999. Photobioreactors. In: Flickinger MC, Drew SW (editors). Encyclopedia of Bioprocess Technology: Fermentation, Biocatalysis and Bioseparation. New York, NY: John Wiley & Sons, Inc. pp. 395–419.
Tredici MR. 2004. Mass production of microalgae: photobioreactors. In: Richmond A (editor). Handbook of Microalgal Cultures. Oxford, UK: Blackwell Publishing Ltd., pp. 178–214.
Tredici MR, Materassi R. 1992. From open ponds to vertical alveolar panels: the Italian experience in the development of reactors for the mass cultivation of phototrophic microorganisms. Journal of Applied Phycology 4:221–231.
US DOE. 2010. National Algal Biofuels Technology Roadmap. US Department of Energy, Office of Energy Efficiency and Renewable Energy, Biomass Program.
Van Heukelem L, Thomas C. 2001. Computer assisted high-performance liquid chromatography method development with applications to the isolation and analysis of phytoplankton pigments. Journal of Chromatography A. 910:31–49.
Volkman JK. 1986. A review of sterol markers for marine and terrigenous organic matter. Organic Geochemistry 9:83–100.
Volkman JK. 1999. Australasian research on marine natural products: chemistry, bioactivity and ecology. Marine and Freshwater Research 50:761–779.
Volkman JK. 2005. Sterols and other triterpenoids: source specificity and evolution of biosynthetic pathways. Organic Geochemistry 36:139–159.
Volkman JK. 2006. Lipid markers for marine organic matter. In: Volkman JK (editor). Marine Organic Matter: Biomarkers, Isotopes and DNA. Berlin: Springer Verlag. pp. 27–70.
Volkman JK, Brown MR. 2006. Nutritional value of microalgae and applications. In: Subba Rao DV (editor). Algal Cultures, Analogues of Blooms and Applications. Enfield, UK: Science Publishers. 1:407–457.
Volkman JK, Brown MR, Dunstan GA, Jeffrey SW. 1993. Biochemical composition of marine microalgae from the class Eustigmatophyceae. Journal of Phycology 29:69–78.
Volkman JK, Barrett SM, Blackburn SI, Mansour MP, Sikes EL, Gelin F. 1998. Microalgal biomarkers: A review of recent research developments. Organic Geochemistry 29:1163–1179.
Von Harmelen T, Oonk H. 2006 Microalgae biofixation processes: applications and potential contributions to greenhouse gas mitigation options. In: International Network on Biofixation of CO_2 and Greenhouse Gas Abatement with Microalgae Operated Under the International Energy Agency Greenhouse Gas R & D Program. The Netherlands: TNO.
Webb KL, Chu F-LE. 1983. Phytoplankton as a food source for bivalve larvae. In: Pruder GD, Langdon CJ, Conklin DE (editors). Proceedings of the Second International Conference on Aquaculture Nutrition. Special Publication No. 2. Baton Rouge, LA: Louisiana State University. pp. 272–291.
Wei D, Chen F, Chen G, Zhang X, Liu L, Zhang H. 2008. Enhanced production of lutein in heterotrophic *Chlorella protothecoides* by oxidative stress. Science of China Series C–Life Sciences 51:1088–1093.
Wiffjels RH, Barbosa MJ. 2010. An outlook on microalgal biofuels. Science 329:796–799.

Wright SW, Jeffrey SW. 1987. Fucoxanthin pigment markers of marine phytoplankton analysed by HPLC and HPTLC. Marine Ecology Progress Series 38:259–266.

Wright SW, Jeffrey SW. 2005. Pigment markers for phytoplankton production. In: Volkman JK (editor). Marine Organic Matter: Biomarkers, Isotopes and DNA. Berlin: Springer Verlag. pp. 71–104.

Wright SW, Jeffrey SW, Mantoura RFC, *et al.* 1991. Improved HPLC method for the analysis of chlorophylls and carotenoids from marine phytoplankton. Marine Ecology Progress Series 77:183–196.

Yamaguchi K. 1997. Recent advances in microalgal bioscience in Japan, with special reference to utilization of biomass and metabolites: a review. Journal of Applied Phycology 8:487–502.

Zhang C-W, Zmora O, Kopel R, Richmond A. 2001. An industrial-size flat plate glass reactor for mass production of *Nannochloropsis* sp. (Eustigmatophyceae). Aquaculture 195:35–49.

10 Bioprocessing Approaches to Synthesize Bio-based Surfactants and Detergents

Douglas G. Hayes

10.1 BIO-BASED SURFACTANTS: OVERVIEW

The main use of surfactants and detergents is to lower interfacial surface free energy at liquid–vapor or liquid–liquid interfaces *via* their interfacial adsorption, with the main goal being to increase the miscibility of two immiscible liquid phases, or to modify the polarity of the interface. The molecular structure of surfactants typically consists of separate hydrophilic and lipophilic domains, which promotes their partition to interfaces. Surfactants, an industry of approximately 10 million metric tons (tonnes) and US$20 billion in size (Hayes, 2009), are found in many commercial products: foods, cosmetics, personal care products, cleaners (e.g., laundry detergents), pharmaceuticals, paints and coatings, and lubricants to name a few. A comprehensive review of bio-based surfactants in foods and their specific roles has been published elsewhere, with "bio-based" referring to derivation from natural and renewable resources (Kralova and Sjoblom, 2009).

In the early twentieth century and before, surfactants were derived from nature, with the most common source being fatty acid salts, or soaps, used in personal care products. But as the availability and utilization of petroleum increased in the 1920s and beyond, this source became the main feedstock for surfactants. However, due to several motivators: the decreased availability of petroleum and other fossil fuels, the increased demand, hence price, for petroleum ($140/bbl reached in 2008), national security concerns about dependency upon non-domestic petrochemical suppliers, environmental concerns with obtaining fossil fuels (due to their increasingly difficult accessibility, evidenced by the environmental damage caused by the leakage of the "Deepwater Horizon" off-shore oil well in the Gulf of Mexico in 2010, 5600 m below sea level) and the generation of carbon dioxide and other greenhouse gases (and their impact upon climate change), interest in bio-based surfactants has increased (Hayes, 2009; Patel, 2004).

Several of these motivators have translated to increased consumer demand for bio-based and eco-friendly products, motivating surfactant producers and suppliers to seek certification for the labels "bio-based" or "green" or "eco-friendly" or "sustainable" (McCoy, 2008; Hayes, 2009). Of note, the feedstock costs will serve as the main economic motivator for manufacturing bio-based surfactants, since the production cost per mass for bio-based is roughly equivalent to petroleum-based (US Department of Energy, 1999). Furthermore, bio-based feedstocks for surfactants must possess reduced price fluctuations and increased

Food and Industrial Bioproducts and Bioprocessing, First Edition. Edited by Nurhan Turgut Dunford.

reliability in their supply to be robust (Hayes, 2009). As a final note, sustainability is not necessarily assured for all bio-based feedstocks. A particular concern is the loss of animal habitat that results from increased acreage of oil-bearing crops and/or increased deforestation, with increased acreage of palm oil in Malaysia and Indonesia serving as a case in point (McCoy, 2007).

Since 2008, the price for crude petroleum has decreased due to the economic recessions in many countries worldwide, which has led to decreased demand, leading to a wane in the interest for bio-based surfactants. However, petroleum prices are expected to increase again in the near future as economic recovery progresses. Therefore, it is anticipated that bio-based surfactants will be economically viable in the near future.

10.2 FEEDSTOCKS FOR BIO-BASED SURFACTANTS

Both the hydrophile and lipophile of surfactants can be derived from bio-based feedstocks. Figure 10.1 depicts the utilization of oleochemical feedstocks to manufacture food products, biofuels, chemicals, and materials as one would envision occurring within an "oleochemical biorefinery." The latter term refers to the concept where an oleochemical (bio-based) feedstock is utilized in a chemical facility, which through various processing and separations schemes produces energy or fuel, chemical intermediates, materials, and other co-products,

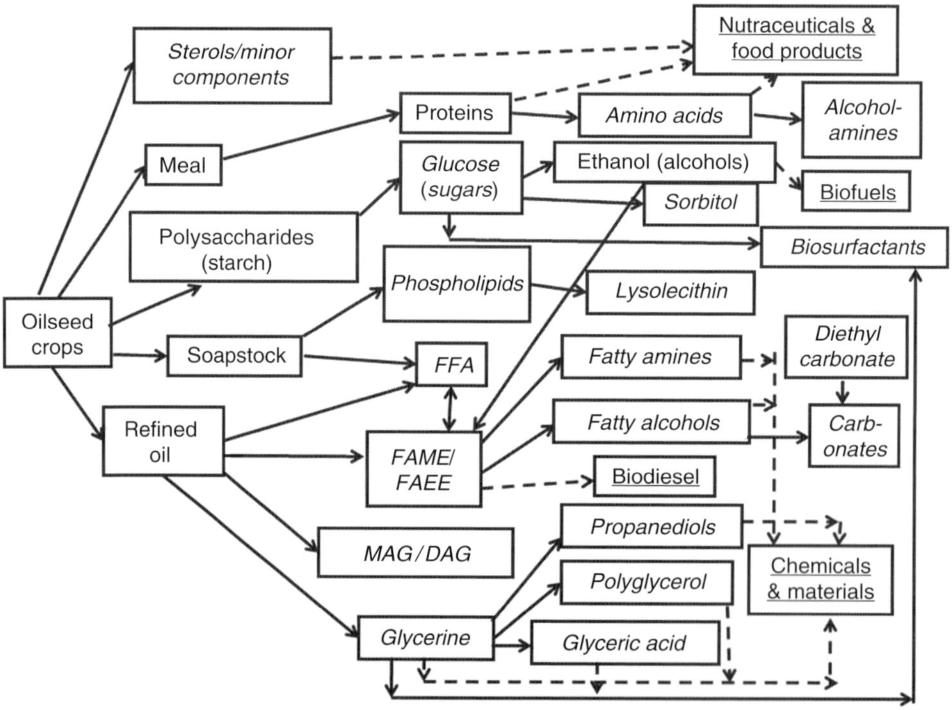

Figure 10.1 Process flow diagram of a possible oleochemical biorefinery to produce targeted chemicals, materials, and fuels (underlined), indicating potential chemical intermediates useful for synthesis of bio-based surfactants (italicized).

akin to a petrochemical refinery, where petroleum is fractionated to produce home and transportation fuels, chemical intermediates, lubricants, asphalt, and so on (Hatti-Kaul *et al.*, 2007; Hill, 2007). As shown in the figure, several chemical intermediates (italicized in the figure) can serve as components of bio-based surfactants.

At the heart of bio-based surfactant preparation is the utilization of long-chain fatty acids found primarily in seed oil in the form of triacylglycerols (TAGs) for the lipophile. The ideal fatty acyl group for industrial, non-food utilization is C_{12}–C_{16} saturates, while for food-based products monounsaturates are more acceptable. Feedstocks enriched in medium-chain saturates include palm, palm kernel particularly palm stearine, a palmitic acyl-rich by-product from the purification and fractionation of palm kernel oil (Santosa, 2008), and coconut oils (Edser, 2004). Cuphea is under investigation in the United States as a future feedstock (Pandey *et al.*, 2000). Other potentially valuable sources of fatty acyl groups for surfactants and detergents are inexpensive feedstocks such as tallow, used cooking oils, and algal oils and oils from jatropha and soapnut (Scheibel, 2007; Chetri *et al.*, 2008; Chisti, 2007). These sources consist mainly of C_{16} and C_{18} saturates and C_{18} monoenes and dienes.

Fatty acyl groups can also be derived from oleochemical co-products enriched in free fatty acids and/or phospholipids, such as soapstock. Many surfactants employ oxygenated fatty acids, particularly hydroxy acids, such as ricinoleic acid (R-18:1-9*cis* OH-12) from castor oil, lesquerolic acid (R-20:1-11*cis* OH-14) from lesquerella oil, and dimorphecolic acids (S-18:2 10 *trans*,12*trans* OH-9) from dimorphotheca oil (and their oligomers) (Hayes, 2004), and epoxidized fatty acids, which are readily prepared from common oils such as soybean (Doll and Erhan, 2009). In addition to fatty acyl groups, other potential bio-based lipophiles include sterols (Svensson and Brinck, 2003) and lignin (Johansson and Svensson, 2001; Holladay *et al.*, 2007).

Typically the fatty acyl groups obtained from TAGs are converted into free fatty acids (FFA) or fatty acid methyl or ethyl esters (FAME and FAEE, respectively) *via* hydrolysis and transesterification, respectively (Figure 10.1). Methyl and ethyl esters are anticipated to be abundant materials in an oleochemical biorefinery due to their use in biodiesel (Ahmad *et al.*, 2007). Methanol and ethanol would be common at biorefineries, readily produced from gasification and fermentation of polysaccharides after their saccharification, respectively. FFA, FAME, and FAEE are typically joined to hydrophiles *via* ester linkages, which enhance biocompatibility and biodegradability, but are readily hydrolyzed under acidic and basic conditions. FAMEs and FAEEs are readily converted to fatty alcohols to fatty amines *via* heterogeneous catalytic reactions (Giraldo *et al.*, 2010; Egan, 1968). Also, bio-based long-chain carbonates have been recently produced, through a catalytic or biocatalytic reaction between fatty alcohol and diphenyl or diethyl carbonate (Kenar *et al.*, 2004; Banno *et al.*, 2007, 2010; Matsumura, 2002). The latter, known as a non-toxic replacement for hazardous chemicals such as phosgene and dimethyl sulfate, is formed from the catalytic oxidation of ethanol (Rudnick, 2006).

As illustrated in Figure 10.1, hydrophiles can also be derived from natural resources. Many of the bio-based hydrophiles are polyhydric alcohols, or polyols, useful for covalent attachment of fatty acyl groups *via* ester bonds or fatty alcohol groups *via* ether bonds. Glycerine, an inexpensive co-product from the manufacture of biodiesel, can be conjugated to fatty acyl groups to produce monoacylglycerols (MAGs), which are common emulsifiers. (Alternatively, MAGs can be obtained from hydrolysis of seed oils.) In addition, glycerine can be converted into other "head group" moieties: 1,2-propanediol, or propylene glycol, *via* a catalytic process, a technological approach under development by Dow, Huntsman, and others; 1,3-propanediol through a fermentation process developed by DuPont Tate and

Lyle LLC (Loudon, TN) (Shelley 2007); glyceric acid *via* fermentation (Habe *et al.*, 2009); and polyglycerol, through homogeneous or heterogeneous catalysis (Barrault *et al.*, 2004).

Mono-, di-, and oligo-saccharides derived from starches, cellulose, and other naturally-derived polysaccharides, such as glucose, xylose, N-acetylglucosamine, maltose, and cellobiose are additional inexpensive choices. Monosaccharide derivatives are also important sources of hydrophiles. Sugar alcohols (e.g., sorbitol and xylitol) are readily produced from monosacharides *via* hydrogenation (Wen *et al.*, 2004), electrochemical reduction (Tang *et al.*, 2004), and microbial transformation utilizing oxidoreductases (Liu *et al.*, 2010). Sorbitan, molecules typically consisting of a five-member ring that contain an oxygen atom, formed from the dehydration of sorbitol, are commonly used to prepare non-ionic surfactants. Oxidation of sacchharides produces furfuryl (De Jong *and Marcotullio*, 2010) and levoglucosanyl (Lakshmanan and Hoelscher, 1970) compounds, also potentially valuable hydrophiles. Glycols, components of alkyl glycosides and esters, are produced *via* fermentation of saccharides (Werpy and Petersen, 2004).

Polyol-based surfactants serve as an example of "non-ionic" surfactants, amphiphiles with polar "head" groups that are uncharged. Some "anionic" and "cationic" surfactants, amphiphiles possessing negatively and positively charged polar groups, are also bio-based (Hayes, 2009). However, surfactants described herein, obtained through bioprocessing, are non-ionic surfactants. Phospholipids serve as bio-based "zwitterionic" surfactants.

Amino acids can serve as surfactant hydrophiles *via* covalent attachment to their carboxylic acid and/or amino functional groups (Infante *et al.*, 2009a, 2009b; Husmann, 2008). In addition, amino acids can be converted into ethanolamine and isopropylamine (from serine and threonine, respectively) (Scott *et al.*, 2007), which serve as hydrophiles for cationic surfactants (Otero, 2009). DNA-derived bio-based surfactants have also been synthesized (Leal *et al.*, 2006; Xu *et al.*, 2005; Bilalov *et al.*, 2004).

Many nonionic surfactants contain poly(ethylene glycol), or equivalently, poly(ethylene oxide), as their hydrophile, through covalent attachment to fatty acyl, fatty alcohol, or fatty amine groups. This hydrophile is typically formed from petroleum-derived ethylene *via* synthesis of ethylene oxide, a material that is carcinogenic, mutagenic, highly flammable, volatile, and reactive. However, in Brazil ethylene has now been derived from bioethanol formed from fermentation of sugar cane (Gielen *et al.*, 2008).

In addition to derivation *via* covalent attachment of bio-derived lipophiles and hydrophiles, bio-based surfactants can be more directly synthesized from bio-renewable resources. Mono- and di-acylglycerols are readily obtained through hydrolysis of TAG. Phospholipids are directly obtained from soapstock, gums, and other oleochemical processing co-products. Lysophospholipids, common emulsifiers in foods that also play important physiological roles in phospholipid metabolism and treatment of arteriosclerosis, are obtained from chemical or enzymatic hydrolysis of phospholipids (D'Arrigo and Servi, 2010). As described in the following, several naturally-occurring surfactants, "biosurfactants," can be produced from fermentation.

10.3 INDUSTRIAL BIO-BASED SURFACTANTS

Several bio-based surfactants are being produced on an industrial scale today. Methyl ester sulfonates (MES), derived from coconut or palm FAME (Figure 10.2), are bio-based replacements for petroleum-derived linear alkylbenzene sulfonates (LAS) in cold-water laundry detergents found in commercial products such as "Care Coldwash" (Danlind,

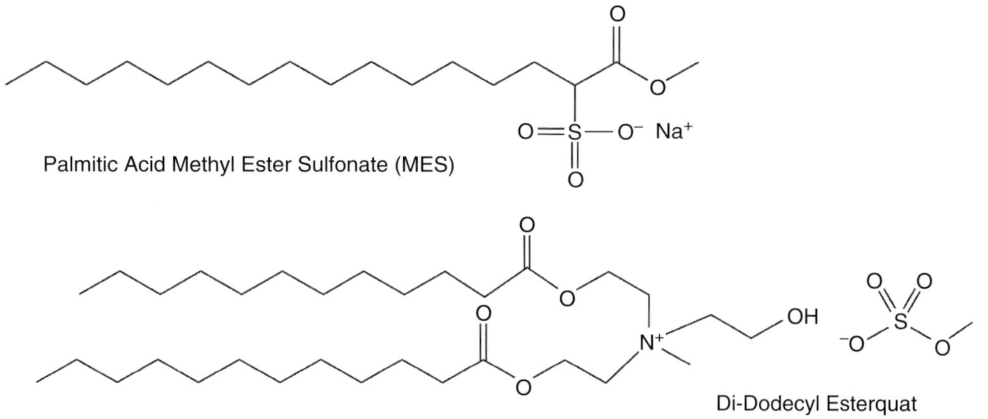

Figure 10.2 Molecular structure of industrial bio-based surfactants that are typically synthesized chemically.

Holstebro, Denmark) and "All/Surf" from Unilever, and are produced by Lion (Johore, Malaysia), Cognis (Selenagor, Malaysia), Stepan (Northfield, IL; 50 000 tonnes/year) and Huish Detergents (Salt Lake City, UT; 80 000 tonnes/year), among others (Roberts *et al.*, 2008; McCoy, 2008; Hayes, 2009). Esterquats, cationic, quaternary ammonium, surfactants derived from FAME (Figure 10.2), are important biodegradable rinse-added fabric softeners and ingredients of hair conditioners, dyes, and bacterial and sanitizer products, including swimming pool biocides, produced by Undessa (Barcelona, Spain) among others (Hayes, 2009; Mishra and Tyagi, 2007; Overkempe *et al.*, 2003). A novel bio-based surfactant series has also been reported, where bio-based hydroxylated diacids, malic and tartaric acids, are covalently attached to fatty acid chlorides under solvent-free conditions to produce fatty acyl-conjugated anhydrides, then covalently attached to triethylene glycol to produce Gemini surfactants that are biodegradable and perform outstandingly as surfactants (Altenbach *et al.*, 2010).

The examples given above are bio-based surfactants that are currently and most probably in the future will be produced chemically. There are several commercial surfactants that can be potentially produced, or partially produced, *via* bioprocessing, particularly fatty acid-polyhydric alcohol (or polyol) esters and fatty alcohol–polyol ethers, employed in foods, cosmetics, and pharmaceuticals (Figure 10.3). Alkyl polyglycosides, or APGs (e.g., β-dodecyl maltoside, β-$C_{12}G_2$; Figure 10.3), have numerous applications in pharmaceuticals and cosmetics due to their high biocompatibility, and are manufactured worldwide, with a capacity of nearly 85 000 tonnes per year (e.g., Glucopon®, Plantacare®, and Plantaren® product lines from Cognis, Cincinnati, OH) (Savic *et al.*, 2010; Hill, 2009). Saccharide–fatty acid esters are useful emulsifiers in foods, cosmetics, and pharmaceuticals which also possesses antimicrobial activity (Otomo, 2009).

Sorbitan esters (known as Span®) and their derivatives containing poly(ethylene glycol), that is, poly(ethylene oxide), or equivalently, "ethoxylate" groups (known as Tween®; Figure 10.3), also known collectively as polysorbates, have numerous uses. These three surfactants can be tailored to possess a specific hydrophilic–lipophilic balance, to direct their employment for oil-in-water (greater hydrophilicity) or water-in-oil (greater lipophilicity) type emulsification, through adjusting the fatty acyl or alcohol tail length and degree of unsaturation, the chemical structure and degree of polymerization of the polyol

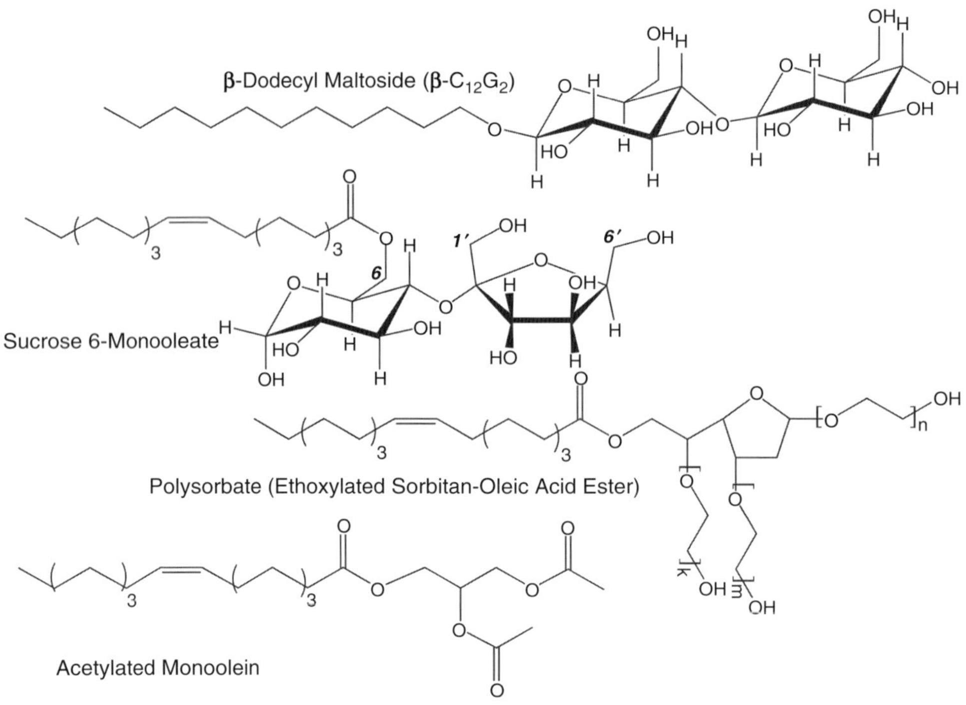

Figure 10.3 Molecular structure of bio-based surfactants that can be produced *via* bioprocessing.

moiety, and the degree of ethoxylation. For example, surfactants to be employed for dispersing water into oil will need to be relatively lipophilic to solubilize in the oil, and will consist of perhaps two or more long alkyl chain length (e.g., $\geq C_{16}$) and a relatively small polar or "head" group, such as glycerol or a monosaccharide. Surfactants for dispersing oil in water will consist of a single medium-chain alkyl group ($\leq C_{14}$) attached to a very polar "head" group, such as a di- or oligo-saccharide and/or a polyol covalently attached to ethoxylate groups.

MAGs, with the hydroxyls of its glycerol moiety being free or esterified with acetic or lactic acid (Figure 10.3), are highly employed emulsifiers. N-acylated amino acids (Perlastan®, Struktol Company of America, Stow, OH) have utility in cosmetics and personal care products due to high biocompatibility and biodegradability, pH insensitivity, and hard water tolerance (Husmann, 2008). A recent review, and its cross-references, provides specific applications and manufacturers of these surfactants (Hayes, 2009).

10.4 ADVANTAGES OF BIOPROCESSING TO PREPARE BIO-BASED NON-IONIC SURFACTANTS

As stated above, bio-based surfactants are typically prepared chemically under harsh conditions. For example, as reviewed elsewhere (Honydonckx *et al.*, 2004), fatty acid–polyol esters are typically prepared using heterogeneous or homogeneous catalysts. MAGs are prepared *via* esterification at 100–200 °C for 10–24 h, or *via* glycerolysis of TAG or FAME at 220–250 °C. Yields of MAGs are low (35–60%) and undesirable by-products frequently form, requiring the need for an additional purification steps, such as molecular distillation (Freitas *et al.*, 2010). Span® is typically prepared by first conducting acid-catalyzed dehydration of sorbitol to obtain sorbitan, followed by alkali (e.g., $NaOCH_3$-) catalyzed transesterification between FAME and sorbitan at 200–250 °C. Sucrose–fatty acid esters are prepared *via* transesterification of FAME at elevated temperatures (>100 °C) and reduced pressure for several hours, often in the presence of toxic solvents such as dimethylformamide (DMF) or dimethylsulfoxide (DMSO). APGs, although readily produced under solvent-free conditions *via* formation of an acetal linkage between fatty alcohol and saccharide under mild reaction temperatures, require molecular distillation, an energy-intensive method, to remove excess fatty alcohol reactant; the process employs significant amounts of energy to operate (Hill, 2009).

Therefore, although chemical-based syntheses provide polyol surfactants at high yields and reaction rates, they possess many disadvantages when examined from a sustainability or life cycle assessment (LCA) based perspective. LCA refers to a quantitative analysis of the costs involved with a product "from cradle to grave", which includes cost factors for their environmental "footprints" (e.g., production of greenhouse gases and toxicants which persist in ecosystems) (Patel, 2004; Hatti-Kaul *et al.*, 2007; Cowan *et al.*, 2008). Firstly, the harsh conditions can lead to degradation of double bonds present in the acyl donor, and to other by-products, which can lead to discoloration at minimum, among other undesired results. The latter may require additional purification steps to be employed. This issue is very important for surfactants containing polyunsaturated acyl groups; for instance, omega-6 fatty acyl-enriched MAGs have been reported recently as effective anti-cancer agents (Fortin, 2010). Moreover, the double bonds of polyunsaturated fatty acids are known to undergo several undesired thermal and oxidative reactions at elevated temperatures (>100 °C), including the formation of polymers that can adversely affect human health, alkehydes, and ketones (Chang, 1988). Secondly, the high temperatures require excessive energy usage, leading to increased production of the greenhouse gas carbon dioxide. Thirdly, the use of heterogeneous catalysts, acid/bases, and/or toxic solvents yields waste products that can lead to environmental harm and also present safety problems for workers. Fourthly, often these reactions produce broad product distributions of desired and undesired products, and, moreover, have lower selectivity, which can impair product performance and impact the products' biocompatibility and biodegradability.

The production of bio-based surfactants via bioprocessing – the use of enzymes in non-aqueous media and fermentation – provides a significant improvement in process sustainability.

10.5 PREPARATION OF BIO-BASED SURFACTANTS VIA ENZYMES IN NON-AQUEOUS MEDIA

Enzymes can, potentially, play an important role in the manufacture of many bio-based surfactants (Hayes, 2004; Karmee, 2008) (Table 10.1). The use of enzymes provides many advantages compared to chemical processing, particularly for the upgrading of process sustainability: lower energy use (due to the use of lower temperatures), lower amounts of waste products and by-products, and the absence of toxic metal catalysts or acids /bases, and safer operating conditions (Cowan *et al.*, 2008). The major disadvantages are the prohibitive costs for enzymes compared to chemical catalysts (although this effect is reduced when enzymes are reused, which is enabled by immobilization, and when potential enzyme inhibitors are absent) and the lower reaction rates that accompany many enzymatic reactions.

Table 10.1 Enzymes used for bioprocessing of bio-based surfactants, and the reactions and products they catalyze.

Enzyme	Surfactant Type	Biocatalytic Role	Reference
Alcohol Dehydrogenase	Surfactants with acetal or ketal linkages	Oxidation of fatty alcohol to long-chain aldehyde or ketone	Orlich *et al.*, 2000
α-Amylase	Alkyl polyglucosides	Simultaneous hydrolysis of starch and conjugation with fatty alcohol	Larsson *et al.*, 2005
Glucosidase	Alkyl glucosides	Formation of ether linkage between monosaccharide and fatty alcohol	Hansson and Adlercreutz, 2002; van Rantwijk *et al.*, 1999
Glucosyl transferases	Alkyl polyglucosides	Addition of monosaccharide units to an alkyl glucoside surfctant	Svensson *et al.*, 2009a, 2009b
Lipase	Mono- and Di-acylglycerol (MAG and DAG, respectively)	Partial hydrolysis of triacylglycerol; esterification or transesterification between glycerol and FFA or FAME, respectively	Hayes, 2004
Lipase	Polyol–Fatty Acid Esters	Formation of ester bonds between fatty acyl group and hydroxyl(s) of polyol, such as MAGs and saccharide–fatty acid esters	Hayes, 2004; Watanabe and Shimada, 2009; Pyo and Hayes, 2009
Lipase	Lysophospholipid	Hydrolysis of phospholipid	Han and Rhee, 1998; Xu *et al.*, 2008; Guo *et al.*, 2005
Lipase	Amino acid surfactants	Formation of amide bond between ε-amine group of lysine and fatty acyl group	Gardossi *et al.*, 1991; Montet *et al.*, 1990; Soo *et al.*, 2003
Lipase	Amino acid surfactants	Conjugation of ε-OH group of homoserine and fatty acyl group	Nagao and Kito, 1989
Lipase	Amino acid surfactants	Ester bond formation between free fatty acids and hydroxyls of arginine esterified to glycerol	Infante *et al.*, 2009
Lipase	Polyglycerol polyricinoleate	Oligomerization of ricinoleic acid and esterification of oligo (ricinoleic acid) and polyglycerol	Bodalo *et al.*, 2009; Montiel Morte and Hayes, 2009
Lipase	Polyol–fatty acid carbonates	Alkyl exchange between bio-based diethyl carbonate and fatty alcohol + polyol	Banno *et al.*, 2007, 2010; Matsumura 2002; Lee *et al.*, 2010
Lipase	Surfactants with amide linkages	N-acylation of alkanolamines and dialkanolamines	Otero, 2009
Papain (and lipase)	Amino Acid surfactants	Ester or amide bond formation between arginine's α-carboxylic acyl group and fatty alcohols, glycerol (polyols), or fatty amines, respectively	Infante *et al.*, 2009
Phospholipase A_1 and A_2	Lysophospholipid	Hydrolysis of phospholipid to create lysophospholipid	Daimer and Kulozik, 2009; Xu *et al.*, 2008; Guo *et al.*, 2005
Phospholipases A_1, A_2, and D (and lipase)	Phospholipids	Tailor-making of phospholipids with specific structure	Xu *et al.*, 2008; Guo *et al.*, 2005; Hossen and Hernandez, 2005

In addition, due to the need to reduce any inhibitory agents, the starting materials must be purified; for instance, fatty acyl-containing material must not contain phospholipids, aldehydes/ketones, peroxides, and other contaminants. But, as energy costs increase (as anticipated), the importance of sustainability increases (due to government regulation and/or consumer demand), and the capabilities of enzymes and their production systems increase (due to improved technology in screening, mutagenesis, protein engineering, recombinant DNA technology, immobilization, bioreactor design, etc.), enzymatic bioprocessing is anticipated to become more cost competitive.

For most of the applications listed in Table 10.1, the main role of the enzyme is the covalent attachment of hydrophile and lipophile. Formation of the ester bond is the most readily achieved using enzymes (particularly hydrolases, such as lipases). Bio-based surfactants with ester bonds are particularly common for food and pharmaceutical applications due to their high biocompatibility. They also have excellent biodegradability (Stjerndahl *et al.*, 2003). However, ester bonds possess relatively low temperature stability and are hydrolyzed under alkaline conditions (Stjerndahl and Holmberg, 2003). Carbonate linkages provide improved stability under alkaline conditions compared to esters (but possess lower stability than amides), and are also highly biodegradable (Stjerndahl and Holmberg, 2005a). Amide bonds provide higher stability toward acidic and basic conditions, important for industrial applications such as laundry detergents, yet retain excellent biodegradability (Stjerndahl and Holmberg, 2005b). The ether linkage provides improved stability compared to ester bonds in alkaline media, but undergoes degradation in acidic media; and, alkyl ethoxylates and APGs possess excellent biodegradability (Steber and Wierich, 1985; Petrovic and Barcelo, 2000; Bozetine *et al.*, 2008; Hill, 2009). Oxidation of fatty alcohols to long-chain aldehydes and ketones *via* horse or yeast alcohol dehydrogenase may be useful for the chemo-enzymatic synthesis of surfactants with acetal or ketal linkages (Orlich *et al.*, 2000).

Of the enzymes listed in Table 10.1, lipases are the "workhorses." The employment of lipases in non-aqueous media is an established art, with over 25 years of research serving as a foundation. Lipases are abundant and relatively inexpensive enzymes that require no co-factors and are easily immobilized. Lipases from several thermophilic organisms have been isolated, cloned, and mass produced *via* recombinant DNA technology in common vectors such as *Escherichia coli*. Some of the examples in Table 10.1 are surfactants formed from enzymatic hydrolysis of oleochemical feedstocks, such as MAG formed from lipase-catalyzed hydrolysis of TAG, and lysophospholipids *via* hydrolysis by lipases or phospholipase A. In the following sections some specific examples from the literature are given of enzyme-catalyzed synthesis of bio-based surfactants. Other examples not described, such as the oxidation of fatty alcohols to aldehydes (Orlich *et al.*, 2000) and the covalent attachment of fatty alcohols and bio-based diethyl carbonate (Banno *et al.*, 2007, 2010; Matsumura 2002; Lee *et al.*, 2010) are covered in the references provided.

10.5.1 Lipase-catalyzed synthesis of monoacylglycerols (MAGs)

Preparation of MAGs using lipases is currently well known and can occur *via* several different routes (Hayes, 2004; Watanabe and Shimada, 2009). [In addition to glycerol, other glycols can serve as acyl acceptor for the preparation of bio-based surfactants (Hayes, 2004).] Key aspects for bioprocessing are the need to enable miscibility between acyl donor and glycerol substrates and retain low water concentrations for reactions involving ester bond formation, to increase the yield of ester upon the approach of thermodynamic equilibrium. The three recent examples discussed encompass the current state-of-the-art. One group reported the formation of MAG *via* esterification between glycerol and lauric (12:0) acid at an 8:1 mole ratio in the absence of solvent at 45 °C, above the melting point of lauric acid (Freitas *et al.*, 2010). The biocatalyst was an immobilized form of *Penicillum cambertii* lipase ("Lipase G" from Amano Enzyme Inc., Nagoya, Japan), known for its selectivity to form partial glycerides, present at 5 wt-%. The reaction was conducted at a 40 gram scale under stirred batch mode. Under these conditions, 92% of fatty acid was consumed, yielding 60% MAG and 27% diacylglycerol (DAG) in a 6 h period. Many lipases are regio-selective towards formation of ester bonds at the 1- and 3- glycerol positions, with isomerization, that is, "acyl migration," occurring non-biocatalytically to convert 1-(3-) MAG into 2-MAG (and *vice versa*) and 1,2-(2,3-)DAG into 1,3-DAG, with 1-(3-) MAG and 1,3-DAG typically predominating (Hayes, 2004).

In a second example, MAGs were produced *via* simultaneous glycerolyis and hydrolysis of Camellia oil, an oleic (18:1-9*cis*) acid-rich TAG, using a 4:1 mole ratio of glycerol-to-TAG at 50 °C in stirred batch mode on a 5–50 gram scale (Zeng *et al.*, 2010). *tert*-Butanol was employed as a co-solvent, at a volume ratio of 60:40 liter of *tert*-butanol per liter of substrates. The reaction was catalyzed by a commercially available immobilized thermophilic lipase from *Candida antarctica* (Novozym 435, Novozymes, Inc., Franklinton, NC). Under these conditions, >95% TAG was consumed, yielding approximately 65% MAG and 11% DAG in 24 h. The predominant isomers were 1-MAG and 1,3-DAG, respectively, reflecting the occurrence of acyl migration (i.e., for 2-MAG formed through hydrolysis).

In a third example, MAGs were obtained by hydrolysis of TAG in oil-in-water emulsions at 50 °C using a 1,3-selctive lipase, which released FFA from the 1- and 3-glycerol positions (Hwang *et al.*, 2009). To make the reaction irreversible, the medium was enriched in calcium cations, to enable the formation of soaps (with the optimal pH being 10.0, further promoting soap formation).

10.5.2 Lipase-catalyzed synthesis of saccharide–fatty acid esters

Covalent attachment of fatty acyl groups and saccharides *via* ester bonds using lipases has been reviewed (Pyo and Hayes, 2009). Use of lipases in reactions utilizing mono- or di-saccharides (e.g., fructose, glucose, maltose, ribose, sucrose, and xylose) or sugar alcohols (e.g., mannitol and sorbitol) as acyl acceptor substrate yields primarily mono- or di-esters due to the regio-selectivity of lipases. Moreover, ester bonds are formed almost exclusively from primary –OH groups, with regio-selectivity toward a particular –OH group when multiple primary hydroxyls exist. For instance, for the lipase-catalyzed esterification of sucrose, the 6-OH group of the latter's pyranose ring is selectively acylated over the 1' and 6' primary hydroxyls of its furanose ring (Figure 10.3). As described above for MAG manufacture, the key goals for the successful biotransformation are to achieve miscibility between the acyl donor and acceptor and to remove water from the reaction system. As reviewed, several creative attempts have been employed to enhance miscibility, including the derivatization of saccharides with protective groups (e.g., isopropylidene) or complexation agents (e.g., phenylboronic acid); but, in most cases, a polar co-solvent has been employed, such as *tert*-butanol, DMF, or recently, ionic liquids. However, the employment of solvent or derivitization is not desirable due to increased cost, the incompatibility of solvents and ionic liquids with food processing regulations, and decreased sustainability.

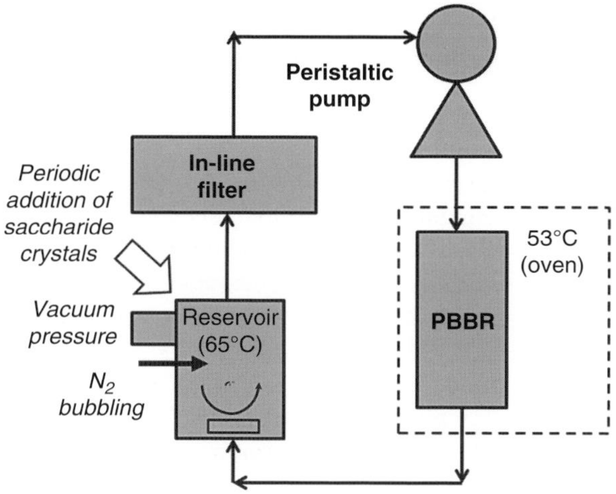

Figure 10.4 Bioreactor system operated under continuous recirculation, for the solvent-free lipase-catalyzed synthesis of saccharide–fatty acid esters (PBBR refers to "packed-bed bioreactor").

Ye and Hayes have employed solvent-free media: 10–100 μm-sized suspensions of saccharide (fructose, glucose, sucrose, or xylose) in crystalline form dispersed in mixtures of acyl donor (e.g., oleic, myristic, lauric, or caprylic acid) and the main reaction product, monoester, with the latter being present initially at 5–25 wt-% (depending on the nature of the acyl acceptor crystals) (Ye *et al.*, 2010; Ye and Hayes, 2011). The presence of monoester greatly increased the solubility of saccharide crystals in the solvent-free media. For instance, for the reaction between fructose and oleic acid, the maximum fructose concentration increased from 0.7 wt-% up to 2.5 wt-% as the ester concentration increased from 5 wt-% (initially) to 93 wt-% (Ye and Hayes, 2011).

The solvent-free reactions were operated in a bioreactor system operated on a 10–20 gram scale under continuous recirculation that contained the following units in series: a reservoir open to the atmosphere maintained at 60 °C (to enable removal of water), an in-line nylon filter of 100 μm nominal pore size, a peristaltic pump, a unit for further removal of water (either a packed column of molecular sieves or a vessel that employed simultaneously vacuum pressure and bubbling of nitrogen gas) and a packed-bed bioreactor containing immobilized *Rhizomucor miehei* lipase, 0.05 g per g of reaction medium (Lipozyme IM, Novozymes, Inc.), the latter maintained at 53 °C (Figure 10.4). Saccharide was added periodically to the reservoir. During the initial phase of the reaction, removal of water *via* free evaporation in the reservoir (at 60 °C) was deemed to be adequate. But when the percent conversion of acyl donor reached ~60%, additional water removal was required to enhance ester formation through lowering the liquid-phase water content, *via* use of a molecular sieve column or application of vacuum pressure plus nitrogen bubbling (Figure 10.4) (Ye and Hayes, 2011).

Under these conditions, for the reaction of fructose and oleic acid, the final product consisted of 85–90% esters (of which 85% is monoester and the remainder diester), representing a productivity of 0.297 mmol per hour per gram of lipase (Ye and Hayes, 2011). The final product may not need of further purification; moreover, it would serve as a technical-grade product.

10.5.3 Lipase-catalyzed synthesis of polyglycerol polyricinoleate

Polyglycerol polyricinoleate, or PGPR, is an important emulsifier in foods, for improved texture and enhanced volume of bakery products, in salad dressings, and in chocolate to decrease the occurrence of fat bloom (Bodalo *et al.*, 2009). It is conventionally formed *via* a chemical process: polycondensation of ricinoleic acid at high temperature, followed by covalent attachment to polyglycerol in the presence of alkali. This approach yields undesired by-products and requires large energy expenditures, hence leads to production of carbon dioxide (Bodalo *et al.*, 2009). In contrast, the enzymatic approach involves relatively low temperatures (75 °C or less), and does not produce by-products. In one approach, polyricinoleate, produced previously by bulk oligomerization of ricinoleic acid catalyzed by lipases, was reacted with polyglycerol of average degree of polymerization 3.0 at a 5:1 mass ratio with immobilized *Rhizopus arrhizus* lipase in a stirred tank reactor operated at 40 °C (solvent-free conditions) (Bodalo *et al.*, 2009). Acid value (AV), a measure of the free COOH groups present in the reaction, was employed to monitor the reaction. Under these conditions, a final AV of 8.7 mg of KOH per g of reaction mixture was achieved, which does not quite reach the maximum allowable AV for PGPR as specified by the European Union (EU), 6 mg KOH per g (Bodalo *et al.*, 2009). However, this requirement was met if vacuum pressure was employed: 5.21 mg KOH per g (Montiel Morte and Hayes, 2009).

In an alternate approach, 33 g (110 mmol) of ricinoleic acid was mixed with 2.2 g (approximately 9 mmol) of the same polyglycerol source in the presence of 12 wt-% Novozym 435 (Montiel Morte and Hayes, 2009). The bulk polymerization was conducted under stirring at 75 °C. This approach yielded an AV of 20 mg KOH per g, thus suggesting vacuum pressure is required to lower the AV to meet EU specifications. Gel permeation chromatographic analysis suggested the number-averaged molecular weight and polydispersity index were 5430 and 3.24, respectively. NMR analyses suggested that of triglycerol's five hydroxyl groups, only the two primary OH groups could serve as acyl acceptor sites.

10.5.4 Enzyme-catalyzed synthesis of alkylpolyglucosides (APGs)

Chemical synthesis of alkyl polyglycosides, or APGs, yields a mixture of α- and β-pyranoside isomers. Anomerically pure glycosides may lead to more consistent and reliable performance of APGs from batch to batch; their chemical synthesis is complex requiring several protection and deprotection steps (van Rantwijk *et al.*, 1999).

Enzymatic synthesis of APGs requires two steps: the synthesis of alkyl (mono)glycoside *via* glycosidases and the extension of the oligosaccharide group catalyzed by cyclodextrin glycosyl transferase. The first step has been reviewed (van Rantwijk *et al.*, 1999). Three examples are provided to describe the current state-of-the-art. Firstly, β-glycosidase from the thermophilic organism *Pyrococcus furiosus* was employed to catalyze the reaction between the disaccharide lactose (β-D-galactopyranosyl-(1→4)-D-glucose) and hexanol (Hansson and Adlercreutz, 2002). Immobilized enzyme (*via* adsorption onto celite) was added to a solution of 93 mM lactose in *n*-hexanol, operated at 75 °C, with its water activity maintained at 0.79 *via* equilibration with saturated salt aqueous solution. This approach did not yield lactoside, due to the enzymatic cleavage of lactose's β-1,4 glycosidic bond.

Therefore, the two major products were hexyl galactoside and glucoside, at a total yield of 46%. The galactose monosaccharide unit was preferably catalyzed by the enzyme over the glucose monosaccharide unit. Major issues for this approach were the limited miscibility of the substrates, requiring that a medium- but not long-chain alcohol be used (C_8 or less), and the higher water activities needed by glycosidases compared to lipases, which reduces the yield at the approach of thermodynamic equilibrium.

Secondly, APGs were produced starting with alkylglycoside and a source of additional saccharide units, α-cyclodextrins, using cyclodextrin glycosyl transferase from *Bacillus macerans* (Svensson *et al.*, 2009a, 2009b). The reaction took place in aqueous media at pH 5.2 and 60 °C, using an eightfold molar excess of cyclodextrin to dodecyl β-D-maltoside. This reaction yielded mainly the dodecyl β-D-maltooctaoside with an overall yield of 50%. The eight glucosidyl units result from two units present in the substrate's maltose unit plus six present in the glucosidyl donor α-cyclodextrin.

An alternative approach to produce APGs was to react soluble starch (a homopolymer of glucose with α-1,4-glycosidic linkages) with 1-butanol in aqueous media in the presence of α-amylase at 50 °C (Larsson *et al.*, 2005). This approach produced a mixture of APGs, with the average degree of polymerization of the oligosaccharides residing between two and four. Yields were generally low. In summary, significant progress has been achieved for the enzymatic synthesis of APGs in recent years; however, the enzymatic route will require significant improvement to be competitive with chemical APG synthesis.

10.5.5 Enzyme-catalyzed synthesis of amino acid derivatives

Enzymes, particularly hydrolases, are capable of covalent attachment of fatty alkyl or acyl groups onto the amino and carboxylic acid functional groups of amino acids. Two examples will be described initially; these involve the covalent modification of arginine. For the first, papain, a protease from papaya latex, catalyzed the attachment of the carboxylic acid moiety of arginine to either a fatty amine or a fatty alcohol, producing an amide or ester bond, respectively (Figure 10.5) (Infante *et al.*, 2009b). The starting material contained the N-benzyloxy ("Cbz") protective group attached to the α-amino group; and, its carbonyl group was esterified to methanol, the latter serving as a good leaving group. Amide bond formation occurred using a 50-fold mole excess of n-C_8–C_{12} alkyl amine at room temperature using a mixture of acetonitrile and 0.1 M borate buffer, pH = 8.2. Papain immobilized *via* adsorption onto polyamide was employed as biocatalyst. Typically, 70–90% conversion of Cbz-argine into the amide occurred. Fatty alcohols of chain length 8–12 were esterified to arginine's hydroxyl group using a solvent-free reaction medium conducted at 50–65 °C, where the alcohol served as substrate and solvent. Ester yields of up to 80–90% occurred. The surfactants have potential application as antimicrobial agents and for DNA compaction. By a similar approach, the carboxylic acyl moieties of several N- (and for cysteine S-) Cbz-protected amino acids (glycine, lysine, serine, aspartate, phenylalanine, tyrosine, and cystine) were conjugated with fatty alcohols (to yield esters), α,ω-diols (yielding diesters containing two amino acids), and α,ω-diamines (yielding diamides) (Valivety *et al.*, 1998).

Covalent attachment of the α-carboxylic acid moiety of CBz-protected amino acids with alcohols or amines has been the most successful approach for enzymic transformations. The attempt to add fatty acyl groups to free amino acids has not been successful (Soo *et al.*, 2003; Valivety *et al.*, 1997). The most successful addition of acyl groups to unprotected amino acids has been reported for the ε-amino group of lysine, the ε-OH group of homoserine,

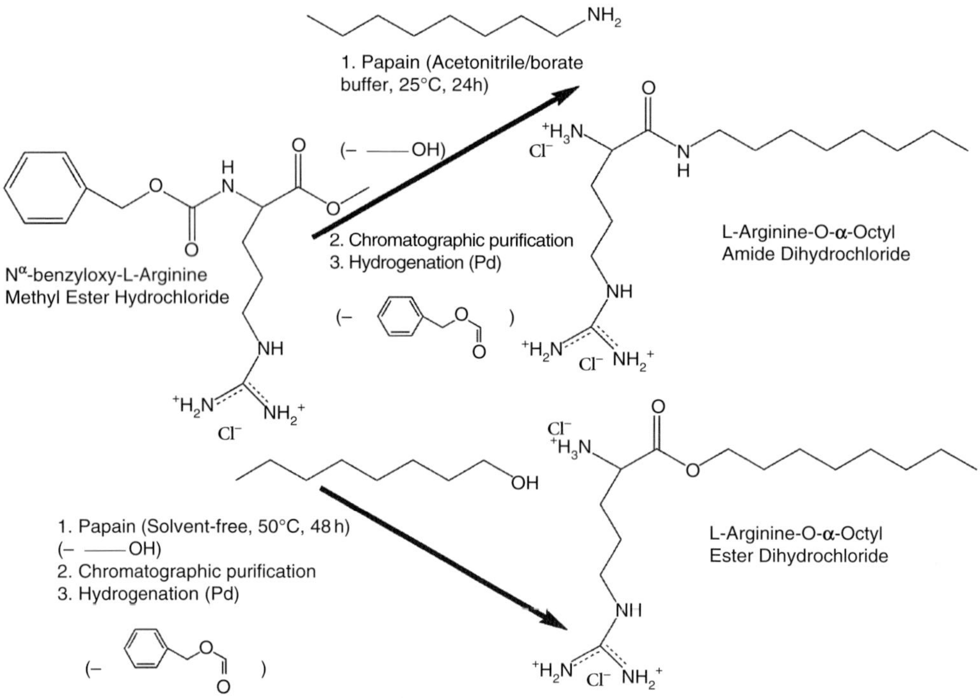

Figure 10.5 Synthesis of long-chain amide and ester of the α-carboxylic acid moiety of arginine using papain as biocatalyst and fatty amine and alcohol as reactant, respectively (adapted from Infante *et al.*, 2009a).

and the unprotected ε-OH group of N-Cbz-protected serine as acyl acceptors, using lipases as catalysts in the presence of apolar solvent, using a molar excess of acyl donor free fatty acid or TAG (Soo *et al.*, 2003; Valivety *et al.*, 1997; Nagao and Kito, 1989; Gardossi *et al.*, 1991; Montet *et al.*, 1990).

An additional example is the synthesis of acylated glycerol–argnine conjugates using lipases (Figure 10.6) (Infante *et al.*, 2009a). In the first, the α-carboxylic acid group of N^{α}-acetylated L-arginine was conjugated with glycerol using immobilized papain or *C. antarctica* lipase as biocatalyst at 50 °C in the presence of 0.2 M borate buffer, pH = 8.2, yielding an ester at 80% conversion. The majority of the product contained an ester bond at the 1- and 3- glycerol position; but, a small, 10% yield of ester at the 2-glycerol position occurred. The second step involved immobilized *R. miehei* lipase-catalyzed esterification of the glycerol moiety's free –OH groups under solvent-free conditions at 65 °C, with the acyl donor, typically lauric acid, also serving as solvent. Di-acylated products occurred at a high yields, 50–75%, with the product distribution containing a mixture of positional isomers (Figure 10.6), resulting from the occurrence of acyl migration.

10.5.6 Enzymatic production of lysophospholipids and structured phospholipids

The enzymatic cleavage of an acyl group from phospholipids yields lysophospholipids, which are more water soluble than phospholipids and, therefore, useful as emulsifiers for oils in water. Three enzymes have been employed to conduct hydrolysis, including phospholipases A_1 and A_2, which cleave the fatty acyl group from the 1- and 2-glycerol

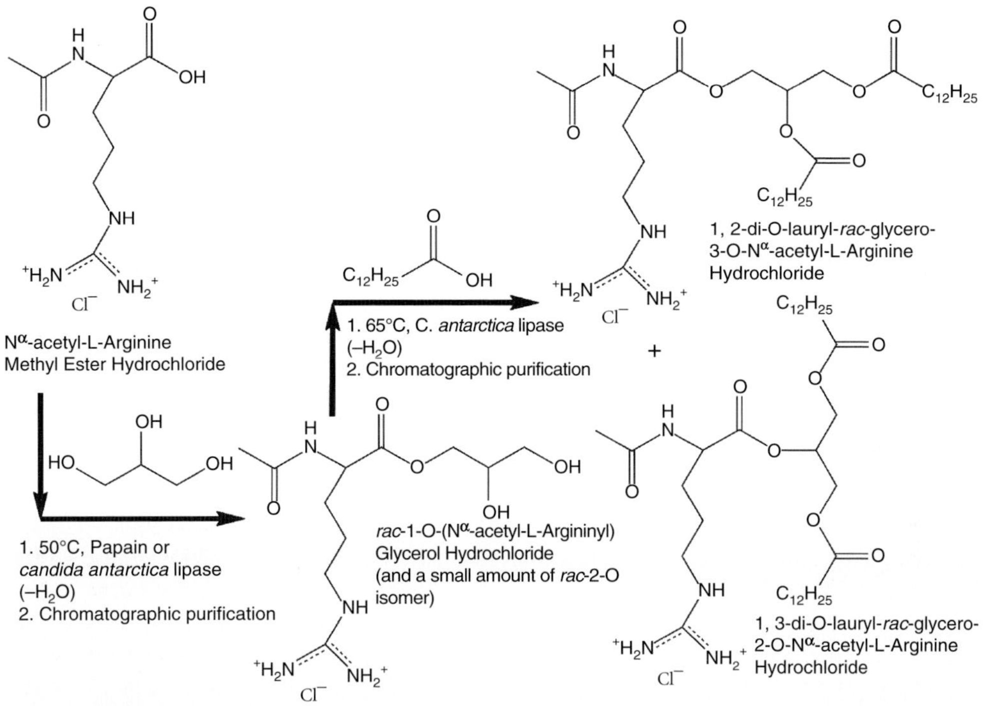

Figure 10.6 Lipase-catalyzed synthesis of di-O-alkyl, O-(N$^{\alpha}$-acetyl-L-argininyl) glycerol (adapted from Infante *et al.*, 2009a).

position of phospholipids, respectively (Xu *et al.*, 2008; Guo *et al.*, 2005). In addition, many lipases catalyze the hydrolysis of acyl groups from the 1-acyl position (Xu *et al.*, 2008; Guo *et al.*, 2005). To convey the current state-of-the-art, a recent paper showed that egg yolk suspended in aqueous solution that contained Lecitase 10L, a commercial preparation of porcine pancreatic phospholipase A_2 developed by Novozymes, Inc., exhibited improved capability for emulsification in aqueous media (Daimer and Kulozik, 2009).

Lipases and phospholipases are also employed to tailor-make "structured" phospholipids," with the goal being to control the surface activity of the bio-based surfactant, and/or incorporate nutritionally beneficial acyl groups, such as omega-3 fatty acids. A potential application of structured phospholipids is lung surfactants, used in the treatment of acute and neonatal distress syndrome (Acosta *et al.*, 2009). Moreover, phospholipids are the main component of synthetic lung surfactant products: Exosurf® (Glaxo Wellcome, Research Triangle Park, NC), Surfaxin® (Discovery Laboratories, Warrington, PA), Venticute® (Altana Pharma AG, Wesel, Germany), and ALEC® (Britannia Pharmaceuticals, Newberry, UK). Acyl groups esterified to the 1- and 2-glycerol position can be selectively incorporated using lipases and phospholipases A_1 and A_2. In addition, phospholipase D *via* hydrolysis and transphosphatidylation can be used to selectively incorporate specific "head groups" attached to the phosphate group contained at position 3.

This topic has been thoroughly reviewed (Xu *et al.*, 2008; Guo *et al.*, 2005). To provide an example, medium-chain (caprylic) acyl groups were incorporated into the acetone-soluble fraction of lecithin, a relatively inexpensive source of phospholipids enriched in phosphatidylcholine (Hossen and Hernandez, 2005). For one series of reactions, lecithin and caprylic (8:0) acid, the latter at 20-fold excess, were dissolved in hexane and treated with immobilized phospholipase A_2 or immobilized *R. miehei* lipase at 40 °C. The former enzyme produced a 45% yield of the desired product, phospholipid that contained one caprylic acyl group per molecule, and 52% of the undesired product, phospholipids containing two caprylic acyl groups per molecule, formed as a result of acyl migration of incorporated caprylic acyl group from the 2- to the 1-glycerol position. The product distribution using lipase as biocatalyst was similar, and the yield of desired product was slightly less than that achieved using phospholipase A_2, 39%. Therefore, this result suggests the preparation of structured phospholipids must be accompanied by selective separations schemes, since multiple products are formed for each step of a synthesis scheme, consistent with previous examples in the literature (Xu *et al.*, 2008; Guo *et al.*, 2005).

10.6 PREPARATION OF BIOSURFACTANTS VIA FERMENTATION

In addition to employing enzymes for bio-based surfactant *via in vitro* systems, microorganisms can produce bio-based surfactants *in vivo* (Kitamoto *et al.*, 2009; Lang, 2003; Muthusamy *et al.*, 2008; Rahman and Gakpe, 2008; Pinzon *et al.*, 2009). Fermentative synthesis of bio-based surfactants, referred to as "biosurfactants," possess many of the advantages described above for enzymatic synthesis, particularly enhanced sustainability (lower energy utilization, the absence of solvents, etc.). Biosurfactants are typically composed of lipid, protein, and carbohydrate moieties and often are associated with cell walls or membranes. Initial interest stemmed from their high biocompatibility and biodegradability. However, in recent years, research has shown their unique behavior in self-assembly systems, suggesting their potential applications in biomedicine and nanotechnology (Kitamoto *et al.*, 2009). Applications of biosurfactants in food products was recently reviewed (Nitschke and Costa, 2007).

Biosurfactants are divided into four categories: glycolipid-type, fatty acid-type (including phospholipids, which were discussed above), lipopeptide-type, and polymer-type (Kitamoto *et al.*, 2009). Glycolipid biosurfactants include rhamnolipids, sophorolipids, mannosylerythritol lipids, and trehalose esters (Figure 10.7). Rhamnolipids, produced from *Pseudomonas* sp. or *Burholderia* sp. bacteria, may have a single rhamnose monosaccharide unit, as depicted in Figure 10.7, or two, with a single product often containing a mixture of mono- and di-rhamnolipid, with variability in acyl chain length as well. Also, the lipophilic moiety can consist of a chain of two 2-β-hydroxy acids esterified together. Sophorolipids consist of sophorose (2-O-β-D-glucopyranosyl, β-D-glucopyranose) with an ether linkage between its reducing end and the hydroxyl of an ω or ω-1- hydroxyl fatty acyl group of 16-18 carbons. The carboxylic moiety of the fatty acid component can be free, or can form an ester bond with the 4′ hydroxyl of sophorose, as depicted in Figure 10.7. Also, the 6 and/or 6′ hydroxyl groups of the sophorose unit can be acylated, as depicted in Figure 10.7, or free. Typically, the product obtained from fermentation is a mixture of both types.

Rhamnolipids are secondary metabolites produced during a stationary growth stage, often when deprived of a nutrient such as nitrogen, and often utilize long-chain lipophilic or non-traditional carbon-energy sources, such as seed oils and alkanes (Pinzon *et al.*, 2009). Their recovery from the fermentation can be challenging due to the frequent occurrence of foaming (Pinzon *et al.*, 2009).

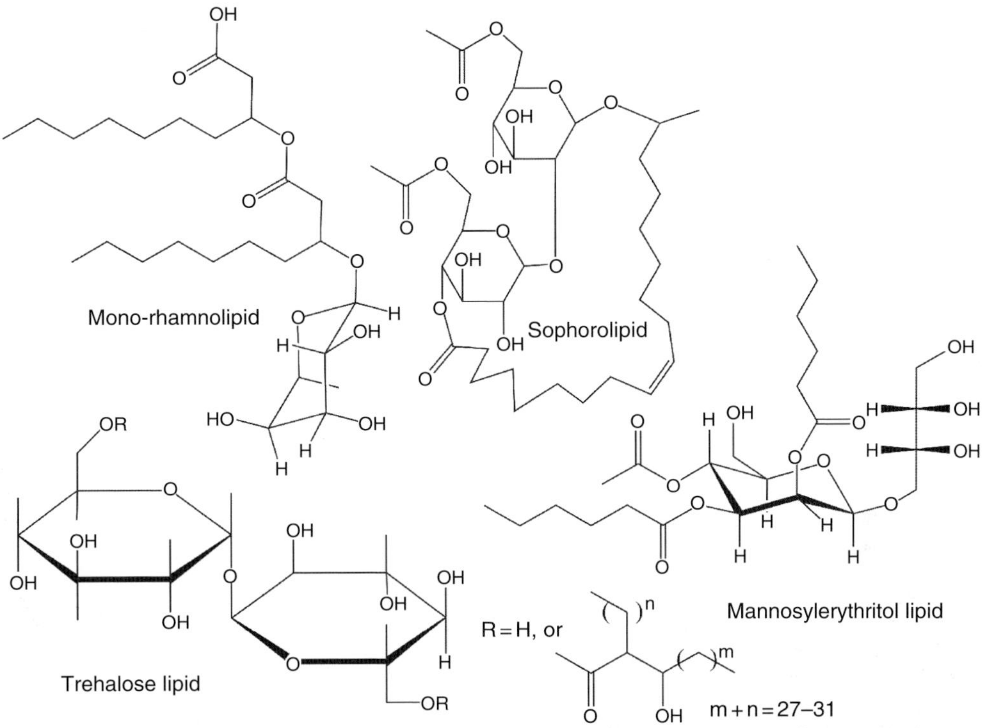

Figure 10.7 Glycolipid-type biosurfactants.

Sophorolipids possessing lactone rings are generally more lipophilic than rhamnolipids and are less sensitive to pH, and often have a lower surface activity per mass than rhamnolipids. They are typically obtained from bacteria or yeasts, particularly from *Candida* sp. Fermentation is often conducted under nitrogen starvation conditions, with abundant dissolved oxygen being present (Pinzon *et al.*, 2009). Sophorolipids can undergo chemical or enzymatic modification to further tune their surface-active properties (Ashby *et al.*, 2009).

As shown in Figure 10.7, mannosylerythritol lipids (MELs), reviewed elsewhere (Rau and Kitamoto, 2009; Arutchelvi *et al.*, 2008), contain mannose, erythritol, and fatty acyl groups as building blocks, with *meso*-erythritol conjugated to the reducing end of mannose, fatty acyl groups of varying chain length esterified to hydroxyls 2′ and 3′ of mannose and often acetyl groups at 4′ and/or 6′ hydroxyls of mannose. MELs possessing acetyl groups at positions 4′ and 6′, 6′, 4′, and the absence of acetyl groups are designated as MEL- A, B, C, and D, respectively. They are typically produced by yeasts using a fatty acyl-enriched carbon-energy source, such as soybean oil or biodiesel, and are recovered from the fermentation by solvent extraction (Rau and Kitamoto, 2009). Typically a mixture of MEL-A, B, C, and D is formed, with MEL-A being the most common.

Trehalose lipids, consisting of the disaccharide trehalose (α-D-glucopyranosyl-(1→1)-α-D-glucopyranoside) esterified to a branched, hydroxy fatty acid at its 6 and/or 6’ hydroxyl group (Figure 10.7), are typically produced from the *Rhodococcus* sp. bacteria. Other trehalose lipids contain mixtures of succinoleyl and fatty acyl groups esterified to trehalose. Their fermentation, chemistry, and applications are described elsewhere (Franzetti *et al.*, 2010).

The biological role of glycolipid biosurfactants appears to be for emulsification of apolar carbon-energy sources, adhesion of cells to hydrophobic surfaces, energy storage, and perhaps to provide resistance to high osmotic pressures (Kitamoto *et al.*, 2009). Recent emphasis has been placed on utilization of low cost, including co-product, streams as carbon-energy sources for glycolipid biosurfactants production, such as sophorolipids from glycerine, soy molasses, and whey, and rhamnolipid from used cooking oil and soapstock and molasses (Nitschke and Costa, 2007; Ashby *et al.*, 2009; Pinzon *et al.*, 2009).

Glycolipid-type biosurfactants are employed in a variety of applications, due to their high surface activity, biodegradability, and biocompatibility. Rhamnolipids are useful in bioremediation and enhanced oil recovery (Pinzon *et al.*, 2009). Sophorolipids are employed in food encapsulation and, more recently, in dishwasher detergents, and have biomedical applications (Kitamoto *et al.*, 2009; Ashby *et al.*, 2009). Mannosylerythritol lipids have several biomedical applications, including the treatment of tumors and as antimicrobial agents (Arutchelvi *et al.*, 2008). Mannosylerythritol lipids, sophorolipids, and several other glycolipid biosurfactants have numerous applications in cosmetics (Lourith and Kanlayavattanakul, 2009). Trehalose lipids have several environmental and biomedical applications (Franzetti *et al.*, 2010).

Lipopeptide surfactants, produced primarily by the *Pseudomonas* and *Bacillus* bacteria, typically consist of polypeptides that contain 5–12 D- or L-amino acid residues and a β-hydroxy fatty acyl group of chain length C_{13}–C_{18}, typically covalently attached *via* amide bond formation at the amino terminus of the polypeptide. Also, many lipopeptides contain a second linkage, an ester bond between the β-hydroxy group of the fatty acyl moiety and the carboxylic acid terminus of the polypeptide. Surfactin, an excellent surface-active agent that can lower the surface tension of water from 72 to 29 mN m^{-1} at a concentration of only 50 μM, is depicted in Figure 10.8. Surfactin contains a mixture of molecules similar in chemical structure to that depicted in the figure with a slight variation in the fatty acyl chain length. Reviewed elsewhere (McInerney *et al.*, 2009), lipopeptides possess many applications, including biomedical employment as antibiotics (e.g., surfactin from *B. subtilis*) and inhibitors of blood clotting. Their thermostability and insensitivity to pH and salinity makes them excellent candidates for oil recovery. Lipopeptides are formed *via* fermentation, using carbon and nitrogen starvation conditions. Recent work has involved utilizing low cost co-product streams as carbon-energy sources, such as molasses (Nitschke and Costa, 2007).

Polymer-type surfactants, typically produced from bacteria or yeast, are primarily polysaccharides. One example is "emulsan," an anionic lipoheteropolysaccharide produced from *Acinetobacter venetianus* when utilizing lipids or long-chain alkanes as carbon-energy sources (Mercaldi *et al.*, 2008). Fatty acyl groups are linked to the polysaccharide backbone *via* ester or amide linkages. Its molecular weight is 98 000. Emulsan's main application is for environmental remediation and oil recovery, with recent investigations showing potential utility in drug delivery. Alasan, a high molecular weight complex of polysaccharide containing covalently attached alanine and protein obtained from *A. radioresistens*, has similar applications as emulsan (Ahmed *et al.*, 2009). Liposan, obtained from fermentation of *Candida lipopytica* using hexadecane as carbon-energy source, contains mainly (83%) polysaccharide (glucose, galactose, galactosamine, and galacturonic acid units) and the remainder protein, and has utility as a gelling agent, emulsifier, stabilizer, flocculant, lubricate, or dispersing agent (Cirigliano and Carman, 1985).

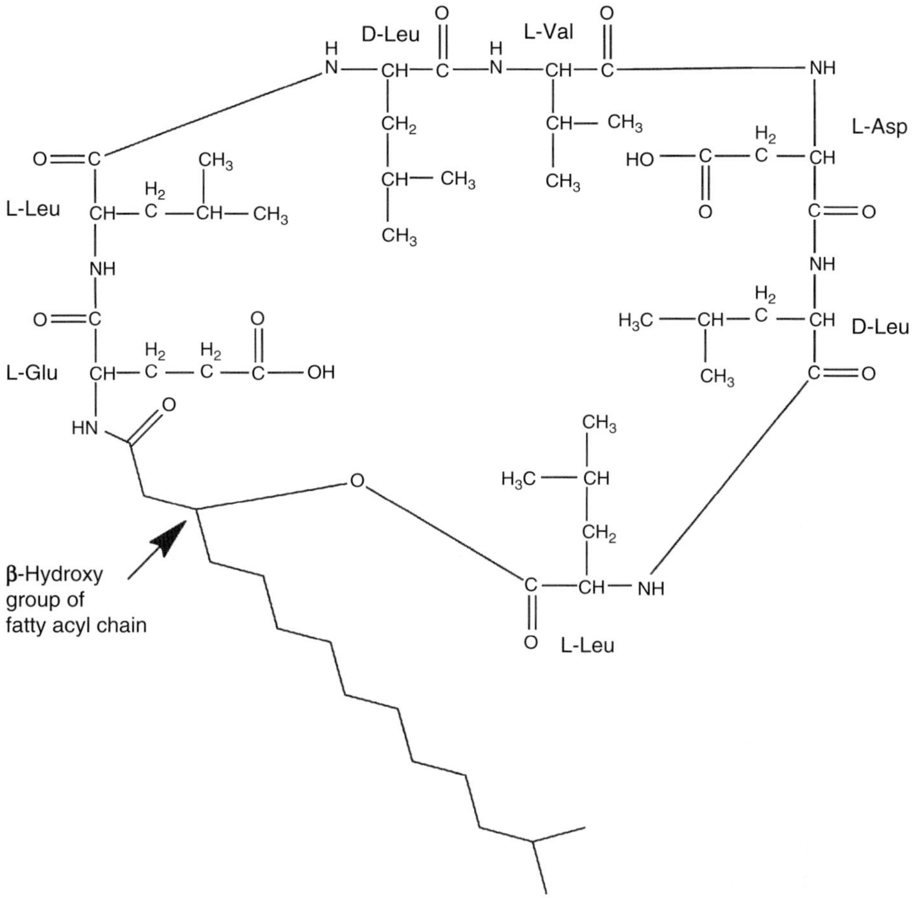

Figure 10.8 Molecular structure of surfactin, a liposaccharide-type biosurfactant.

10.7 CONCLUSIONS

Bio-based surfactants will account for an increasing share of the surfactants and detergents market worldwide as the cost of petrochemical feedstocks increases and reaches and/or surpasses the cost of bio-based feedstocks; this is anticipated to occur in the near future. Bio-based feedstocks possess the capacity to meet worldwide demand for surfactants and detergents. Bioprocessing of bio-based surfactants *via* use of enzymes or microorganisms (product from the latter referred to as "biosurfactant") will become increasingly attractive due to their excellent sustainability profile and their high biocompatibility and biodegradability. Bioprocessing will become competitive with chemical approaches in terms of operating costs in the future due to their reduction of energy usage and their narrower selectivity, reducing the costs of downstream separations. Bioprocessing is particularly robust for niche applications where highly pure surfactants are needed, such as biomedicine. Further research is needed in developing more robust biocatalysts and chemo-enzymatic processing methodologies.

REFERENCES

Acosta EJ, Saad SMI, Kang N, Policova Z, Hair ML & Neumann AW. 2009. Lung surfactants: formulation, evaluation, and polymeric additives. In: Hayes, D.G., Kitamoto, D., Solaiman, D.K.Y. & Ashby, R.D. (editors). Bio-based Surfactants and Detergents: Synthesis, Properties, and Applications. Champaign, IL: American Oil Chemists' Society Press. pp. 191–229.

Ahmad S, Siwayanan P, Abd Murad Z, Abd Aziz H & Seng Soi H. 2007. Beyond biodiesel: methyl esters as the route for the production of surfactants feedstock. INFORM 18(4):216–220.

Ahmed MMM, Hafez EE & El-Saadani MA. 2009. Biodegradation of polyaromatic hydrocarbons by recombinant bacteria containing Alasan gene. International Journal of Environment and Pollution 38(4):415–424.

Altenbach H-J, Ihizane R, Jakob B, *et al.* 2010. Synthesis and characterization of novel surfactants: combination products of fatty acids, hydroxycarboxylic acids and alcohols. Journal of Surfactants and Detergents 13(4):399–407.

Arutchelvi JI, Bhaduri S, Uppara PV & Doble M. 2008. Mannosylerythritol lipids: a review. Journal of Industrial Microbiology & Biotechnology 35(12):1559–1570.

Ashby RD, Solaiman DKY & Zerkowski JA. 2009. Production and modification of sophorolipids from agricultural feedstocks. Bio-based Surfactants and Detergents: Synthesis, Properties, and Applications. Champaign, IL: American Oil Chemists' Society Press. pp. 29–49.

Banno T, Toshima K, Kawada K & Matsumura S. 2007. Synthesis and properties of biodegradable and chemically recyclable cationic surfactants containing carbonate linkages. Journal of Oleo Science 56(9):493–499.

Banno T, Kawada K & Matsumura S. 2010. Creation of novel green and sustainable Gemini-type cationics containing carbonate linkages. Journal of Surfactants and Detergents 13(4):387–398.

Barrault J, Clacens JM & Pouilloux Y. 2004. Selective oligomerization of glycerol over mesoporous catalysts. Topics in Catalysis 27(1–4):137–142.

Bilalov A, Leal C & Lindman B. 2004. Mixing oil and water by a DNA-based surfactant. Journal of Physical Chemistry B 108(39):15408–15414.

Bodalo A, Bastida J, Maximo MF, Montiel MC, Gomez M & Ortega S. 2009. Screening and selection of lipases for the enzymatic production of polyglycerol polyricinoleate. Biochemical Engineering Journal 46(2):217–222.

Bozetine I, Ahmed Zaid T, Chitour CE & Canselier JP. 2008. Optimization of an alkylpolyglucoside-based dishwashing detergent formulation. Journal of Surfactants and Detergents 11(4):299–305.

Chang SS. 1988. Challenges of the Deodorization of Fish Oil. n-3 News 3(3):1–4.

Chetri AB, Tango MS, Budge SM, Watts KC & Islam MR. 2008. Non-edible plant oils as new sources for biodiesel production. International Journal of Molecular Sciences 9:169–180.

Chisti Y. 2007. Biodiesel from microalgae. Biotechnology Advances 25(3):294–306.

Cirigliano MC & Carman GM. 1985. Purification and characterization of liposan, a bioemulsifier from *Candida lipolytica*. Applied and Environmental Microbiology 50(4):846–850.

Cowan D, Oxenbøll KM & Holm HC. 2008. Enzymatic bioprocessing of oils and fats. INFORM 19(4):210–212.

D'Arrigo P & Servi S. 2010. Synthesis of lysophospholipids. Molecules 15:1354–1377.

Daimer K & Kulozik U. 2009. Oil-in-water emulsion properties of egg yolk: effect of enzymatic modification by phospholipase A_2. Food Hydrocolloids 23(5):1366–1373.

De Jong W & Marcotullio G. 2010. Overview of biorefineries based on co-production of furfural, existing concepts and novel developments. International Journal of Chemical Reactor Engineering 8:A69 (available at http://www.bepress.com/ijcre/vol8/A69).

Doll KM & Erhan SZ. 2009. Polyol and amino acid-based biosurfactants, builders, and hydrogels. In: Hayes, D.G., Kitamoto, D., Solaiman, D.K.Y. & Ashby, R.D. (editors). Bio-based Surfactants and Detergents: Synthesis, Properties, and Applications. Champaign, IL: American Oil Chemists' Society Press. pp. 425–448.

Edser C. 2004. Growing use of oleochemicals. Focus on Surfactants (Nov):1–2.

Egan RR. 1968. The preparation and properties of amines and cationic surfactants from fatty acids. Journal of the American Oil Chemists' Society 45(7):481–486.

Fortin S. 2010. Polyunsaturated fatty acid monoglycerides, derivatives, and uses thereof. US Patent 2009-535048, 20100160261.

Franzetti A, Gandolfi I, Bestetti G, Smyth TJP & Banat IM. 2010. Production and applications of trehalose lipid biosurfactants. European Journal of Lipid Science and Technology 112(6):617–627.

Freitas L, Paula AV, dos Santos JC, Zanin GM & de Castro HF. 2010. Enzymatic synthesis of monoglycerides by esterification reaction using *Penicillium camembertii* lipase immobilized on epoxy SiO_2-PVA composite. Journal of Molecular Catalysis B: Enzymatic 65(1–4):87–90.

Gardossi L, Bianchi D & Klibanov AM. 1991. Selective acylation of peptides catalyzed by lipases in organic solvents. Journal of the American Chemical Society 113(16):6328–6329.

Gielen D, Newman J & Patel MK. 2008. Reducing industrial energy use and CO_2 emissions: the role of materials science. MRS Bulletin 33(4):471–477.

Giraldo L, Camargo G, Tirano J & Moreno-Pirajan JC. 2010. Synthesis of fatty alcohols from oil palm using a catalyst of Ni-Cu supported onto zeolite. E-Journal of Chemistry 7(4):1138–1147.

Guo Z, Vikbjerg AF & Xu X. 2005. Enzymatic modification of phospholipids for functional applications and human nutrition. Biotechnology Advances 23(3):203–259.

Habe H, Shimada Y, Yakushi T, *et al.* 2009. Microbial production of glyceric acid, an organic acid that can be mass produced from glycerol. Applied and Environmental Microbiology 75(24):7760–7766.

Hansson T & Adlercreutz P. 2002. Enzymatic synthesis of hexyl glycosides from lactose at low water activity and high temperature using hyperthermostable β-glycosidases. Biocatalysis and Biotransformation 20(3):167–178.

Hatti-Kaul R, Toernvall U, Gustafsson L & Boerjesson P. 2007. Industrial biotechnology for the production of bio-based chemicals – a cradle-to-grave perspective. Trends in Biotechnology 25(3):119–124.

Hayes DG. 2004. Enzyme-catalyzed modification of oilseed materials to produce eco-friendly products. Journal of the American Oil Chemists' Society 81(12):1077–1103.

Hayes DG. 2009. Bio-based surfactants, overview and industrial state-of-the-art. In: Hayes, D.G., Kitamoto, D., Solaiman, D.K.Y. & Ashby, R.D. (editors). Bio-based Surfactants and Detergents: Synthesis, Properties, and Applications. Champaign, IL: American Oil Chemists' Society Press. pp. 3–25.

Hill K. 2007. Industrial Development and Application of Bio-based Oleochemicals. Pure and Applied Chemistry 79(11):1999–2011.

Hill K. 2009. Alkyl polyglycosides – where "green" meets performance. SOFW Journal 135(1/2):6, 8–12, 14.

Holladay JE, Bozell JJ, White JF & Johnson D. 2007. Top value-added chemicals from biomass. Volume II. Results of screening for potential candidates from biorefinery lignin. US Department of Energy, Pacific Northwest National Laboratory. http://www1.eere.energy.gov/biomass/pdfs/pnnl-16983.pdf (accessed 17 December 2010).

Honydonckx HE, De Vos DE, Chavan SA & Jacobs PA. 2004. Esterification and transesterification of renewable chemicals. Topics in Catalysis 27(1–4):83–96.

Hossen M & Hernandez E. 2005. Enzyme-catalyzed synthesis of structured phospholipids with conjugated linoleic acid. European Journal of Lipid Science and Technology 107(10):730–736.

Husmann M. 2008. Surfactants derived from naturally-occurring amino acids. SOFW Journal 134(3):34–36, 38–40.

Hwang S, Lee S, Ahn I-S & Jung J-K. 2009. Highly efficient production of monoglycerides by the continuous removal of fatty acids from lipase-catalyzed oil hydrolysis. Biocatalysis and Biotransformation 27(5–6):290–295.

Infante MR, Perez L, Moran C, Pons R & Pinazo A. 2009a. Synthesis, aggregation properties, and applications of biosurfactants derived from arginine. In: Hayes, D.G., Kitamoto, D., Solaiman, D.K.Y. & Ashby, R.D. (editors). Bio-based Surfactants and Detergents: Synthesis, Properties, and Applications. Champaign, IL: American Oil Chemists' Society Press. pp. 351–387.

Infante MR, Perez L, Moran MC, *et al.* 2009b. Biocompatible surfactants from renewable hydrophiles. European Journal of Lipid Science and Technology 112(1):110–121.

Johansson I & Svensson M. 2001. Surfactants based on fatty acids and other natural hydrophobes. Current Opinion in Colloid & Interface Science 6(2):178–188.

Karmee SK. 2008. Lipase catalyzed synthesis of ester-based surfactants from biomass derivatives. Biofuels, Bioproducts & Biorefining 2(2):144–154.

Kenar JA, Knothe G & Copes AL. 2004. Synthesis and characterization of dialkyl carbonates prepared from mid-, long-chain, and Guerbet alcohols. Journal of the American Oil Chemists' Society 81(3):285–291.

Kitamoto D, Morita T, Fukuoka T & Imura T. 2009. Self-assembling properties of glycolipid biosurfactants and their functional developments. In: Hayes, D.G., Kitamoto, D., Solaiman, D.K.Y. & Ashby, R.D. (editors). Bio-based Surfactants and Detergents: Synthesis, Properties, and Applications. Champaign, IL: American Oil Chemists' Society Press. pp. 231–272.

Kralova I & Sjoblom J. 2009. Surfactants used in food industry: a review. Journal of Dispersion Science and Technology 30(9):1363–1383.

Lakshmanan CM & Hoelscher HE. 1970. Production of levoglucosan by pyrolysis of carbohydrates. Pyrolysis in hot inert gas stream. Industrial & Engineering Chemistry Product Research and Development 9(1):57–59.

Lang S. 2003. Surfactants produced by microorganisms. In: Holmberg, K. (editor). Novel Surfactants: Preparation, Applications, and Biodegradability, 2nd Edn. New York: Marcel Dekker, Inc. pp. 279–315.

Larsson J, Svensson D & Adlercreutz P. 2005. α-Amylase-catalysed synthesis of alkyl glycosides. Journal of Molecular Catalysis B: Enzymatic 37(1–6):84–87.

Leal C, Bilalov A & Lindman B. 2006. Phase behavior of a DNA-based surfactant mixed with water and *n*-alcohols. Journal of Physical Chemistry B 110(34):17221–17229.

Lee KH, Park C-H & Lee EY. 2010. Biosynthesis of glycerol carbonate from glycerol by lipase in dimethyl carbonate as the solvent. Bioprocess and Biosystems Engineering 33(9):1059–1065.

Liu C, Dong H, Zhong J, Ryu DDY & Bao J. 2010. Sorbitol production using recombinant *Zymomonas mobilis* strain. Journal of Biotechnology 148(2–3):105–112.

Lourith N & Kanlayavattanakul M. 2009. Natural surfactants used in cosmetics: glycolipids. International Journal of Cosmetic Science 31(4):255–261.

Matsumura S. 2002. Biodegradable surfactants having carbonate bond and their manufacture. Japan Patent 2000–316398 2002128744.

McCoy M. 2007. A new kind of clean. Chemical & Engineering News 85(12):29–31.

McCoy M. 2008a. Converging pathways: chemical companies and environmentalists edge closer together in the pursuit of sustainability. Chemical & Engineering News 86(33):47–46.

McCoy M. 2008b. Greener cleaners. Chemical & Engineering News 86(3):15–23.

McInerney MJ, Youssef N & Nagle DP. 2009. Lipopeptide biosurfactants and their use in oil recovery. In: Hayes, D.G., Kitamoto, D., Solaiman, D.K.Y. & Ashby, R.D. (editors). Bio-based Surfactants and Detergents: Synthesis, Properties, and Applications. Champaign, IL: American Oil Chemists' Society Press. pp. 129–153.

Mercaldi MP, Dams-Kozlowska H, Panilaitis B, Joyce AP & Kaplan DL. 2008. Discovery of the dual polysaccharide composition of Emulsan and the isolation of the emulsion stabilizing component. Biomacromolecules 9(7):1988–1996.

Mishra S & Tyagi VK. 2007. Ester quats: the novel class of cationic fabric softeners. Journal of Oleo Science 56(6):269–276.

Montet D, Servat F, Pina M, *et al.* 1990. Enzymic synthesis of N-ε-acyllysines. Journal of the American Oil Chemists' Society 67(11):771–774.

Montiel Morte MC & Hayes DG. 2009. Biocatalytic synthesis of polyglycerol polyricinoleate, a value-added, bio-based, food emulsifier. 16th Annual BioEnvironmental Polymer Society (BEPS) Meeting (Polymers and the Environment – Emerging Technologies and Science. Chicago, IL.

Muthusamy K, Gopalakrishnan S, Ravi TK & Sivachidambaram P. 2008. Biosurfactants: properties, commercial production and application. Current Science 94(6):736–747.

Nagao A & Kito M. 1989. Synthesis of O-acyl-L-homoserine by lipase. Journal of the American Oil Chemists' Society 66(5):710–713.

Nitschke M & Costa SGVAO. 2007. Biosurfactants in food industry. Trends in Food Science & Technology 18(5):252–259.

Orlich B, Berger H, Lade M & Schomacker R. 2000. Stability and activity of alcohol dehydrogenases in W/O-microemulsions: enantioselective reduction including cofactor regeneration. Biotechnology and Bioengineering 70(6):638–646.

Otero C. 2009. Selective enzymatic synthesis of N-acylated alkanolamine emulsifiers. In: Hayes, D.G., Kitamoto, D., Solaiman, D.K.Y. & Ashby, R.D. (editors). Bio-based Surfactants and Detergents: Synthesis, Properties, and Applications. Champaign, IL: American Oil Chemists' Society Press. pp. 299–322.

Otomo N. 2009. Basic properties of sucrose fatty acid esters and their applications. In: Hayes, D.G., Kitamoto, D., Solaiman, D.K.Y. & Ashby, R.D. (editors). Bio-based Surfactants and Detergents: Synthesis, Properties, and Applications. Champaign, IL: American Oil Chemists' Society Press. pp. 275–298.

Overkempe C, Annerling A, Van Ginkel CG, Thomas PC, Boltersdorf D & Speelman J. 2003. Esterquats. In: Holmberg, K. (editor). Novel Surfactants: Preparation, Applications, and Biodegradability New York: Marcel Dekker, Inc. pp. 347–384.

Pandey V, Banerji R, Dixit BS, Singh M, Shukla S & Singh SP. 2000. Cuphea, a rich source of medium-chain triglycerides: fatty acid composition and oil diversity in *Cuphea procumbens*. European Journal of Lipid Science and Technology 102(7):463–466.

Patel M. 2004. Surfactants based on renewable raw materials: carbon dioxide reduction potential and policies and measure for the European Union. Journal of Industrial Ecology 7(3/4):47–62.

Petrovic M & Barcelo D. 2000. The stability of non-ionic surfactants and linear alkylbenzene sulfonates in a water matrix and on solid-phase extraction cartridges. Fresenius' Journal of Analytical Chemistry 368(7):676–683.

Pinzon NM, Zhang Q, Koganti S & Ju L-K. 2009. Advances in bioprocess development of rhamnolipid and sophorolipid production. In: Hayes, D.G., Kitamoto, D., Solaiman, D.K.Y. & Ashby, R.D. (editors). Bio-based Surfactants and Detergents. Champaign, IL: American Oil Chemists' Society Press. pp. 77–105.

Pyo S-H & Hayes DG. 2009. Synthesis of saccharide fatty acid ester biosurfactants catalyzed by lipase. In: Hayes, D.G., Kitamoto, D., Solaiman, D.K.Y. & Ashby, R.D. (editors). Bio-based Surfactants and Detergents: Synthesis, Properties, and Applications. Champaign, IL: American Oil Chemists' Society Press. pp. 323–350.

Rahman PKSM & Gakpe E. 2008. Production, characterization and applications of biosurfactants-review. Biotechnology (Faisalabad, Pakistan) 7(2):360–370.

Rau U & Kitamoto D. 2009. Mannosylerythritol lipids: production and downstream processing. In: Hayes, D.G., Kitamoto, D., Solaiman, D.K.Y. & Ashby, R.D. (editors). Bio-based Surfactants and Detergents: Synthesis, Properties, and Applications. Champaign, IL: American Oil Chemists' Society Press. pp. 51–75.

Roberts DW, Giusti L & Forcella A. 2008. Chemistry of methyl ester sulfonates. INFORM ("Surfactants from Biorenewable Sources: A Special Supplement", Biorenewable Resources #5)(March):2–9.

Rudnick LR. 2006. Dialkyl carbonates. In: Rudnick, L.R. (editor). Synthetics, Mineral Oil, and Bio-Based Lubricants: Chemistry and Technology. Boca Raton, FL: CRC Press. pp. 235–245.

Santosa SJ. 2008. Palm oil boom in Indonesia: from plantation to downstream products and biodiesel. Clean: Soil, Air, Water 36(5–6):453–465.

Savic S, Tamburic S & Savic MM. 2010. From conventional towards new – natural surfactants in drug delivery systems design: current status and perspectives. Expert Opinion on Drug Delivery 7(3):353–369.

Scheibel J. 2007. The impact of feedstocks on future innovation in surfactant technology for the detergent market. 98th American Oil Chemists' Society Annual Meeting. Quebec City, Canada.

Scott E, Peter F & Sanders J. 2007. Biomass in the manufacture of industrial products – the use of proteins and amino acids. Applied Microbiology and Biotechnology 75(4):751–762.

Shelley S. 2007. A renewable route to propylene glycol. Chemical Engineering Progress 103(8):6–9.

Soo EL, Salleh AB, Basri M, Rahman RNZRA & Kamaruddin K. 2003. Optimization of the enzyme-catalyzed synthesis of amino acid-based surfactants from palm oil fractions. Journal of Bioscience and Bioengineering 95(4):361–367.

Steber J & Wierich P. 1985. Metabolites and biodegradation pathways of fatty alcohol ethoxylates in microbial biocenoses of sewage treatment plants. Applied and Environmental Microbiology 49(3):530–537.

Stjerndahl M & Holmberg K. 2003. Synthesis and chemical hydrolysis of surface-active esters. Journal of Surfactants and Detergents 6(4):311–318.

Stjerndahl M & Holmberg K. 2005a. Hydrolyzable nonionic surfactants: Stability and physicochemical properties of surfactants containing carbonate, ester, and amide bonds. Journal of Colloid and Interface Science 291(2):570–576.

Stjerndahl M & Holmberg K. 2005b. Synthesis, stability, and biodegradability studies of a surface-active amide. Journal of Surfactants and Detergents 8(4):331–336.

Stjerndahl M, van Ginkel CG & Holmberg K. 2003. Hydrolysis and biodegradation studies of surface-active esters. Journal of Surfactants and Detergents 6(4):319–324.

Svensson D, Ulvenlund S & Adlercreutz P. 2009a. Efficient synthesis of a long carbohydrate chain alkyl glycoside catalyzed by cyclodextrin glycosyltransferase (CGTase). Biotechnology and Bioengineering 104(5):854–861.

Svensson D, Ulvenlund S & Adlercreutz P. 2009b. Enzymatic route to alkyl glycosides having oligomeric head groups. Green Chemistry 11(8):1222–1226.

Svensson M & Brinck J. 2003. Surfactants based on sterols and other alicyclic compounds. In: Holmberg, K. (editor). Novel Surfactants : Preparation, Applications, and Biodegradability, 2nd Edn. New York: Marcel Dekker, Inc. pp. 217–239.

Tang Y, Chen H, Jiang J, Tang Z, Huang B & Roesky HW. 2004. Application of hydrogen-storage alloy electrode in electrochemical reduction of glucose. Journal of Power Sources 130(1–2):56–60.

US Department of Energy. 1999. The Technology Roadmap for Plant /Crop-Based Renewable Resources 2020. www1.eere.energy.gov/biomass/pdfs/technology_roadmap.pdf (accessed 17 December 2010).

Valivety R, Jauregi P, Gill I & Vulfson E. 1997. Chemo-enzymic synthesis of amino acid-based surfactants. Journal of the American Oil Chemists' Society 74(7):879–886.

Valivety R, Gill IS & Vulfson EN. 1998. Application of enzymes to the synthesis of amino acid-based bola and gemini surfactants. Journal of Surfactants and Detergents 1(2):177–185.

van Rantwijk F, Woudenberg-van Oosterom M & Sheldon RA. 1999. Glycosidase-catalyzed synthesis of alkyl glycosides. Journal of Molecular Catalysis B: Enzymatic 6(6):511–532.

Watanabe Y & Shimada Y. 2009. Production of monoacylglycerols through lipase-catalyzed reactions. In: Hou, C.T. & Shaw, J.F. (editors). Biocatalysis and Agricultural Biotechnology. Boca Raton, FL: CRC Press. pp. 181–197.

Wen J-P, Wang C-L & Liu Y-X. 2004. Preparation of sorbitol from D-glucose hydrogenation in gas-liquid-solid three-phase flow airlift loop reactor. Journal of Chemical Technology and Biotechnology 79(4):403–406.

Werpy T & Petersen G. 2004. Top Value Added Chemicals from Biomass, Volume I, Results of Screening for Potential Candidates from Sugars and Synthesis Gas. US Department of Energy. http://www1.eere.energy.gov/biomass/pdfs/35523.pdf (accessed 17 December 2010).

Xu C, Taylor P, Fletcher PDI & Paunov VN. 2005. Adsorption and hybridization of DNA-surfactants at fluid surfaces and lipid bilayers. Journal of Materials Chemistry 15(3):394–402.

Xu X, Vikbjerg AF, Guo Z, Zhang L & Acharya AK. 2008. Enzymatic modification of phospholipids and related polar lipids. In: Gunstone, F.D. (editor). Phospholipid Technology and Applications. Bridgwater, UK: Oily Press. pp. 41–82.

Ye R & Hayes DG. 2011. Optimization of the solvent-free lipase-catalyzed synthesis of fructose-oleic acid ester through programming of water removal. Journal of the American Oil Chemists' Society 88:1351–1359.

Ye R, Pyo S-H & Hayes DG. 2010. Lipase-catalyzed synthesis of saccharide-fatty acid esters using suspensions of saccharide crystals in solvent-free media. Journal of the American Oil Chemists' Society 87(3):281–293.

Zeng F-K, Yang B, Wang Y-H, Wang W-F, Ning Z-X & Li L. 2010. Enzymatic production of monoacylglycerols with camellia oil by the glycerolysis reaction. Journal of the American Oil Chemists' Society 87(5):531–537.

11 Biopolymers

Oğuz Türünç and Michael A. R. Meier

11.1 INTRODUCTION

It has been many years since human beings began using plant and animal products for purposes other than nutrition, for example as paints or clothing. Since that time, the methods and knowledge-base used to produce these products from such resources has steadily developed. Eventually, the discovery of fossil oil and its consequent exploitation, including the improved knowledge about petroleum chemistry and chemical engineering, resulted in a "plastic revolution", which first seemed to be very cheap, easily formed, and a long-lasting technology. However, due to the recent awareness of not only the green house effect but also accumulation of plastic waste in almost every part of the world, as well as environmentally unsound production technologies, mankind has started to search for sustainable, environmentally friendly feedstocks and technologies to improve the ecological impact of the chemical industry. The fact that fossil oil is a limited feedstock only further clarifies the need to develop more sustainable alternatives. Today, carbohydrate- and plant oil-based polymers and synthesis technologies are being extensively researched and even commercialized to some extent. This chapter summarizes current knowledge and applications of some carbohydrate- and plant oil-based biopolymers derived from natural resources and discusses possible future applications/products as well as properties of these materials.

11.2 CARBOHYDRATE-BASED POLYMERS

11.2.1 Polymers from starch

Starch is a highly abundant polysaccharide and is produced by plants to store energy. It is a semi-crystalline polymer composed of amylose and amylopectin. Amylose is an almost linear polymer consisting of $\alpha(1 \rightarrow 4)$ linked glucose molecules, whereas amylopectin has a similar (1–4)-linked polymer backbone with an additional 5% (approximately) $\alpha(1 \rightarrow 6)$ linkages with branches typically occurring every 24 to 30 glucose units (Figure 11.1) (Voet *et al.*, 2006). The relative amounts of amylose and amylopectin depend on the plant source and affect the final properties of the material. Corn starch granules for instance contain approximately 70% amylopectin and 30% amylose (Avella *et al.*, 2005). Starch is one of the

Food and Industrial Bioproducts and Bioprocessing, First Edition. Edited by Nurhan Turgut Dunford.

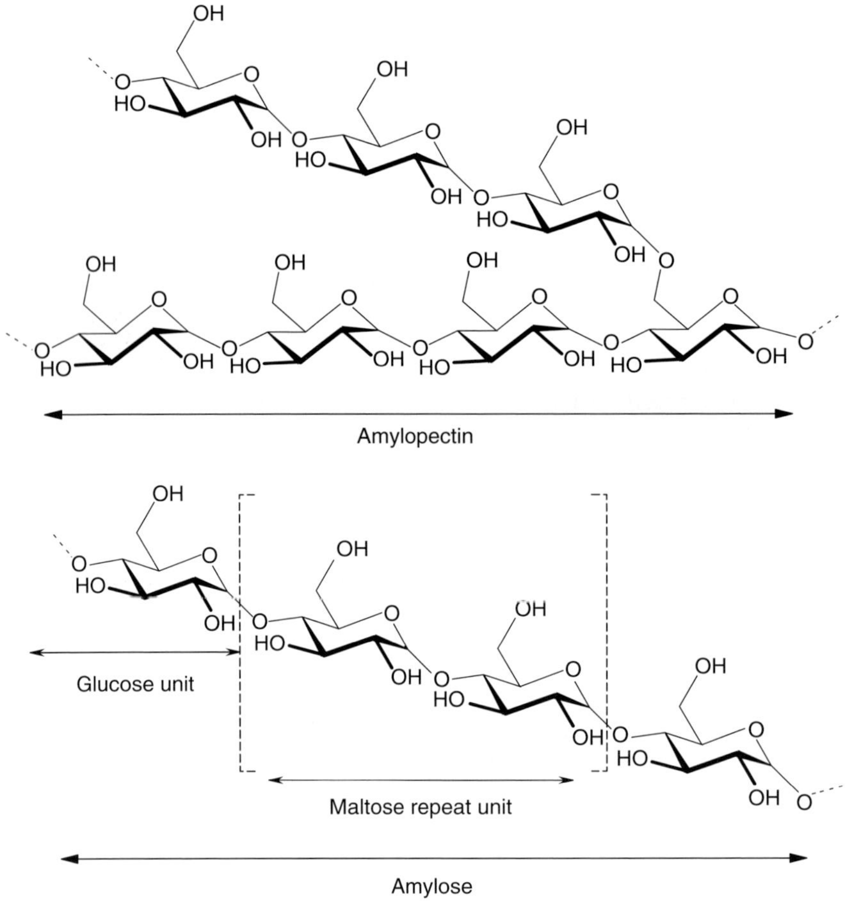

Figure 11.1 Chemical structure of starch.

cheapest and most abundant natural polymers (Mecking, 2004). The abundance and degradability make starch interesting for several industrial applications.

Although starch is itself a thermoplastic polymer (thermoplastic starch, TPS) (Li *et al.*, 2008) and can be processed *via* extrusion or molding with the aid of a plasticizer (Chuayuljit *et al.*, 2009), this natural polymer is a rapidly degrading material with minimal moisture resistance. Application possibilities of TPS are as fillers or blends with other thermoplastics (Arvanitoyannis *et al.*, 1997, 1998). Chemical modification of starch, on the other hand (Gandini, 2008; Lee *et al.*, 2007), can be used to adjust degradation properties as well as the water stability of the resulting materials.

For instance, blends of starch with polyethylene or other polyesters are manufactured by Novamont on a scale of about 36 300 metric tons (tonnes) per year for packaging applications (Mooney, 2009). Moreover, foam materials of starch blends are also used in packaging applications. Firstly, the blend is prepared *via* the extrusion of TPS and a biodegradable (Nabar *et al.*, 2006) or non-biodegradable (Mihai *et al.*, 2007; Biswas *et al.*, 2006) polymer. Subsequently, the blends are blown into foams in order to improve the hydrophobic character and mechanical properties of starch foams and films (Yoon *et al.*, 2007; Avella *et al.*, 2005).

In 2006, Yoon *et al.* studied the effect of additives such as glycerol, succinic acid, malic acid, and tartaric acid on the physical properties of starch/PVA blend films; the results showed that the flexibility and strength of the films could be improved with alcohol and carboxylic acid functionalities.

Flieger and co-workers (2003), in their review, classify starch composites according to their starch content: low (~10–20%), medium (~40–60%), and high amounts (~90%) of starch, each class demonstrating distinct material properties. Addition of low amounts of starch to, for example, polyethylene (Lim and Jane, 1992) accelerates the disintegration of the polymer and the breaking apart of the material in such a way that the fast degradation of starch produces pores in the remaining plastic film, which leads to a better interaction between oxygen, microorganisms and the plastic molecules. In some applications photo initiators, such as benzophenone derivatives, are used as pro-oxidants to speed up the degradation progress *via* generation of free radicals, which results in polymer oxidation and, finally, chain scission (Bagheri and Naimian, 2002).

Jansson and co-workers (2006) studied free films made of starch and PVA mixtures (produced by casting) with respect to their mechanical and moisture uptake in order to establish their potential as spray-dried powders. The moisture uptake was substantially higher when glycerol was added and the glycerol-containing films had a lower failure stress under tension than the corresponding glycerol-free films. Nevertheless, the films containing 20% glycerol showed almost double the water uptake compared to the glycerol-free films. The observed swelling behavior indicated that some of the powders were good candidates for plastisol coatings, which are suspensions of polymer particles in a plasticizer (Persico *et al.*, 2009) (commonly dioptyl adipate) used mainly in textile inks.

Starch can also be chemically modified *via* esterification with fatty acids to synthesize completely renewable and biodegradable materials. Sagar and Merrill (1995) esterified high amylose starch with different chain length fatty acid chlorides in the presence of pyridine and then evaluated the rheological, thermal, and mechanical properties of the resulting materials. It was shown that ester groups within the structure act as internal plasticizers, making the material more easily processable and more ductile. The starch esters of pentanoic and hexanoic acid were semi-crystalline. The report showed that fatty acid modified starch could be used for some applications where biodegradability is required but mechanical property requirements are not too demanding. Moreover, it was shown that the degradation rate of starch could be controlled by the degree of acetylation (Fringant *et al.*, 1998).

A great deal of attention has been given to the incorporation of polysaccharides into polyurethane (PU) foams (Donnelly *et al.*, 1991). Attempts to form PUs from starch and its degradation products, polyol glucoside ethers, suitable for polyurethane were reported in the early 1960s (Otey and Miehltretter, 1962). Some methods for depolymerizing polysaccharides are well known. For instance, starch can be enzymatically degraded to yield saccharide mixtures (containing dextrin, maltose, maltotriose and glucose), high maltose syrups, or high glucose syrups *via* the use of α-amylase, pullulanase, and amyloglucosidase enzymes (Donnelly *et al.*, 1991). Polyethers were derived from starch degradation products by alkoxylation of the carbohydrate to a degree necessary to give a liquid of the desired viscosity and hydroxyl number. Ge and co-workers (2000) utilized starch and wattle tannin of the bark of *Acacia mearnsii* (a fast-growing tree native to Australia) and toluene diisocyanate to synthesize flexible PU foams for car cushions, which need a high cross-linking density to maintain a high-resilient property. On the other hand, Kim and co-workers (2007) used starch as the main polyol component, with the amount of 30–50 wt-%, for polyurethane film formation. Two endothermic peaks were observed by DSC analysis, resulting from urethane and starch domains. From tensile testing, the stress at break and modulus were observed to increase significantly with the starch content and –NCO/–OH molar ratio. Much research has also been carried out to enhance the fire resistance of PU by the incorporation of phosphorus, halogens, or nitrogen-containing compounds, or by the use of rigid molecules either as polyol or as isocyanate (Donnelly *et al.*, 1991).

Because of their low immunogenic potential, potential bioactive behavior, capability of interacting with the host's tissue, chemical versatility, and almost unlimited source, starch-based polymers are considered to be ideal materials for tissue engineering (Cunha *et al.*, 2004). Different degradable porous architectures have thus been developed using polymers based on blends of corn starch with poly(ethylene-co-vinyl alcohol)(SEVA-C, Novamont, Italy) and cellulose acetate (SCA, Novamont, Italy)(Gomes *et al.*, 2001). These materials present a non-cytotoxic behavior and are under consideration for a wide range of biomedical applications, such as scaffolds for bone-tissue engineering (Salgado *et al.*, 2004).

Another future application of starch might be in the field of electro-active polymers or polymer electrolytes. Although starch is an insulator and its proton mobility is low, upon doping with MX (NaCl, NaI, LiCl, LiI), it exhibits conductance between 10^{-5} and 10^{-6} S/cm (compared to 10^{-9}–10^{-11} S/cm for starch which contains 30% water) (Finkenstadt, 2005). Recently, Khiar and Arof (2010) studied an electrolyte based on starch and different amounts of ammonium nitrate. The study revealed that the conductivity was increased from 6.28×10^{-10} to 2.83×10^{-5} upon addition of 25% ammonium nitrate at room temperature.

One of the most recent and interesting application possibilities of starch is shape-memory materials. Non-modified starch has been shown to present very efficient shape-memory capabilities, with a recovery ratio of $R_r > 90\%$ for a fixed deformation of 200%, which was permanently shaped by extrusion, and this shape was stabilized below T_g. A second shape was gained *via* thermo-molding above T_g and shape recovery was triggered by water sorption at 20 °C. The study stimulates the use of starch as environmentally friendly smart material (Véchambre *et al.*, 2010).

11.2.2 Polymers from cellulose

Cellulose, next to chitin, is the most abundant biopolymer on earth. It is the major constituent of plant cell walls and more than half of the organic carbon on earth is fixed in cellulose. It is composed of unbranched, linear chains of d-glucose molecules linked through 1,4-β-d glycosidic bonds (Figure 11.2).

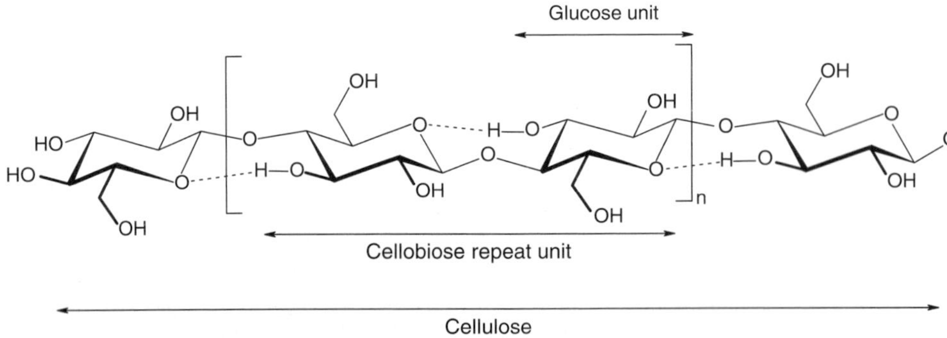

Figure 11.2 Chemical structure of cellulose.

Interestingly, no vertebrate has the capacity to digest cellulose enzymatically. Herbivores subsist largely on cellulose, not because they can digest it themselves, but because their digestive tracks contain microbes that produce cellulose-hydrolyzing cellulases; various cellulase types are also synthesized by fungi. When compared to starch, cellulose is relatively resistant to biodegradation. Each cellulose molecule is an unbranched polymer of 10^3–10^6 d-glucose units (Flieger *et al.*, 2003).

Although paper has been known for more than 2000 years, the sensitivity of the mechanical properties towards water limits the use of cellulose as a material in many applications. Moreover, cellulose cannot be thermally processed into plastics because of its hydrogen-bonded structure. However, microfibrils with diameters ranging from 2 to 10 nm and lengths reaching several micrometers can be isolated from native cellulose fibers through a combination of chemical and mechanical methods. The viscose process has been known for more than 100 years and although it is far from being environmentally friendly (Figure 11.3) it dominates other methods for isolating cellulose fibers. Cellulose processed in this way finds applications from textile fibers to technical fibers for cords of high performance tires (rayon) or for film (cellophane) production, which is suitable for food packaging (Guerra *et al.*, 2005). An environmentally friendly method, the Lyocell process, was developed in the 1980s as an alternative to the viscose process, in which N-methylmorpholine-N-oxide (NMMO) monohydrate is used as a more convenient solvent system, resulting in a process practically free from emissions since solvent recovery is almost complete (Klemm *et al.*, 2005).

Their mechanical properties, with a Young's modulus of ~ 140 GPa, close to that of Kevlar, make cellulose fibers ideally suited as reinforcing elements in composite materials (Gandini, 2008). Moreover, cellulose ultrathin films are also of great interest for their extreme wetting behavior with aqueous systems and for being extremely stable against oxidation and thermal degradation (Schaub *et al.*, 1993; Kontturi *et al.*, 2006). These features make cellulose interesting for two distinct types of materials, cellulose composites and thermoplastically processable cellulose derivatives.

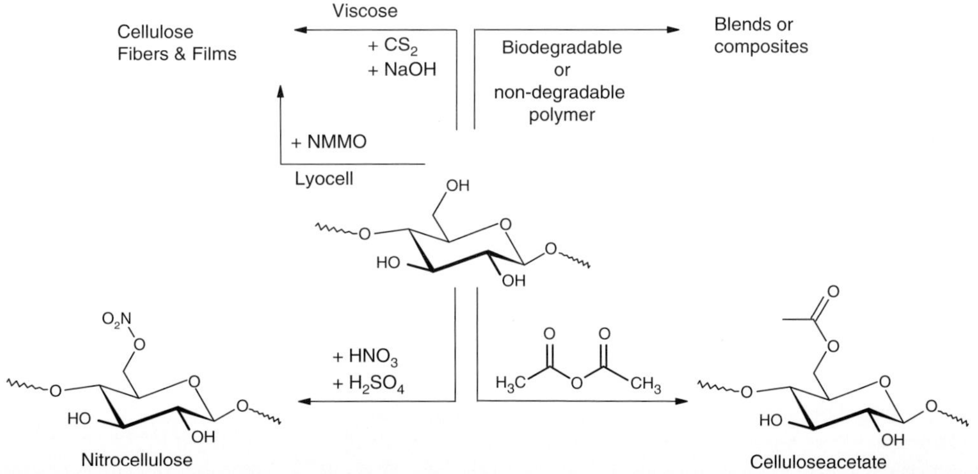

Figure 11.3 Modification possibilities of cellulose.

The introduction of cellulose into a graft-copolymer matrix and common thermoset polymers (Gindl and Jeronimidis, 2004) was shown to improve the mechanical properties of the composite materials (Mohanty *et al.*, 2000). Fiber surface modification by physical and chemical means (Felix and Gatenholm, 1993), and the use of coupling agents to overcome the incompatibility between cellulose and a hydrocarbon polymer matrix as well as to improve interfacial interactions were applied (Nair and Thomas, 2003). When cellulose is used as a natural fiber in a biodegradable polymer matrix, so called eco-composite materials are formed; these have drawn increasing attention due to environmental considerations. The processes required to prepare eco-composites are very similar to fiber glass composites and have been reviewed by Bogoeva-Gaceva and co-workers (2007). A brief review covering composites of biodegradable and non-degradable polymers with natural fibers was given by Netravali and Chabba in 2003.

Chemical modification of cellulose is another important research topic, as is industrial application area. The high number of hydroxyl groups present on each repeating unit of cellulose gives potential to be chemically modified through all possible alcohol-involved organic reactions, typically esterifications and etherifications. Moreover, click chemistry is currently being investigated to functionalize cellulose (Liebert *et al.*, 2006; Zhao *et al.*, 2010). Cellulose acetate (Figure 11.3) is the most widely known and used example of esterified cellulose derivative. It is used for potentially biodegradable fibers and films. The degradation rates of cellulose acetate can be controlled with the degree of substitution (Buchanan *et al.*, 1993). Parkesin™, a moldable nitro derivative of cellulose (Figure 11.3), has been used as a replacement for ivory. Moreover, celluloid, invented in the 1860s, is a nitro-cellulose derivative that uses camphor as a softener to improve its flexibility (Mooney, 2009).

The crystal structure and three-dimensional network prevent cellulose from behaving as a polyol for PU syntheses. To overcome this problem a liquefaction process in the presence of organic solvents was developed and resulted in products suitable for PU synthesis. Yan and co-workers (2008) liquefied corn stalk, an agricultural by-product, and tested it for the synthesis of PU foams blown by water. The report revealed that such polyurethane foams had excellent mechanical and thermal properties and could be used as heat insulating materials.

In addition to its renewability and satisfactory mechanical properties, cellulose presents biocompatibility, which makes it possible to use in some pharmaceutical applications. Different chemical modifications allow different applications of cellulose, such as oxycellulose for controlled drug delivery matrices, sodium carboxymethyl cellulose as emulsifying agents, and cellulose acetate phthalate for tablet coatings (Kamel *et al.*, 2008).

Just like starch, cellulose has also been shown to exhibit some important electro-active properties. For instance, cellophane was found to be a piezoelectric material that transforms electrical energy into mechanical energy. Nevertheless, some cellulosic blends, for example the blend of cellulose xanthate and propylene oxide-grafted hydroxyethyl cellulose with PEG, showed conductivities around 10^{-5} and 10^{-4} S/cm (Finkenstadt, 2005).

11.2.3 Polymers from lactic acid and lactide

An important feature of starch is its potential enzymatic hydrolysis into glucose and subsequent fermentation into lactic acid. Poly(lactic acid) (PLA) can be obtained from this fermentation product *via* direct condensation or *via* its cyclic lactide form (l-, d-, or meso-lactide) (Mecking, 2004; Lichtenthaler, 2006) (Figure 11.4). PLA chemistry was investigated by Carothers in 1932 and since then it has been extensively studied by means of efficient

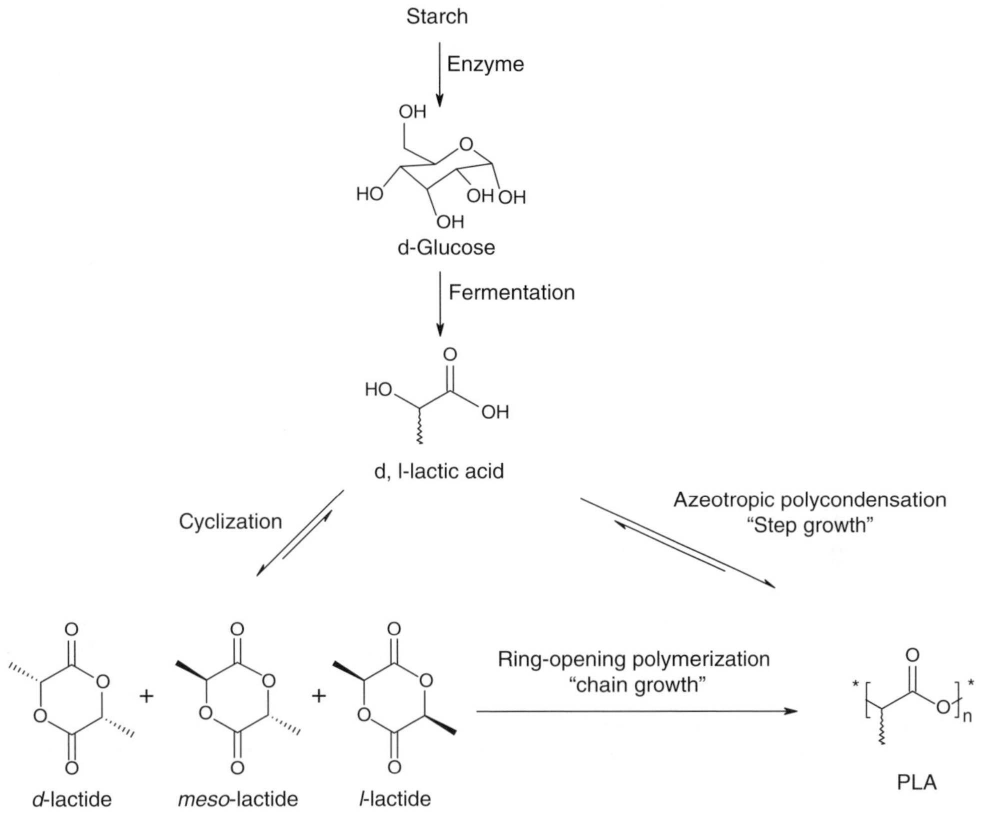

Figure 11.4 Synthesis of lactic acid, lactide and poly(lactic acid).

glucose fermentation, enantiomerically pure monomer and polymer synthesis, a wide range of catalyst systems, and final product properties and enhancements.

PLAs have been industrially fabricated into fibers, films, and surgical implants and sutures. Currently, most PLA is produced by Natureworks® (Dow-Cargill) in an amount of 136 000 tonnes per year in its plant in Nebraska, USA (Mooney, 2009). The advantageous properties of PLA include being renewable, biodegradable, recyclable, compostable, biocompatible, processable, and energy saving. Nevertheless, PLA has poor toughness, slow degradation, hydrophobicity, and lack of reactive side-chain groups, which needs to be improved (Rasal *et al.*, 2010).

Currently PLA is industrially synthesized *via* chemical pathways (e.g. ring-opening polymerization of lactide or solvent-based azeotropic condensation). Although azeotrotopic distillation overcomes the limitation of the molecular weight, a typical drawback of step-growth polymerization kinetics, to some extent, the ring-opening polymerization of lactide allows better control of the polymerization and remains by far the most widely used method for the synthesis of well-defined materials (Dechy-Cabaret *et al.*, 2004). In contrast, Yang and co-workers (2010) studied one-step fermentative *in vivo* production of PLA in *E. coli*. The results led to the biosynthesis of PLA and its copolymers with 3-hydroxybutyrate containing various lactate fractions within the range 8.7–64.4 mol%.

The polymerization rate, molecular weight control, and stereo control are the key parameters for PLA synthesis (Williams and Hillmyer, 2008). The physical, mechanical, and degradation properties of PLAs are strongly dependent on the chain stereochemistry. Isotactic PLLA is a highly crystalline polymer with a T_m of 170 °C (Zhong *et al.*, 2003). It has excellent mechanical properties and degrades rather slowly. Atactic poly(*rac*-lactic acid) is amorphous and degrades faster. Therefore, it is very important to control the chain stereochemistry for PLA homopolymers and copolymers. Much effort has focused on the development of new catalysts for the controlled polymerization of lactide, including chiral aluminum isopropoxides based on enantiopure or racemic cyclohexylsalen ligands (Jacobsen ligand), imidazole (Kricheldorf *et al.*, 2008), aluminum-methyl complexes supported by tetradentate phenoxy-amine ligands (Tang and Gibson, 2007), Group 3 metal complexes supported by dianionic alkoxy-amino-bisphenolate ligands (Amgoune *et al.*, 2006), primary and secondary alcohol adducts of 1,3-dimesitylimidazolin-2-ylidene (Csihony *et al.*, 2005), as well as tin(II) complexes supported by β-diketiminate ligands (Dove *et al.*, 2006). Moreover, Lohmeijer and co-workers (2006) reported some guanidine and amidine organocatalysts for ring-opening polymerization of lactide, which avoid heavy metal contaminants from catalyst residues in the polymer. It was shown that 1,5,7-Triazabicyclo[4.4.0]dec-5-ene (TBD), *N*-methyl-TBD (MTBD), and 1,8-diazabicyclo[5.4.0]-undec-7-ene (DBU) are very efficient catalysts for ring-opening polymerization with excellent control of molecular weight and polydispersity.

Since PLAs are generally very brittle and exhibit lower elongation at break, they need to be toughened for some applications. Anderson and co-workers (2008) reviewed the toughening methods as adjusting stereochemistry, crystallinity, and processing, plasticization, and blending with other degradable or non-degradable polymers. For instance, when poly(l-lactic acid)(PLLA) is toughened with polymerized soybean oil, the blends exhibit a tensile toughness as high as four times and strain at break values as high as six times greater than those of unmodified PLLA (Robertson *et al.*, 2010).

In 2008 Pounder and co-workers took advantage of thiol-ene reactions, which are considered to be one of the click reactions, and used a furan protected alcohol functional maleimide as an initiator for the polymerization of lactide. After polymerization and deprotection, some aromatic and aliphatic thiol compounds with various functionalities were reacted to yield PLAs with different chain-end functionalities under very mild conditions without any degradation. Another interesting approach to synthesize PLA grafted with poly(ethylene oxide)(PEO), containing various repeating ethylene oxide units, was reported by Jiang and co-workers in 2008. Mono-methoxy PEGs were functionalized with a lactic acid moiety through a three-step process and the subsequent cyclization of these lactic acid moieties into lactide was carried out with the aid of an organic acid. Finally, this adduct was subjected to ring-opening polymerization using $Sn(2\text{-ethylhexanoate})_2$, a very common catalyst for PLA synthesis. The resulting polymers, containing one and two PEO repeat units in the pendant chain, were hydrophilic but not water soluble. Those having longer PEO chains were water soluble. Moreover, a lower critical solution temperature was detected for the two polymers (19 and 37 °C, respectively) indicating their thermo responsive behavior.

PLA can also be used for medical applications owing to its biocompatibility and biodegradability. PLA was shown to degrade *via* hydrolytic erosion (Suggs *et al.*, 2007), in which the mechanical properties remain similar during the degradation progress. The degradation rate and mechanism, as well as other properties of PLA, depend on molecular weight and distribution, pH, T_g, crystallinity, and temperature. For further enhancement of the physical and mechanical properties, PLA was copolymerized with glycolic acid, resulting in a more hydrophilic and highly crystalline polymer. Interestingly, no linear relationship between the composition and degradability was reported; hence, copolymers with high or low comonomer ratios are less sensitive to hydrolysis than copolymers with equal ratio, due to their greater crystallinity (Reed and Gilding, 1981). Another medical application of PLA and its copolymers, owing to their tissue compatibility, is the preparation of scaffolds for tissue engineering (Kayaman-Apohan *et al.*, 2001).

11.2.4 Polyhydroxyalkanoates

Polyhydroxyalkanoates (PHA) are a form of carbon and energy storage in the cytoplasm of most of the bacteria belonging to the Halobacteriaceae family (Steinbüchel and Füchtenbusch, 1998; Flieger *et al.*, 2003). After the destruction of the cell walls, up to 90% poly(3-hydroxybutyrate) (PHB) polymer, the most known class of PHAs, can be isolated (Mecking, 2004). These polymers are biodegraded by microbes within five to six weeks. During this period they can be metabolized, both in the presence and absence of oxygen, into carbon dioxide and water or methane, respectively (Shimao, 2001).

PHAs can be used for the manufacture of films, coated paper, compost bags and can also be molded into bottles and razors (Mooney, 2009). Additionally, since they are biocompatible, they can also be used as implants without causing inflammation (Ramakrishna *et al.*, 2001).

Copolymers of PHAs are more useful for the industry, since they exhibit lower crytallinity, easy processability, and more flexible final product properties. The final PHA polymer properties can thus be adjusted from semi-crystalline to elastomeric plastics. The most common instance is poly(3-hydroxybutyrate-co-3-hydroxyvalerate) (PHBV). The relative content of hydroxybutyrate and hydroxyvalerate monomers can be adjusted by the amount of the valine added to the glucose-based culture medium (Eschenlauer *et al.*, 1996). The melting point of PHB (180 °C) is lowered to 137 °C by 25 mol% of hydroxyvalerate units (Mecking, 2004). PHBV was first commercialized by ICI (Imperial Chemical Industries plc) in the late 1980s as Biopol® and its production was stopped in 1998 after it was sold to Monsanto in 1996. Currently, the Biomer Company in Münich, Germany, produces about 100 kg of PHB per cubic meter of fermentation medium; a white powder with 98% purity can be collected after extraction (Figure 11.5).

PHAs require expensive bacterial fermentation and isolation processes resulting in more expensive production costs if compared to other petroleum-derived polymers. Thus, in the beginning of the 2000s, an alternative strategy for lowering the production costs proposed to

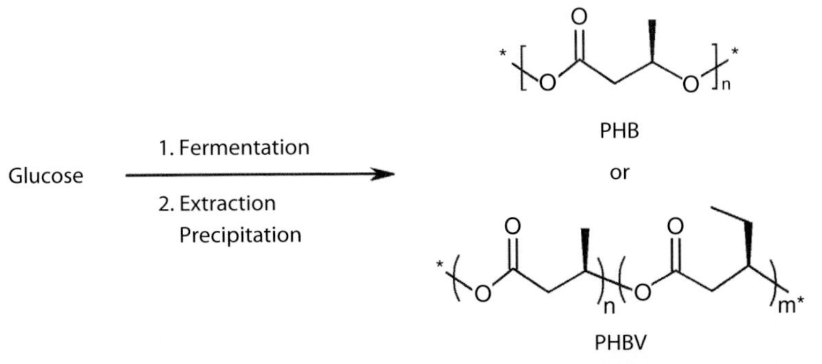

Figure 11.5 Synthesis of PHAs.

develop transgenic plants (genetically modified plants) that produce PHAs (Mittendorf *et al.*, 1999; Snell and Peoples, 2002). Presently, a variety of PHAs can be synthesized in many transgenic plants such as tobacco, rapeseed, cotton, alfalfa, flax, and oil palm (Yunus *et al.*, 2008).

Moreover, alternative carbon sources for bacterial fermentations to synthesize PHAs are also under investigation. These include fatty acids (Tan *et al.*, 1997; Chakraborty *et al.*, 2009) and glycerol (Ibrahim and Steinbüchel, 2009). With the aid of genetically-modified bacterial strains a direct utilization of plant oils, for example bacteria with lipase activity to utilize triacylglycerols, has also been described (Marsudi *et al.*, 19988). Furthermore, it was shown that by varying the fatty acid feed composition, the properties of the plastic material can be tuned (Lemos *et al.*, 2006). Additionally, low-grade coal liquefaction products could also be used as carbon source, since low-grade coal exists in high amounts and has no value for common industrial processes (Füchtenbusch and Steinbüchel, 1999).

11.2.5 Polymers from chitin or chitosan

Chitin is a polysaccharide found in the outer skeleton of crustaceans, insects, crabs, shrimp, lobsters, in the internal structures of other invertebrates, and in the cell walls of fungi. Chitin (poly-N-acetyl-d-glucosamine, 2-acetamido-2-deoxy-1,4-β-d-glucan) (Figure 11.6) is one of the three most abundant polysaccharides in nature, is biodegradable, non-toxic, and readily biocompatible.

Chitosan is also a natural product and can be obtained *via* (partial) deacetylation of chitin, most commonly by aqueous alkali, although this N-deacetylation is almost never complete

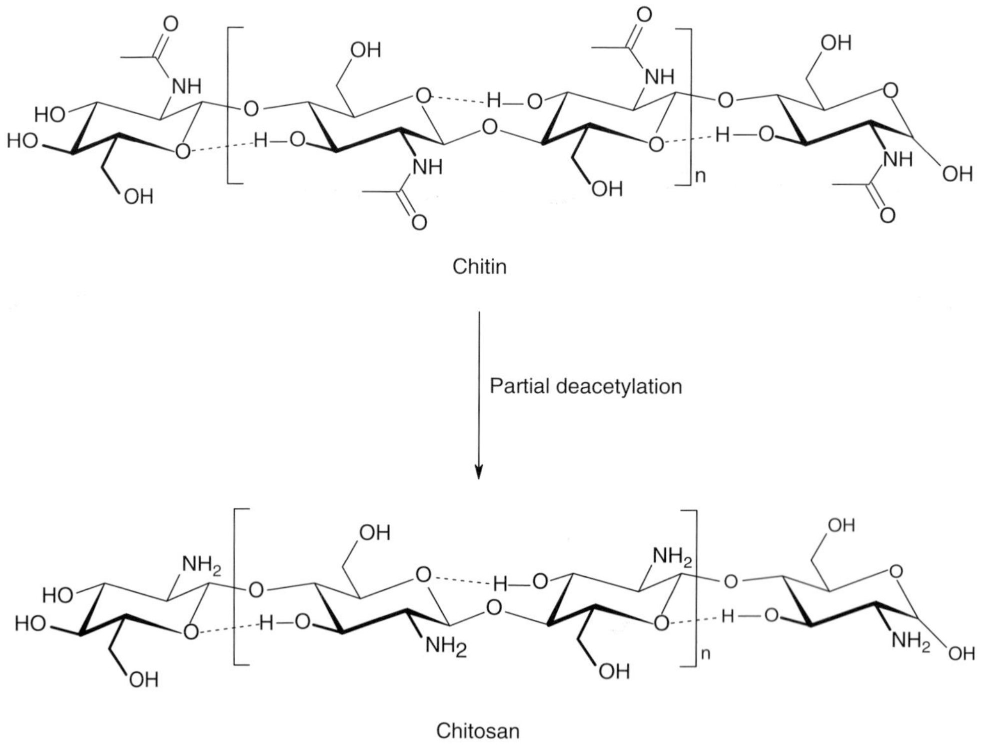

Figure 11.6 Chemical structures of chitin and chitosan.

(Figure 11.6). Chitin is insoluble in aqueous solutions at neutral pH, but N-deacetylation increases the aqueous solubility of the polymer also providing reactive primary amines for chemical modification as the molecular weight reduces from 1000–2500 to 100–500 kDa during the deacetylation process.

The improved solubility of chitosan enables the synthesis of biomaterial conjugates by grafting of synthetic hydrophobic polymers, which induce amphiphilic pH sensitive and thermally sensitive properties (Baldwin and Kiick, 2009).

Preparation methods, physicochemical properties, and applications of chitosan have been reviewed by Wani and co-workers (2010). Moreover, chitooligosaccharides, which are chemically or enzymatically derived chitosan oligomers, have been found to exhibit various bioactivities, although the mechanisms remain poorly understood (Aam *et al.*, 2010). Chitin is degraded into its monomer, N-acetyl glucosamine, by chitinase enzymes derived from plants, fungi, bacteria, insects, and fish (Flieger *et al.*, 2003). The production of microbial chitinases, for example the production of chitinases by genetically engineered organisms and by incorporating modern fermentation techniques, were summarized by Felse and Panda (2000). More recently Sahoo and co-workers (2009) reviewed the chemical modification, depolymerization processes, and the use of chitin and chitosan in biomedical, food/nutritional, material, microbiological, immunological, and other miscellaneous applications. In addition, just like the aforementioned carbohydrates, chitosan films also show electro-activity when doped with $NaClO_4$, NaI, $LiCF_3SO_3$, and $LiCH_3CO_2$ at varying concentrations depending on salt type and amount, and hydrated films with 10^{-4} S/cm conductance (Finkenstadt, 2005). On the other hand, the potential of chitin and chitosan for arsenic removal from groundwater and the use of these natural polymers for arsenic trioxide delivery in tumor therapy were described. The efficacy was attributed to the hydroxyl and amine functionalities acting as metal scavengers (Da Sacco and Masotti 2010).

11.3 FAT- AND OIL-BASED POLYMERS

The relatively low cost, ready availability, renewability, and the potential biodegradability of materials derived from plant oils make vegetable oils advantageous starting materials for many applications. This class of renewable raw materials possesses great potential as a sustainable resource for the polymer industry, since naturally occurring fatty acids can be exploited for monomer and polymer synthesis without many reaction steps. The direct use of plant oils without chemical modification or further functionalization most often leads to cross-linked structures (thermosets, coatings, resins). Moreover, thermoplastic materials can be prepared from fatty acids and their derivatives (obtained by transesterification of the triglycerides) having linear and hyperbranched architectures and resulting in polymers with tunable properties.

11.3.1 Polymers from triglycerides

Plant oils mainly comprise triacylglycerides that can be directly used for the synthesis of a variety of polymers. For instance, they have been used in the synthesis of coatings, often avoiding additional costs and time associated with the modification of the starting materials (Derksen *et al.*, 1995). A wide range of polymerization methods, including condensation, radical, cationic, and metathesis procedures, have been investigated. The scope, limitations, and possibility of utilizing these methods for polymer production from triacylglycerides have been reviewed by Güner and co-workers (2006).

11.3.1.1 *Polymers from soybean oil*

Soybean oil is a vegetable oil dominating today's food oil market. About 80% of the soybean oil produced each year is used for human food; another 6% is used for animal feed, while the remainder (14%) serves non-food uses (soaps, fatty acids, lubricants, coatings, etc.)(Valverde *et al.*, 2008). Another increasingly important non-food use of soybean oil is biodiesel production (in 2007 about 80% of the vegetable oil used to produce biodiesel was soybean oil in the United States).

Soybean oil is less expensive than corn, safflower, and sunflower oils. It also has a higher level of unsaturation (typical composition: 53% linoleic, 21% oleic, 8% linolenic, 10% palmitic, and 5% stearic) (Meier *et al.*, 2007) compared to some other vegetable oils. Crude soybean oil contains approximately 95–97% triacylglycerides, making it a potential candidate as a renewable macromonomer for the polymer industry.

The double bonds present on the fatty acid chains can undergo cationic or radical polymerization processes. The reactivity towards different polymerization techniques depends on the number and position of the double bonds; hence conjugated double bonds are more reactive. Henna and co-workers (2007) prepared a copolymer of conjugated low-saturation soybean oil, acrylonitrile, and either divinylbenzene or dicyclopentadiene *via* free-radical polymerization. The resulting transparent yellow polymers exhibited 10% weight loss in the dicyclopentadiene and divinylbenzene at 402–428 and 370–391 °C, respectively. Complete conversion of conjugated low-saturation soybean oil was achieved when the oil concentration was kept between 40 and 65%.

An important application of soybean oil is its conversion to polyols and use for PU foam synthesis (Figure 11.7).

Soybean oil was converted into polyols to improve its reactivity with isocyanate and used to synthesize PUs *via* hydroformylation and subsequent hydrogenation reactions. John and co-workers (2002) used three different polyols made from soybean oil triacylglycerides for PU synthesis with TDI and MDI and studied their reactivities and foam formation, and

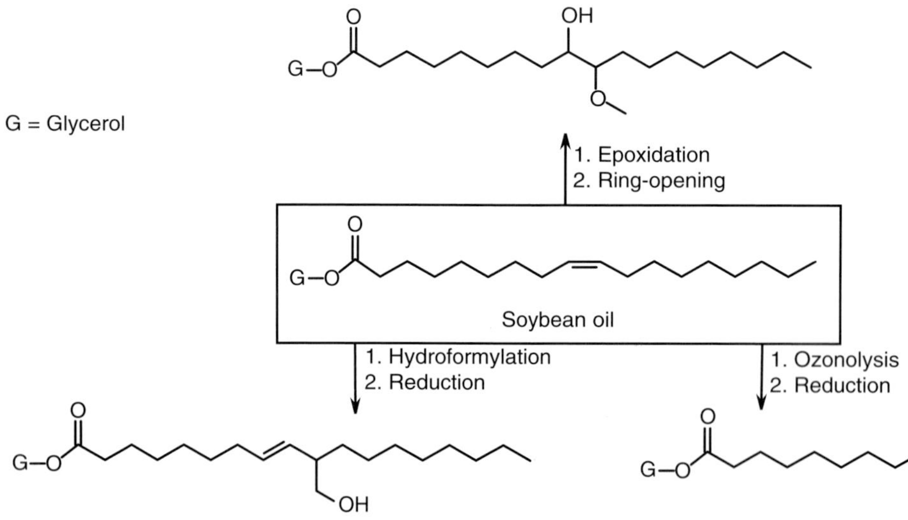

Figure 11.7 Different ways to prepare polyols for PU synthesis from soybean oil (G: glycerol moiety with two more fatty acid derivatives attached).

physical properties. The results revealed that soy polyols react similarly to synthetic polyols, and the foams prepared with MDI were more rigid than those synthesized with TDI. Guo and co-workers (2002) reported that the rhodium-catalyzed hydroformylation resulted in a polyol with 95% conversion producing rigid polyurethane, while the cobalt-catalyzed reaction gave a polyol with 67% conversion leading to a material with lower mechanical strength. Recently, a similar route, but with partial esterification of the hydroxyl groups with formic acid, was also examined (Petrović *et al.*, 2008). In addition to hydroformylation, soybean oil epoxidation followed by ring opening with methanol (Guo *et al.*, 2006; Lu and Larock, 2008) or phosphoric acid (Dwan'Isa *et al.*, 2003) and ozonolysis (Petrović *et al.*, 2005) are other methods for synthesizing polyols.

The soybean oil-based polyols were studied in detail in order to establish structure-property relationships of the polyurethanes produced thereof. The polyols synthesized through hydroformylation, bearing primary alcohols, showed higher reactivity towards PU formation and resulted in less rigid PU materials than those obtained through epoxidation. On the other hand, the polyols synthesized from soybean oil *via* ozonolysis, free from long pendant alkyl chains, resulted in PUs which displayed better mechanical properties and higher T_g. A different approach to synthesize PUs from soybean oil excluding isocyanates is the cyclic carbonation of epoxidized soybean oil (ESBO) (Türünç *et al.*, 2008) (Figure 11.8), a successfully commercialized product, and curing with primary amines to yield β-hydroxy urethanes (Tamami *et al.*, 2003; Javni *et al.*, 2008). Additionally, thermal and mechanical properties of these polyurethanes were enhanced by incorporation of silica nano-particles *via* nano-composite formation (Türünç *et al.*, 2008).

Eren and co-workers (2003) modified soybean oil *via* anhydride functionalization of the double bonds and polymerized the resulting monomers by polycondensation with low molecular weight polyols and long diols (Figure 11.8). These polyesters, which were resilient soft rubbers at room temperature, could find application as adhesives, film formers, textile and paper sizes, and tackifiers.

In 2008, Biswas and co-workers reviewed nitrogen containing monomers, for example fatty amines, fatty amides, fatty imidazolines, and polymers (PUs) made from soybean oil-based

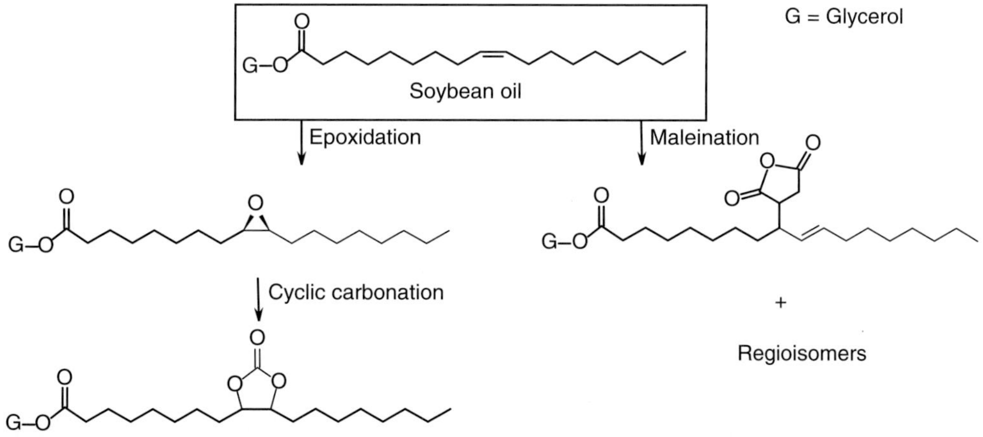

Figure 11.8 Modification possibilities of soybean oil (G: glycerol moiety with two more fatty acid derivatives attached).

triacylglyceride oils, fatty acid esters, and methyl esters of triacylglyceride oils. Among these products, fatty amines and amides are industrially significant and widely used.

11.3.1.2 *Polymers from castor oil*

The castor oil plant (*Ricinus communis*) is a native of tropical Asia and Africa. The castor oil extraction and refining processes have been reviewed by Ogunniyi (2006) and, more recently, by Mutlu and Meier (2010). The fatty acids generally consist of up to 90% ricinoleic, 4% linoleic, 3% oleic, 1% stearic, and less than 1% linolenic acids. The high content of ricinoleic acid is the reason for the high value of castor oil and its versatile application possibilities in the chemical industry. Moreover, the hydroxyl functionality of ricinoleic acid, which makes castor oil a natural polyol, provides oxidative stability to the oil, and a relatively high shelf life compared to other oils by preventing the formation of peroxides. Furthermore, the hydroxyl group allows a variety of chemical reactions to be performed on ricinoleic acid and its derivatives, including dehydration, halogenation, alkoxylation, esterification, and sulfation. Castor oil is already widely used for industrial chemical products like paints, coatings, inks, lubricants, and so on.

Castor oil is often used as a polyol and reacted with various diisocyanates in order to produce PU products, ranging from coatings, cast elastomers, thermoplastic elastomers, rigid foams, semi-rigid foams, sealants, adhesives to flexible foams. Moreover, Moeini (2007) reported production of novel green polyether-ester urethane and insulating coatings based on polyols derived from glycolyzed polyethylene terephtalate (PET), castor oil, and adipic acid, which exhibited excellent mechanical properties and electrical insulation characteristics. Since castor oil has a low hydroxy functionality, transesterification with glycerol can improve the properties and yield rigid PU foams with good physicomechanical properties (Petrović *et al.*, 2009). Furthermore, with the addition of trimethylol propane and pentaerythritol, an even higher thermal stability was achieved (Valero *et al.*, 2008). Karak and co-workers (2009) reported the synthesis of castor oil-modified hyperbranched thermoplastic PU. These materials were derived from castor oil, a macroglycol (e.g. poly-ε-caprolactone, (PCL)-diol or PEG), and MDI with or without a chain extender. Castor oil-derived PUs are also of interest for composite materials. For instance, castor oil/graphite composites were studied as electrode materials (Mendes *et al.*, 2002). Moreover, when doped with sulfonated polyaniline, electrically conductive PUs were synthesized from castor oil (Amado *et al.*, 2006).

Castor oil has found application in the synthesis of interpenetrating polymer networks (IPNs). These materials can be defined as a combination of two polymer networks, at least one of which is synthesized and/or cross-linked in the immediate presence of the other. They are called semi-IPN if just one of the polymers is a network (Athawale *et al.*, 2003). Early reports on castor oil IPNs appeared in 1977 by Yenwo and co-workers. The report discussed the synthesis possibilities *via* cross-linking of double bonds with sulfur, reaction of hydroxyl groups with diisocyanates, and emulsion polymerizations with saponified ricinoleic acids as emulsifier. Moreover, the IPNs from acrylic polymers, such as polymethyl methacrylate and poly-2-ethoxyethyl methacrylate, and castor oil-based polyurethanes were reported to contribute to the final properties of the material (Cunha *et al.*, 2004; Sanmathi *et al.*, 2004). Incorporation of acrylic moieties into the PU networks increased toughness and thermal properties. In contrast, IPN polyesters derived from castor oil and dibasic acids (e.g. malonic, succinic, glutaric, adipic, suberic, and sebacic acid) were obtained as soft and opaque elastomers (Suthar *et al.*, 2003).

Mohapatra and co-workers (1996) synthesized a castor-oil based PU *via* isocyanates and a novolac resin to produce semi-IPNs. Various PUs from diphenylmethane diisocyanate, hexamethylene diisocyanate, and isophorone diisocyanate, and novolac resins from cardanol, furfural, and *p*-aminobenzoic acid were copolymerized to yield thermally stable semi-IPNs. It was shown that the stability of the networks increased with the increase of the novolac resin content. The semi-IPNs prepared from castor oil-based PUs from iso-phorone as well as diphenylmethane diisocyanates and the cardanol-based resins were reported to withstand very high temperatures before decomposition.

The double bonds present on ricinoleic acid triglycerides are inactive to free-radical polymerization. To overcome this, Wang and co-workers (2008) prepared a half-ester of castor oil *via* maleation and copolymerized this monomer with styrene as reactive diluent and a free radical initiator. Polymers with different castor oil/styrene ratio and curing methods were prepared and optimized to yield biodegradable plastic foams with tuned rigidity. In addition to being more cost effective and bio-based, the foam plastics had a comparable compressive stress at 25% strain to commercial semi-rigid PU foams based on fossil oil. It was found that higher styrene content resulted in higher compressive stress but decreased biodegradation rate of the product.

11.3.1.3 *Polymers from other oils*

Although plant oil-based polymers are mainly derived from soybean and castor oil, an ever increasing number of publications are devoted to the use of other plant oils, such as linseed oil, tung oil, canola oil, high oleic sunflower oil, and many others.

Kundu and Larock (2005) copolymerized conjugated (87%) linseed oil, which is more reactive towards free radical reactions, with styrene and divinyl benzene (3–5%) by gradual heating. Soxhlet extractions of the products showed that only 35–85% of the oil could be incorporated into the cross-linked thermosets. The dynamic mechanical analysis of these polymers indicated that they were phase separated, as two separate T_gs appeared; these materials presented a soft rubbery phase with a sharp T_g at –50 °C and a hard brittle plastic phase with a broadened T_g at 70–120 °C. These polymers were shown to be stable below 350 °C. Moreover, the same group (Henna *et al.*, 2007) used 100% conjugated linseed oil and copolymerized it with acrylonitrile and DVB using the thermal initiator 2,2' -azobis isobutyronitrile (AIBN). Much better oil incorporation (61–96%) was revealed by the soxhlet extractions. Carter and co-workers (2008) investigated commercially available epoxidized linseed oil (ELO) as an attractive alternative to poly(vinyl chloride) (PVC) flooring products. Upon screening of the process parameters and extensive flooring investigations, it was concluded that the ELO system was suitable as a flooring product with compatible performance to commercial PVC-based flooring material.

Ring-opening metathesis was also applied to vegetable oil derivatives. A commercially available monomer containing a five-membered cyclic mono-ene structure, Dilulin, is produced *via* Diels-Alder reactions by subjecting linseed oil with cyclopentadiene to high temperature and pressure. Henna and co-workers (2008) performed the ring-opening metathesis-copolymerization of Dilulin with dicyclopentadiene with the aid of Grubbs second generation catalyst. The resulting bio-based polymers offer unique and promising properties, encouraging the replacement of petrochemical-based materials in some applications, although sometimes fiber reinforcement or other fillers are needed to further enhance the mechanical properties. Casado and co-workers (2009) utilized tung oil *via* epoxidation-ring opening and increased hydroxyl functionality with triethanol amine insertion, yielding a highly reactive tung oil-based polyol for PU synthesis. Incorporation of 10–15% pine wood flour to prepare PU composites led to a stronger material that not only showed improved tensile strength, but also improved impact behavior, attributed to very good interfacial adhesion. A different approach to synthesize polyols for PU synthesis from canola oil was utilized by Petrović and co-workers (2005). Ozonolysis of canola oil yielded polyols with terminal primary hydroxyl functionalities in a similar fashion as already discussed for soybean oil.

High oleic sunflower oil (HOSO) contains a minimum of 80% (up to 93%) oleic acid that makes it especially attractive for certain synthetic applications. Lluch and co-workers (2010) synthesized a triacylglyceride derivative containing allylic alcohol groups *via* direct photo-peroxidation of HOSO in the presence of singlet oxygen followed by reduction with sodium borohydride. This derivative was acrylated and cross-linked *via* radical polymerization. The extensive hydrogenation of this derivative, prior to acrylation and radical polymerization, led to an improvement of the thermal stability of the cross-linked polymers. In both cases, highly cross-linked thermosets were obtained.

11.3.2 Polymers from fatty acids

Today, plant oils represent the major renewable resource for the chemical industry, offering broad possibilities for polymer production (Metzger and Bornscheuer, 2006). Plant oils offer a number of fatty acids with different chain lengths and functional groups as well as different numbers and positions of C=C double bonds that can be transesterified with an alcohol, mainly methanol (Figure 11.9). Since plant oils consist mainly of triacylglycerides of various saturated and unsaturated fatty acids, it is difficult to synthesize structurally well-defined and property-tailored polymers. As a result, fatty acids serve as valuable

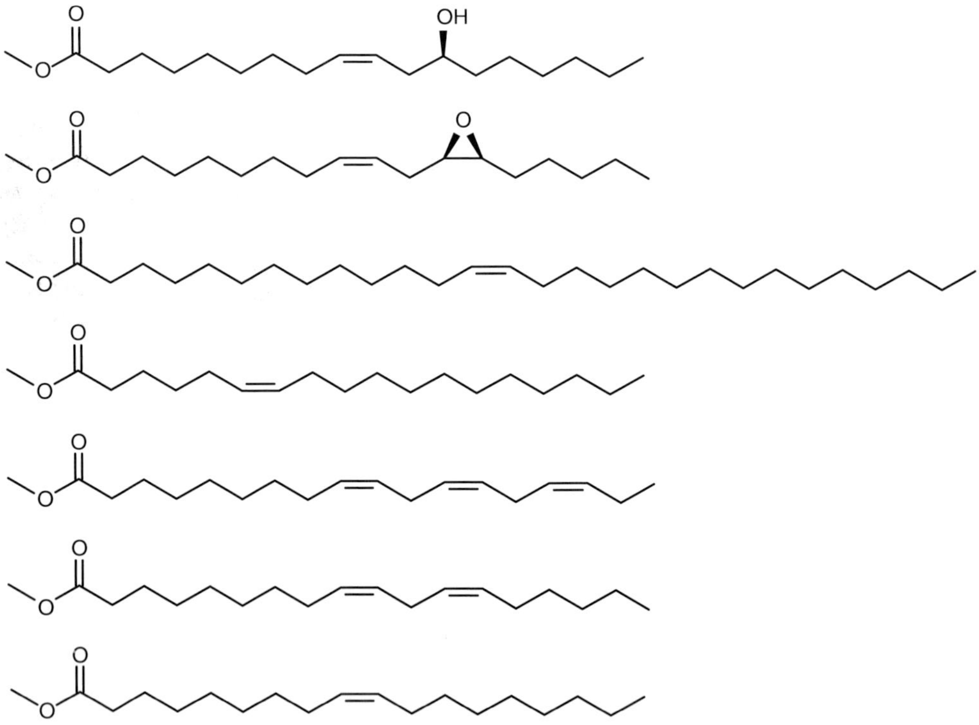

Figure 11.9 Chemical structure of the most widely applied fatty acid methyl esters.

monomers in the formation of linear monomers, (co)polymers, cross-linked and interpenetrating polymers and composite materials for the chemical industry.

Fatty acids and fatty acid methyl esters (FAME) can be used directly or after functionalization as monomers for the synthesis of a variety of polymeric materials. The most important functionalization possibilities of the double bonds and the ester groups have been extensively reviewed in the literature (Biermann *et al.*, 2001, 2007; Biermann and Metzger 2004).

In the case of ricinoleic acid, which has hydroxyl functionality on the aliphatic chain, it is easy to introduce functional groups suitable for polymerization leading to fatty acid-based biodegradable polymers. Domb and Nudelman (1995) utilized succinic and maleic anhydride (Figure 11.10) to modify ricinoleic acid and subsequently copolymerize the obtained monomers with sebacic acid to yield a polyanhydride *via* condensation with acetic anhydride. These polymers were shown to be useful as drug carriers and they are biocompatible and degrade within 4–6 weeks both *in vivo* and *in vitro*. Ricinoleic acid was also used for macrolactone formation, but subsequent ring-opening copolymerization with lactide to produce 100% renewable polyesters was found to be difficult, yielding low molecular weight polymers and oligomers (Figure 11.10). This behavior was attributed to the low reactivity of ricinoleic acid macrolactones (Slivniak and Domb, 2005).

In 2008, Sokolsky-Papkov and co-workers prepared copolymers of ricinoleic acid lactones and lactide then evaluated them as biomaterials for *in situ* formation of injectable gels. Copolymerization with lactide resulted in molecular weights in the range of 5–16 kDa with melting temperatures between 100 and 130 °C for copolymers containing 10–50% ricinoleic acid.

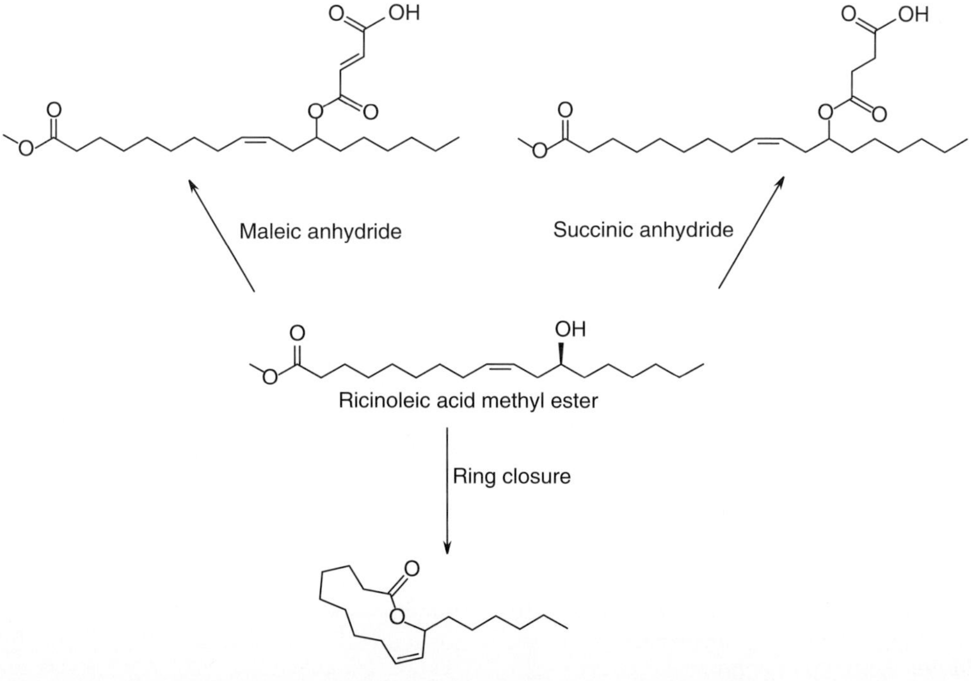

Figure 11.10 Modification possibilities of ricinoleic acid.

Since fatty acids have only one carboxylic acid functionality, different functionalization or polymerization routes were studied. Warwel and co-workers (2001) synthesized ω-olefinic fatty acids *via* cross-metathesis of ethylene and FAME catalyzed by heterogeneous rhenium or homogeneous ruthenium catalysts. These ω-olefinic fatty acids were then converted to various polymers including polyolefins, polyesters, polyethers, and polyamides. Moreover, a number of metathesis catalysts and reaction conditions were investigated in order to optimize the cross-metathesis of FAMEs for high conversion and selectivity, with the goal to synthesize new polycondesation monomers from FAMEs (Jacobs *et al*., 2009; Rybak and Meier, 2007, 2008). Furthermore, dimers of fatty acids have been exploited as renewable di-functional monomers for linear polymer synthesis and have become commercially available products, although they are ill-defined and also contain monomers and trimers to some extent. A class of these polymers is dimer-acid derived polyamides. Wang and co-workers (2009) polymerized a commercial dimer acid synthesized from C-18 fatty acids and different aromatic diamines *via* melt polycondensation. The physical and mechanical properties of the resulting polyamides were found to be similar. The heat resistance and mechanical properties of poly (4,4′-diphenylsulfone dimeramide) and poly(4,4'-diphenyl dimeramide) were found to be comparable to those of PA 12,12. Moreover, modified FAMEs were also used as reactive diluents for the replacement of petroleum-based compounds. Campanella and co-workers (2009) acrylated an epoxidized FAME (AFAME) and utilized this compound as a replacement for styrene and studied the thermomechanical properties of the resulting materials. The study revealed that the use of AFAME introduced the advantage of reducing hazardous air pollutants (HAPs) and volatile organic compounds (VOCs), but this process was not industrially viable due to cost issues, viscosity, and thermo-mechanical properties. Instead, blends of styrene and AFAME (17.5% of each) resulted in viscosities that met the industry requirements for composites, T_g (~100 °C), and storage modulus (at 30 °C ~1200 MPa), comparable with petroleum-based commercial resins.

10-Undecenoic acid (10-UA), a platform chemical obtained *via* pyrolysis of ricinoleic acid, offers various opportunities as polymer building blocks (Mutlu and Meier, 2009). The terminal double bond of this fatty acid is successfully used for thiol-ene addition reactions (Hoyle *et al*., 2004) as well as acyclic diene metathesis (ADMET) reactions (Baughman and Wagener, 2005). Recently, methyl 10-undecenoate was functionalized with 1-thioglycerol, mercaptoethanol, and butane-1,4-dithiol. The resulting monomers were further polymerized *via* polycondensation with TBD, a guanidine-type organic base catalyst (Türünç and Meier, 2010) (Figure 11.11). The synthesized linear and hyperbranched polyesters did not exhibit any important weight-loss below 300 °C and showed thermal properties similar to commercially available polyesters. Lluch and co-workers (2010) synthesized the allyl ester of undecenoic acid and polymerized it *via* UV-initiated thiol-ene addition reactions. The one-pot reaction yielded a series of well-defined telechelics with molecular weights in the range of 1000–3000 g/mol and hydroxyl, carboxyl, or trimethoxysilyl end-groups. The telechelics with hydroxy end groups were then reacted with a diisocyanate compound producing bio-based multi-block polyester urethanes.

The use of metathesis reactions as a versatile tool in fatty acid-based monomer synthesis was studied in detail by Meier's group (Rybak *et al*., 2008; Meier, 2009) (Figure 11.11). A comparative study of dimerization of methyl 10-undecenoate *via* self-metathesis (Djigoué and Meier, 2009) and subsequent polycondensation with different diamines was reported by Mutlu and Meier in 2009.

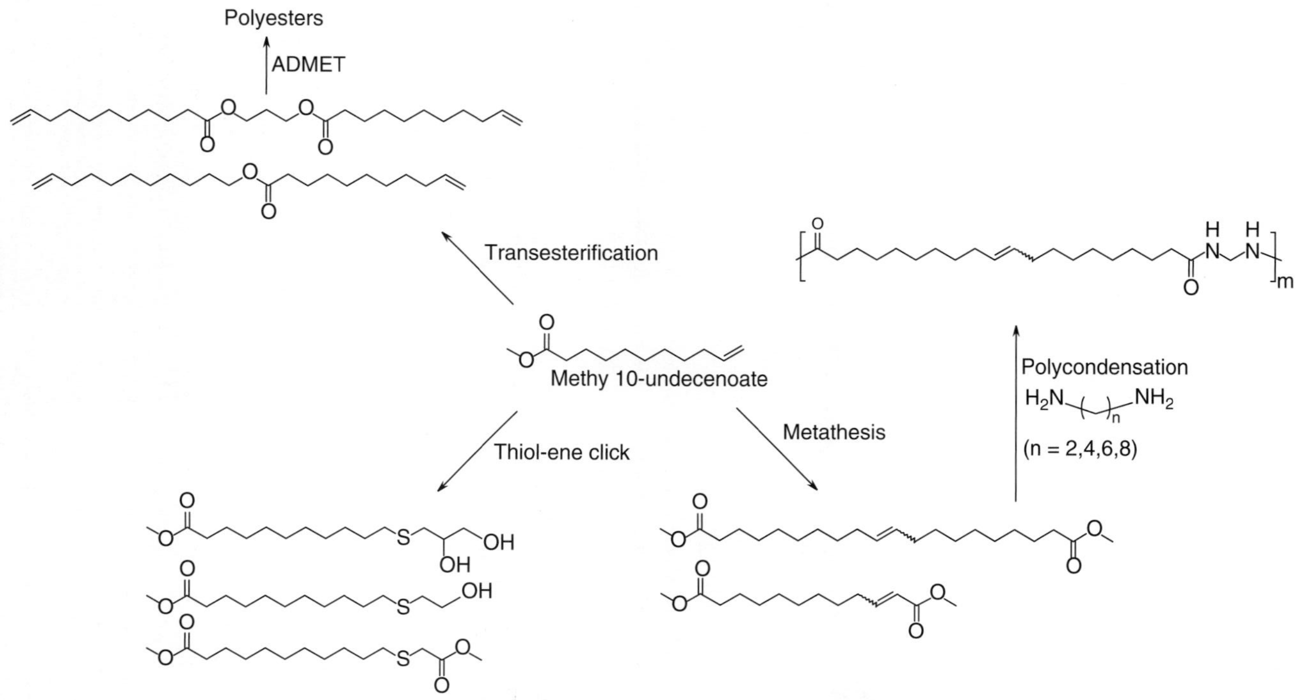

Figure 11.11 Use of methyl 10-undecenoate as a sustainable building block for the polymer industry.

The synthesis of α,ω-diene monomers is suitable for the synthesis of fatty acid-based biodegradable polyesters *via* ADMET polymerization (Rybak and Meier, 2008; Fokou and Meier, 2009, 2010). Since the material properties of the polymers strongly depend on the polymer structure, it is important to establish and suppress the olefin isomerization trend observed during ADMET polymerizations. Recently, an α,ω-diene monomer from methyl 10-undecenoate and 1,3-propanediol was first ADMET polymerized; then transesterification (with MeOH) products of this polymer were analyzed by GC-MS to establish the degree of isomerization (Fokou and Meier, 2010). The results revealed that the use of benzoquinone can prevent olefin isomerization efficiently, leading to the preparation of polymers with well-defined repeating units *via* ADMET of functional group containing monomers. Moreover, ADMET polymerization of fatty acid-derived dienes is also applicable to phosphorous-containing monomers, leading to flame retardant renewable polymers (Montero de Espinosa *et al.*, 2009, 2010). Linear polymers with different phosphorous contents were prepared by copolymerization of a phosphorous and a non-phosphorous containing monomer, revealing that approximately 3% phosphorous content resulted in a limiting oxygen index (LOI) of 23.5, whereas an LOI of 19.0 was obtained for the polymer without phosphorous. Thus, the flame retardancy was significantly increased. In case of the cross-linked polymers, the LOI values increased from 18.4 to 25.7 with increasing phosphorous content from 0 to 3.8%.

11.4 CONCLUSION

As highlighted in this chapter there are encouraging carbohydrate- and plant oil-based polymers that could substitute, at least partially, mineral oil-based materials which are in the market today. Although some renewable polymeric materials are already commercialized, others are still not economically feasible for large scale production. However, the declining fossil oil reserves might change the price advantage of petroleum-based polymers soon. Moreover, the advances in pure scientific and engineering knowledge on generating platform chemicals, polymer building blocks, and plastic materials from renewable resources will definitely help with use of sustainable resource utilization globally in the future. Considering the depleting fossil fuel reservoirs, the switch from petroleum-based products to renewable ones must be realized soon and these efforts should led by universities, industrial R&D departments, and local and federal governments.

REFERENCES

Aam BB, Heggset EB, Norberg AL, Sørlie M, Vårum KM, Eijsink VGH. 2010. Production of Chitooligosaccharides and Their Potential Applications in Medicine. Mar Drugs 8:1482–1517.

Amado FDR, Rodrigues LF, Forte MMC, Ferreira CA. 2006. Properties evaluation of the membranes synthesized with castor oil polyurethane and polyaniline. Polym Eng Sci 46:1485–1489.

Amgoune A, Thomas CM, Roisnel T, Carpentier JF. 2006. Ring-Opening Polymerization of Lactide with Group 3 Metal Complexes Supported by Dianionic Alkoxy-Amino-Bisphenolate Ligands: Combining High Activity, Productivity, and Selectivity. Chem Eur J 12:169–179.

Anderson KS, Schreck KM, Hillmyer MA. 2008. Toughening Polylactide. Polym Rev 48(1):85–108.

Arvanitoyannis I, Psomiadou E, Biliaderis CG, Ogawa H, Kawasaki N, Nakayama A. 1997. Biodegradable Films Made from Low Density Polyethylene (LDPE), Ethylene Acrylic Acid (EAA), PolyCaprolactone (PCL) and Wheat Starch for Food Packaging Applications: Part 3. Starch/Stärke 49(7/8):306–322.

Arvanitoyannis I, Biliaderis CG, Ogawab H, Kawasakib N. 1998. Biodegradable films made from low-density polyethylene (LDPE), rice starch and potato starch for food packaging applications: Part 1. Carbohyd Polym 1998 36:89–104.

Athawale VD, Kolekar SL, Raut SS. 2003. Recent Developments in Polyurethanes and Poly(acrylates) Interpenetrating Polymer Networks. J Macromol Sci Part C-Polym Rev 43(1):1–26.

Avella M, De Vlieger JJ, Errico ME, Fischer S, Vacca P, Volpe MG. 2005. Biodegradable starch/clay nanocomposite films for food packaging applications. Food Chem 93(3):467–474.

Bagheri R and Naimian F. 2002. Effect of photo-initiators on melt degradation of polypropylene and starch-filled polymer. Polym Int 51:907–911

Baldwin AD, Kiick KL. 2009. Polysaccharide-Modified Synthetic Polymeric Biomaterials, Peptide Sci 94(1):128–140.

Baughman TW, Wagener KB. 2005. Recent Advances in ADMET Polymerization. In: Buchmeiser M (editor). Advances in Polymer Science, Volume 176. Germany: Springer-Verlag GmbH. pp. 1–42.

Biermann U, Metzger JO. 2004. Catalytic C,C-Bond Forming Additions to Unsaturated Fatty Compounds. Topics in Catalysis 38:3675–3677.

Biermann U, Fürmeier S, Metzger JO. 2001. New Chemistry of Oils and Fats. In Oleochemical Manufacture and Applications (eds F.D. Gunstone, R.J. Hamilton). Sheffield Academic Press and CRC Press, ISBN 1-84127-219-1, pp. 266–299.

Biermann U, Butte W, Eren T, Haase D, Metzger JO. 2007. Regio- and Stereoselective Diels-Alder Additions of Maleic Anhydride to Conjugated Triene Fatty Acid Methyl Esters. Eur J Org Chem 2007(23): 3859–3862.

Biswas A, Willet JL, Gordon SH, Finkenstadt VL, Cheng HN. 2006. Complexation and blending of starch, poly(acrylic acid), and poly(N-vinyl pyrrolidone). Carbohyd Polym 65(4):397–403.

Biswas A, Sharma BK, Willett JL, Erhan SZ, Cheng HN. 2008. Soybean oil as a renewable feedstock for nitrogen-containing derivatives. Energy Environ Sci 1:639–644.

Bogoeva-Gaceva G, Avella M, Malinconico M, *et al.* 2007. Natural Fiber Eco-Composites. Polym Composite 28(1):98–107.

Buchanan CM, Gardner RM, Komarek R. 1993. Aerobic Biodegradation of Cellulose Acetate. J Appl Polym Sci 47(10):1709–1719.

Campanella A, La Scala JJ, Wool RP. 2009. The Use of Acrylated Fatty Acid Methyl Esters as Styrene Replacements in Triglyceride-based Thermosetting Polymers. Polym Eng Sci 49(12):2384–2392.

Carter DT, Stansfield N, Mantle RJ, France CM, Smith PA. 2008. An investigation of epoxidised linseed oil as an alternative to PVC in flooring applications. Ind Crop Prod 28:309–319.

Casado U, Marcovich NE, Aranguren MI, Mosiewicki MA. 2009. High-Strength Composites Based on Tung Oil Polyurethane and Wood Flour: Effect of the Filler Concentration on the Mechanical Properties. Polym Eng Sci 49(4):713–721.

Chuayuljit S, Hosililak S, Athisart A. 2009. Thermoplastic Cassava Starch/Sorbitol-Modified Montmorillonite Nanocomposites Blended with Low Density Polyethylene: Properties and Biodegradability Study. J Mater Min 19(1):59–65.

Chakraborty P, Gibbons W and Muthukumarappan K. 2009. Conversion of volatile fatty acids into polyhydroxyalkanoate by Ralstonia eutropha. J Appl Microbiol 106(6):1996–2005.

Csihony S, Culkin DA, Sentman AC, Dove AP, Waymouth RM, Hedrick JL. 2005. Single-Component Catalyst/Initiators for the Organocatalytic Ring-Opening Polymerization of Lactide. J Am Chem Soc 127:9079–9084.

Cunha FOV, Melo DHR, Veronese VB, Forte MMC. 2004. Study of castor oil polyurethane-poly(methyl methacrylate) semi-interpenetrating polymer network (SIPN) reaction parameters using a 23 factorial experimental design. Mater Res 7:539–543.

Da Sacco L and Masotti A. 2010. Chitin and Chitosan as Multipurpose Natural Polymers for Groundwater Arsenic Removal and As_2O_3 Delivery in Tumor Therapy. Mar Drugs 8:1518–1525.

Dechy-Cabaret O, Martin-Vaca B, Bourissou D. 2004. Controlled Ring-Opening Polymerization of Lactide and Glycolide. Chem Rev 104:6147–6176.

Derksen JTP, Cuperus FP, Kolster P. 1995. Paints and coatings from renewable resources. Ind Crop Prod 3:225–236.

Djigoué GB, Meier MAR. 2009. Improving the Selectivity for the Synthesis of two Renewable Platform Chemicals *via* Olefin Metathesis. Appl Catal A 368:158–162.

Domb AJ, Nudelman R. 1995. Biodegradable Polymers Derived from Natural Fatty Acids. J Polym Sci Part A: Polym Chem 33: 717–725.

Donnelly MJ, Stanford JL, Still RH. 1991. The Conversion of Polysaccharides into Polyurethanes: A Review. Carbohyd Polym 14(3):221–240.

Dove AP, Gibson VC, Marshall EL, Rzepa HS, White AJP, Williams DJ. 2006. Synthetic, Structural, Mechanistic, and Computational Studies on Single-Site β-Diketiminate Tin(II) Initiators for the Polymerization of rac-Lactide. J Am Chem Soc 128:9834–9843.

Dwan'Isa JPL, Mohanty AK, Misra M, Drzal T, Kazemizadeh M. 2003. Novel Biobased Polyurethanes Synthesized from Soybean Phosphate Ester Polyols: Thermomechanical Properties Evaluations. J Polym Environ 11(4):161–168.

Eschenlauer AC, Stoup SK, Srienc F, Somers DA. 1996. Production of Heteropolymeric Polyhydroxyalkanoate in *Eschericha Coli* from a single carbon source. Int J Biol Macmol 19:121–130.

Eren T, Küsefo lu S, Wool R. 2003. Polymerization of Maleic Anhydride-Modified Plant Oils with Polyols. J Appl Polym Sci 90:97–202.

Felix JM, Gatenholm P. 1993. Controlled Interactions in Cellulose-Polymer Composites. I: Effect on Mechanical Properties. Polym Composite 14(6):449–457.

Felse PA, Panda T. 2000. Production of microbial chitinases – A revisit. Bioprocess Eng 23:127–134.

Finkenstadt VL. 2005. Natural polysaccharides as electroactive polymers. Appl Microbio Biot 67(6): 735–745.

Flieger M, Kantorová M, Prell A, ezanka T, Votruba J. 2003. Biodegradable Plastics from Renewable Sources. Folia Microbiol 48(1):27–44.

Fokou PA, Meier MAR. 2009. Use of a Renewable and Degradable Monomer to Study the Temperature-Dependent Olefin Isomerization during ADMET Polymerizations. J Am Chem Soc 131:1664–1665.

Fokou PA, Meier MAR. 2010. Studying and Suppressing Olefin Isomerization Side Reactions during ADMET Polymerizations. Macromol Rap Com 31:368–373.

Fringant C, Rinaudo M, Gontard N, Guilbert S, Derradji H. 1998. A biogradable starch based coating to waterproof hydrophilic materials. Starch/Stärke 50(7):292–296.

Füchtenbusch B, Steinbüchel A. 1999. Biosynthesis of polyhydroxyalkanoates from low-rank coal liquefaction products by Pseudomonas oleovorans and Rhodococcus ruber. Appl Microbiol Biotechnol 52:91–95.

Gandini A. 2008 Polymers from Renewable Resources: A Challenge for the Future of Macromolecular Materials. Macromolecules 41(24):9491–9504.

Ge J, Zhong W, Guo Z, Li W, Sakai K. 2000. Biodegradable Polyurethane Materials from Bark and Starch. I. Highly Resilient Foams. J Appl Polym Sci 77(12):2575–2580.

Gindl W, Jeronimidis G. 2004. Wood pulp fiber reinforced melamine-formaldehyde composites. J Mater Sci 39(9):3245–3247.

Gomes ME, Ribeiro AS, Malafaya PB, Reis RL, Cunha AM. 2001. A new approach based on injection moulding to produce biodegradable starch-based polymeric scaffolds: morphology, mechanical and degradation behaviour. Biomaterials 22(9):883–889.

Guerra NP, Macías CL, Agrasar AT, Castro LP. 2005. Development of a bioactive packaging cellophane using Nisaplin® as biopreservative agent. Lett Appl Microbiol 40(2):106–110.

Güner FS, Ya cı Y, Erciyes AT. 2006. Polymers from triglyceride oils. Prog Polym Sci 31(7):633–670.

Guo A, Demydov D, Zhang W, Petrović ZS. 2002. Polyols and Polyurethanes from Hydroformylation of Soybean Oil. J Polym Environ 10(1–2):49–52.

Guo A, Zhang W, Petrović ZS. 2006. Structure-property relationships in polyurethanes derived from soybean oil. J Mater Sci 41(15):4914–4920.

Henna PH, Andjelkovic DD, Kundu PP, Larock RC. 2007. Biobased Thermosets from the Free-Radical Copolymerization of Conjugated Linseed Oil. J Appl Polym Sci 104:979–985.

Henna PH, Kessler MR, Larock RC. 2008. Fabrication and Properties of Vegetable-Oil-Based Glass Fiber Composites by Ring-Opening Metathesis Polymerization. Macromol Mater Eng 293:979–990.

Hoyle CE, Lee TY, Roper T. 2004. Thiol-enes: Chemistry of the past with promise for the future. J Polym Sci Part A: Polym Chem 42(21):5301–5338.

Ibrahim MHA, Steinbüchel A. 2009. Zobellella denitrificans strain MW1, a newly isolated bacterium suitable for poly(3-hydroxybutyrate) production from glycerol. J Appl Microbiol 108:214–225.

Jacobs T, Rybak A, Meier MAR. 2009. Cross-metathesis reactions of allyl chloride with fatty acid methyl esters: Efficient synthesis of α,β-difunctional chemical intermediates from renewable raw materials. Appl Catal A-Gen 353:32–35.

Jansson A, Järnström L, Rättö P, Thuvander F. 2006. Physical and Swelling Properties of Spray-Dried Powders made from Starch and Poly(vinyl alcohol). Starch/Stärke 58(12):632–641.

Javni I, Hong DP, Petrović ZS. 2008. Soy-Based Polyurethanes by Nonisocyanate Route. J Appl Polym Sci 108(6):3867–3875.
Jiang X, Smith MR, Baker GL. 2008. Water-Soluble Thermoresponsive Polylactides. Macromolecules 41:318–324.
John J, Bhattacharya M, Turner RB. 2002. Characterization of Polyurethane Foams from Soybean Oil. J Appl Polym Sci 86:3097–3107.
Kamel S, Ali N, Jahangir K, Shah SM, El-Gendy AA. 2008. Pharmaceutical significance of cellulose: A review. eXPRESS Polym Lett 2(11):758–778.
Karak N, Rana S, Cho JW. 2009. Synthesis and characterization of castor-oil-modified hyperbranched polyurethanes. J Appl Polym Sci 112:736–743.
Kayaman-Apohan N, Karal-Yılmaz O, Baysal K, Baysal BM. 2001. Poly(dl-lactic acid)/triblock PCL-PDMS-PCL copolymers: synthesis, characterization and demonstration of their cell growth effects *in vitro*. Polymer 42:4109–4116.
Khiar ASA, Arof AK. 2010. Conductivity studies of starch-based polymer electrolytes. Ionic 16(2):123–129.
Kim DH, Kwon OJ, Yang SR, Park JS. 2007. Preparation of Starch-based Polyurethane Films and Their Mechanical Properties. Fibers Polym 8(3):249–256.
Klemm D, Heublein B, Fink HP, Bohn A. 2005. Cellulose: Fascinating Biopolymer and Sustainable Raw Material. Angew Chem Int Ed 44:3358–3393.
Kontturi E, Tammelin T, Österberg M. 2006. Cellulose-model films and the fundamental approach. Chem Soc Rev 35:1287–1304.
Kricheldorf HR, Lomadze N, Schwarz G. 2008. Cyclic Polylactides by Imidazole-Catalyzed Polymerization of L-Lactide. Macromolecules 41:7812–7816.
Kundu PP, Larock RC. 2005. Novel Conjugated Linseed Oil-Styrene-Divinylbenzene Copolymers Prepared by Thermal Polymerization. 1. Effect of Monomer Concentration on the Structure and Properties. Biomacromolecules 6:797–806.
Lee WJ, Youn YN, Yun YH, Yoon SD. 2007. Physical Properties of Chemically Modified Starch(RS4)/PVA Blend Films – Part 1. J Polym Envir 15(1):35–42.
Lemos PC, Serafim LS, Reis MAM. 2006. Synthesis of polyhydroxyalkanoates from different short-chain fatty acids by mixed cultures submitted to aerobic dynamic feding. J Biotechnol 122:226–238.
Li G, Sarazin P, Favis BD. 2008. The Relationship between Starch Gelatinization and Morphology Control in Melt-Processed Polymer Blends with Thermoplastic Starch. Macromol Chem Physic 209(10):991–1002.
Lichtenthaler FW. 2006. Biorefineries – Industrial Processes and Products. Status Quo and Future Directions. Vol. 2. Weinheim, Germany: Wiley-VCH Verlag GmbH & Co. KGaA.
Liebert T, Häsch C, Heinze T. 2006. Click Chemistry with Polysaccharides. Macromol Rapid Commun 27:208–213.
Lim S, Jane J. 1992. Effect of Starch Granule Size on Physical Properties of Starch-Filled Polyethylene Film. Biotechnol Progr 8(1):51–57.
Lluch, Ronda JC, Galiá M, Lligadas G, Cádiz V. 2010. Rapid Approach to Biobased Telechelics through Two One-Pot Thiol-Ene Click Reactions. Biomacromolecules 11(6):1646–1653.
Lohmeijer BGG, Pratt RC, Leibfarth F, *et al.* 2006. Guanidine and Amidine Organocatalysts for Ring-Opening Polymerization of Cyclic Esters. Macromolecules 39(25):8574–8583.
Lu Y, Larock RC. 2008. Soybean-Oil-Based Waterborne Polyurethane Dispersions: Effects of Polyol Functionality and Hard Segment Content on Properties. Biomacromolecules 9(11):3332–3340.
Marsudi S, Unno H, Hori K. 1998. Palm oil utilization for the simultaneous production of polyhydroxyalkanoates and rhamnolipids by *Pseudomonas aeruginosa*. Appl Microbiol Biotechnol 78:955–961.
Mecking S. 2004. Nature or Petrochemistry? – Biologically Degradable Materials. Angew Chem Int Ed 43(9):1078–1085.
Meier MAR. 2009. Metathesis with Oleochemicals: New Approaches for the Utilization of Plant Oils as Renewable Resources in Polymer Science. Macromol Chem Phys 210:1073–1079.
Meier MAR, Metzger JO, Schubert US. 2007. Plant oil renewable resources as green alternatives in polymer science. Chem Soc Rev 36:1788–1802.
Mendes RK, Claro-Neto S, Cavalheiro ETG. 2002. Evaluation of a new rigid carbon-castor oil polyurethane composite as an electrode material. Talanta 57:909–917.
Metzger JO, Bornscheuer UT. 2006. Lipids as renewable resources: Current state of chemical and biotechnological conversion and diversification. Appl Microbiol Biotechnol 71:13–22.
Mihai M, Huneault MA, Favis BD. 2007. Foaming of Polystyrene/ Thermoplastic Starch Blends. J Cell Plast 43(3):215–236.

Mittendorf V, Bongcam V, Allenbach L, Coullerez G, Martini N, Poirier Y. 1999. Polyhydroxyalkanoate synthesis in transgenic plants as a new tool to study carbon flow through beta-oxidation. Plant J 20(1):45–55.

Moeini HR. 2007. Preparation and properties of novel green poly(ether-ester urethane)s insulating coatings based on polyols derived from glycolyzed PET, castor oil, and adipic acid and blocked isocyanate. J Appl Polym Sci 106:1853–1859.

Mohanty AK, Misra M, Hinrichsen G. 2000. Biofibres, biodegradable polymers and biocomposites: An overview. Macromol Mater Eng 276–277(1):1–24.

Mohapatra DK, Das D, Nayak PL, Lenka S. 1996. Polymers from Renewable Resources. XX. Synthesis, Structure, and Thermal Properties of Semi-Interpenetrating Polymer Networks Based on Cardanol-rmaldehyde-Substituted Aromatic Compounds Copolymerized Resins and Castor Oil Polyurethanes. J Appl Polym Sci 70:837–842.

Montero de Espinosa L, Ronda JC, Galià M, Meier MAR. 2009. Fatty Acid Derived Phosphorus-Containing Polyesters *via* Acyclic Diene Metathesis Polymerization. J Polym Sci Part A: Polym Chem 47:5760–5771.

Montero de Espinosa L, Meier MAR, Ronda JC, Galià M, Cádiz V. 2010. Phosphorus-Containing Renewable Polyester-Polyols *via* ADMET Polymerization: Synthesis, Functionalization, and Radical Crosslinking. J Polym Sci Part A: Polym Chem 48:1649–1660.

Mooney BP. 2009 The second green revolution? Production of plant-based biodegradable Plastics. Biochem J 418(2):219–232.

Mutlu H, Meier MAR. 2009. Unsaturated PA X,20 from Renewable Resources *via* Metathesis and Catalytic Amidation. Macromol Chem Phys 210:1019–1025.

Mutlu H, Meier MAR. 2010. Castor oil as a renewable resource for the chemical industry. Eur J Lipid Sci Technol 112(1):10–30.

Nabar YU, Draybuck D, Narayan R. 2006. Physicomechanical and Hydrophobic Properties of Starch Foams Extruded with Different Biodegradable Polymers. J Appl Polym Sci 102(1):58–68.

Nair KCB, Thomas S. 2003. Effect of Interface Modification on the Mechanical Properties of Polystyrene-Sisal Fiber Composites. Polym Composite 24(3):332–343.

Netravali AN, Chabba S. 2003. Composites get greener. Mater Today 6(4):22–29.

Ogunniyi DS. 2006. Castor oil: A vital industrial raw material. Bioresource Technol 97(9):1086–1091.

Otey FH, Miehltretter. 1962. Degraded starch polyoxyalkylene ether compositions and process for producing the same. US Patent 3,165,508.

Persico P, Ambrogi V, Acierno D, Carfagna C. 2009. Processability and Mechanical Properties of Commercial PVC Plastisols Containing Low-Environmental-Impact Plasticizers. J Vinyl Addit Techn 15(3):139–146.

Petrović ZS, Zhang W, Javni I. 2005. Structure and Properties of Polyurethanes Prepared from Triglyceride Polyols by Ozonolysis. Biomacromolecules 6(2):713–719.

Petrović ZS, Guo A, Javni I, Cvetkovi I, Hong DP. 2008. Polyurethane networks from polyols obtained by hydroformylation of soybean oil. Polym Int 57:275–281.

Petrović ZS, Ionescu M, Javni I. 2009. US Patent Application 20090082483.

Pounder RJ, Stanford MJ, Brooks P, Richards SP, Dove AP. 2008. Metal free thiol-maleimide "Click" reaction as a mild functionalization strategy for degradable polymers. Chem Commun 7(41):5158–5160.

Ramakrishna S, Mayer J, Wintermantel E, Leong KW. 2001. Biomedical Applications of Polymer-composite Materials: a review. Compos Sci Technol 61:1189–1224.

Rasal RM, Janorkar AV, Hirt DE. 2010. Poly(lactic acid) modifications. Prog Polym Sci 35(3):338–356.

Reed AM, Gilding DK. 1981. Biodegradable polymers for use in surgery – poly(glycolic)/poly(lactic acid) homo and copolymers: 2. *In vitro* degradation. Polymer 22(4):494–498.

Robertson ML, Chang K, Gramlich WM, Hillmyer MA. 2010. Toughening of Polylactide with Polymerized Soybean Oil. Macromolecules 43:1807–1814.

Rybak A, Meier MAR. 2007. Cross-metathesis of fatty acid derivatives with methyl acrylate: renewable raw materials for the chemical industry. Green Chem 9:1356–1361.

Rybak A, Meier MAR. 2008a. Cross-Metathesis of Oleyl Alcohol with Methyl Acrylate: Optimization of reaction conditions and comparison of their environmental impact. Green Chem 10:1099–1104.

Rybak A, Meier MAR. 2008b. Acyclic Diene Metathesis with a Monomer from Renewable Resources: Control of Molecular Weight and One-Step Preparation of Block Copolymers. ChemSusChem 1:542–547.

Rybak A, Fokou PA, Meier MAR. 2008. Metathesis as a versatile tool in oleochemistry. Eur J Lipid Sci Technol 110:797–804.

Sagar AD, Merrill EW. 1995. Properties of Fatty-Acid Esters of Starch. J Appl Polym Sci 58(9):1647–1656.
Sahoo D, Sahoo S, Mohanty P, Sasmal S, Nayak PL. 2009. Chitosan: a New Versatile Bio-polymer for Various Applications. Des Monomers Polym 12:377–404.
Salgado AJ, Coutinho OP, Reis RL. 2004. Bone Tissue Engineering: State of the Art and Future Trends. Macromol Biosci 4(8):743–765.
Sanmathi CS, Prasannakumar S, Sherigara BS. 2004. Interpenetrating polymer networks based on polyol modified castor oil polyurethane and poly(2-ethoxyethyl methacrylate) synthesis, chemical, mechanical, thermal properties, and morphology. J Appl Polym Sci 94:1029–1034.
Schaub M, Wenz G, Wegner G, Stein A, Klemm D. 1993. Ultrathin Films of Cellulose on Silicon Wafers. Adv Mater 5(12):919–922.
Shimao M. 2001. Biodegradation of Plastics. Curr Opin Biotechnol 12:242–247.
Slivniak R, Domb AJ. 2005. Macrolactones and Polyesters from Ricinoleic Acid. Biomacromolecules 6:1679–1688.
Snell KD, Peoples OP. 2002. Polyhydroxyalkanoate Polymers and Their Production in Transgenic Plants. Metab Eng 4:29–40.
Sokolsky-Papkov M, Shikanov A, Kumar N, Vaisman B, Domb AJ. 2008. Fatty acid based biodegradable polymers-synthesis and applications. Bull Israel Chem Soc 23:12–17.
Steinbüchel A, Flüchtenbusch B. 1998. Bacterial and other biological systems for polyester production. Trends Biotechnol 16:419–427.
Suggs LJ, Moore SA, Mikos AG. 2007. Physical Properties of Polymers Handbook. New York: Springer. pp. 939–950.
Suthar B, Dave M, Jadav K. 2003. Sequential-interpenetrating polymer networks from castor oil-based polyesters XXVI. J Appl Polym Sci 50:2143–2147.
Tamami B, Sohn S, Wilkes GL. 2003. Incorporation of Carbon Dioxide into Soybean Oil and Subsequent Preparation and Studies of Nonisocyanate Polyurethane Networks. J Appl Polym Sci 92(2):883–891.
Tan IKP, Kumar KS, Theanmalar M, Gan SN, Gordon III B. 1997. Saponified palm kernel oil and its major free fatty acids as carbon substrates for the production of polyhydroxyalkanoates in *Pseudomonas putida* PGA1. Appl Microbiol Biotechnol 47: 207–211.
Tang Z, Gibson VC. 2007. rac-Lactide polymerization using aluminum complexes bearing tetradentate phenoxy-amine ligands. Europ Polym J 43:150–155.
Türünç O, Meier MAR. 2010. Fatty acid derived monomers and related polymers *via* thiol-ene (click) additions. Macromol Rap Com 31(20): pp. 1822–1826.
Türünç O, Kayaman-Apohan N, Kahraman MV, Mencelo lu Y, Güngör A. 2008. Nonisocyanate based polyurethane/silica nanocomposites and their coating performance. J Sol-Gel Sci Technol 47(3):290–299.
Valero MF, Pulido JE, Ramírez Á, Cheng Z. 2008. Polyurethanes synthesized of polyols obtained from castor oil modified by transesterification with pentaerythritol. Quím Nova 31:2076–2082.
Valverde M, Andjelkovic D, Kundu PP, Larock RC. 2008. Conjugated Low-Saturation Soybean Oil Thermosets: Free-Radical Copolymerization with Dicyclopentadiene and Divinylbenzene. J Appl Polym Sci 107:423–430.
Véchambre C, Chaunier L, Lourdin D. 2010. Novel Shape-Memory Materials Based on Potato Starch. Macromol Mater Eng 295(29):115–122.
Voet D, Voet JG, Pratt CW. 2006. Fundamentals of Biochemistry: Life at the Molecular Level, 2nd Edn. New York: John Wiley & Sons, Inc.
Wang HJ, Rong MZ, Zhang MQ, Hu J, Chen HW, Czigány T. 2008. Biodegradable Foam Plastics Based on Castor Oil. Biomacromolecules 9:615–623.
Wang W, Zhang Z, Xu M and Zhang Y. 2009. Synthesis and characterization of polyamides based on dimer acid. J Wuhan Univ Technol 24(3):367–370.
Wani MY, Hasan N, Malik MA. 2010. Chitosan and Aloe Vera: Two Gifts of Nature. J Disper Sci Technol 31(6):799–811.
Warwel S, Brüse F, Demes C, Kunz M, Rüsch GEN, Klaas M. 2001. Polymers and surfactants on the basis of renewable resources. Chemosphere 43:39–48.
Williams CK, Hillmyer MA. 2008. Polymers from Renewable Resources: A Perspective for a Special Issue of Polymer Reviews. Polym Rev 48(1):1–10.
Yan Y, Pang H, Yang X, Zhang R, Liao B. 2008. Preparation and Characterization of Water-Blown Polyurethane Foams from Liquefied Cornstalk Polyol. J Appl Polym Sci 110(2):1099–1111.
Yang TH, Kim TW, Kang HO, *et al.* 2010. Biosynthesis of Polylactic Acid and Its Copolymers Using Evolved Propionate CoA Transferase and PHA Synthase. Biotechnol Bioeng 105(1):150–160.

Yenwo GM, Manson JA, Pulido J, Sperling LH, Conde A, Devia N. 1977. Castor-oil-based interpenetrating polymer networks: Synthesis and characterization. J Appl Polym Sci 21:1531–1541.

Yoon SD, Chough SH, Park HR. 2006. Effects of Additives with Different Functional Groups on the Physical Properties of Starch/PVA Blend Film. J Appl Polym Sci 100(5):3733–3740.

Yoon SD, Chough SH, Park HR. 2007. Preparation of Resistant Starch/Poly(vinyl alcohol) Blend Films with Added Plasticizer and Crosslinking Agents. J Appl Polym Sci 106(4):2485–2493.

Yunus AMM, Parveez GKA, Ho C-L. 2008. Production of Polyhydroxyalkanoates in Transgenic Plants. AsPac J Mol Biol Biotechnol 16(1):1–10.

Zhao GL, Deiana L, Córdova A. 2010. Heterogeneous "Organoclick" Derivatization of Polysaccharides: Photochemical Thiol-ene Click Modification of Solid Cellulose. Macromol Rapid Commun 31:740–744.

Zhong Z, Dijkstra PJ, Feijen J. 2003. Controlled and Stereoselective Polymerization of Lactide: Kinetics, Selectivity, and Microstructures. J Am Chem Soc 125:11291–11298.

12 Lignocellulosic Biomass Processing

Fei Yu and Jonathan Y. Chen

12.1 INTRODUCTION

Currently there is a great concern about the sustainability of the environment and energy sources and consumers are demanding biomass-based products. Therefore, scientists are putting intensive efforts into exploring renewable biomass sources and expanding the utilization of these resources for bioproduct development. A lot of attention is drawn to lignocellulosic biomass, one of the major renewable energy sources, due to its many advantages over other energy sources. Biomass-based energy and materials lessen greenhouse gas effects and can be generated locally. Some biomaterials, such as lignocellulosic ethanol, polylactic acid (PLA) and polyhydroxyalkanoate (PHA) can be produced through biological processes, for example fermentation, while some biofuels are obtained through thermochemical processes, such as biopolyols, bio-oils and synthesis gas.

The United States federal government, state governments and industries are stepping up efforts to shift the current fossil-oil-based economy to a renewable bio-based economy. The non-renewable resource and fossil-oil-based economy has adverse impacts on our environment and is non-sustainable. In contrast, an economy based on renewable resources, such as lignocellulosic biomass, will supplement energy and material production from fossil resources, and offers a strong possibility of replenishing or displacing the use of petroleum. Firstly, biomass resources are domestic, reducing dependence on foreign oil and reducing concerns over the stability and sustainability of that supply. The conversion processes from lignocellulosic biomass sources also have a major advantage over petroleum refining processes in that they release far fewer pollutants to the atmosphere and generate less toxic waste during production. Perhaps most importantly to the agricultural sector, producing chemicals and materials from renewable biomass resource means a new market for agriculture and forestry. Exploring new uses for agricultural and forest residues and their derivatives can reduce the effect of food commodity market limits on agricultural product prices, delivering higher profits to producers.

12.2 AVAILABILITY OF LIGNOCELLULOSIC BIOMASS

Agricultural and forest processing always generates a tremendous amount of biomass that can be utilized to produce bioproducts, biopolymers or biofuels. Forestry and crop residues have been traditionally used as animal feed directly or, after pretreatments, to improve

Food and Industrial Bioproducts and Bioprocessing, First Edition. Edited by Nurhan Turgut Dunford.

digestibility. It is only in recent years that researchers increased their effort to convert biomass to industrial products. One of the ways of utilizing biomass is to directly convert it into biofuels or biopolymers. Another major way of utilizing biomass is to convert it into smaller molecules as chemical building stocks (CBS) for making chemical end products.

12.2.1 Southern pine wood

Woody biomass contains plant roots, bark, leaves, branches, limbs, trunks, and vines from trees and shrubs. Just as trees are grown to produce lumber and other products for paper and furniture industries, trees can be grown for chemicals, polymers as well as energy. In the southeastern United States, pine plantation is the most abundant source of woody biomass.

Plantation pine silvicultural practices have been adopted for a rapidly increasing share of timberlands in the South as shown in Figure 12.1 by the increased annual acreage of trees planted over time. From 1952 to 1996, 57 million acres of pines were planted in the southern United States (Smith *et al.*, 2009). In 1994, total privately owned plantation pine acreage in the South was 30 million acres or about 47% of the total privately-owned forested acres (Siry and Bailey, 2003). Modern plantation pine cultural practices call for planting of genetically improved seedlings that grow 16% faster than the traditional stock. Wide, between-seedling spacing followed by early thinning has resulted in rapidly increased growth rates (Zobel and Jett, 1995). Siry and Bailey (2003) estimate that pine growth in the southern United States increased at an average rate of 2.6% annually from 1987 to 1994, and increased at an average rate of 2.8% annually from 1994 to 2004. Fox and co-workers (2007) had the same prediction and reported the pine growth rate in intensively managed plantations in the South may exceed 400 ft^3 /ac/y).

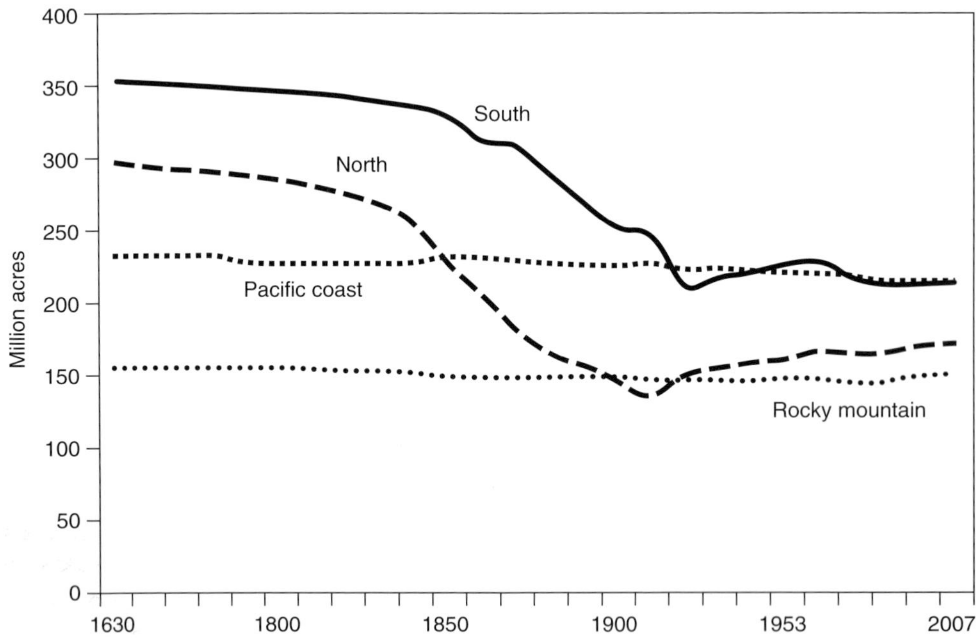

Figure 12.1 Forest area trends in the United States by major region, 1630–2007 (Smith et al., 2009).

12.2.2 Corn stover

Published calculations indicate that the potential for corn stover in the United States could be 75 million dry tons (US) per year (Perlack *et al.*, 2005). Corn stover consists of stalks, leaves, and cobs after the corn kernels are harvested. More than 90% of the stover is left in the fields, less than 1% is collected for industrial processing, and about 5% is baled for animal feed and bedding. Much of the remaining 90% or so must be plowed. The plowing operation can cause organic carbon and nitrogen losses due to oxidation, and increases the amount of fertilizer and chemicals that need to be applied. The decay of the stover increases the release of carbon dioxide, a greenhouse gas. Although some residue (<30%) is required to protect the soil from erosion, most residues (>60%) can be safely taken out of the fields for valuable utilization, which has the potential to be a win-win situation for the producer, the processor, and the consumer. The farmer wins from the stover sales, reduced cultivation costs and possible carbon credits for the greenhouse gas offset. The processors make value-added industrial products that are previously produced from petroleum. The consumer benefits from productive agricultural practices and an improved environment.

12.2.3 Bast fiber crops

Bast fiber crops are a group of plants that can produce natural cellulose fibers from plant stem skin. In history, cultivation of bast fiber crops is the oldest method to produce natural fibers for meeting clothing needs and other daily necessaries. The most important bast fiber crops are ramie, flax, hemp, jute, and kenaf, based on their production capacity and consumption quantity. To date, the production of bast fiber crops still primarily aims at the textile market. Facing fierce competition from synthetic fibers that have increased productivity and a steadily expanded end-use market, the world capacity of bast fiber production continues to decline. Currently, the production volume of the major bast fibers in the world is about 4.8 million metric tons, equivalent to 14% of the global production of manufactured fibers (Fibersource, 2002).

Bast fiber crops are also a group of lignocellulosic biomass that can be used for bioenergy production. A basic fact is that the bast fiber extraction rate from the bast fiber biomass is only about 10–30%. The remaining material after the fiber extraction process is a large portion of residues (woody core and short fiber). While the traditional textile application of bast fibers is well established, the potential of using bast fiber residues for biofuel production is still undervalued. Furthermore, some subtropical fiber crops, like sunn hemp, can grow in summer with a short cultivation period and large amount of biomass yield (IFAS, 2009). The production of these fiber crops requires no use of extra land, no use of pesticide and nitrogen fertilizer, has no competition with food crops, and no conflict with the current infrastructure for industrial crops. Therefore, the bast fiber crops can be developed as a type of sustainable feedstock viable for bioenergy and bio-based product diversification, rural economic development, and efficient use of land resources.

Bast fiber crops have a similar biological stem structure. They are typically composed of a bark layer, a bast layer and a stem core. The bark layer is called cuticle, a thin skin to form a protective stem surface and to hold bast fiber bundles. The bast layer includes a primary fiber layer where bast fibers can be extracted. The stem core usually consists of woody tissue called xylem and pith. Figure 12.2 shows the cross-sectional view of sunn hemp stem. The height of bast fiber crops varies depending on crop species. Overall, it is in the range of 60 to 600 cm (Krishnan *et al.*, 2000; Salmon-Minotte and Franck, 2000; Sponner *et al.*, 2000; Kozlowski *et al.*, 2000).

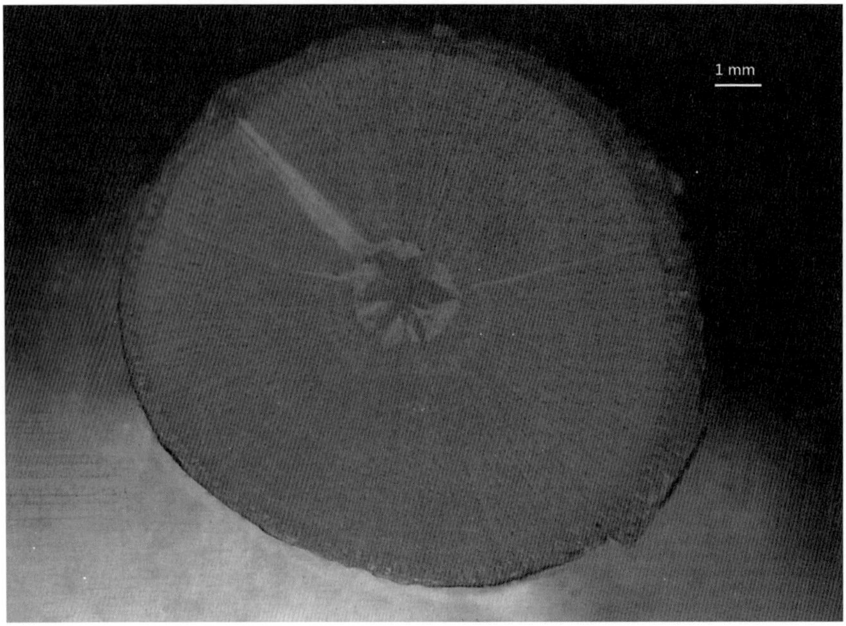

Figure 12.2 Cross-section of sunn hemp stem.

12.2.4 Other lignocellulosic feedstocks

Switchgrass (*Panicum virgatum*) has been the model herbaceous species targeted for development as a dedicated energy crop (McLaughlin, 1993), while bermudagrass (*Cynodon dactylon*) is the most widely used forage species in the southern United States (Wu and Taliaferro, 2009). More recently, Schmer and co-workers (2008) reported that switchgrass produced 5.4 times more energy than the non-renewable energy input needed for its cultivation in large scale (3–9 ha) field trials on marginal land in the mid-continental United States. In the southern area, unfertilized tall- and mixed-grass prairies typically produce 2.47–7.41 tons of dry biomass/ha/year from a single mid-summer harvest. Mean annual dry biomass yields in the range of 7.41 to 14.83 tons/ha are realized from tame pasture species like switchgrass and bermudagrass under typical management (Ball *et al.*, 2002; Dalrymple, 1990; Dabo *et al.*, 1987). Current breeding, genetic and cultural research with switchgrass and bermudagrass will continuously provide benefits in the form of high yielding varieties and improved cultural/management strategies.

Several grass species have shown potential for use in biomass production in the humid Southeast and there is sufficient land available to produce any of several biomass crops that could be used for conversion to energy. However, of all the crops and crop residues evaluated, switchgrass would be the feedstock of choice (Baldwin and Graham, 2005). Several factors combine to modify the quality of biomass feedstocks. Genetic variability, management practices, and storage and handling techniques will all affect the compositional traits, which are important during biomass processing. To maintain a continuous supply of herbaceous feedstocks to a processing plant, the biomass must be placed into long-term storage. The key to sustainable biomass production is maximizing yield with minimal inputs. Feasibility and ease of establishment also affect acceptance of a species. Continuous screening of potential grass species and sustainability studies should be performed from time to time.

12.3 PROCESSING

Many polymers conventionally derived from petrochemicals can be produced from renewable resources such as lignocellulosic biomass. Tremendous efforts have been made to produce biological substitutes for petrochemical feedstocks. Many technically feasible approaches are available to convert biomass to biopolymers. However, no large-scale commercial facility is operating to date. This situation can be attributed to a combination of the following three factors: technical inadequacy, lack of economic competitiveness, and lack of understanding of the industrial need.

The market potential for bio-based products is very promising. The Roadmap for Biomass Technologies in the United States (USDOE, 2007) forecasts a large increase in bio-based products. Production of chemicals and materials from biomass increases substantially from approximately 5.67 billion kg or 5% of the current production of target United States chemical commodities in 2001, to 12% in 2010, 18% in 2020 and 25% in 2030.

There are several biological processes for converting lignocellulosic biomass, such as anaerobic digestion and fermentation. However, many technical and economic barriers remain. There also exist various thermochemical conversion mechanisms to breakdown biomass feedstocks, including combustion, liquefaction, fast pyrolysis and gasification. Future bio-based production technologies have the opportunity to increase the attractiveness of scale-up and commercialization by improving the conversion of biomass to viable chemicals and biofuels.

12.3.1 Biological conversion

Ethanol produced from corn is a major renewable transportation fuel in the United States. The amount of ethanol produced in 2009 was in excess of 38 billion liters and the produced amount anticipated by the end of 2015 is expected to be around 60 billion liters. However, the main drawback of modern ethanol technologies is that the corn-based ethanol is mostly starch based. There are also non-starch materials that can be utilized to produce ethanol. Agricultural and forest residue and dedicated energy crops are the major sources of lignocellulosic feedstock available for ethanol production. Rooney (1998) estimate that the United States annually produces 306 million metric dry weight tons (tonnes) of agricultural residues like sugarcane bagasse, corn stover, wheat straw, and rice hulls that could be used as biofuel feedstock. Arkenol, BC International, Iogen/Petro Canada, and the Masada Resources Group are working on these low- to no-cost feedstocks in current commercial bioenergy development projects in North America (Glassner, 1999). However, large scale production of ethanol and/or the dynamics of providing ethanol production plants with sustainable supplies of feedstock will require dedicated energy crops as part of a supply system.

12.3.2 Thermochemical conversion

12.3.2.1 *Liquefaction*

Liquefaction is an effective way to convert lignocellulosic materials into bio-based chemicals. Liquefaction is a process in which solid biomass is liquefied in organic solvents with or without catalysts, resulting in a liquid containing smaller solvent-soluble chemicals that can be used directly or from which other products can be derived. Liquefied materials have been made into reactive adhesives, foams, moldings, fibers, carbon fibers, and so on.

The basic conditions for liquefaction are temperature and solvent, and sometimes catalyst. The conventional liquefaction of lignocellulosics requires one to several hours of treatments at 300–400 °C with or without catalysts. More recently developed liquefaction processes can be conducted in organic solvents at temperatures of 240–270 °C without catalysts or at temperatures from 80 to 150 °C with acidic catalysts. The newer processes can result in a yield around 90–95% compared with only 40–60% for the conventional processes. Much research has been centered on the selection of liquefying reagents and catalysts for particular biomass.

Pu and Shiraishi (1993a, 1993b) studied the liquefaction of air-dried wood chips or meals in the presence of phenols at about 250 °C. They found that the biomass could be converted to dioxane-soluble products within a few hours. Heating temperature greatly affected the liquefaction rate; for example, it took less than 30 min at 280 °C and more than 1.5 days at 200 °C for complete liquefaction. The presence of water accelerated the liquefaction, and even the existence of water content equivalent to that in green wood (100–150% by dry mass) facilitated the liquefaction (Pu and Shiraish, 1993b). They also found that stronger acids were more effective than weak acids in accelerating the liquefaction process while alkali or alkaline salt retards the reaction (Pu and Shiraish, 1994). The converted wood-phenol liquid was fairly fluid even at room temperature. In the initial stage of the liquefaction (about 10 min at 250 °C), the majority of lignin was liquefied rapidly if the amounts of phenol were enough. The cellulosic component was most resistant to liquefaction.

Yamada and Ono (1999) used ethylene carbonate (EC) or propylene carbonate (PC) in the presence of acid catalyst at elevated temperature (120–150 °C) in their liquefaction process converting wood and cellulose waste into chemicals. It was found that the rate of the EC liquefaction of cellulose was approximately 10 times faster than that of polyhydric alcohol liquefaction. However, not all biomass responds to liquefying reagent the same way. Satisfactory liquefaction seems to be dependent on the type of lignin, that is hardwood lignin or softwood lignin. For example, when applied to softwood (Japanese cedar and Japanese cypress), liquefaction could not be accomplished. Yamada and Ono (1999) solved this problem by blending ethyleneglycol (EG) with EC. ^{13}C-NMR revealed that the EC liquefaction products from cellulose include levulinic acid compounds, which also result from EG liquefaction of cellulose. Beldman and co-workers (1982) found that lignin effects liquefaction as well.

Other research revealed that liquefying reagent blends are more effective than a single reagent. Yao and co-workers (1994) concluded after studying liquefaction of biomass in dilute solvents that, in most cases, no single solvent could dissolve all of the liquefied components completely. The most effective diluent solvents were binary systems composed of solvents with different polarities, one an electron donor moderately polar solvent, such as dioxane, tetrahydrofuran or acetone, and the other an hydroxyl-containing electron donor-acceptor, highly polar solvent, such as methanol, ethylene glycol or water. The binary of dioxane and water was found to be suitable for a wide range of liquefaction solvents, and a ratio of 4:1 dioxane:water was recommended as a universal diluent for liquefied biomass.

Co-liquefaction is another approach to the effective use of liquefying reagents. Yao and co-workers (1993) investigated the co-liquefaction of a wood and starch system. It is generally very difficult to obtain a liquid with a large wood concentration. On the other hand, starch is very easy to liquefy in the PEG/glycerine system, even at a very small liquid ratio and catalyst concentration. In their study, a combined liquefaction procedure for wood meal (*Betula* sp.) and starch was designed for preparing large biomass content liquids. However, it was found that the extent of the wood liquefaction in a simultaneous liquefaction process, in which the wood and starch were added at the same time, was independent of the biomass composition (wood:starch ratio) but dependent on the liquid ratio and the catalyst concentration. When the liquid ratio was decreased close to one (no matter what the wood:starch ratio was), about 40% of the wood became difficult to liquefy. Based on these findings, a stepwise liquefaction procedure was proposed in which the wood could be preliquefied alone at a relatively large liquid ratio followed by the addition and liquefaction of the starch. By this procedure, a large biomass-content liquid was prepared with a relatively small amount of unliquefied residue. Similar results were observed by Cho and co-workers (1998), who, taking a similar approach, studied the effect of addition of liquefied starch to a corn stover liquefaction system and found that the liquefaction process was promoted and a higher corn stover to liquefying reagent ratio could be used.

It appears that for each type of biomass, liquefying conditions must be developed. These conditions may include liquefying reagents, catalysts, temperature and time. Different types of biomass contain different functional groups, which determine the potential use for the liquefied biomass. Currently, most research has been centered on how effective the liquefaction process is in terms of yield, time and energy use. Little effort has been made to understand the processes and the chemical profiles of the liquefied biomass, not to mention controlling the processes to produce liquefied biomass with desirable chemical profiles.

Liquefied biomass may contain low molecular weight polymers. These low molecular weight polymers can be directly used as final products such as adhesive (glue) (Chen and Chen, 1996). They also can be copolymerized to produce other polymeric materials such as polyurethane, polyester, fibers, and so on. If they are blended with other biomass particles, such as lignocellulosic particles or distiller's dried grains with solubles (DDGS), high density panels or molded articles can be produced.

Pu and Shiraishi (1994) reported that water-soluble adhesives were prepared from a phenol liquefied wood solution. The adhesives revealed satisfactory waterproof quality. Addition of alkylresorcinol as a cure promoter to the adhesives resulted in a medium temperature, curable, aqueous phenol resin adhesive.

The large quantity of hydroxyl groups in liquefied biomass is perfect for the production of polyurethane (PU) through copolymerization. The use of PU foams continues to grow at a rapid pace throughout the world, attributed to their light weight, excellent strength:weight ratio, energy absorbing performance, and comfort features. Currently, PU foams are produced primarily from non-renewable, fossil-origin chemicals. Interest in making PU from renewable resources is increasing. Kurimoto and co-workers (2001) prepared polyurethane (PU) films by solution-casting through copolymerization between liquefied wood and polymeric methylene diphenylene diisocyanate (PMDI) at [NCO]:[OH] ratios of 1.0 and 1.2. The PU films prepared from liquefied wood with high viscosity were found to be more rigid than the films prepared from the liquefied wood with low viscosity. The increase in the viscosity contributed to increases in the cross-link density of the PU films. Varying the viscosity is a way to control the mechanical properties of PU films at a constant [NCO]:[OH] ratio. In a separate study, Kurimoto and co-workers found that the properties of PU films were affected by the isocyanate:hydroxyl group ([NCO]:[OH]) ratio and wood content in PU film. The increase of wood content at a [NCO]:[OH] ratio of 1.0 significantly enhanced the Young's modulus and reduced the maximum elongation of PU film. The rigid mechanical properties were due to the increase of dissolved wood fragments.

PU foams can also be prepared from liquefied woods or starch, diphenylmethane diisocyanate (MDI), catalyst, foaming stabilizer, and viscosity aids (Yao *et al.*, 1995; Cho *et al.*, 1998). The polyol content, isocyanate:hydroxyl group (NCO:OH) ratio, and water content were found to affect the foaming and water absorption of the PU foams. The foaming rates increased with increasing moisture contents of liquefied wood. Moisture content of 7% resulted in a more than 30 times increase of foaming rate comparing with moisture content of 0.6%. But an increase in water content may result in a decrease in cross-links between wood polyol and isocyanate, because the NCO:OH ratio is constant. Increasing moisture contents have significantly decreased the density of PU foams. The optimum water content should be about 2.5% or less in this preparation condition.

Polyols for PU foam production are required to have hydroxyl levels in the range of 300 to 500. Many biomass feedstocks have a hydroxyl value greater than 15 000. For PU foam production, liquefaction conditions must allow the preparation of polyols having hydroxyl values in the range of 300 to 500. Research has found that liquefying agents, temperature and processing time have a strong influence on the hydroxyl values of polyols. Sometimes, the usability of PU foams may be determined by their rigidity/flexibility and hydrophilic water absorbing capability (Yao *et al.*, 1995).

Since biopolyols can act as reactive adhesives, blending solid materials and liquefied materials with or without copolymerization may present another way of making useful materials from liquefied biomass. Lin and co-workers (1996) blended soy protein isolate (SPI) and polyether polyols to make water-blown rigid PU foam. Foams containing SPI exhibited thermal conductivity values similar to, or slightly higher than, foams containing no SPI. The density, compressive strength, compressive modulus, and dimensional stability of foams with or without SPI decreased as the initial water content increased. An idea inspired by this research is to incorporate DDGS (as filler) into liquefied biomass to produce PU foams (Yu *et al.*, 2008).

Sellers and co-workers (2000) used a high-pressure flow-molding press to make flat and/or molded panels from recycled powdered thermoplastics (polyethylene and polystyrene) and various lignocellulosic materials (wood and kenaf). The blended materials had a ratio of 1:1, and the panels had a relative density range from 600 to 900 kg/m^3. The modulus of elasticity values for the polystyrene/wood panels were significantly higher compared to the polyethylene/wood panels. All plastic/fiber panels had low thickness swell values. The reactive adhesiveness of liquefied biomass makes liquefied biomass a perfect candidate for making high density materials by mixing them with solid particles, such as lignocellulosic meals or DDGS meals, and then molding the mixture to high strength panels or other forms of products. Some copolymers may be added to the mixture to enhance the performance of the materials.

12.3.2.2 *Fast pyrolysis*

Fast pyrolysis processes to convert cellulosic materials to produce value-added chemicals or materials (such as sugars, acids, phenol, etc.) generally referred to as bio-oil, have been developed. The fast pyrolysis process requires the reduction of the biomass fuel to approximately sawdust size. Particles are heated to between 400 and 550 °C very rapidly in the absence of oxygen followed by cooling to condense the pyrolysis product. This treatment fractures the molecular bonds converting the biomass to the final bio-oil. The charcoal by-product of the process is used as fuel to produce the required high pyrolyzation temperatures, so that the process is nearly energy neutral. The yield of bio-oil is relatively high, at about 60% dry weight basis or higher depending on the production process (Bridgwater *et al.*, 1999).

Bio-oil chemical properties vary with the feedstock but woody biomass typically produces a mixture of 30% water, 30% phenolics, 20% aldehydes and ketones, 15% alcohols, and 10% miscellaneous compounds. As a fuel, bio-oil has environmental advantages when compared to fossil fuels, producing half the NO_x and no SO_x. As a fuel derived from a renewable resource, bio-oil is considered carbon dioxide neutral (Mulraney *et al.*, 2002).

Bio-oil can be burned directly in engines or mixed with diesel oil. Electricity has been produced by diesel engines and turbines have been specially modified to burn bio-oil (Mulraney *et al.*, 2002). However, technical issues, such as acidity, immiscibility, viscosity change over time and other problems, must be solved prior to widespread application for use in engines (Bridgwater *et al.*, 1999).

In addition to the bio-oil product produced in the condensation stage, the fast pyrolysis process produces combustible gases as exit gas. The amount produced is less than 10% by weight of the dry feedstock. These exit gases are most often entrained and utilized as process gas for feedstock pyrolyzation. Some bio-oil reactors, however, do not utilize this gas as process gas but remove it from the pyrolysis system (Bridgwater *et al.*, 1999).

Almost all pyrolysis reactors employed worldwide are fluidized-bed reactors. However, these reactors are known to have relatively high capital requirements. The Department of Forest Products at Mississippi State University (MSU) houses a 10 kg/h auger pyrolysis reactor. Auger reactors are estimated to require approximately 30% of the capital required for fluidized-bed technology for equivalent production capacity. The MSU auger reactor utilizes a novel pyrolysis vapor condensation system that produces very high quality bio-oil in that the water content is approximately 20% or lower, versus 25–30% for bio-oils produced by fluidized bed reactors. Analysis of the chemical composition of bio-oils produced by this reactor is currently underway for both southern pine and red oak feedstocks produced from bole wood and bark at several temperature and residence time regimes.

Bio-oil can be burned directly in engines and electricity has been produced by bio-oil fueled diesel engines; turbines have been specially modified to successfully burn bio-oil. However, some properties of bio-oil, such as lower octane, acidity, immiscibility, viscosity change over time and a distinctive odor, have prevented its commercial use to date for other than pilot and demonstration projects (Bridgwater *et al.*, 1999).

Bio-oil can contain up to 45% oxygen, which is responsible for various negative properties described. Hydrotreatments (hydrogen treatments) of bio-oils to reduce the oxygen content have been investigated using various catalysts. They have resulted in reducing or eliminating some of the negative properties of bio-oils due to the elimination of oxygen and/or reduction of double bonds and aldehydic and keto groups. Two methods have been investigated: (1) catalytic treatment of the pyrolysis vapors prior to condensation into bio-oil, and (2) catalytic treatment of the condensed bio-oils with hydrogen or hydrogen precursors using typical petroleum refining catalysts, including zeolites. The first method has an advantage of not using a reducing gas such as hydrogen and the overall cost of the treatment is lower. The method also produces liquid hydrocarbons directly suitable as fuels but the yields are low and the catalysts are deactivated relatively rapidly due to the high percentages of coking of bio-oil components (Bridgwater and Cottam, 1992; Czernik *et al.*, 2002).

The second method, catalytic hydrogen treatment of the condensed bio-oil liquids, has been investigated using standard petroleum hydrogenating procedures and catalysts by employing slightly modified technologies currently utilized for petroleum refining and infrastructure. Fuel grade materials and selected chemicals may be produced depending on the extent of the hydrogen treatments. Raw bio-oils, due to the high oxygen contents, cannot be refined using the current petroleum refining processes.

Commonly, the hydrogen treatment for reducing or eliminating oxygen to the extent of transforming bio-oils to a liquid hydrocarbon mixture with properties very similar to petroleum crude oil is called hydrodeoxygenation (HDO) (Scholze, 2002). HDO treatments of raw bio-oils can be carried out very similarly to the hydrodesulfurization process used in the petroleum refining industry to reduce the sulfur content of petroleum crude oil. Both hydrodesulfurization and HDO require catalysts combined with some level of heat and pressure. Researchers have hypothesized that only minor changes are required for the current hydrodesulfurization processes and infrastructure of the petroleum industry to be applied to the HDO of bio-oil. Likewise, HDO treated bio-oils can potentially be refined in existing petroleum refineries, again with only minor adjustments to the current petroleum industry refinery infrastructure (Bridgwater and Cottam, 1992).

The carbohydrates, cellulose and hemicellulose components of wood contain oxygen as primary and secondary alcohols. Oxygen is also present in the phenolic and methyl ether groups of lignin. During pyrolysis, carbohydrates are dehydrated and broken to form molecules having various aldehyde, keto and hydroxyacid groups in bio-oil products. Phenol and a wide array of alkyl and methoxy substituted phenols are derived from the lignin component of wood.

The combined chemical characteristics of these functional groups are the cause of many of the undesirable attributes of bio-oil. Immiscibility with petroleum fuels, odor of burned wood, low heat content, instability, and corrosiveness are examples of some of the more undesirable properties. The chemical reactivity of different functional groups provides numerous possible approaches for modifying and improving the characteristics of bio-oil for fuel or chemical use. Oxygen is also a linkage in the higher molecular weight carbohydrate-derived polymeric components of bio-oils and removal of the linkage oxygen will decrease product viscosity from reduction of polymeric components' molecular weights (Czernik *et al.*, 2002).

Char particles entrained into bio-oil during pyrolysis are known to catalyze polymerization reactions, leading to viscosity increases over time. Filtering the bio-oil immediately after production has been proven to improve bio-oil stability and color (Czernik *et al.*, 2002).

The approximately 25% of water contained in crude bio-oils has been shown to foul the catalysts applied during the hydrogen treatment step. Previous researchers have fractionated the bio-oils into water and organic fractions and often the water fractions were discarded prior to hydrogenating the organic fractions. Researchers have also shown that bio-oils when mildly hydrogenated do not increase in viscosity with application of heat as compared to non-hydrogenated bio-oils (Baker and Elliott, 1988; Elliott and Neuenshwander, 1996).

Hydrogenation followed by esterfication will eliminate most of the undesirable characteristics of bio-oils previously listed. Acetic acid is one of the most common carboxylic acids present in bio-oil and is a primary cause of the corrosiveness characteristic. Acetic acid can be reacted with methanol to form methyl acetate. When methyl acetate is hydrogenated using $CuOCuCr_2O_4$ as the catalyst, methanol and ethanol are the products. These chemicals have value as fuel or can be introduced into another product stream. About 10–20% of the bio-oil being produced in the pyrolysis unit at MSU can be converted to esters.

The dehydration of carbohydrates also yields furfural, furanone, and other similar compounds. Mild hydrogenation will reduce the alkenes to alkanes and the aldehyde and ketone groups to alcohols. The overall positive effect of this reaction is that the heat of combustion of the products is higher than that of the starting material. The aliphatic and aromatic carbonyl compounds are partially responsible for the odor of bio-oil and the reduction of aldehydes to alcohols will reduce the odor of bio-oils. Aldehydes and ketones can be hydrogenated with a wide variety of catalysts.

Substituted phenols make up a larger portion of the bio-oils. Phenols from the bio-oils of pine contain 3-methoxy groups along with a wide variety of other functional groups in the 1- position. Phenols and substituted phenols yield the corresponding cyclohexanes when hydrogenated and the pungent odor associated with phenols will not be present after hydrogenation. Hydrogenation of phenolic compounds will have the greatest impact on the chemical and physical properties of bio-oils.

Only about one-third to one-half of the compounds in crude bio-oils can be analyzed by gas chromatography/mass spectrometry (GC/MS). The non-volatile compounds are dimers, trimers and oligimers of carbohydrate and lignin monomeric units. These high molecular weight compounds will also be hydrogenated and with some of the large molecules will be cleaved to monomeric units by removing the oxygen linkage between the monomeric units.

Catalytic HDO research has, to date, been based on application of traditional petroleum catalysts. Moderate successes in applying these HDO catalysts on bio-oils have been attained. However, specialized catalysts for bio-oil application, as opposed to petroleum applications, have yet to be developed. The complex chemistry of bio-oils presents a special challenge compared to that of processing crude petroleum oils that are mainly composed of hydrocarbons. A promising approach identified by Bridgwater and Cottam (1992) is the identification of a modified zeolite that is more selective toward bio-oil components and desired products.

Zeolites are various types of crystalline aluminosilicates that occur in nature and can also be synthesized (Augustine, 1996). More than 100 varieties of zeolites are currently available with new candidate zeolites added monthly. Zeolites, such as ZSM-5, HZSM-5 and MCM-41 in various forms, have been shown to be effective in catalyzing pyrolysis vapors *via* catalytic cracking to produce liquid hydrocarbons suitable for fuels (Adam *et al.*, 2005; Diebold and Scahill, 1988; Evans and Milne, 1988; Horne *et al.*, 1997; Sharma and Bakhshi, 1993; Renaud *et al.*, 1988). Catalytic cracking (as opposed to the milder HDO treatment) of the liquid bio-oils themselves has also been performed with zeolites, both by treating revaporized bio-oils (Renaud *et al.*, 1988) or by directly treating liquid bio-oils (Samolada and Vasalos, 1997; Evans and Milne, 1988). Relatively low yields and rapid coking resulting in deactivation of catalysts have both been reported as problems from the cracking of bio-oil liquids or revaporized bio-oils.

HDO treatments of bio-oils with metallic catalysts adopted from the petroleum industry have been performed (Baker and Elliott, 1988; Bridgwater and Cottam, 1992; Czernik *et al.*, 2002; Centeno *et al.*, 1995, 1997, 1999; Conti *et al.*, 1997; Ferrari *et al.*, 2002; Puente *et al.*, 1999; Oasmaa and Boocock, 1992; Zhang *et al.*, 2003). Researchers have shown that a two-stage process is required (Baker and Elliott, 1988; Gagnon and Kaliaguine, 1988). The first stage applies a mild hydrogenation at relatively low temperatures below about 270 °C. Full HDO of bio-oil requires temperatures above 300 °C, which results in polymerization of highly oxygenated components of raw bio-oils (Elliott and Neuenschwander, 1996). Therefore, currently the HDO stage at higher temperatures must be reserved for application after the first mild hydrogenation stage. HDO of bio-oils with zeolites has not been reported in the literature. Zeolites are effective hydrogenating catalysts as well as proven cracking catalysts. However, zeolites have to be tested as hydrogenation catalysts for bio-oils. The problems of low yields, rapid coking and deactivation of catalysts reported for zeolytic cracking of bio-oils are not expected to occur in the application of mild hydrogenation followed by HDO.

12.3.2.3 *Gasification*

There are several types of gasifiers (such as downdraft, updraft and crossdraft) widely used in syngas production. Syngas is a chemical building block to produce biopolymers, biomaterials, or biofuels. Generally, an updraft gasifier has air passing through feedstocks from the bottom and the combustible gases come out the top of the gasifier, while the downdraft gasifier lets the air pass through in the opposite direction. In a crossdraft gasifier, the air enters from one side and syngas is taken out from the other side. All existing gasifiers fall into the above categories with slight modification.

In the research on gasification of biomass at Mississippi State University, a downdraft gasifier has been used for feedstocks such as hard and soft wood chips and switchgrass pellets (Wei *et al.*, 2008). The study found that the downdraft gasifier produced syngas with similar quality regardless of the three feedstocks used. Additional feedstocks, such as forestry wastes, chicken litter, gin trash, and crop residues, are scheduled to be gasified in the near future.

Syngas from a downdraft gasifier system generally contains high contents of nitrogen, hydrogen, carbon monoxide and carbon dioxide as well as a small amount of impurities. One by-product of the gasification process is tar. Some tar compounds may be toxic to the catalysis process. Tar elimination reactions are known to be kinetically limited, so reaction rates can be increased through the use of catalysts. Many different catalysts have been explored for reducing tars. The most widely used catalysts include dolomite, alkali metals and nickel. Dolomite is a cheap and disposable catalyst that has resulted in significant improvement in tar conversion. Nickel catalysts are widely used in hydrocarbon reforming and are known to be useful in control and adjustment of gas composition. Dual systems have also been proven effective, where dolomite may be used in the reactor bed, followed by further catalytic cleaning of the hot gas downstream with a nickel-based reforming catalyst (Boerrigter *et al.*, 2003). Physical removal of tars and particulates in the syngas is also well studied using different wet and dry scrubbing processes (Devi *et al.*, 2003). Acetone is one of the extensively tested solvents for scrubbing out condensables from biomass syngas, particularly in tar and particulate measurement (Patil *et al.*, 2006; Cateni, 2007, 2008).

Researchers at MSU have also demonstrated the proof of concept that syngas can be converted to chemicals, such as ethylene, alcohols, and hydrocarbons using a combination of commercial and proprietary catalysts and reactor design (Liu *et al.*, 2009). They showed that the enhanced conversions of higher alcohols (e.g. ethanol, propanol, and butanol) to hydrocarbon (30–40% by mass) at pressures to 70 bar compared to that observed for methanol conversion at (1–2% by mass). The analyses of hydrocarbon products showed an octane number of 80–90 with API gravities of 50–55.

Syngas from biomass gasification can also be utilized to produce ethanol or acetic acid *via* fermentation (Ahmed and Lewis, 2006; Cateni, 2007). Virtually all fermentation bioreactors of technical importance deal with heterogeneous systems involving at least two phases. Many challenges exist to optimize synthesis gas fermentation, including overcoming mass transfer limitations, increasing cell densities, reducing the production of unwanted by-products, and minimizing the effects of inhibitory synthesis gas constituents generated *via* biomass gasification (Worden *et al.*, 1997). Work has shown the effects of gas constituents on cell growth, product distribution, and hydrogenase activity (Ahmed and Lewis 2006). In particular, it was shown that nitric oxide can inhibit hydrogenase activity (Ahmed and Lewis, 2007). It has also been observed that the carbon monoxide partial pressure plays a critical role in the cell concentration and the product distribution (Hurst and Lewis, 2010).

Similar findings of partial pressure effects have also been noted by other researchers (Younesi *et al.*, 2005). However, a critical analysis of partial pressure effects on key gas substrate enzymes has not been undertaken. This work will provide critical information towards developing a predictive model that can be used to understand how carbon dioxide partial pressure affects formate dehydrogenase (FDH), which is one of the three gas-substrate enzymes.

The literature on the microbiology of syngas fermentation is limited, although recent reviews point out the great promise of this bioprocess (Henstra *et al.*, 2007; Tanner, 2008). While it has been known that some clostridia can directly reduce organic acids to alcohols (Fraisse and Simon, 1988), this is still an area of interest and active research (Steinbusch *et al.*, 2008; Tashiro *et al.*, 2007). Understanding of the basic biochemistry underlying syngas fermentation should lead to improvement of this bioprocess and is an important component in past and present research in this area (Andersch *et al.*, 1983; Inokuma *et al.*, 2007).

12.3.3 Bast fiber production

Bast fiber production starts with crop cultivation. The crop growers' job includes planting and harvesting. During a harvesting period, the process of mowing, dew retting, and baling is carried out. After harvesting, baled crop stalk is transported to a mechanical processing line for fiber extraction. The processing line is designed in different ways that requires a specific input bale shape (round or rectangular) and determines output fiber quality. Through this mechanical process, bast fiber with a certain quality is produced together with woody core by-product with different sizes (Figures 12.3 and 12.4). For some fiber crops, like flax and hemp, seed is also produced as a by-product. Bast fibers are traditionally used in the production of textiles, fabrics and apparel. They are particularly suitable for making

Figure 12.3 Extracted bast fiber. (Courtesy of TEMAFA.)

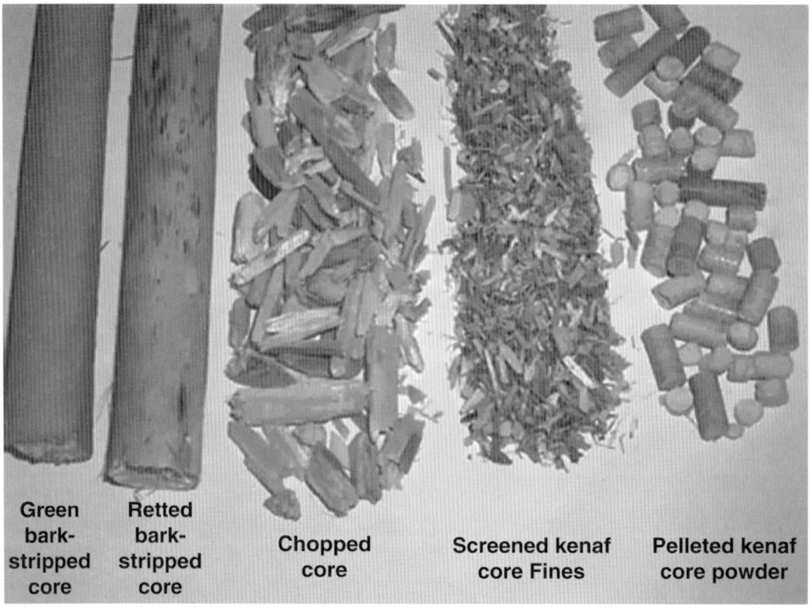

Figure 12.4 Woody core by-product after fiber extraction. (Courtesy of TEMAFA.)

non-woven fabrics that are increasingly employed in filtration, insulation, environmental clean-up and vehicle interiors. Woody core and short (trash) fiber can be used as biomass feedstock for biofuel and biochar conversion. Fiber crop seeds can be used for producing food ingredients, nutraceutical products, cosmetic products, and technical products. Although the harvesting technologies influence the produced fiber quality, the fiber extracting process is a key step in the bast fiber production. Two important technologies currently used in bast fiber production are described here.

One technology is a mechanical scutching line for producing line bast fiber, woody core (shives) and seed. This processing line keeps bast fiber in its stalk length in each process step, so that the extracted fiber is long staple fiber and can be cut into short staple fiber, depending on the different cutting lengths required by different end uses, such as blended textiles, non-woven felts, composites or paper. The automatic scutching line evolved from a traditional manual operation of scutching flax using wooden blades. The whole machine system of a scutching line usually consists of the following units: bale opening, stalk breaking, scutching, and cutting and baling. Figure 12.5 illustrates a typical flax scutching line.

The bale opening unit is to open delivered bales, remove seed, and make a continuous and even layer of fiber stalk on the feed belt. These mechanical actions are performed by a bale opener and a stalk divider. To assure processing efficiency, harvested fiber crop stalks need to be baled parallel to each other, with stalk top on one side and stalk bottom on the other side. The stalk breaking unit performs woody core crushing without fiber damage. This enables the broken woody core chips (large size shives) to be separated from the fiber bundles. The breaking unit has two modules, one for breaking fiber stalk bottom and the other for breaking fiber stalk top. The scutching unit is used to remove woody core particles (small size shives) remaining in the fiber bundles after the stalk breaking. In the flax scutching line as shown in

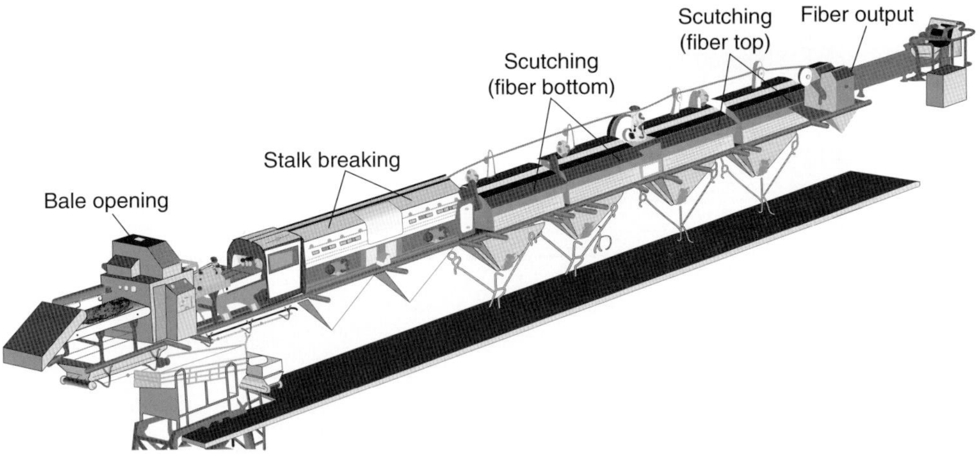

Figure 12.5 Flax scutching line. (Courtesy of Van Dommele Engineering.)

Figure 12.5, the scutching is performed by a turbine with its axis along the fiber moving direction. There are two scutching steps. In the first step, two turbines are used for scutching bast fiber bottoms. The second scutching step uses two other turbines to beat on bast fiber tops. The removed trash particles and short fiber are transported to a collecting unit through funnels and a pneumatic transportation system. Some scutching lines use a cylinder to perform scutching process. Bast fiber layer, with a fiber-laying orientation perpendicular to its moving direction, is moved by feeding rolls into a scutching area, where the bast fiber is beaten by the spring steel blades mounted on the surface of a rotating cylinder to scrape off woody core particles and short fiber (called tow fiber). In this case, the cylinder axis is parallel to the fiber-laying orientation. Two scutching units are needed, one for rough scutching and another for fine scutching. Additionally, this cylinder scutching is often followed by a separating unit. This unit is composed of shaking screens that shake up fiber bundles and let loose shives and short fiber fall into an vacuum chamber underneath the shaking screens. Thus, clean bast fiber with shorter stalk length is produced. This line fiber product can be packed and stored directly, or can be cut and baled by an automatic hydraulic baling press.

Another technology is a decorticating line, which produces short staple bast fiber, woody core, and seed. Figure 12.6 exhibits the major components that constitute this bale-to-bale fiber processing line (Temafa GmbH, 2005). In comparison with the above scutching line, these critical components include bale opening, stalk breaking, fiber breaking and separating, fiber opening and cleaning and bale press. After a fiber stalk bale is delivered to the feeding point of this line, the bale is opened and fiber stalk is spread out on a moving belt to form a continuous layer using an opener and divider. Because the decorticating line is designed for short staple fiber production, fiber stalk laid on the moving belt does not necessarily keep parallel. The fiber stalk breaking unit uses a large-size cylinder surfaced with scattered beaters to crush the woody core inside the fiber. Two stalk breaking units can be arranged to achieve a better efficiency of woody core breaking. The crushed fiber stalk is fed through retaining rollers to a fiber breaking and separating unit to break fiber bundles and small size shives. This is performed by a set of crushing rollers. To separate the crushed shives from fiber bundles, horizontal or angled shaking screens are used for removing short fiber and shives during the transportation of fiber bundles.

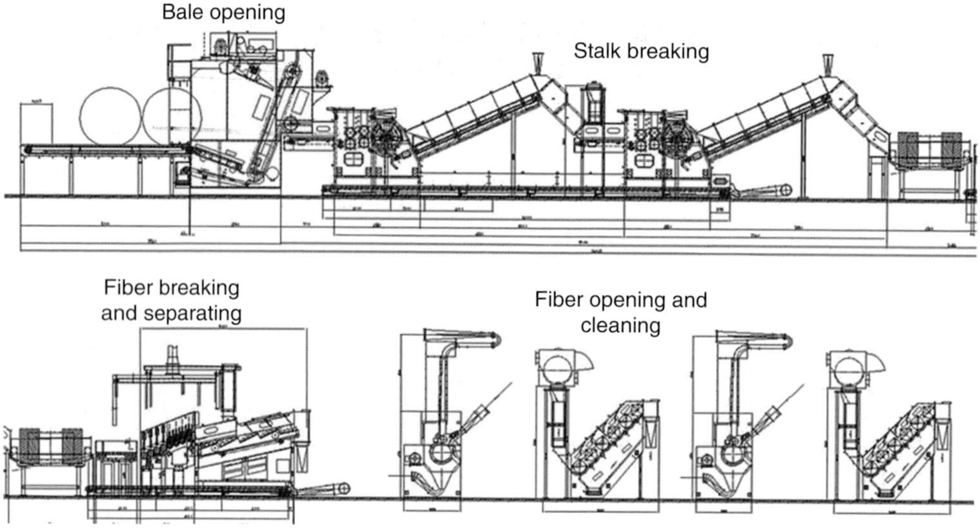

Figure 12.6 Bast fiber decorticating line. (Courtesy of TEMAFA.)

The fiber opening unit is designed for shortening and refining the bast fiber to meet different end-use requirements. The fiber bundles are fed onto a rotating cylinder for opening and cleaning through a feeding table or a volumetric feeding chute. The cylinder surface can be covered either by pins or by saw-tooth wires, depending on length and fineness of the short staple fiber to be produced. Following this fiber opening procedure is a step cleaner for further removing loose shives and short fiber encapsulated in the fiber bundles. This cleaning unit is constructed with a set of beater rollers and filter grids. After this cleaning process, staple fiber is produced and can be transported to a bale press for packing. Shive content of the output fiber is within 5%. Length of the produced short staple fiber is about 50–120 mm. Fiber fineness is between 4 and 10 dtex. A typical decorticating line can have a stalk feeding speed of 3000 kg/h. Extracted fiber yield is in the range of 500 to 700 kg/h (Temafa GmbH, 2003).

REFERENCES

Adam J, Blazso M, Meszaros E, *et al.* 2005. Pyrolysis of biomass in the presence of Al-MCM-41 type catalysts. Fuel 84:1494–1402.

Ahmed A, Lewis RS. 2006. Effects of biomass-generated syngas constituents on cell growth, product distribution and hydrogenase activity of Clostridium carboxidivorans P7T. Biomass and Bioenergy 30(7):665–672.

Andersch W, Bahl H, Gottschalk G. 1983. Levels of enzymes involved in acetate, butyrate, acetone and butanol formation by Clostridium acetobutylicum. Eur. J. Appl. Microbiol. Biotechnol. 18:327–332.

Augustine RL. 1996. Heterogeneous Catalysis for the Synthetic Chemist. New York: Marcel Dekker, Inc.

Baldwin BS, Graham JW. 2006. Effect of population density and row spacing on dry matter yields and bork:core ratio of kenaf (*Hibiscus cannabinus L.*). Industrial Crops and Prod 23:244–248.

Baker EG, Elliott DC. 1988. Catlytic hydrotreating of biomass-derived oils. In: Soltes EJ, Milne TA (editors). Pyrolysis Oils from Biomass. American Chemical Society Symposium Series 376. American Chemical Society.

Ball DM, Hoveland CS, Lacefield GD. 2002. Southern Forages: Modern Concepts for Forage Crop Management, 3rd edn. Norcross, GA: Potash and Phosphate Institute/Foundation for Agronomic Research.

Beldman G, Voragen AGJ, Rombouts FM, Pilnik W, Demirbas A. 1982. Liquefaction and saccharification of agricultural biomass: Effect of lignin content on aqueous liquefaction products of biomass. Energy from biomass: Proceedings of the EC contractors' meeting, Brussels. 41:1601–1607.
Boerrigter H, Uil H, Calis HP. 2003. Green diesel from biomass via Fischer-Tropsch synthesis: new insights in gas cleaning and process design. In: Bridgwater, AV (editor). Pyrolysis and gasification of biomass and waste. Newbury, UK: CPL Press. pp. 371–383.
Bridgwater AV, Cottam ML. 1992. Opportunities for biomass pyrolysis liquids production and upgrading. Energy and Fuels 6:113–120.
Bridgwater AV, Czernik C, Diebold J, Mekr D, Radlein P. 1999. Fast Pyrolysis of Biomass: A Handbook. Newbury, UK: CPL Scientific Publishing Services Ltd.
Cateni, B. 2007. Effect of feed composition and gasification parameters on producer gas from a pilot scale fluidized bed gasifier. PhD Thesis. Oklahoma State University, Stillwater, OK.
Centeno A, Laurent E, Delmon B. 1995. Influence of the support of CoMo sulfide catalysts and of the addition of potassium and platinum on the catalytic performances for the hydrodeoxygenation of carbonyl, carboxyl, ant guaiacol-type molecules. J Catalysis 154:288–298.
Centeno A, David O, Vanbellinghen C, Maggi R, Delmon B. 1997. Behaviour of catalysts supported on carbon in hydrodeoxygenation reactions. In: Bridgwater AV, Boocock DGB (editors). Developments in Thermochemical Biomass Conversion, Vol. 1. London: Blackie Academic & Professional.
Centeno A, Maggi R, Delmon B. 1999. Use of noble metals in hydrodeoxgenation reactions. Studies in Surface Science and Catalysis 127:77–84.
Chen C, Chen CM. 1996. State of the art report: adhesives from renewable resources. Holzforschung und Holzverwertung 48:58–60.
Cho N, Seo W, Han G, Cho NS, Seo WS, Han GS. 1998. Development of urethane foams for planting media from wood wastes. J the Korean Wood Science and Technology 26:43–49.
Conti L, Scano G, Boufala J, Mascia S. 1997. Bio-crude oil hydrotreating in a continuous bench-scale plant. In: Bridgwater AV, Boocock DGB (editors). Developments in Thermochemical Biomass Conversion, Vol. 1. London: Blackie Academic & Professional.
Czernik S, Maggi R, Peacoke GVC. 2002. Review of methods for upgrading biomass-derived fast pyrolysis oils. In: Bridgwater, AV (editor). Fast Pryolysis of Biomass: A Handbook, Vol. 2. Newbury, UK: CPL Press.
Dabo SM, Taliaferro CM, Coleman SW, Horn FP, Claypool PL. 1987. Yield and digestibility of old world bluestem grasses as affected by cultivar, plant part, and maturity. J Range Management 40:10–15.
Dalrymple RL. 1990. Old world bluestem. Agricultural. Division, The Samuel Roberts Noble Foundation, Inc. Publication NFOUB-1.
Devi L, Ptasinski KJ, Janssen FJJG. 2003. A review of the primary measures for tar elimination in biomass gasification processes, revisited. Biomass and Bioenergy 24(2):125–140.
Diebold J, Scahill J. 1988. Biomass to gasoline. In: Soltes EJ, Milne TA (editors). Pyrolysis Oils from Biomass. American Chemical Society Symposium Series 376. American Chemical Society.
Elliott DC, Neuenschwander GG. 1996. Liquid fuels by low-severity hydrotreating of biocrude. In: Bridgwater AV, Boocock DGB (editors). Developments in Thermochemical Biomass Conversion, Vol. 1. London: Blackie Academic & Professional. pp. 611–621.
Evans RJ, Milne T. 1988. Molecular-beam, mass-spectrometric studies of wood vapor and model compounds over an HZSM-5 catalyst. In: Soltes EJ, Milne TA (editors). Pyrolysis Oils from Biomass. American Chemical Society Symposium Series 376. American Chemical Society.
Ferrari M, Delmon B, Grange P. 2002. Influence of the impregnation order of molybdenum and cobalt in carbon supported catalysts for hydrodeoxygenation reactions. Carbon 40:497–411.
Fibersource. 2002. Annual World Surveys of Manufactured Fiber Production [Internet]. Fiber Economics Bureau, The Fiber Organon. [Accessed 27 December 2010]. Available at: http://www.fibersource.com/F-Info/fiber%20production.htm.
Fox TR, Jokela EJ, Allen HL. 2007. The development of pine plantation silviculture in the southern United States. J For 105:337–447.
Fraisse L, Simon H. 1988. Observations on the reduction of non-activated carboxylates by *Clostridium formicoaceticum* with carbon monoxide or formate and the influence of various viologens. Arch. Microbiol. 150:381–386.
Gagnon J, Kaliaguine S. 1988. Catalytic hydrotreatment of vacuum pyrolysis oils from wood. Industrial Engineering and Chemical Research 27:1783–1788.
Glassner D, Hettenhaus J, Schechinger T 1999. Corn stover potential: recasting the corn sweetner industry. In: J. Janick (editor). Perspectives on new crops and new uses. Alexandria, VA: ASHS Press. pp. 74–82.

Henstra AM, Sipma J, Rinzema A, Stams AJM. 2007. Microbiology of synthesis gas fermentation for biofuel production. Current Opinion in Biotechnology 18:200–206.

Horne PA, Nugranad N, Williams PT. 1997. The influence of steam on the zeolite catalytic upgrading of biomass pyrolysis oils. In: Bridgwater AV, Boocock DGB (editors). Developments in Thermochemical Biomass Conversion, Vol. 1. London: Blackie Academic & Professional.

Hurst KM, Lewis RS. 2010. Carbon monoxide partial pressure effects on the metabolic process of syngas fermentation. Biochemical Engineering 48: 159–165.

IFAS: Sunn Hemp – A Cover Crop in Florida [Internet]. Homestead, FL: Institute of Food and Agricultural Sciences; c2000–2009 [Accessed 27 December 2010]. Available at: http://edis.ifas.ufl.edu/tr003.

Inokuma K, Nakashimada Y, Akahoshi T, Nishio N. 2007. Characterization of enzyme involved in the ethanol production of Moorella sp. HUC22-1. Arch. Microbiol. 188:37–45.

Kozlowski R, Rawluk M, Barriga-Bedoya J. 2000. Ramie. In: Franck RR (editor). Bast and other plant fibres. Cambridge, UK: Woodhead Publishing Limited. pp. 207–227.

Krishnan KB, Doraiswamy I, Chellamani KP. 2000. Jute. In: Franck RR (editor). Bast and other plant fibres. Cambridge, UK: Woodhead Publishing Limited. pp. 24–93.

Kurimoto Y, Koizumi A, Doi S, Tamura Y, Ono H. 2001. Wood species effects on the characteristics of liquefied wood and the properties of polyurethane films prepared from the liquefied wood. Biomass and Bioenergy 21:381–390.

Lin Y, Hsieh F, Huff HE, Iannotti E. 1996. Physical, mechanical, and thermal properties of water-blown rigid polyurethane foam containing soy protein isolate. Cereal Chem 73: 189–196.

Liu S, Gujar A, Thomas P, Toghiani H, White M. 2009. Synthesis of Gasoline-Range Hydrocarbons over Mo/HZSM-5 Catalysts. Applied Catalysis A: General 357:18–25.

McLaughlin SB. 1993. New switchgrass biofuels research program for the southeast. Proceedings of the 1992 Annual Automotive Technol. Dev. Contractors Coordinating Meeting, 2–5 November 1992; Dearborn, MI. pp. 111–115.

Mulraney H, Farang IH, LaClaire CE, Barrett CJ. 2002. Technical, environmental and economic feasibility of Bio-oil in New Hampshire's north country. www.unh/p2/bio-oil/bounhif.pdf.

Oasmaa A, Boocock DGB. 1992. The catalytic hydrotreatment of peat pyrolysate oils. The Canadian Journal of Chemical Engineering. 70:294–300.

Patil KN, Huhnke RL, Bellmer DD. 2006. Performance of a unique downdraft gasifier. ASABE Annual International meeting, 9–12 July 2006, Portland, Oregon.

Perlack RD, Wright LL, Turhollow AF, Graham RL, Stokes BJ, Erbach DC. 2005. Biomass as feedstock for a bioenergy and bioproducts industry: The technical feasibility of a billion-ton annual supply (DOE/GO-102005-2135). Oak Ridge, TN: US Department of Energy.

Pu S, Shiraishi N. 1993a. Liquefaction of wood without a catalyst I. Time course of wood liquefaction with phenols and effects of wood/phenol ratios. J the Japan Wood Research Society 39:446–452.

Pu S, Shiraishi N. 1993b. Liquefaction of wood without a catalyst II. Weight loss by gasification during wood liquefaction, and effects of temperature and water. J the Japan Wood Research Society 39:453–458.

Pu S, Shiraishi N. 1994. Liquefaction of wood without a catalyst IV. Effect of additives, such as acid, salt, and neutral organic solvent. J the Japan Wood Research Society 40:824–829.

Puente G, Gil A, Pis JJ, Grange P. 1999. Effects of support surface chemistry in hydrodeoxygenation reactions over CoMo/activated carbon sulfided catalysts. Langmuir 15:5800–5806.

Renaud M, Grandmaison JL, Roy C, Kaliaguine S. 1988. Low pressure upgrading of vacuum-pyrolysis oils from wood. In: Soltes EJ, Milne TA (editors). Pyrolysis Oils from Biomass. American Chemical Society Symposium Series 376. American Chemical Society.

Rooney T. 1998. Lignocellulosic feedstock resource assessment. Report SR-580-24189. Golden, CO: National Renewable Energy Laboratory. p. 123.

Salmon-Minotte J, Franck RR. 2000. Flax. In: Franck RR (editor). Bast and other plant fibres. Cambridge, UK: Woodhead Publishing Limited. pp. 94–175.

Samolada MC, Vasalos IA. 1997. Catalytic cracking of biomass flash pyrolysis liquids. In: Bridgwater AV, Boocock DGB (editors). Developments in Thermochemical Biomass Conversion, Vol. 1. London: Blackie Academic & Professional.

Schmer MR, Vogel KP, Mitchell RB, Perrin RK. 2008. Net energy of cellulosic ethanol from switchgrass. PNAS 105:464–469.

Scholze B. 2002. Long-term stability, catalytic upgrading, and application of pyrolysis oils – Improving the properties of a potential substitute for fossil fuels. [DPhil dissertation]. Hamburg, Germany: University of Hamburg.

Sellers T, Miller GD, Katabian M. 2000. Recycled thermoplastics reinforced with renewable lignocellulosic materials. Forest Prod J 50:24–28.

Sharma RK, Bakhshi NN. 1993. Catalytic upgrading of fast pyrolysis oil over HZSM-5. Canadian Journal of Chemical Engineering 71:383–391.

Siry JP, Bailey RL. 2003. Increasing southern pine growth and its implications for regional wood supply. Forest Prod J 53:32–37.

Smith WB, Miles PD, Perry CH, Pugh SA. 2009. Forest resources of the United States, 2007. General Technical Report WO-78. Washington, DC: US Department of Agriculture, Forest Service.

Sponner J, Toth L, Cziger S. 2000. Hemp. In: Franck RR (editor). Bast and other plant fibres. Cambridge, UK: Woodhead Publishing Limited. pp. 176–206.

Steinbusch KJJ, Mamelers HVM, Buisman CJN. 2008. Alcohol production through volatile fatty acids reduction with hydrogen as electron donor by mixed cultures. Water Res. 42:4059–4066.

Tanner RS. 2008. Production of ethanol from synthesis gas. In: JD Wall, CS Harwod, A Demain (editors). Bioenergy. Washington, DC: ASM Press. pp. 147–151.

Tashiro Y, Shinto H, Hayashi M, Baba S, Kobayashi G, Sonomoto K. 2007. Novel high-efficient butanol production from butyrate by non-growing Clostridium saccharoperbutylacetonicum N1-4 (ATCC 13564) with methyl viologen. J. Biosci. Bioengineer. 104:238–240.

Temafa GmbH. 2003. Lin Line System-Preparation of Kenaf Fibers for the Use in Technical Products. TEMAFA Maschinenfabrik GmbH, Bergisch Gladbach, Germany.

Temafa GmbH 2005. Excellence in Bast Fiber Preparation. TEMAFA Machines Global Blending Solutions. TEMAFA Maschinenfabrik GmbH, Bergisch Gladbach, Germany.

USDOE: Roadmap for Biomass Technologies in the United States. [Internet]. Biomass Research and Development Initiative, Department of Energy. [Accessed 27 December 2010]. Available at: www1.eere.energy.gov/biomass/pdfs/obp_roadmapv2_web.pdf.

Wei L, Thomason JA, Bricka RM, Sui R, Wooten JR, Columbus EP. 2008. Syn-gas quality evaluation for biomass gasification with a downdraft gasifier. Trans of the ASABE 52:21–37.

Worden RM, Bredwell MD, Grethlein AJ. 1997. Engineering issues in synthesis-gas fermentations. In: Fuels and Chemicals from Biomass. Washington, DC: ACS. pp. 320–335.

Wu YQ, Taliaferro CM. 2009. Bermudagrass. In: Singh RJ (editor). Genetic Resources, Chromosome Engineering, and Crop improvement: Forage Crops. CRC Press. pp. 229–273.

Yamada T, Ono H. 1999. Rapid liquefaction of lignocellulosic waste by using ethylene carbonate. Bioresour Technol 70:61–67.

Yao Y, Yoshioka M, Shiraishi N, Yao YG. 1995. Rigid polyurethane foams from combined liquefaction mixtures of wood and starch. J the Japan Wood Research Society 41:659–668.

Yao Y, Yoshioka M, Shiraishi N. 1993. Combined liquefaction of wood and starch in a polyethylene glycol/glycerin blended solvent. J the Japan Wood Research Society 39:930–938.

Yao Y, Yoshioka M, Shiraishi N. 1994. Soluble properties of liquefied biomass prepared in organic solvents I. The soluble behavior of liquefied biomass in various diluents. J the Japan Wood Research Society 40:176–184.

Younesi H., Najafpour G, Mohamed AR. 2005. Ethanol and acetate production from synthesis gas via fermentation processes using anaerobic bacterium. Biochemical Engineering 27: 110–119.

Yu F, Le Z, Chen P, Liu Y, Lin X, Ruan R. 2008. Atmospheric pressure liquefaction of dried distillers grains (DDG) and its application to polyurethane foams. Applied Biochemistry and Biotechnology 148:235–243.

Zhang SP, Yan YJ, Ren Z, Li T. 2003. Study of hydrodeoxygenation of bio-oil from the fast pyrolysis of biomass. Energy Sources 25:57–65.

Zobel BJ, Jett JB. 1995. Genetics of Wood Production. Springer Series in Wood Science. New York: Springer-Verlag.

13 Recent Developments in Non-thermal Processess

Fernando Sampedro and Howard Q. Zhang

13.1 INTRODUCTION

The need for "fresh", safe and convenient foods with more health benefits is supported by demographic and socioeconomic changes that are occurring in the population. The industry is conscious about the changes in consumer trends. A significant fact is the introduction of a great variety of minimally processed foods that need chilled chain of storage and distribution systems (fruit juices, vegetable-based soups and sauces and "ready to eat" (RTE) products, among others) in the market. Through the years, thermal treatment has been used for producing microbiologically safe food with low enzyme activity and this continues to be the best processing method for mass production of foods. Nevertheless, in certain products containing health beneficial thermolabile bioactive compounds, quality and nutritional content (vitamins, color and flavor, among others) are compromised after thermal processing, even at mild temperatures.

Taking into account the limitations of traditional thermal processes and the trends in consumer demand, scientists have searched for alternative technologies that could ensure the microbial safety while retaining the fresh-like or fresh-prepared quality. These technologies have been called "emerging preservation technologies" and particularly, the ones based on processing parameters other than temperature, "nonthermal technologies" (Barbosa-Cánovas *et al*., 1998). These technologies include pressure-based processing (high pressure processing, high pressure homogenization and high pressure carbon dioxide), electric fields (pulsed and radiofrequency), radiation-based technologies (UV light and irradiation) and high intensity ultrasounds. Some of these technologies are more suitable for solid food processing, such as high pressure processing (HPP), ionizing irradiation and static-UV processes, whereas pulsed electric fields (PEF), high pressur e homogenization (UHPH), high pressure carbon dioxide (HPCD), ultrasounds and flow-through UV treatment are continuous processes used for liquid food pasteurization. In some cases, HPP and irradiation technologies can also be used for processing packaged liquid foods and PEF static devices for extraction of bioactive compounds in solid foods.

Food and Industrial Bioproducts and Bioprocessing, First Edition. Edited by Nurhan Turgut Dunford.

13.2 RECENT ADVANCES IN NON-THERMAL TECHNOLOGIES

13.2.1 High Pressure Processing (HPP)

The potential of using pressure as a food preservation mechanism has been known for over a century (Wilson *et al.*, 2008). High pressure processing (HPP) is mainly based on the "isostatic" principle, which indicates that pressure is transmitted in a uniform and quasi instantaneous manner throughout the whole food independent of its shape and size (Heremans, 2002). The main components of the HPP equipment are: (i) vessel and yoke; (ii) hydraulics (pressure generation system); (iii) temperature control system. HPP technology involves a batch operation. The packaged product to be treated is placed in a vessel and a pressurizing fluid (water or oil) is pumped into the vessel. Once the vessel is filled with the liquid, pressure is built up by a hydraulic pump and is transmitted through the liquid to the vessel until the desired pressure (100–1000 MPa) is attained. The system is held at this pressure for the duration of treatment time (1–20 min) then rapidly decompressed. Temperature is controlled by a thermostated mantle connected to a cryostat that surrounds the vessel (due to adiabatic heating a temperature increase of 3 °C/100 MPa is expected) (Rovere, 2002). The measurement of the temperature is usually carried out by thermocouples placed inside the vessel in contact with the pressure medium.

The mechanism of microbial inactivation by HPP is a combination of different reactions, such as non-covalent bonds breakdown and increase in cell membrane permeability (Rastogi *et al.*, 2007). A review of the latest literature on microbial inactivation by HPP in different food matrices clearly indicates that pressures between 200 and 600 MPa at room or mild temperatures (35–50 °C) are usually sufficient to inactivate the majority of vegetative cells, including molds and yeasts, in several minutes (Wilson *et al.*, 2008; Rastogi *et al.*, 2007). Based on this knowledge, industrial HPP treatment is conducted at a pressure of 600 MPa for a holding time of 1–5 min (Torres and Velazquez, 2005). *Listeria monocytogenes* is used as a HPP target microorganism, especially in RTE products, due to its high resistance to pressure and presence in low acid refrigerated foods. However, in acidic environments, *E. coli* O157:H7 is commonly used as a microorganism of concern.

The primary goal of any non-thermal processing is food preservation while maintaining the quality of fresh product. The inactivation of quality deteriorative enzymes is also extremely important. Covalent bonds are not affected by HPP treatment, thus the primary structure of the enzymes will not be damaged. The hydrogen bonds are also relatively baroresistant, hence secondary enzyme structure will not be affected up 700 MPa. However, HPP treatment affects electrostatic and hydrophobic interactions that maintain the stability of tertiary and quaternary protein structures. HPP treatment also produces structural damage to the active site interfering in the enzyme-substrate interaction (Ludikhuyze *et al.*, 2002). In vegetable-based products, enzyme baroresistance is generally higher than that for the indigenous microorganisms. For that reason, preservation treatment of such products is based on the inactivation of the enzymes responsible for the quality deterioration. Low water activity seems to have a detrimental effect on HPP effectiveness due to the need of a fluid for pressure transmission (Oxen and Knorr, 1993). As commented earlier, HPP processing does not affect covalent bonds and for that reason, most of the bioactive compounds present in food (water-soluble vitamins such as ascorbic acid, vitamin B group and folates, lipid-soluble vitamins such as vitamin E, K, carotenoids (vitamin A) and anthocyanins) will not be affected by HPP processing. However, some factors, such as high oxygen concentration, combined high temperature and high pressure treatment and other chemical reactions, may degrade the bioactive compounds to some extent (Oey *et al.*, 2008a). Regarding sensory properties, HPP has a limited effect on pigments (e.g. chlorophyll, carotenoids, anthocyanins, etc.) responsible for the color of fruits and vegetables (Oey *et al.*, 2008). Numerous studies using both trained panelists and consumers have shown no difference among HPP-treated and fresh samples regarding food flavor (Parish, 1998; Fernández-García *et al.*, 2001; Baxter *et al.*, 2005; Laboissière *et al.*, 2007). However, during shelf-life, sensory properties can be altered due to incomplete inactivation of enzymes and microorganisms, resulting in undesired chemical reactions (both enzymatic and non-enzymatic) in the food matrix. These reactions are also influenced by the type of packaging, oxygen concentration in the package and storage temperature.

HPP was one of the first non-thermal technologies developed and implemented in food industry. At the beginning of the 1990s, two companies in Japan commercialized HPP-treated marmalades, vegetable-based sauces and fruit juices (Thakur and Nelson, 1998). Applications of HPP technology in industry are growing, including a wide range of products in the meat industry (cured jam, bacon and turkey, pork sausages, foie gras, marinated raw meat and "ready to eat" meat products), seafood products (oysters, lobster and crab meat) and vegetable products (fruit juices and smoothies, fruit purees, fresh cut fruits, guacamole, jams, deli salads, dressings, sauces and "ready to eat" vegetable dishes) (Tonello, 2008). Several companies in the United States, Europe and Asia design and market HPP industrial- and laboratory-scale equipment. Avure technologies Inc. in the USA and NC Hyperbaric in the European Union are the world-leading HHP equipment manufacturers. The industrial equipment is designed with a capacity up to 350 liters (single vessel) or 600 liters (two vessels) and provide pressures up to 600 MPa with a maximum overall production around 5000 kg/h, depending on the holding time and package size. Food is usually packaged in a flexible plastic material with minimal headspace to minimize the volume variation (15%) during the treatment (van der Berg *et al.*, 2002). The development of industrially reliable equipment has lowered the estimated cost of HPP to US$0.05–0.5 per liter (Balasubramian, 2003). Recently, the US Food and Drug Administration (FDA) accepted pressure-assisted thermal-sterilization (PATS) as a process for commercial application in low acid foods (National Center for Food Safety and Technology, 2009). Combined high temperature and high pressure treatment brings a new challenge for the equipment manufacturers in the design of a vessel that is able to resist compression and heating forces in a large volume. The scientific community needs to assess the impact of this new process on the quality and nutritional content of the HPP-sterilized food. Only pilot-scale equipment was available for the PATS process in 2010.

13.2.2 Ultra High Pressure Homogenization (UHPH)

Homogenization is a well-known fluid mechanical process that involves the division of particles or droplets into micron sizes (0.2–2 μm) by passing the fluid through a minute gap in a homogenization valve, creating a stable dispersion or emulsion (Diels and Michiels, 2006). Homogenization at relatively low pressure (50 MPa) has been extensively used in many industries, such as the pharmaceutical, cosmetic, chemical and dairy industries (i.e. to reduce the fat globule size) (Diels *et al.*, 2005a). Recently developed robust homogenization equipment has been able to reach pressure of up to 400 MPa, the so-called dynamic high pressure (DHP) or ultra high pressure homogenization (UHPH) (López-Pedemonte *et al.*, 2006). Unlike HPP technology, UHPH is a continuous process with treatment time lasting several seconds, which makes it more suitable for liquid food processing.

The microbial inactivation exerted by UHPH is believed to be due to the mechanical destruction of cell integrity caused by different mechanisms, including spatial pressure and velocity gradients, turbulence, impingement and cavitation, which occur in liquids during high pressure homogenization (Donsì *et al.*, 2009). Different studies have shown significant microbial inactivation of pathogenic and spoilage organisms (*E. coli* O157:H7, *L. monocytogenes*, *S. aureus*, *Penicillium* ssp, *S. cerevisiae*, *L. plantarum* and *L. mesenteroides*, among others) by applying pressure up to 400 MPa in a single cycle (Briñez *et al.*, 2006a, 2006b, 2007; Campos and Cristianini, 2007) or in consecutive passes (Tahiri *et al.*, 2006; Bevilacqua *et al.*, 2009; Capra *et al.*, 2009; Donsì *et al.*, 2009). Fluid viscosity has been shown to be a major environmental parameter affecting bacterial inactivation by UHPH, as opposed to water activity and product composition, whereas high cellular concentration seems to exert a protective effect on microbial cells, which are less exposed to the effects of pressure (Diels *et al.*, 2005; Tahiri *et al.*, 2006). In addition, several authors observed the lack of sublethal injury after UHPH (Wuytack *et al.*, 2002, 2003; Diels *et al.*, 2005a; Briñez *et al.*, 2006a, 2006b, 2007; Roig-Sagués *et al.*, 2009). As a recent technology, little scientific information on the effects of dynamic pressure on quality parameters and bioactive compounds is found in the literature. The potential of UHPH to inactivate food quality-deteriorative enzymes is still unknown. Pectinmethylesterase (PME) seems to be highly resistant to dynamic pressure (25 and 80% inactivation after a single pass at 150 MPa and five passes at 250 MPa, respectively); despite the low inactivation achieved, the homogenization process seems to decrease the size of juice particles in suspension and avoid destabilization and precipitation (Welti-Chanes *et al.*, 2009; Carreño *et al.*, 2010). Regarding sensory properties, several studies have shown the potential of UHPH to produce milk and soy derived products such as yoghurt, cheese and ice cream with enhanced sensory and rheological properties (Vannini *et al.*, 2008; Cruz *et al.*, 2009; Innocente *et al.*, 2009; Serra *et al.*, 2009).

The UHPH system consists of two positive displacement high pressure pumps in series with two-stage homogenization valves. The first stage valve, which is made of ceramic, is used for the homogenization process and usually set at a pressure between 50 and 350 MPa. The second stage pneumatic valve, which is located after the first one, is used for micronization and usually maintained at 50 MPa. A manometer connected to the circuit displays the pressure of the system (Kumar *et al.*, 2009). A flow-through micrometric gap between the valve and the seat causes the acceleration of the fluid in concurrence with the pressure drop, generating significant mechanical stress, which then causes a significant temperature increase depending on the fluid viscosity (estimated as 1.5–2.5 °C per 10 MPa) (Popper and Knorr, 1990). A tubular heat exchanger (set at 4 °C) connected immediately after the homogenization valve is used to minimize the thermal effects. Several companies in the European Union (Stansted Fluid Power Inc. and GEA Niro Soavi) commercialize UHPH bench- and pilot plant-scale equipment (up to 400 MPa and production rates around 200 l/h). Promising results have been achieved using UHPH pilot plant-scale equipment. However, further research on the design of large-scale equipment and scientific knowledge on the effect of dynamic pressure on quality parameters and nutritional content of food is needed for the scale-up of this technology to industrial production.

13.2.3 High Pressure Carbon Dioxide (HPCD)

The inhibitory effect of carbon dioxide (CO_2) and its combination with pressure on the growth of microorganisms has been known for more than 100 years (Liao *et al.*, 2007). However, the application of pressurized CO_2, so-called high pressure carbon dioxide (HPCD), is relatively new (Damar and Balaban, 2006). HPCD processing involves contacting food with either subcritical or supercritical CO_2 in a batch (using a pressure vessel), semi-batch or continuous system. Besides the environmentally benign nature of the HPCD process (CO_2 is non-toxic and easily removed by simple depressurization and degassing), the CO_2 pressures applied for preservation purposes are much lower (generally 20 MPa) than the pressures employed in HPP (García-González *et al.*, 2009).

Kincal and co-workers (2005) and García-González co-workers (2007) have hypothesized that the elimination of oxygen in the system during HPCD reduces growth rate of aerobic bacteria. Furthermore, when CO_2 is dissolved in water carbonic acid is produced. Presence of carbonic acid in the medium increases cellular acidity and decreases intracellular pH, affecting metabolic activities. Carbon dioxide can also rapidly penetrate into the cells and may enhance its chemical activity on the internal metabolic processes. Finally, carbon dioxide has an inhibitory effect on certain enzyme systems. Higher temperatures and agitation increase the diffusivity of CO_2 and the fluidity of the cell membrane, making CO_2 penetration into the product easier (Hong *et al.*, 1997; Hong and Pyun, 1999). Treatments conducted with supercritical CO_2 have proven to be more effective in inactivating microbial cells than CO_2 under subcritical conditions due to liquid-like density and gas-like viscosity and diffusivity of supercritical CO_2, which enhances its penetration into cells resulting in increased cell disruption (Tomasula, 2003; Gunes *et al.*, 2005). Vegetative cells in a low water content system seem to be more resistant to HPCD processing, maybe due to the direct result of a decreased CO_2 solubility and higher cellular concentrations, which seem to exert a protective effect on microbial cells because of the decreased exposure of cells to the effects of pressurized CO_2 (García-González *et al.*, 2009). Usually pressures between 4 and 50 MPa at room temperature or mild temperatures (35 –C) for 5–120 min are enough to inactivate or cause at least five-log reduction of several spoilage and pathogen microorganisms (García-González *et al.*, 2007). Yuk and co-workers (2010) reported inactivation of *E. coli* K-12 in apple cider as a function of CO_2 concentration, process temperature and residence time. A seven-log reduction was achieved at 42 °C and 8% CO_2 in the cider.

The enzyme inactivation exerted by pressurized CO_2 is believed to be due to pH lowering, conformational changes of the enzyme, and inhibitory effects of molecular CO_2 on enzyme activity (Damar and Balaban, 2006). As opposed to microorganisms, some enzymes, such as peroxidase (POD) and polyphenoloxidase (PPO), are less sensitive to pressure changes under supercritical conditions than subcritical CO_2 conditions. This may be due to the ability of subcritical CO_2 to better interact with the molecules of the enzymes and thereby affect the rate of denaturation (Liu *et al.*, 2008). It is believed that HPCD processing has a minimal impact on some bioactive compounds found in fruit juices, such as ascorbic acid and carotenoids, due to the exclusion of O_2 from the system and pH lowering (Boff *et al.*, 2003; Zhou *et al.*, 2009). The pH decrease observed after HPCD processing is related to CO_2 dissolution, which further dissociates into bicarbonate and carbonate and formation of hydrogen ions (Liu *et al.*, 2008). Color (L^* and a^* values) of the product is slightly affected by HPCD (Park *et al.*, 2002; Kincal *et al.*, 2006). No difference in sensory properties among HPCD-treated and fresh samples has been detected after two weeks of storage at low refrigeration temperatures (1.5 °C). The CO_2 concentration threshold at which panelists could notice the "carbonation" effect of HPCD processing was 230 ppm of CO_2 in the product (Kincal *et al.*, 2006). An enhanced cloudiness in citric and carrot juices has been observed, due to both precipitation of calcium as calcium carbonate and reduced particle size by pressure, which delays the destabilization of colloidal suspension (Park *et al.*, 2002; Boff *et al.*, 2003; Kincal *et al.*, 2006; Lim *et al.*, 2006; Zhou *et al.*, 2009).

The HPCD continuous system is more suitable for liquid food processing and industrial scale-up. It is designed to continuously mix and pressurize a stream of product with carbon dioxide. The product feed tank is connected to a pump and the pump stroke length is adjusted based on desired pressure, CO_2:product (w/w) ratio, and fluid flow rate. A reciprocating intensifier pump is used to pressurize CO_2. Proper mixing of CO_2 and product is achieved by using a reactor tank or a membrane reactor. The CO_2/product mixture is held in a tube for a particular residence time. Thermocouples are used to measure temperature in the holding tube. The system also has equipment for depressurization and separation of CO_2 and product (Kincal *et al.*, 2006). Praxair Inc. (USA) has designed HPCD units for processing at 37, 74, and 148 l/min continuous product flow rates (Damar and Balaban, 2006; García-González *et al.*, 2007). Further developments in large-scale high pressure equipment manufacturing at a lower cost and development of niche products for this application are challenges for the commercial success of HPCD technology.

13.2.4 Pulsed Electric Fields (PEF)

For many years high intensity electric fields have been applied to induce electroporation, a phenomenon used to promote bacterial DNA interchange by perforating microbial membranes (Zhang *et al.*, 1995). Recent advances in the design of the treatment chamber led to the development of a new processing technology known as pulsed electric fields (PEF) (Vega-Mercado *et al.*, 1997, Yin *et al.*, 1997) and radio frequency electric fields (RFEF) (Geveke and Brunkhorst, 2003). The principle of PEF technology is the treatment of a biological material or food placed between two electrodes installed 0.1–1.0 cm apart in a treatment chamber separated by an insulator with short pulses (1–10 μs) that are generated by a high voltage (5–20 kV) pulse generator. A power source is used to charge a capacitor bank and a switch is used to discharge energy to the treatment chambers (Zhang *et al.*, 1995). Two principle types of chambers are used in the PEF technology: continuous chambers for pasteurization of liquid food and static chambers more suitable for studies in solid food (extraction of bioactive compounds) (Barbosa-Cánovas *et al.*, 1998). New designs for uniform electric field distribution and geometry, such as the co-field continuous chamber, applying the electric field in the same direction as the fluid flow (Yin *et al.*, 1997), have made it possible to apply a uniform electric field in large-scale equipment. New manufacturing materials such as graphite are much more resistant to electrolysis and metal migration (Toepfl *et al.*, 2007; Huang and Wang, 2009).

The main processing parameter is the electric field (kV/cm), which is defined by the relationship between the electrical potential difference applied to two electrodes and the distance between them. The main pulse waveform used in a PEF treatment is the square wave, which can be applied in monopolar or bipolar mode (Ho *et al.*, 1995). Treatment time (μs) is defined by the pulse width (μs), the pulse frequency (Hz), flow rate and chamber dimensions. An oscilloscope is used to monitor voltage, current and pulse width during the treatment. PEF treatment produces a temperature rise due to ohmic heating and a cooling system is used to minimize thermal effects. Increase in temperature generally produces a synergetic effect on microbial and enzyme inactivation (Wouters *et al.*, 2001a; Toepfl *et al.*,

2007). High temperatures also increase the electrical conductivity of the product, lowering the maximum electric field strength viable in the treatment chamber before undergoing a "dielectric breakdown" of the system, evidenced by an electric spark. In addition, air bubbles or suspended particles may also produce "dielectric breakdown" (Wouters *et al.*, 2001a; Góngora-Nieto *et al.*, 2002, 2003).

Most of the studies on PEF microbial inactivation agree that high voltage treatment produces a series of structural and functional changes in the cellular membrane that lead to microorganism death (Mañas and Pagán, 2005). The well-known theory exposed by Zimmermann (1986) states that when an external electric field is applied to a system it induces a potential difference across the cellular membrane. The maximum potential difference that the membrane can withstand is known as the "critical transmembrane potential" or "critical electric field" (E_c). When the external electric field exceeds E_c, membrane breakdown occurs leading to the formation of pores. The magnitude of E_c principally depends on the membrane characteristics of the microorganism (Gram-positive and Gram-negative), size and shape (smaller cells and rod-shaped cells require higher E_c values) (Qin *et al.*, 1998; Heinz *et al.*, 2002; Toepl *et al.*, 2007). In general, "reversible" pores are formed under conditions where the electric field is lower than E_c, and the membrane returns to its initial form after the treatment. These pores may cause certain sublethal damage, leading to cellular death in stress conditions such as low pH and refrigeration temperature. On the contrary, if the electric field is much higher than E_c, "irreversible" pores are produced, leading to the death of the microorganism. Besides structural changes in the membrane, other changes have been observed inside the cell, such as leakage of intracellular material and alterations in cell proteins (Wouters *et al.*, 2001b; Aronsson *et al.*, 2005; García *et al.*, 2007). A high electric field and short time treatment (HEST) (30–40 kV/cm for 20–500 μs) combined with mild temperature (40–55 °C) has been proposed as optimal to achieve the highest degree of inactivation of some pathogen microorganisms, such as *E. coli* O157:H7, *Listeria monocytogenes*, *Listeria innocua*, *S. aureus* and different *Salmonella* strains (Mosqueda-Melgar *et al.*, 2008). Each microorganism may behave differently under PEF treatment. Gurtler and co-workers (2010) conducted a comparison study with 23 strains of pathogenic and spoilage microorganisms and identified *E. coli* 35218 as a surrogate for *E. coli* O157:H7.

Several studies on inactivation of enzymes during PEF processing have reported conformational changes and alteration of protein helix alignment due to the movement of charges produced by the electric field, making difficult the active site-substrate interaction (Bendicho *et al.*, 2002; Zhang *et al.*, 1995). PEF processing showed some effects on bioactive compounds, such as carotenoids, and water-soluble vitamins, such as ascorbic acid in fruit juices, thiamin, riboflavin, biotin, folic acid and pantothenic acid in milk, other lipid-soluble vitamins (retinol, cholecalciferol and α-tocopherol) and the fatty acids in milk (Bendicho *et al.*, 2002; Rivas *et al.*, 2007; Zulueta *et al.*, 2007; Riener *et al.*, 2008; Soliva-Fortuny *et al.*, 2009). On the other hand, sensory properties such as flavor, aroma and color are not significantly affected by electric field treatment in fruit juices (Jia *et al.*, 1999; Min and Zhang, 2003; Min *et al.*, 2003a, 2003b). Applications of this technology are mainly focused on preservation of liquid food. PEF treatment has also been applied to improve the extraction of different bioactive components from diverse foodstuff by increasing the permeability of plant cells (Ade-Omawaye *et al.*, 2001), thereby improving the juice yield and quality (Guderjan *et al.*, 2007).

Several pilot plant-scale PEF systems have been developed and a great number of bench-scale equipment is spread around the world in different research groups. The use of

PEF by the juice processing industry became a reality in the USA where several types of fruit juices treated by PEF were commercialized (Clark, 2006). In addition, several companies manufacturing PEF processing units have appeared in recent years. Among them, Diversified Technologies Inc. and PurePulse Technologies Inc. in the USA and ScandiNova Systems AB in Sweden manufacture PEF units with an overall flow rate of 400–8000 l/h. The cost of a PEF-processed product by these units is estimated in $0.04 per liter (Huang and Wang, 2009). It is expected that the high number of research studies and industrial efforts will lead to new industrial applications of PEF technology in the near future.

13.2.5 Ultraviolet Light (UV)

The use of ultraviolet light (UV) for air and water treatment and surface decontamination has been well established for nearly 50 years (Masschelein, 2002). As a physical preservation method, UV irradiation has a positive consumer image (non-ionizing radiation) and is of interest to the food industry as a low cost technology. UV irradiation of foods does not produce chemical residues, by-products or radiation and it is a simple, dry and low temperature process (Bachmann, 1975; Morgan, 1989). The UV light wavelength range used for food processing varies from 200 to 280 nm, which is known as UV-C or germicidal range, since it effectively inactivates bacteria and viruses (Guerrero-Beltrán and Barbosa-Cánovas, 2004). Typically, three general types of mercury UV lamps are used: low pressure mercury lamps (LPM), low pressure high output (LPMHO), and medium pressure mercury lamps (MPM). These terms are based on the vapor pressure of mercury when the lamps are operating. The FDA has approved the use of LPM lamps for juice processing and the process has already been successfully commercialized (Koutchma, 2009).

Liquid foods absorb UV radiation. The presence of dissolved organic and suspended solids and high initial microbial populations (such as yeast cells) in liquid foods leads to strong UV attenuation effects (penetration of only 1 mm is required for absorption of 90% of the light) (Sizer and Balasubramaniam, 1999). Penetration of light through the system decreases as color or turbidity increases because of high light absorption (Tran and Farid, 2004). The UV-C dose emitted from a lamp is usually measured using UV sensors in W/m^2 units. It has been lately proposed that a turbulent flow increases UV light penetration (Koutchma, 2009). UV-C light is also applied to fresh fruits, vegetables and roots before being stored to reduce the initial count of microorganisms on the surface of the product and to induce host resistance to the microorganisms (Guerrero-Beltrán and Barbosa-Cánovas, 2004).

Microorganisms that are exposed to UV light are affected at the DNA level. The UV-C light absorbed by DNA causes a physical shifting of electrons to render splitting of the DNA bonds, delay of reproduction or cell death (Wright *et al.*, 2000). In addition, DNA transcription and replication are blocked, compromising cellular functions and eventually leading to cell death (Sastry *et al.*, 2000). The highest germicidal effect against bacteria is obtained at wavelengths near 254 nm (UV-C, generated by LPM lamps) (Guerrero-Beltrán and Barbosa-Cánovas, 2006). Photoreactivation is a phenomenon that must be taken into account when the UV-C injured cells are subsequently exposed to wavelengths within the range of 330–480 nm (Liltved and Landfald, 2000). The damage occurring at the DNA level could be repaired, increasing the number of viable microorganisms during storage if light-transmitting packages are used. The variation of microbial resistance is believed to be due to the presence of UV absorbing proteins or differences in the structure of the nucleic acids as well as the ability of the microorganisms to repair UV damage (Guerrero-Beltrán and Barbosa-Cánovas, 2004). Up to now, few studies have shown the effects of UV on quality-deteriorative enzymes, and variable resistance has been found among them. For example, while UV light caused 80% inactivation in PPO, the effect of UV on texture-related enzymes such as pectinmethylesterase (PME) and polygalacturonase (PG) were negligible (Barka *et al.*, 2000; Tran and Farid 2004; Guerrero-Beltrán and Barbosa-Cánovas, 2006). Some food components are known to be "light sensitive": carotenes, tryptophan, certain food pigments, folic acid, vitamins B12, A, D, K, B2 and E, and unsaturated fatty acid residues in oils, solid fats, and phospholipids (Spikes, 1981). Riboflavin (vitamin B2) and beta-carotene seem to be the most sensitive in terms of degradation; however, effects of UV light strongly depend on the wavelength used in the treatment (Tran and Farid, 2004; Koutchma, 2009). It is well known that milk and milk products are highly light sensitive and UV irradiation produces undesirable changes in the sensory and chemical properties (Matak *et al.*, 2007) whereas in other products, such as apple cider, no differences in sensory properties have been found (Tandon *et al.*, 2003; Donahue *et al.*, 2004).

Recent advances in science and engineering of UV irradiation have led to the development of continuous processing equipment, which rendered UV irradiation a viable alternative to thermal pasteurization for a range of liquid foods and ingredients (fresh juices, soft drinks, raw milk, liquid eggs, liquid sugars and sweeteners, etc.) (Koutchma, 2009). Currently, several companies market continuous UV equipment, such as Dill Instruments Inc (USA), manufactures an irradiator with a rotating cylinder with four LPM lamps that are able to process up to 100 l/h of product, and Oesco Inc. (USA), which manufactures a UV processor with 8 LPM lamps specifically designed to treat apple cider at around 1000 l/h.

13.2.6 Irradiation

Food irradiation is a physical means of food processing that involves exposing the pre-packaged or bulk foodstuffs, such as tuber and bulb crops, stored grains, dried ingredients, meat, poultry, fish and fruits, to gamma rays, X-rays, or electrons (Barbosa-Cánovas *et al.*, 1998). Foodstuffs are generally irradiated with gamma radiation from a radioisotope source, energetic electrons from particle accelerators and X-rays emitted by high energy electron beams (Cleland, 2006). The potential application of ionizing radiation in food processing is based mainly on the fact that ionizing radiation damages the DNA very effectively so that living cells become inactivated, therefore microorganisms, insect gametes, and plant meristems are prevented from reproducing, resulting in various preservative effects as a function of the absorbed radiation dose (Thayer, 1990). The choice of radiation source for a particular food will depend on thickness and density of the material, dose uniformity, ratio, minimum dose, processing rate and economics (Cleland, 2006). Gamma rays and X-rays have high penetrating characteristics, thus they can be used to treat food even in pallet-size containers (Farkas, 2006). These types of radiation do not induce radioactivity in foods or packaging materials and are available in quantities and at costs that allow commercial use of the irradiation processes which are considered safe by the World Health Organization (WHO) and Food and Agriculture Organization (FAO) (El-Samahy *et al.*, 2000; Farkas, 2004). The most important parameter in the irradiation process is the absorbed dose, which is proportional to the ionizing energy absorbed per unit mass of irradiated material measured in gray (Gy) or 1 J/kg (Cleland, 2006). Radiation treatment causes practically no temperature rise in the product and can be used as a terminal treatment in the packaged product (Borsa, 2006).

The effect of irradiation on microbial cells is mainly due to the damage of nucleic acids and generation of reactive oxygen species (mainly hydrogen peroxide), which result in oxidative damage to cell membrane (Abn and Lee, 2006). The moisture content of the product, temperature during irradiation, presence or absence of oxygen, the fresh or frozen state (increasing resistance in frozen state) influence the microbial resistance to radiation, particularly in the case of vegetative cells (Farkas, 2006). *Salmonella* serotypes seem to be the most radioresistant bacteria and the dose to eliminate salmonellae in food would also be sufficient to inactivate all non-sporeforming pathogens (Farkas, 1998). Recommended dose of radiation varies with the type of food to be irradiated, for example: frozen and chilled poultry, 3–5 and 1.5–2.5 kGy, respectively; 4 kGy to control pathogens in frozen fish, shrimp, prawn and frog-legs and *Vibrio vulnificus* in oysters; 4–6 kGy to kill food-borne parasites and 3–10 kGy to guarantee an adequate microbial safety level in spices, dried vegetables and herbal teas (Farkas, 1998; O'Bryan *et al.*, 2008; Arvanitoyannis *et al.*, 2009a, 2009b). Irradiation does not inactivate viruses, enzymes and microbial toxins within the doses recommended for foods (Farkas, 1998). Proteins, fats and carbohydrates are not notably altered by irradiation although higher doses may cause slight aroma and color changes in beef, pork and poultry (Wood and Bruhn, 2000; Abn and Lee, 2006). Lipid oxidation is accelerated by irradiation by generating hydroxyl radicals and off-odors. The presence of oxygen plays an important role in increasing lipid oxidation whereas in frozen foods free radical mobility is compromised, lowering the rate of lipid oxidation. In addition, sulfur compounds that are part of protein structure may generate off-odors during aerobic storage of irradiated foods. Undesirable changes in the final product can be minimized by using double packaging and antioxidant addition (Abn and Lee, 2006). New advances in lowering the irradiation dose that is needed to inactivate pathogenic bacteria, such as minimizing overdose zones or inducing sensitivity to microorganisms, are necessary to abate the adverse effects of irradiation on sensory properties of some products (Borsa, 2006).

At present there are approximately 60 commercial irradiation facilities operating in the United States and in over 40 countries irradiation is allowed (Sommers, 2006). Irradiation has been approved by the FDA to eliminate insects and bacteria from wheat, flour, spices, and fruits, to control sprouting of potatoes and onions and ripening of fruits and vegetables, control trichinosis in pork and inactivate pathogenic bacteria in red and poultry meat (Abn and Lee, 2006; Sommers, 2006). The FDA has now approved irradiation to eliminate harmful bacteria in lettuce and spinach leaves (FDA, 2008). Several companies across the United States supply irradiated ground beef and poultry meat and irradiation has been successfully used for shipping horticultural products from Hawaii to the United States and from Australia to New Zealand. In addition, authorities in Brazil have approved irradiation of all foods as a quarantine method (Borsa, 2006; Sommers, 2006). More than 200 gamma-ray facilities are being used for food radiation. Cobalt-60 (approved by the FDA) is the most widely used radiation source due to its ability to treat large pallet loads (30 tonnes per hour) of low density packages (0.3 g/cm^3) with a minimum dose of 2 kGy. More than 1000 industrial electron beam accelerators are available for irradiation with energies up to 10 MeV. However, only few are used for food irradiation (Cleland, 2006). X-rays with energies up to 7.5 MeV are also approved by the FDA. X-ray processing facilities that are capable of treating 50 tonnes/h of food in pallets with package density varying from 0.5 to 0.8 g/cm^3 and using a minimum dose of 2 kGy (Cleland, 2006).

13.2.7 High Intensity Ultrasounds

The term "ultrasound" in its most basic definition refers to pressure waves with a frequency of 20 kHz to 10 MHz (Butz and Tauscher, 2002). Power ultrasound, also known as high intensity ultrasound, refers to sound waves with low frequencies (20–100 kHz) and high sound intensity (10–1000 W/cm^2) (Feng *et al.*, 2008). The use of ultrasonics in industrial processes has two main requirements: a liquid medium and a source of high energy vibrations (the ultrasound) (Bates and Patist, 2009). During the sonication process, longitudinal waves are created when a sonic wave meets a liquid medium, thereby creating regions of alternating compression and expansion (Sala *et al.*, 1995). These regions of pressure change cause cavitation to occur and gas bubbles are formed in the medium, creating shock waves. The pressure changes resulting from these implosions are the main bactericidal effect in ultrasound, creating micromechanical shocks that disrupt cellular, structural and functional components up to the point of cell lysis (Butz and Tauscher, 2002). The combination of ultrasound in conjunction with pressure treatment (200–500 kPa) (manosonication), heat treatment (thermosonication) or both (manothermosonication) treatments has been proposed to enhance microbial reduction (Piyasena *et al.*, 2003). The enhanced mechanical disruption of cells by physical damage to the cell envelope in the form of wrinkles, ruptures, and perforations is the reason for the enhanced microbial destruction when ultrasound is combined with heat or pressure (Ugarte *et al.*, 2006, 2007). In recent years, studies into the application of multifrequency sonication to enhance cavitation have received increasing interest. The hypothesis is that multiple frequency sonication will allow bubbles with a wide range of sizes to generate cavitation, and hence increase the cavitation intensity (Feng *et al.*, 2008).

Enzyme inactivation caused by ultrasound has been attributed to the ability of sonication to break large macromolecules or particles by shear and mechanical stress generated by shock waves derived from bubble implosion; this causes the breakdown of hydrogen bonds and Van der Waals interactions, leading to the modification of the enzymes secondary and tertiary structure and loss of biological activity (Sala *et al.*, 1995; Feng *et al.*, 2008). In addition, as a result of intense cavitation, water molecules can be dissociated, generating highly reactive free radicals that can react with enzymes, so modifying their biological activity (Demirdoven and Baysal, 2009). High intensity ultrasound can also denature proteins and produce free radicals that can adversely affect the flavor of fruit based and high fat foods and milk (Sala *et al.*, 1995; Riener *et al.*, 2009). The sonication process seems not to affect to a large extent anthocyanins, ascorbic acid and color values from strawberry, blueberry and orange juice (5–15% loss) (Tiwari *et al.*, 2009a, 2009b; Valdramidis *et al.*, 2010). However, further research is needed to better understand the effects of ultrasound processing on bioactive compounds and sensory properties of foods.

The application of ultrasound as a laboratory-based technique for assisting extraction (Ultrasound Assisted Extraction, UAE) of metabolites from plant material has been widely demonstrated (Knorr 2003; Vilkhu *et al.*, 2008). As examples, commercial extraction systems using high power ultrasound in the food and beverage industry have been used in the extraction of phenolic compounds and pigments from grape marc (solid waste of the wine making process) and blueberries, extraction of lipids and proteins from plant seeds, improvement of oil extraction (palm fruit, corn germ and citrus fruit), extraction of bioactive compounds from coffee beans, tea and carrot (α- and β-carotene), extraction of isoflavones from soybeans and extraction of carbohydrates, polysaccharides and other functional compounds such as hemicelluloses (Bates and Patist, 2009; Demirdoven and Baysal, 2009). Thereby, ultrasound may reduce the dependence on solvents and enable the use of alternative solvents which may provide more attractive economics, environmental and health and safety benefits (Vilkhu *et al.*, 2008).

An ultrasound processing system consists of three basic parts: (i) generator; (ii) transducer; and (iii) coupler. Ultrasonic generators transform electrical energy into ultrasound energy (a type of mechanical energy) via a transducer, with piezoelectric transducers the most commonly used in commercial-scale applications due to their scalability (Bates and Patist, 2009). The intensity of the ultrasound treatments can be measured in terms of power (Demirdoven and Baysal, 2009). The scale-up of an ultrasonic treatment system can be achieved by integrating flow-cell modules, either in series or in parallel, providing 2–16 kW of power, with amplitudes ranging from 1 to 150 micron peak-to-peak displacement, and flows ranging from 1 to 1000 l/min, with an energy efficiency of 90–95% depending on the application (Feng *et al.*, 2008; Bates and Patist, 2009). Companies such as Cavitus Pty Ltd. (Australia) provide a wide range of high power ultrasonic equipment for different food applications, such as extraction, emulsification/homogenization, viscosity alteration, de-foaming and cleaning/sanitation in the wine industry with a payback less than two years (Bates and Patist, 2009).

13.2.8 Hurdle approach

All non-thermal technologies cited in this chapter have several drawbacks. In general, high production costs, the unavailability of industrial-scale equipment, the inability to reduce spores and the lack of effective communication to the consumers about the principles and benefits of these technologies are the main limitations. Taking into account all of these limitations, the hurdle approach has been proposed as the most efficient way to improve overall effectiveness. The combination of temperature and/or low pH with other stress factors, such as pressure, electric field or radiation, has been demonstrated to have a synergistic effect on microbial and enzyme inactivation in numerous studies. Research on high pressure treatment at either elevated temperature for high pressure assisted sterilization or low temperature for high pressure assisted quick freezing and combination of ultrasound and temperature treatment (thermosonication) are some examples. In addition, the use of a combination of different non-thermal processes by applying them consecutively, that is HPP + PEF, HPP + irradiation, and PEF + ultrasound, may represent a new approach to improve the efficacy of the entire process by decreasing the intensity of treatments required for any process alone. Lowering pH, oxygen concentration and water activity during storage have also been used as a barrier to control the growth of microorganisms and preventing the recovery of any sublethally injured cell during storage.

The use of biopreservatives in combination with non-thermal technologies has also been proposed to enhance the effectiveness of these processes against vegetative cells and bacterial spores and help to control microbial growth during storage under chilled conditions. In this sense, bacteriocins and natural antimicrobial compounds, such as nisin, lysozyme, lactoferrin, lactoperoxidase system, chitosan and essential oils from diverse plants, have been successfully applied. Lately, there has been a growing interest in using natural ingredients as biopreservatives. In addition to providing interesting flavors and colors, natural ingredients produced from microorganisms, plants and animals contain substances with antioxidant and antimicrobial properties. To be used in food applications, these antimicrobial agents of natural origin need to be recognized as GRAS (Generally Recognized as Safe) and/or approved as food ingredients by the FDA.

13.3 FUTURE TRENDS

Because of the limitations of traditional thermal and non-thermal technologies, the future of food preservation is moving towards *a-la-carte processing* that involves designing a specific treatment for each type of food. The scientific community will continue to search for new developments and applications of non-thermal technologies to fulfil the new challenges of the twenty-first century with an emphasis on the environmental impact and sustainability of these technologies, the use of non-thermal technologies to reduce food allergies, to produce foods with natural ingredients, eliminate the use of potential carcinogenic additives, and develop healthier and convenient foods for high risk population groups, such as the elderly, children and the inmunocompromised consumers.

Despite all the efforts of the scientific community and equipment manufacturers to develop non-thermal technologies, there are still some important points that need to be addressed. The selection of the target microorganism as well as the non-pathogenic surrogates for each technology is still limited. In addition, the mechanism of microbial inactivation by non-thermal processing needs to be fully understood. Spores inactivation is still a main limitation to the industrial application of most non-thermal technologies. It is clear that new approaches have to be developed to reduce spore counts to a safe level. Regulatory authorities also demand toxicological studies before considering non-thermal technologies safe for the public.

In the last few years great advances have been achieved regarding the understanding of the effects of non-thermal technologies on quality-deteriorative enzymes, sensory properties and bioactive compounds in different food matrices. Many studies have shown the potential to reduce enzyme activity while preserving the fresh-like characteristics of food. However, despite the efforts, deeper scientific knowledge of the mechanism of non-thermal inactivation of enzymes and their quality-related effects is still needed. It appears that non-thermal technologies preserve most of the health beneficial phytochemicals. However, clinical and *in vivo* studies on bioavailability of these compounds after non-thermal processing are still limited. It is apparent that a large gap still exists among industry, the scientific community and consumers. Effective communication strategies are needed to close this gap. Studies on consumer acceptance and preference of non-thermal technologies would also help to steer future research and development efforts.

It is the authors' belief that in the near future new industrial applications of non-thermal technologies to produce healthy, fresh-like and natural foods in a sustainable manner will become common in the market place.

ACKNOWLEDGEMENTS

The authors wish to thank Dr Dave Geveke of the USDA Eastern Regional Research Center for helping in the reviewing process of this manuscript.

Mention of trade names or commercial products in this publication is solely for the purpose of providing specific information and does not imply recommendation or endorsement by the US Department of Agriculture.

REFERENCES

Abn DU, Lee EJ. 2006. Mechanisms and prevention of quality changes in meat by irradiation. In: Sommers CH, Fan X (editors). Food irradiation research and technology. Iowa: Blackwell Publishing Ltd., p. 127.

Ade-Omowaye BIO, Angersbach A, Taiwo KA, Knorr D. 2001. Use of pulsed electric field pre-treatment to improve dehydration characteristics of plant based foods. Trends Food Sci Technol 12:285–295.

Aronsson K, Ronner U, Borch E. 2005. Inactivation of *Escherichia coli*, *Listeria innocua* and *Saccharomyces cerevisiae* in relation to membrane permeabilization and subsequent leakage of intracellular compounds due to pulsed electric field processing. Int J Food Micro 99:19–32.

Arvanitoyannis IS, Stratakos A, Mente E. 2009a. Impact of irradiation on fish and seafood shelf life: a comprehensive review of applications and irradiation detection. Crit Rev Food Sci Nutr 49:68–112.

Arvanitoyannis IS, Stratakos A, Tsarouhas P. 2009b. Irradiation applications in vegetables and fruits: a review. Crit Rev Food Sci Nutr 49:427–462.

Bachmann R. 1975. Sterilization by intense ultraviolet radiation. The Brown Boveri Review 62, 206–209.

Balasubramaniam VM. 2003. High pressure food preservation. In: Heldman D (editor). Encyclopedia of Agriculture, Food and Biological Engineering. New York: Marcel Dekker, Inc. pp. 490–496.

Barbosa-Cánovas GV, Palou E, Pothakamury UR, Swanson BG. 1998. Nonthermal preservation of foods. New York: Marcel Dekker Inc. pp. 9–20, 53–110.

Barka EA, Kalantari S, Makhlouf J, Arul J. 2000. Impact of UV-C irradiation on the cell wall-degrading enzymes during ripening of tomato (*Lycopersicon esculentum* L.) fruit. J Agric Food Chem 48:667–671.

Bates D, Patist A. 2009. Industrial applications of high power ultrasonics in the food, beverage and wine industry. Institute of Food Technology (IFT), Nonthermal Processing Division Newsletter, 11:2–15.

Baxter IA, Easton K, Schneebeli K, Whitfield FB. 2005. High pressure processing of Australian navel orange juices: Sensory analysis and volatile flavor profiling. Innov Food Sci Emerg Technol 6:372–387.

Bendicho S, Espachs A, Arantegui J, Martin O. 2002. Effect of high intensity pulsed electric fields and heat treatments on vitamins of milk. J Dairy Res 69:113–123.

Bevilacqua A, Costa C, Corbo MR, Sinigaglia M. 2009. Effects of the high pressure of homogenization on some spoiling micro-organisms, representative of fruit juice microflora, inoculated in saline solution. Letters Appl Micro 48:261–267.

Boff JM, Truong TT, Min DB, Shellhammer TH. 2003. Effect of thermal processing and carbon dioxide-assisted high-pressure processing on pectinmethylesterase and chemical changes in orange juice. J Food Sci 68:1179–1184.

Borsa J. 2006. Introduction: Food irradiation moving on. In: Sommers CH, Fan X (editors). Food irradiation research and technology. Iowa: Blackwell Publishing Ltd., pp. 6–7.

Bríñez W, Roig-Sagués A, Herrero M. 2006a. Inactivation by ultrahigh-pressure homogenization of *Escherichia coli* strains inoculated into orange juice. J Food Prot 69:984–989.

Bríñez WJ, Roig-Sagués AX, Hernández MM, Guamis B. 2006b. Inactivation of *Listeria innocua* in milk and orange juice by ultrahigh-pressure homogenization. J Food Prot 69:86–92.

Bríñez WJ, Roig-Sagués AX, Hernández MM, Guamis B. 2007. Inactivation of *Staphylococcus* spp. strains in whole milk and orange juice using ultra high pressure homogenization at inlet temperatures of 6 and 20°C. Food Control 18:1282–1288.

Butz P, Tauscher B. 2002. Emerging technologies: chemical aspects. Food Res Int 35:279–284.

Campos FP, Cristianini M. 2007. Inactivation of *Saccharomyces cerevisiae* and *Lactobacillus plantarum* in orange juice using ultra high-pressure homogenization. Innov Food Sci Emerg Technol 8:226–229.

Capra ML, Patrignani F, del Luján Quiberoni A, Reinheimer JA, Lanciotti R, Guerzoni ME. 2009. Effect of high pressure homogenization on lactic acid bacteria phages and probiotic bacteria phages. Int Dairy J 19:336–341.

Carreño JM, Gurrea MC, Sampedro F, Carbonell JV. 2010. Effect of high hydrostatic pressure and high pressure homogenisation on *Lactobacillus plantarum* inactivation kinetics and quality parameters of mandarin juice. Eur Food Res Tech 232:265–274.

Clark JP. 2006. Pulsed electric field processing. Food Tech 1:66–67.

Cleland MR. 2006. Advances in gamma ray, electron beam, and x-ray technologies for food radiation. In: Sommers CH, Fan X (editors). Food irradiation research and technology. Iowa: Blackwell Publishing Ltd., p. 11.

Cruz NS, Capellas M, Jaramillo DP, Trujillo AJ, Guamis B, Ferragut V. 2009. Soymilk treated by ultra high-pressure homogenization: Acid coagulation properties and characteristics of a soy-yogurt product. Food Hydr 23:490–496.

Damar S, Balaban MO. 2006. Review of dense phase CO_2 technology: microbial and enzyme inactivation, and effects on food quality. J Food Sci 71:1–11.

Demirdoven A, Baysal T. 2009. The use of ultrasound and combined technologies in food preservation. Food Rev Int 25:1–11.

Diels AMJ, Michiels CW. 2006. High-pressure homogenization as a non-thermal technique for the inactivation of microorganisms. Crit Rev Micro 32:201–216.

Diels AMJ, De Taeye J, Michiels CW. 2005a. Sensitisation of *Escherichia coli* to antibacterial peptides and enzymes by high-pressure homogenization. Int J Food Micro 105:165–175.

Diels AMJ, Callewaert L, Wuytack EY, Masschalck B, Michiels CW. 2005b. Inactivation of *Escherichia coli* by high-pressure homogenization is influenced by fluid viscosity but not by water activity and product composition. Int J Food Micro 101:281–291.

Donahue DW, Canitez N, Bushway AA. 2004. UV inactivation of *E. coli* 0157:H7 in apple cider: quality, sensory and shelf-life analysis. J Food Process Preserv 28:368–387.

Donsì F, Ferrari G, Lenza E, Maresca P. 2009. Main factors regulating microbial inactivation by high-pressure homogenization: operating parameters and scale of operation. Chem Eng Sci 64:520–532.

El-Samahy SK, Youssef BM, Aaskar AA, Swailam HMM. 2000. Microbiological and chemical properties of irradiated mango. J Food Safety 20:139–156.

Farkas J. 1998. Irradiation as a method for decontaminating food: A review. Int J Food Micro 44:189–204.

Farkas J. 2004. Food irradiation. In: Mozumder A, Hatano Y (editors). Charged particle and photon interactions with matter. New York: Marcel Dekker, Inc. pp. 785–812.

Farkas J. 2006. Irradiation for better foods. Trends Food Sci Tech 17:148–152.

FDA. 2008. Irradiation in the Production, Processing and Handling of Food. Federal Register, 73, number 164, 49593–49603. Food and Drug Administration, Department of Health and Human Services (HHS).

Feng H, Yang W, Hielscher T. 2008. Power Ultrasound. Food Sci Tech Int 14:433–436.

Fernández-García A, Butz P, Bognàr A, Tauscher B. 2001. Antioxidative capacity, nutrient content and sensory quality of orange juice and an orange-lemon-carrot juice product after high pressure treatment and storage in different packaging. Eur Food Res Tech 213:290–296.

García D, Gómez N, Mañas P, Raso J, Pagán R. 2007. Pulsed electric fields cause bacterial envelopes permeabilization depending on the treatment intensity, the treatment medium pH and the microorganism investigated. Int J Food Micro 113:219–227.

García-González L, Geeraerd AH, Spilimbergo S, *et al.* 2007. High pressure carbon dioxide inactivation of microorganisms in foods: The past, the present and the future. Int J Food Micro 117:1–28.

García-González L, Geeraerd AH, Elst K, Van Ginneken L, Van Impe JF, Devlieghere F. 2009. Influence of type of microorganism, food ingredients and food properties on high-pressure carbon dioxide inactivation of microorganisms. Int J Food Micro 129:253–263.

Geveke DJ, Brunkhost C. 2003. Inactivation of *Saccharomyces cerevisiae* using radio frequency electric fields. J Food Prot 66:1712–1715.

Góngora-Nieto MM, Sepúlveda DR, Pedrow P, Barbosa-Cánovas GV, Swanson BG. 2002. Food processing by pulsed electric fields: treatment delivery, inactivation level and regulatory aspects. Lebensm Wiss Tech 35:375–388.

Góngora-Nieto MM, Pedrow PD, Swanson BG, Barbosa-Cánovas GV. 2003. Impact of air bubbles in a dielectric liquid when subjected to high field strengths. Innov Food Sci Emerg Tech 4:57–67.

Guderjan M, Élez-Martínez P, Knorr D. 2007. Application of pulsed electric fields at oil yield and content of functional food ingredients at the production of rapeseed oil. Innov Food Sci Emerg Tech 8:55–62.

Guerrero-Beltrán JA, Barbosa-Cánovas GV. 2004. Review: advantages and limitations on processing foods by UV light. Food Sci Tech Int 10:137–147.

Guerrero-Beltrán JA, Barbosa-Cánovas GV. 2006. Inactivation of *Saccharomyces cerevisiae* and polyphenoloxidase in mango nectar treated with UV light. J Food Prot 69:362–368.

Gunes G, Blum LK, Hotchkiss JH. 2005. Inactivation of yeasts in grape juice using a continuous dense phase carbon dioxide processing system. J Sci Food Agr 85:2362–2368.

Gurtler JB, Rivera RB, Zhang HQ and Geveke DJ. 2010. Selection of surrogate bacterial in place of *E. coli* O157:H7 and *Salmonella Typhimurium* for pulsed electric field treatment of orange juice. International Journal of Food Microbiology. 139:1–8.

Heinz V, Álvarez I, Angersbach A, Knorr D. 2002. Preservation of liquid foods by high intensity pulsed electric fields-basic concepts for process design. Trends Food Sci Tech 12:103–111.

Heremans K. 2002. The effects of high pressure on biomaterials. In: Hendrickx M, Knorr D (editors). Ultra High Pressure Treatments of Foods. New York: Kluwer Academic. pp. 23–25.

Ho S, Cross J, Griffiths M. 1995. Inactivation of *Pseudomonas fluorescens* by high voltage electric pulses. J Food Sci 60:1337–1343.

Hong SI, Pyun YR. 1999. Inactivation kinetics of *Lactobacillus plantarum* by high pressure carbon dioxide. J Food Sci 64:728–733.

Hong SI, Park WS, Pyun YR. 1999. Non-thermal inactivation of *Lactobacillus plantarum* as influenced by pressure and temperature of pressurized carbon dioxide. Int J Food Sci Tech 34:125–130.

Huang K, Wang J. 2009. Designs of pulsed electric fields treatment chambers for liquid foods pasteurization process: A review. J Food Eng 95:227–239.

Innocente N, Biasutti M, Venir E, Spaziani M, Marchesini G. 2009. Effect of high-pressure homogenization on droplet size distribution and rheological properties of ice cream mixes. J Dairy Sci 92:1864–1875.

Jia M, Zhang QH, Min DB. 1999. Pulsed electric field processing effects on flavor compounds and microorganisms of orange juice. Food Chem 65:445–451.

Kincal D, Hill WS, Balaban MO, Portier KM, Wei CI, Marshall MR. 2005. A continuous high pressure carbon dioxide system for microbial reduction in orange juice. J Food Sci 70:249–254.

Kincal D, Hill WS, Balaban MO, *et al.* 2006. A continuous high-pressure carbon dioxide system for cloud and quality retention in orange juice. J Food Sci 71:338–344.

Knorr D. 2003. Impact of non-thermal processing on plant metabolites. J Food Eng 56:131–134.

Koutchma T. 2009. Advances in ultraviolet light technology for non-thermal processing of liquid foods. Food Bioprocess Tech 2:138–155.

Kumar S, Thippareddi H, Subbiah J, Zivanovic S, Davidson PM, Harte F. 2009. Inactivation of *Escherichia coli* k-12 in apple juice using combination of high-pressure homogenization and chitosan. J Food Sci 74:8–14.

Laboissière LHES, Deliza R, Barros-Marcellini AM, Rosenthal A, Camargo LMAQ, Junqueira RG. 2007. Effects of high hydrostatic pressure (HHP) on sensory characteristics of yellow passion fruit juice. Innov Food Sci Emerg Tech 8:469–477.

Liao H, Hu X, Liao X, Chen F, Wu J. 2007. Inactivation of *Escherichia coli* inoculated into cloudy apple juice exposed to dense phase carbon dioxide. Int J Food Micro 118:126–131.

Lim S, Yagiz Y, Balaban MO. 2006. Continuous high pressure carbon dioxide processing of mandarin juice. Food Sci BioTech 15:13–18.

Liltved H, Landfald B. 2000. Effects of high intensity light on ultraviolet-irradiated and non-irradiated fish pathogenic bacteria. Water Res 34:481–486.

Liu X, Gao Y, Peng X, Yang B, Xu H, Zhao J. 2008. Inactivation of peroxidase and polyphenol oxidase in red beet (*Beta vulgaris* L.) extract with high pressure carbon dioxide. Innov Food Sci Emerg Tech 9, 24–31.

López-Pedemonte T, Bríñez WJ, Roig-Sagués AX, Guamis B. 2006. Fate of *Staphylococcus aureus* in cheese treated by ultrahigh pressure homogenization and high hydrostatic pressure. J Dairy Sci, 89:4536–4544.

Ludikhuyze L, Van Loey A, Indrawati, Hendrickx M. 2002. High pressure processing of fruit and vegetables. In: Hendrickx M, Knorr D (editors). Ultra High Pressure Treatments of Foods. New York: Kluwer Academic. pp. 168–171.

Mañas P, Pagán R. 2005. Microbial inactivation by new technologies of food preservation. J Appl Micro 98:1387–1399.

Masschelein WJ. 2002. Ultraviolet light in water and wastewater sanitation. Boca Raton, FL: CRC Press.

Matak KE, Sumner SS, Duncan SE, *et al.* 2007. Effects of ultraviolet irradiation on chemical and sensory properties of goat milk. J Dairy Sci, 90:3178–3186.

Min S, Zhang QH. 2003. Effects of commercial-scale pulsed electric field processing on flavor and color of tomato juice. J Food Sci 68:1600–1606.

Min S, Jin ZT, Min SK, Yeom H, Zhang QH. 2003a. Commercial-scale pulsed electric field processing of orange juice. J Food Sci 68:1265–1271.

Min S, Jin ZT, Zhang QH. 2003b. Commercial scale pulsed electric field processing of tomato juice. J Agr Food Chem 51:3338–3344.

Morgan R. 1989. UV "green" light disinfection. Dairy Ind Int 54:33–35.

Mosqueda-Melgar J, Élez-Martínez P, Raybaudi-Massilia RM, Martín-Belloso O. 2008. Effects of pulsed electric fields on pathogenic microorganisms of major concern in fluid foods: a review. Crit Rev Food Sci Nutr 48:747–759.

National Center for Food Safety and Technology. 2009. NCFST receives regulatory acceptance of novel food sterilization process. Press release, Illinois Institute of Technology.

O'Bryan CA, Crandall PG, Ricke SC, Olson DG. 2008. Impact of irradiation on the safety and quality of poultry and meat products: a review. Crit Rev Food Sci Nutr 48:442–457.

Oey I, Lilleb M, Van Loey A, Hendrickx M. 2008a. Effect of high pressure processing on colour, texture and flavour of fruitand vegetable-based food products: A review. Trends Food Sci Tech 19:320–328.

Oey I, Van der Plancken I, Van Loey A, Hendrickx M. 2008b. Does high pressure processing influence nutritional aspects of plant based food systems? Trends Food Sci Tech 19:300–308.

Oxen P, Knorr D. 1993. Baroprotective effects of high solute concentrations against inactivation of *Rhodotorula rubra*. Lebens Wiss Tech 26:220–223.

Parish ME. 1998. Orange juice quality after treatment by thermal pasteurization or isostatic high pressure. Lebens Wiss Tech 31:439–442.

Park SJ, Lee JI, Park J. 2002. Effects of a combined process of high-pressure carbon dioxide and high hydrostatic pressure on the quality of carrot juice. J Food Sci 67:1827–1834.

Piyasena P, Mohareb E, McKellar RC. 2003. Inactivation of microbes using ultrasound: a review. Int J Food Micro 87:207–216.

Popper L, Knorr D. 1990. Applications of high-pressure homogenization for food preservation. Food Tech 44:84–89.

Qin B, Barbosa-Cánovas GV, Swanson B, Pedrow P. 1998. Inactivating microorganism using a pulsed electric field continuous treatment system. IEEE Trans Ind Appl 34:43–49.

Rastogi NK, Raghavarao KSMS, Balasubramaniam VM, Niranjan K, Knorr D. 2007. Opportunities and challenges in high pressure processing of foods. Crit Rev Food Sci Nutr 47: 69–112.

Riener J, Noci F, Cronin DA, Morgan DJ, Lyng JG. 2008. Effect of high intensity pulsed electric fields on enzymes and vitamins in bovine raw milk. Int J Dairy Tech 62:1–6.

Riener J, Noci F, Cronin DA, Morgan DJ, Lyng JG. 2009. Characterisation of volatile compounds generated in milk by high intensity ultrasound. Int Dairy J 19:269–272.

Rivas A, Rodrigo D, Company B, Sampedro F, Rodrigo M. 2007. Effects of pulsed electric fields on water-soluble vitamins and ACE inhibitory peptides added to a mixed orange juice and milk beverage. Food Chem 104:1550–1559.

Roig-Sagués AX, Velázquez RM, Montealegre-Agramont P, *et al.* 2009. Fat content increases the lethality of ultra-high-pressure homogenization on *Listeria monocytogenes* in milk. J Dairy Sci 92:5396–5402.

Rovere P. 2002. Industrial-scale high pressure processing of foods. In: Hendrickx M, Knorr D (editors). Ultra High Pressure Treatments of Foods. New York: Kluwer Academic. pp. 251–255.

Sala FJ, Burgos J, Condón S, López P, Raso J. 1995. Effect of heat and ultrasound on microorganisms and enzymes, In: Gould GW (editor). New Methods of Food Preservation. London: Blackie Academic & Professional. pp. 176–204.

Sastry SK, Datta AK, Worobo RW. 2000. Ultraviolet light. J Food Sci 65:90–92.

Serra M, Trujillo AJ, Guamis B, Ferragut V. 2009. Flavour profiles and survival of starter cultures of yoghurt produced from high-pressure homogenized milk. Int Dairy J 19:100–106.

Sizer CE, Balasubramaniam VM. 1999. New intervention processes for minimally processed Juices. Food Tech 53:64–67.

Soliva-Fortuny R, Balasa A, Knorr D, Martin-Belloso O. 2009. Effects of pulsed electric fields on bioactive compounds in foods: a review. Trends Food Sci Tech 20:544–556.

Sommers CH. 2006. Preface. In: Sommers CH, Fan X (editors). Food irradiation research and technology. Iowa: Blackwell Publishing Ltd., pp. XV–XVI.

Spikes J. 1981. Photodegradation of foods and beverages. In: Smith KC (editor). Photochemical and Photobiological Reviews. New York: Plenum Press. pp. 39–81.

Tahiri I, Makhlouf J, Paquin P, Fliss I. 2006. Inactivation of food spoilage bacteria and *Escherichia coli* O157:H7 in phosphate buffer and orange juice using dynamic high pressure. Food Res Int 39:98–105.

Tandon K, Worobo RW, Churey JJ, Padilla-Zakour OI. 2003. Storage quality of pasteurized and UV treated apple cider. J Food Process Pres 27:21–35.

Thakur BR, Nelson PE. 1998. High-pressure processing and preservation of food. Food Rev Int 14:427–447.

Thayer DW. 1990. Food irradiation: benefits and concerns. J Food Quality 13:147–169.

Tiwari BK, O'Donnell CP, Patras A, Brunton N, Cullen PJ. 2009a. Stability of anthocyanins and ascorbic acid in sonicated strawberry juice during storage. Eur Food Res Tech 228:717–724.

Tiwari BK, O'Donnell CP, Cullen PJ. 2009b. Effect of sonication on retention of anthocyanins in blackberry juice. J Food Eng 93:166–171.

Toepfl S, Heinz V, Knorr D. 2007. High intensity pulsed electric fields applied for food preservation. Chem Eng Process 46:537–546.

Tomasula PM. 2003. Supercritical fluid extraction of foods. In: Heldman D (editor). Encyclopedia of Agricultural, Food, and Biological Engineering. New York: Marcel Dekker, Inc. pp. 964–967.

Tonello, C. 2008. "All natural" products through high pressure processing. Institute of Food Technology (IFT), Nonthermal Processing Division Newsletter 10:5–6.

Torres JA, Velazquez G. 2005. Commercial opportunities and research challenges in the high pressure processing of foods. J Food Eng 67:95–112.

Tran MTT, Farid M. 2004. Ultraviolet treatment of orange juice. Innov Food Sci Emerg Tech 5:495–502.

Ugarte E, Feng H, Martin ES. 2007. Inactivation of *Shigella* and *Listeria monocytogenes* with power ultrasound at sub-lethal and lethal temperatures. J Food Sci 72:103–107.

Ugarte E, Feng H, Martin ES, Cadwallader KR. 2006. Inactivation of *Escherichia coli* with power ultrasound in apple cider. J Food Sci 71:102–108.

Valdramidis VP, Cullen PJ, Tiwari BK, O'Donnell CP. 2010. Quantitative modeling approaches for ascorbic acid degradation and non-enzymatic browning of orange juice during ultrasound processing. J Food Eng 96:449–454.

van der Berg RW, Hoogland H, Lelieveld HLM, Schepdael LV. 2002. High pressure equipment designs for food processing applications. In: Hendrickx M, Knorr D (editors). Ultra High Pressure Treatments of Foods. New York: Kluwer Academic. pp. 297–304.

Vannini L, Patrignani F, Iucci L, *et al.* 2008. Effect of a pre-treatment of milk with high pressure homogenization on yield as well as on microbiological, lipolytic and proteolytic patterns of "Pecorino" cheese. Int J Food Micro 128:329–335.

Vega-Mercado H, Martín-Belloso O, Qin B-L, *et al.* 1997. Non-thermal food preservation: Pulsed electric fields. Trends Food Sci Tech 8:151–157.

Vilkhu K, Mawson R, Simons L, Bates D. 2008. Applications and opportunities for ultrasound assisted extraction in the food industry-A review. Innov Food Sci Emerg Tech 9:161–169.

Welti-Chanes J, Ochoa-Velasco CE, Guerrero-Beltrán JA. 2009. High-pressure homogenization of orange juice to inactivate pectinmethylesterase. Innov Food Sci Emerg Tech 10:457–462.

Wilson DR, Dabrowski L, Stringer L, Moezelaar R, Brocklehurst TF. 2008. High pressure in combination with elevated temperature as a method for the sterilisation of food. Trends Food Sci Tech 19:289–299.

Wood OB, Bruhn CM. 2000. Position of the American Dietetic Association: food irradiation. J Amer Dietetic Assoc 100:246–253.

Wouters PC, Alvarez I, Raso J. 2001a. Critical factors determining inactivation kinetics by pulsed electric field food processing. Trends Food Sci Tech 12:112–121.

Wouters PC, Bos ADP, Ueckert J. 2001b. Membrane permeabilization in relation to inactivation kinetics of *Lactobacillus* species due to pulsed electric fields. Appl Environ Micro 67:3092–3101.

Wright JR, Sumner SS, Hackney CR, Pierson MD, Zoecklein BW. 2000. Efficacy of ultraviolet light for reducing *Escherichia coli* O157:H7 in unpasteurized apple cider. J Food Prot 63:563–567.

Wuytack EY, Diels AMJ, Michiels CW. 2002. Bacterial inactivation by high-pressure homogenization and high hydrostatic pressure. Int J Food Micro 77:205–212.

Wuytack EY, Duong Thi Phuong L, Aertsen A, *et al.* 2003. Comparison of sublethal injury induced in *Salmonella enterica* serovar Typhimurium by heat and by different nonthermal treatments. J Food Prot 66:31–37.

Yin Y, Zhang QH, Sastry SK. 1997. High voltage pulsed electric field treatment chambers for the preservation of liquid food products. United States Patent 5,690,978.

Yuk HG, Geveke DJ, Zhang HQ. 2010. Efficacy of supercritical carbon dioxide for nonthermal inactivation of *Escherichia coli* K12 in apple cider. Int J Food Micro 138:91–99.

Zimmermann U. 1986. Electrical breakdown, electropermeabilization and electrofusion. Rev Physiol Biochem Pharmacol 105:175–256.

Zhang Q, Barbosa-Cánovas GV, Swanson BG. 1995. Engineering aspects of pulsed electric field pasteurization. J Food Eng 25:261–281.
Zhou L, Wang Y, Hu X, Wu J, Liao X. 2009. Effect of high pressure carbon dioxide on the quality of carrot juice. Innov Food Sci Emerg Tech 10:321–327.
Zulueta A, Esteve MJ, Frasquet I, Frigola A. 2007. Fatty acid profile changes during orange juice-milk beverage processing by high-pulsed electric field. Eur J Lipid Sci Tech 109:25–31.

14 Enzymes as Biocatalysts for Lipid-based Bioproducts Processing

Ling-Zhi Cheong, Zheng Guo, Sergey N. Fedosov, Bena-Marie Lue, Ram C.R. Jala, Gündüz Güzel, and Xuebing Xu

14.1 INTRODUCTION

Bioproducts are materials that are derived from renewable biological resources such as agriculture, forestry, and biological waste. The use of enzymes as biocatalysts for lipid-based bioproducts processing has been discussed and reviewed both in the academia and industry over the last five years (Hou, 2005; Xu *et al.*, 2006, 2007a, 2008; Divakar and Manchar, 2007; Naik *et al.*, 2010; Fernandez-lafuente, 2010). This chapter provides insights into distinct enzyme characteristics essential in industrial processing, including enzyme kinetics. A brief review of the industrial applications of enzymes is also given with special focus on enzymatic processing of four different lipid-based bioproducts, namely: partial acylglycerols, bioactive compounds, phospholipids and fatty acid alkyl esters.

14.2 ENZYME CHARACTERISTICS

Biocatalysis refers to the applications of either isolated enzymes or whole cells as catalysts (Schmid *et al.*, 2001). One of the remarkable characteristics of enzymes is their chemical precision or selectivity. Enzymes selectivity can be categorized into substrate, stereo, region and functional group selectivity. Enzymes with substrate selectivity, such as lipases, can be used to selectively remove, separate or enrich specific fatty acids for nutritional and pharmaceutical applications (Xu, 2004a). Stereoselectivity is the enzyme's capability to selectively act on a single enantiomer or diastereomer. Aspartase has been used industrially to stereoselectively add ammonia to fumarate producing highly pure L-aspartic acid at low cost (Fusee, 1987). Regioselectivity refers to the enzyme's ability to act on one location in a molecule. This has been widely used for preparation of structured lipids (Xu, 2004a). Meanwhile functional group selectivity is defined as enzyme selectivity towards a certain functional group in the presence of other equally or more reactive functional groups. For example, Nitrilase is capable of selectively hydrolyzing nitrile group in the presence of ester or amide, which is almost impossible through traditional chemical means (Rozzell, 1999).

Another remarkable characteristic of enzymes is their high catalytic efficiency. Enzyme-catalyzed reactions usually display high reaction rate accelerations from 10^5 to 10^{17} fold,

Food and Industrial Bioproducts and Bioprocessing, First Edition. Edited by Nurhan Turgut Dunford.

which are remarkable taking into consideration these reactions normally occur at ambient temperature and pressure. For example, 1 kg of immobilized aspartate ammonia lyase can produce more than 100 000 kg of L-aspartic acid, making it one of the most efficient biocatalytic processes known (Rozzell, 1999).

Enzymes were once dictated as requiring water to maintain their active conformation for catalytic activity. Hence, enzymatic reactions were initially conducted in aqueous solution. This has greatly undermined the capabilities of enzymes and limited its applications. In the 1960s, certain enzymes, namely chymotrypsin (Dastoli *et al.*, 1966), xanthine oxidase (Dastoli and Price, 1967) and porcine pancreatic lipase (Zaks and Klibanov, 1985) were found to be catalytically active even when suspended in organic solvents. This has led to the finding that only a minute amount of water is required to maintain the activity of enzyme. Hence, it is now possible to apply enzymes as catalysts in non-aqueous reactions such as organic synthesis (Zaks and Klibanov, 1985; Schmid *et al.*, 2001).

In addition, enzymes are able to act on a broad range of substrates (Rozzell, 1999; Schmid *et al.*, 2001). Most importantly, unlike chemical catalysts, enzymes function at mild reaction conditions, ambient temperature, atmospheric pressure, and neutral pH.

14.3 ENZYME KINETICS IN INDUSTRIAL APPLICATIONS

Principles of enzyme kinetics have been reviewed in textbooks by Marangoni (2003) and Cornish-Bowden (2004). The rate of any chemical reaction follows strict mathematical rules, and understanding of the rules is important for correct predictions of the process timing, particularly in industrial scale processing. The following section briefly introduces some basic aspects of the enzyme kinetics with particular focus on lipases.

Enzyme (E) accelerates conversion between a substrate (S) and a product (P) by formation of the intermediate binding complexes ES and EP.

$$E+S \leftrightarrow ES \leftrightarrow EP \leftrightarrow E+P \tag{14.1}$$

Initially, the reaction proceeds forward in the absence of product. The rate of the reaction at this stage ($v=v_+$) depends only on the substrate concentration and follows the Michaelis–Menten equation.

$$v=\frac{V_+}{1+\frac{K_{mS}}{s}};\quad v\approx\frac{V_+}{K_{mS}}s\,(s<<K_{mS});\quad v\approx V_+\,(s>>K_{mS}) \tag{14.2}$$

where $V_+=k_+\cdot e_0$ is the maximal v_+ at the current enzyme concentration e_0; s is the concentration of free substrate (usually equated to the total substrate $s_0=s+es\approx s$ if $s_0>>e_0$); K_{mS} is the Michaelis constant or $s_{½}$ (i.e. the concentration of substrate giving a half-maximal rate). The Michaelis constant is often identical to the dissociation constant K_s, which describes the equilibrium $E+S\leftrightarrow ES$.

The most common biochemical reactions typically involve two substrates (A and B) and one or two product (P and Q):

$$A+B+E \leftrightarrow P+Q+E \tag{14.3}$$

The initial reaction rate ($v \approx v_+$ at $p, q \approx 0$) obeys the general equation below with two variables:

$$v = \frac{V_+}{\left(1 + \frac{K_{mA}}{a} + \frac{K_{mB}}{b} + \frac{K_{sA}K_{mB}}{ab}\right)} \tag{14.4}$$

where V_+ is the maximal rate of forward reaction; K_{mA}, K_{mB} are the Michaelis constants (which refer to the respective equilibriums of $A+EB \leftrightarrow EAB$ and $B+EA \leftrightarrow EAB$ for a pseudo-equilibrium binding mechanism); K_{sA} describes the dissociation of $A+E \leftrightarrow EA$; a and b are the concentrations of substrates A and B, respectively.

Bi-substrate reactions follow one of the specific binding mechanisms, for example random, sequential or ping-pong. The scheme below describes the equilibrium random binding of A and B to E as:

$$E + A \leftrightarrow EA;\ E + B \leftrightarrow EB;\ EA + B \leftrightarrow EAB;\ EB + A \leftrightarrow EAB;\ EAB \rightarrow E + P + Q \tag{14.5}$$

where the rate of the reaction depends on a and b in accordance with Equation 14.4. Sequential binding follows a different pattern:

$$E + A \leftrightarrow EA;\ EA + B \leftrightarrow EAB;\ EAB \rightarrow E + Q \tag{14.6}$$

where A does not dissociate from EAB. This results in $K_{\mathrm{mA}} = 0$ considering a pseudo-equilibrium binding mechanism. Description of a ping-pong reaction requires two sequential steady state steps:

$$E + A \leftrightarrow EA \rightarrow P + EX;\ EX + B \leftrightarrow EXB \rightarrow E + Q \tag{14.7}$$

This model results in $K_{sA} \cdot K_{mB} = 0$ at p and $q = 0$.

Evaluation of parameters of a bisubstrate reaction is relatively complex and can be done using three different approaches. In the first approach, the concentration of one substrate is kept constant and included into the apparent values of V_{app} and $K_{m,app}$. Two separate plots of v as a function of a (b = constant) and b (a = constant) permit evaluation of all coefficients in Equation 14.4. In the second approach, a 3D regression analysis can be performed to obtain a v-surface dependent on both a and b. The third approach requires a fixed ratio between the two substrates, $b = \gamma \cdot a$, to transform the two-variable Equation 14.4 to a function with only one variable, a.

$$v = \frac{V}{1 + \frac{1}{a}\left(K_{mA} + \frac{K_{mB}}{\gamma}\right) + \frac{1}{a^2}\frac{K_{sA}K_{mB}}{\gamma}} \tag{14.8}$$

Analysis of v data for two or more γ allows calculation of the constants in Equation 14.8. Parameters of all functions discussed here can be calculated by regression fitting on a computer, which is a routine procedure in modern science.

Lipases are a family of enzymes which catalyze cleavage of ester bonds between organic alcohols and fatty acids (Lowe, 2002; Reis *et al.*, 2009). They have distinctive structural characteristics responsible for different kinetic behavior. For example, human pancreatic lipase (hPL) (Reis *et al.*, 2009), Candida antarctica lipase A (CALA) (Ericsson *et al.*, 2007) and lipase from Thermomyces lanuginosum (TLL) (Brzozowski *et al.*, 2000) contain an active site covered by a peptide loop known as the lid. All of these enzymes require interfacial activation to open the active site (Reis *et al.*, 2009). Upon formation of a micelle, the covering lid interacts with lipid–aqueous interface and exposes the active site (Winkler *et al.*, 1990; Lowe, 2002). An "open" lipase is additionally stabilized by protein colipase, which helps to remove the lid from the active site (Thomas *et al.*, 2005). Other lipases, such as PLRP2 from guinea pig, Candida antarctica lipase B (CALB) (Uppenberg *et al.*, 1994) and lipase A from bacteria Bacillus subtilis (van Pouderoyen *et al.*, 2001), have no lid and do not require interfacial activation.

All lipases, regardless of their structure and necessity of interfacial activation, follow the same mechanism of catalysis (Haeffner and Norin, 1999; Reis *et al.*, 2009). The triad of amino acids, Ser153, Asp177 and His264 (numbering according to hPL, mature protein), initiate conversion of substrates. The charge-transfer residues Asp and His assist in partial ionization of Ser by abstracting its proton. The ionized residue $E-Ser-O^{\delta-}$ attacks the ester bond of a substrate, for example triacylglycerol (TAG), and cleaves it into two fragments. The produced organic alcohol, for example diacylglycerol (DAG), is liberated into the medium, whereas fatty acid (FA) becomes covalently bound to Ser. At the next stage, the second substrate R–OH removes fatty acid from the Ser-residue by a substitutive mechanism. In aqueous solution, R–OH corresponds to water which liberates free fatty acid. In non-aqueous solution, R–OH can be an alcohol, carboxylic acid or acylglycerol. All of them attack the Ser–FA bond to produce a novel ester.

Reaction kinetics of lipases is relatively complex and generally depends on the reaction conditions, which can be aqueous, interfacial or non-aqueous. In an aqueous medium, lidless enzymes preferentially hydrolyze soluble substrates such as p-nitrophenyl acetate, triacetin and tributyrin. This type of lipolytic reaction follows the mono-substrate Michaelis–Menten equation (Reis *et al.*, 2009). Concentration of the second substrate (water, 55.5 M) is assumed as a constant and becomes a part of parameters V and K_{mS}.

Lipases with a lid demonstrate a more complex behavior in aqueous medium. They react preferentially with micellar substrates (S^*). Figure 14.1 shows the mechanism for simultaneous hydrolysis of S and S^*. The velocity of the reaction is expressed as the sum of Michaelis and Hill equations where a steep increase of activity is observed after formation of S^* ($s \approx K*$).

$$v = \frac{V_s}{1+\frac{K_S}{s}\left(1+\frac{S^n}{K_*^n}\right)} + \frac{V_*}{1+\frac{K_*^n}{S^n}\left(1+\frac{S}{K_s}\right)} \qquad (14.9)$$

V_s and $V*$ are the maximal rates of hydrolysis of S and S^*; K_s and $K* = (K_n \cdot K_s^*)^{1/n}$ are the half-effect constants of the corresponding catalytic reactions (K_n and K_s^* are the dissociation

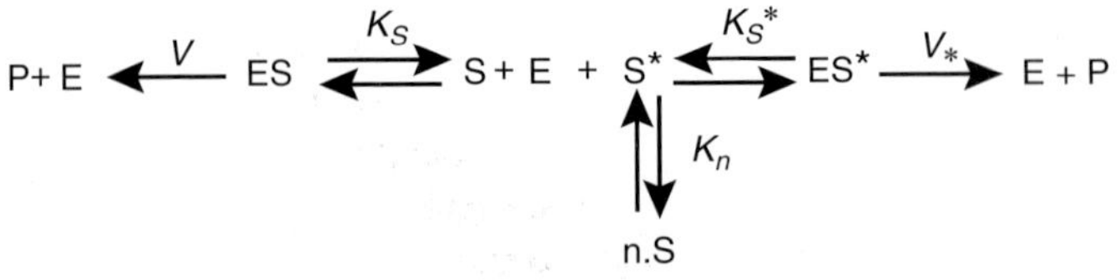

Figure 14.1 Mechanism for simultaneous hydrolysis of substrate and micellar substrate.

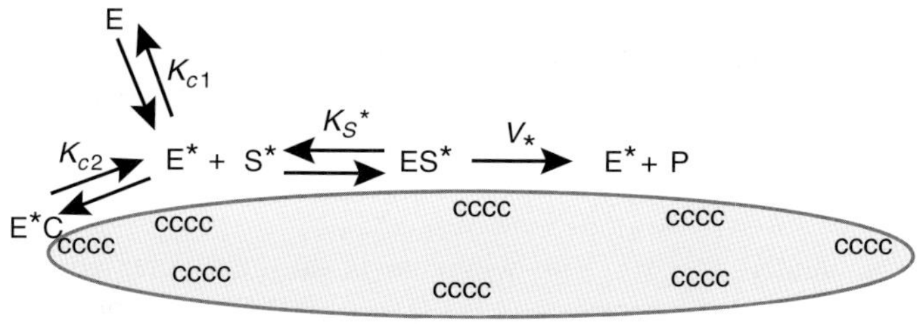

Figure 14.2 Enzymatic reactions involving water-insoluble substrates.

constants of micelle and micelle-enzyme complexes); s is the concentration of monomeric substrate in the water phase (it can be calculated from the total concentration s_0 using the expression $s_0 = s + n{\cdot}s^n/K_n$); n is the average number of S molecules in micelle. The concentration of free monomers (s) is frequently substituted by the total concentration $s_0 = s + n\,s^* \approx s$, though this assumption is valid only at a small s_0. At higher substrate concentrations, increasing amounts of S^* are produced due to the association $n{\cdot}S \leftrightarrow S_n$, and the expression $s_0 \approx n\,s^*$ becomes more appropriate.

Reactions with water-immiscible substrates are even more complex (Figure 14.2). Thus, an emulsifier is required to form the interface (active capturing surface C), where the surface-bound forms of the enzyme (E^*) and the substrate (S^*) react. Some emulsifiers, such as bile salts, might occlude lipase and form the inactive complex E^*C (Lowe, 2002, Pan and Bahnson, 2007). Other surfactants, for example Triton X-100, do not demonstrate such inhibiting effects (Salis *et al.*, 2003). The initial velocity of the reaction at interface can be written as:

$$v = \frac{V_*}{1 + \frac{K_{S^*}}{s^*}\left(1 + \frac{K_{c1}}{c} + \frac{c}{K_{c2}}\right)} = \frac{V_*}{1 + \frac{K_{mS^*}}{s^*}} \qquad (14.10)$$

where s^* is the concentration of substrate at the interface (usually treated as its bulk concentration); V^* is the maximal rate of forward reaction, c is the capturing surface per volume unit (it is often equated to the bulk concentration of emulsifier), K_{c1} stands for the equilibrium dissociation constant of E^* from the surface $CCCC$ (Figure 14.2), K_{c2} describes dissociation of the encapsulated inactive complex E*C. If c is kept constant, it can be incorporated into the apparent constant $K_{mS}{}^* = K_s{}^*\,(1 + K_{c1}/c + c/K_{c2})$, whereupon v becomes a function of one variable, s^*. Dependence on another variable c (capturing surface = emulsifier) at a constant s^* should be analysed using another form of the same equation:

$$v = \frac{V}{1 + \frac{K_{c1}^*}{c} + \frac{c}{K_{c2}^*}};\ V = \frac{V_*}{1 + \frac{K_{s^*}}{s^*}};\ K_{c1}^* = \frac{k_{c1}}{1 + \frac{s^*}{Ks^*}};\ K_{c2}^* = K_{c2}\left(1 + \frac{s^*}{K_{s^*}}\right) \qquad (14.11)$$

This function reaches its maximum at $c = (K_{c1}{}^*{\cdot}K_{c2}{}^*)^{1/2}$ and approaches zero at $c \to \infty$ (in the case of C-caused inhibition). If the emulsifier causes no inhibition ($K_{c2} \to \infty$), C acts as an essential activator. Hence, the dependence of v on c becomes a simple hyperbole (Martin *et al.*, 1994). Presence of residual enzymatic activity without emulsifier means that the capturing

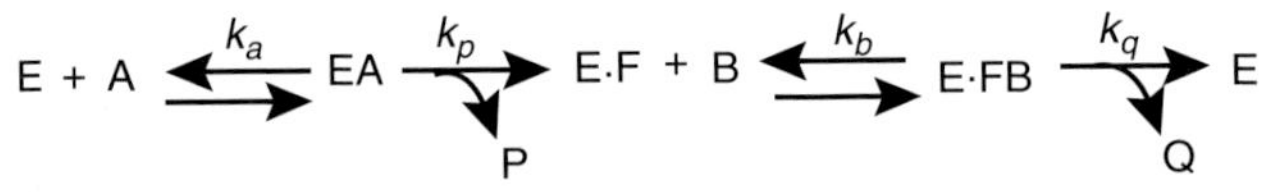

Figure 14.3 Enzymatic reactions in anhydrous medium.

surface should be treated as $c = c_0 + m$, where m is the emulsifier concentration and c_0 represents the endogenous active surface at $m = 0$ (no emulsifier present).

Lipase-catalyzed reactions do not necessarily require water as the second reactant. An immobilized enzyme is often stable in anhydrous media (Sharma *et al.*, 2001) and can use different alcohols as the secondary acceptors of fatty acids. The concentration of alcohol is relatively low and cannot be treated as a constant. Consequently, the reaction obeys the bisubstrate ping-pong mechanism (Cheirsilp *et al.*, 2008; Mitchell *et al.*, 2008; Xiong *et al.*, 2008) as shown in Figure 14.3. The corresponding equation of the reaction rate in the absence of products is:

$$v = \frac{V}{1 + \frac{K_{mA}}{a} + \frac{K_{mB}}{b}}; \quad V = \frac{k_p . k_q}{k_p + k_q}; \quad K_{mA} = \frac{K_a k_q}{k_p + k_q}; \quad K_{mB} = \frac{K_b k_p}{k_p + k_q} \qquad (14.12)$$

where a corresponds to the concentration of acyl substrate, for example monoacylglycerol (MAG), and b is the concentration of alcohol, for example methanol. It should be emphasized that substitution of one substrate (b=methanol) by its analogue (b=butanol) changes not only K_{mB} but also K_{mA} of the other substrate due to a different value of k_q. At the same time, the ratios V/K_{mA} and V/K_{mB} remain the same after substitution of either B or A.

The current and most used method to analyze kinetics of lipase-catalyzed transesterification involves regression analysis of the time-dependent conversions of TAG, DAG, MAG and fatty acid to the corresponding alcohol esters (Cheirsilp *et al.*, 2008; Mitchell *et al.*, 2008). Yet, the reliability of these calculations is not very high because a reasonably good fit of 30–50 experimental points is not necessarily stipulated by one unique combination of 15–20 parameters. Other combinations, not explored by the authors, can prove equal within the same significance level.

14.4 ENZYMES IN INDUSTRIAL APPLICATIONS

Applications of enzyme technology in various industries have seen a rapid growth in the last decade. Table 14.1 shows some of the enzymes that are used in various industries.

The food industry is a traditional market for enzymes with the frequent discovery of new enzymes applications. Pazczola (2001) provided a comprehensive review on the latest enzyme applications in the food industry. Lipase-catalyzed modifications of oils and fats, including lipase-catalyzed production of margarine, coco butter equivalents and structured lipids, represent some of those technologies that have been industrialized (Xu, 2004a).

In the pharmaceutical industry, enzymes have significantly reduced the number of steps required for syntheses of desired compounds, by-product and waste formation in drugs manufacturing, which subsequently decreased the overall production cost. For example, the company DSM has developed a four-step enzymatic semi-synthesis of cephalexin from cephalosporin C, which is a reduction from the ten-step conventional manufacturing process.

Table 14.1 Enzymes and their industrial applications.

Enzymes	Substrates	Catalyzed Reactions	Industries
Lipases	Lipids, esters	Hydrolysis, esterification and transesterification	Food, effluent treatment, detergents, fine chemicals and pharmaceuticals
Pectinase	Pectins	Clarification of fruit juices	Food and beverage
Proteases	Proteins	Proteolysis	Detergents, food, pharmaceutical, chemicals synthesis
Carbohydrases	carbohydrates	Hydrolysis	Food, feed additives, pulp and paper, sugar, textiles, detergents
Cellulases	Cellulose	Hydrolysis	Pulp, textile, feed, detergents
Amylases	Polysaccharides	Hydrolysis	Food and feed additives

This process generates 15 kg of waste per kg product, in comparison to the conventional approach which generates 50 kg of waste per kg product (Gavrilescu and Christi, 2005).

Progress in non-aqueous biocatalysis opens up new opportunities for enzyme technology to penetrate the well established chemical industry. For example, acrylamide is conventionally produced chemically from acrylonitrile by sulfuric acid or copper-catalyzed hydrolysis. This process is high in cost, low in selectivity and product quality, and results in environmental pollution. Mitsubishi Rayon Co. Ltd (Japan) developed an enzymatic acrylamide process using immobilized nitrile hydrase for the production of acrylamide. About 100 000 tons of acrylamide has been produced annually by this process with greater single-pass conversion, higher final product concentration, near quantitative conversion and 400% energy saving (Gavrilescu and Christi, 2005).

Biofuels are produced from renewable resources, such as biomass, which are constantly replenished as a sustainable alternative energy source to conventional fossil fuel. Enzymes such as cellulase can be used to hydrolyze cellulosic biomass, which can be converted to bioethanol. Meanwhile, lipase can be used in a similar manner for the production of fatty acids' alkyl esters of the lipids/oils used as the feedstock. Currently, intensive investigations are being carried out to develop efficient protocols for enzymatic production of biofuels.

The aforementioned are brief examples of successful industrial applications of biocatalysts. In the following section, four detailed case studies of enzymatic processing of lipid-based bioproducts, namely partial acylglycerols, bioactive compounds, phospholipids and fatty acid alkyl esters are presented.

14.4.1 Enzymatic processing of partial acylglycerols

Partial acylglycerols (PG), namely MAG and DAG, are esters of glycerol (GLY) with either one or two of the hydroxyl groups esterified with fatty acids, respectively (Yasukawa and Katsuragi, 2004). With hydroxyl groups in the structure, partial acylglycerols have marked surface activity, which enables them to reduce the surface tension at the oil–aqueous interface. Thus, they are prominently used as emulsifiers in the food industry. Besides that, partial acylglycerols have also been used as basic building blocks for various organic compounds in the pharmaceutical and nutraceutical industries (Cheong and Lai, 2009). In the late 1990s, DAG has attracted attention for its' potential as a healthful form of oil, capable of reducing postprandial lipid level, increasing β-oxidation of fat, as well as managing and preventing obesity (Takase, 2007). Due to their promising applications in various industries, there has been a surge in research on enzymatic processing for the synthesis of partial acylglycerols.

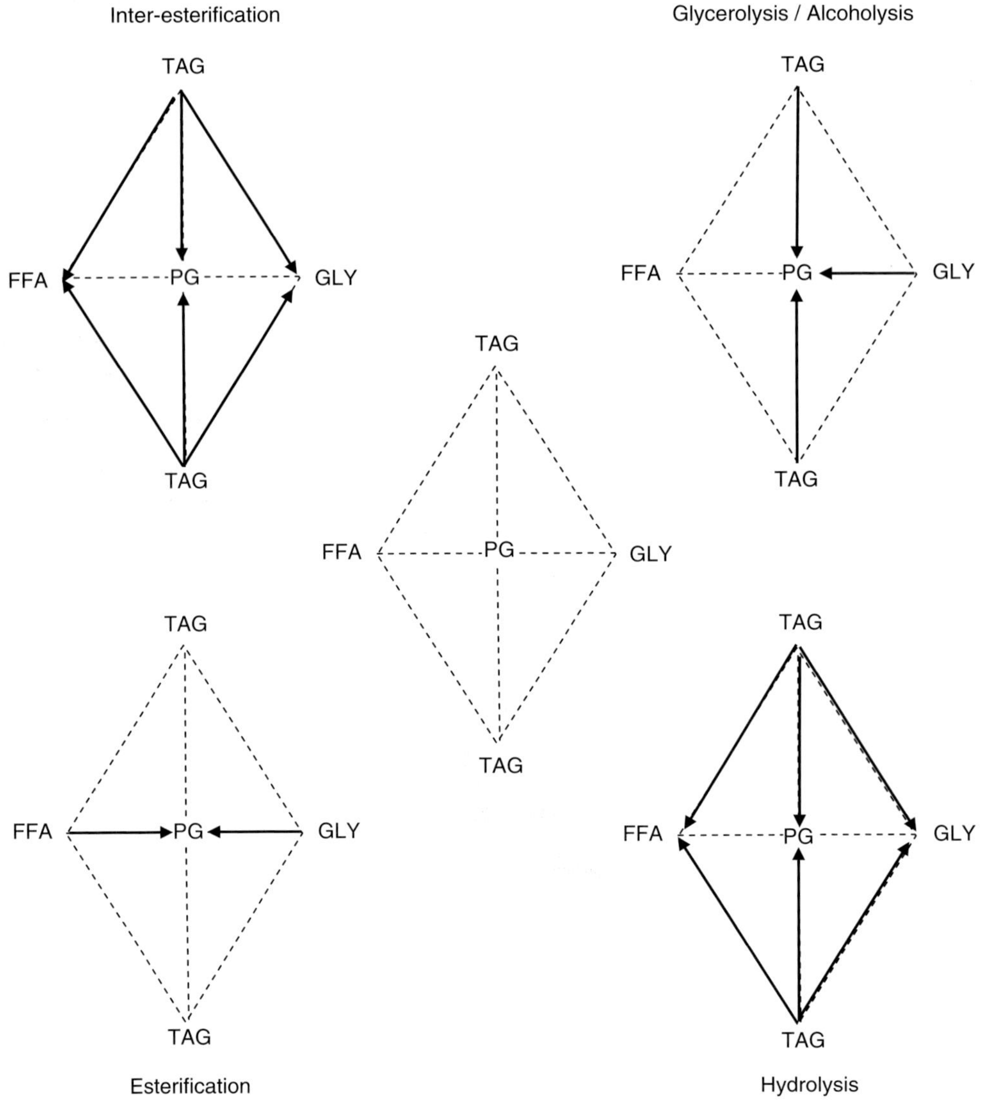

Figure 14.4 Enzymatic processing routes for production of partial acylglcyerols.

There are a number of enzymatic processing routes for production of partial acylglcyerols, namely esterification, acidolysis, glycerolysis, alcoholysis, partial hydrolysis and interesterification (Figure 14.4). Among these, extensive research has been carried out on glycerolysis, esterification and partial hydrolysis, both in academia and industry, during the last five years (Table 14.2). These processes include lipolysis (hydrolysis of ester bonds), liposynthesis (synthesis of ester bonds) or a combination of both. As partial acylglycerols are intermediates in the lipolysis or liposynthesis process, careful control of the process parameters is of the upmost importance to direct the reaction equilibrium towards production of partial acylglycerols.

As a whole, enzymatic synthesis of partial acylglycerols is governed by several common process parameters, namely enzyme type and dosage, ratio of the reactants, reaction media,

Table 14.2 Literature compilation on enzymatic processing for production of partial acylglycerols.

Enzymatic processing description	Partial glycerides yield (wt-%)	Reference
Enzymatic glycerolysis of polyunsaturated oil in *tert*-butanol medium using batch and packed-bed bioreactor (PBR) directed towards high yield of MAG	70% MAG (2 h in batch system/30 min of residence time in PBR)	Yang *et al.* (2005)
Enzymatic alcoholysis of fish oils in the presence of acetone using stirred tank (STR) and PBR directed towards high yield of MAG.	~65% MAG (STR discontinued mode)	Esteban *et al.* (2005)
Enzymatic glycerolysis of sunflower oil in the presence of solvent for MAG production	68–82% MAG (*tert*-butanol/*tert*-pentanol systems)	Damstrup *et al.* (2005)
Solvent-free enzymatic glycerolysis of rapeseed and sunflower oil for production of DAG	~65% DAG (Similar yield in 20 kg pilot batch reactor)	Kristensen *et al.* (2005)
Enzymatic glycerolysis of fats and oils in ionic liquid medium for production of MAG	~90% MAG (cocosalkyl pentaethoxi methyl ammonium methosulfate medium)	Guo and Xu (2005)
Enhancing kinetics of esterification for production of DAG by addition of purified MAG into subsequent production cycle	~38–58% DAG	Minoru *et al.* (2006)
Lipase-catalyzed esterification of glycerol and oleic acid in organic solvent system for production of partial acylglycerols	~90% partial acylglycerols (hexane system after 72 h)	Tripathi *et al.* (2006)
Enzymatic partial hydrolysis of fats and oils for production of DAG	~40% DAG	Lai *et al.* (2007)
Enzymatic glycerolysis of palm olein in organic solvent medium using PBR for production of MAG	~65% MAG (Acetone/isooctane system. Similar yield in 20 kg pilot batch reactor)	Kittikun *et al.* (2007)
Lipase-catalyzed esterification of palmitic and oleic acid with glycerol for production of dioleopalmitin	48% DAG (Similar yield in 10 litre pilot packed-bed reactor)	Lo *et al.* (2007)
Solvent-free enzymatic partial hydrolysis of palm olein for production of DAG	33% DAG (Similar yield in 10 litre pilot packed-bed reactor reactor)	Cheong *et al.* (2007)
Solvent free enzymatic glycerolysis of olive oil in AOT surfactant for MAG production	55% MAG	Fiametti *et al.* (2008)
Enzymatic glycerolysis of triolein in ionic liquid medium for production of DAG	73% DAG (trioctylmethyl ammonium trifluoroacetate/tetraalkyl ammonium sulfate system)	Guo *et al.* (2009)
Enzymatic glycerolysis of olive oil in compressed n-butane and AOT for production of partial acylglycerols.	~80% partial glycerides (70% MAG and 10% DAG)	Valerio *et al.* (2009)
Lipase-catalyzed partial hydrolysis of soybean oil under ultrasound irradiation for production of DAG	~40% DAG (1 h and 30 min under ultrasound irradiation)	Babicz *et al.* (2010)

temperature and type of reactor used. Lipase, depending on the amount of water present, catalyzes either the hydrolysis or synthesis of ester bonds in acylglycerols (Angkawidjaja and Kanaya, 2006). Non-regiospecific lipases have been shown to catalyze often the lipolysis and liposynthesis to completion, forming either triacylglycerol (TAG) or free fatty acids (FFA). Meanwhile, 1,3 regiospecific lipases, such as lipases from *Rhizomucor meihei* (RML), *TLL* and *Rhizopus sp.*, are often used to catalyze the synthesis of partial acylglycerol (Plou *et al.*, 1996). Optimum lipase dosage is also equally important in producing partial acylglycerols at reasonable yield. Although enzyme dosage has a positive correlation with yield, it is not usually kept at too high a concentration due to mass transfer limitations between enzymes and substrates. Furthermore, high enzyme concentration is not economically feasible for industrial application.

The ratio of the reactants has also significant effect on the yield of the desirable product. For example, in enzymatic glycerolysis to produce DAG, the ideal molar stoichiometric ratio of oil to glycerol is 2:1. On the other hand, a molar stoichiometric ratio of oil to glycerol of 1:2 will favor MAG production:

$$2\ \text{TAG} + 1\ \text{glycerol} \rightarrow 3\ \text{DAG} \tag{14.13}$$

$$1\ \text{TAG} + 2\ \text{glycerol} \rightarrow 3\ \text{MAG} \tag{14.14}$$

Various approaches have been used to modify the ratio of reactants during the course of reactions to shift the reaction equilibrium towards higher partial acylglycerols yield. For example, Rosu and co-workers (1997) performed a stepwise temperature reduction during enzymatic glycerolysis, which resulted in a high yield of more than 90% MAG. During the course of the temperature reduction, the MAG formed, which had a higher melting temperature than that of the bulk reaction mixture, gradually precipitated and eventually solidified. This shifted the reaction equilibrium towards production of MAG. Nevertheless, this approach was found to be ineffective in increasing production of DAG (Kristensen *et al.*, 2005). Kristensen and co-workers (2005) deduced that the carrier of the immobilized lipase might be susceptible to blockage by the paste-like DAG, which hindered contact among the reactants and lipase. The presence of a significant amount of DAG crystals in the reaction mixture might have also prompted DAG to act as a seed for co-crystallization with TAG, which led to the reaction rate stagnating (Cheong and Lai, 2009). Another approach currently adopted in industrial continuous production of DAG is addition of purified MAG into the second and subsequent production cycles to drive the esterification forward resulting in high DAG yield (Minoru *et al.*, 2006).

The reaction media is an important part of enzymatic processing. Conventionally, solvents such as *n*-hexane, *n*-heptane, acetone, *tert*-butanol and diethyl ether are sometimes used to improve the miscibility of reaction mixtures and subsequently increasing overall reaction rate. Tertiary alcohols, namely *tert*-butanol, *tert*-pentanol or mixtures of them with hexane were found to be efficient reaction media for reasonably high yield of MAG (Damstrup *et al.*, 2005). Despite that, solvent engineering has been deemed undesirable, as it poses problems in waste management and, thus, will be environmentally unfriendly. Ionic liquids, which are considered as eco-friendly solvents due to their negligible vapor pressure, have emerged as new reaction media. Benefits of ionic liquids include adjustable solubility, enhanced stability of lipases, positive effects on specificity of enzymes and facilitation of recoverability and recyclability of lipases. Kahveci and co-workers (2009) found that binary ionic liquid systems serve as good reaction media with at least 70% DAG yield.

Reaction temperature is well known for its effect on enzyme activity. Most lipases are active in the temperature range 40–75 °C. Within this temperature range, higher temperature usually increases the reaction rate. Nevertheless, lipases are deactivated faster at higher reaction temperature due to protein denaturation. Cheong and co-workers (2007) found that 1,3-specific lipase from RML was quite stable at 65 °C and could be reused in at least ten continuous batches of enzymatic partial hydrolysis (120 h) to produce DAG without significant loss of catalytic activity.

Reactor design has a significant effect on the enzymatic production efficiency of partial acylglycerols. For example, enzymatic glycerolysis involves the reaction between oil and glycerol, which are immiscible. A stirred tank batch reactor that provides efficient mechanical stirring is preferable in such a reaction to enhance mass transfer and reaction rate. Nevertheless, enzyme support material may be broken down by the strong mechanical stirring, resulting in lower recoverability and reusability of the enzyme (Xu *et al.*, 2007b). Packed-bed reactors, which usually comprise of pumping the reaction substrates through a column containing a bed of enzyme, are beneficial in terms of recoverability and reusability of enzyme. However, it may not be suitable for use in enzymatic glycerolysis, as glycerol may coat the enzyme bed, resulting in decrease of mass transfer and reaction rate. Nonetheless, it can be used in other enzymatic reactions where strong mechanical stirring is not required, such as enzymatic partial hydrolysis (Cheong *et al.*, 2007).

Enzymatic processing for synthesis of partial acylglycerols has been a regular feature in lipid technology. In fact, numerous enzymatic processes have been patented for industrial scale DAG production (Lai *et al.*, 2004). Although it may seem that the subject has been heavily and thoroughly investigated, a few reaction-related problems still persist. One such problem is low product conversion due to immiscible reaction substrates. A few studies have been conducted to improve the miscibility of reaction substrates, such as employing novel reaction media, namely compressed n-butane and sodium dioctyl sulfosuccinate (AOT) surfactant (Valerio *et al.*, 2009), and also introducing sonochemical irradiation to the enzymatic reaction (Babicz *et al.*, 2010). Besides that, purification of the partial acylglycerol product is also worth studying, as the final products from enzymatic processing usually contain a mixture of glycerol, FFA, MAG, DAG and TAG. Xu (2004b) has conducted extensive studies on purification of partial acylglycerol products using short path distillation technology. In fact, this technology has been used industrially for production of DAG oil. Nevertheless, other problems persist in regards to coloration and possible development of carcinogenic glycidol fatty acid esters during the heat treatment. The coloration of oil can be easily solved through an adsorption/bleaching process (Minoru *et al.*, 2010). Thus, it will be beneficial to investigate other purification methods or elimination of the glycidol fatty acid esters.

14.4.2 Enzymatic processing of bioactive compounds

Described as non-nutritional microconstituents of plants with significant biological activity (Patil *et al.*, 2009; Saura-Calixto and Goñi, 2009), bioactive compounds have been widely linked to the maintenance and improvement of human health, such as prevention of cardiovascular disease and various cancers. Current research suggests that the ability to protect against oxidative damage is yet another key attribute of bioactive compounds (Balsano and Alisi, 2009; Madhuri and Pandey, 2009). Polyphenols, ascorbic acid, carotenoids and limonoids are just some examples of bioactive compounds present in foods such as fruits and vegetables, wine, tea, oilseeds and cocoa (Rice-Evans *et al.*, 1996). Due largely to their natural origin and healthy properties, the overall demand for these compounds continues to expand.

Still more widespread use of bioactive compounds is limited due to issues such as lower solubility and miscibility in more hydrophobic environments. As such, strategies meant to alter the physicochemical properties of hydrophilic bioactives have been developed. One relatively successful approach has been the enzymatic acylation of bioactive compounds with more hydrophobic groups, whereby the addition of such groups impacts on partitioning and even on the emulsification properties of the resulting products (Sasaki *et al.*, 2010; Villeneuve, 2007).

Enzymatic acylation to yield bioactive compounds with additional properties and/or altered functionalities generally takes place via esterification or transesterification reactions. In simple esterification, the bioactive compound reacts with a fatty acid (or alcohol, based on its structure) to yield an ester and a molecule of water. In contrast, transesterification reactions involve acyl exchange between an ester and alcohol to yield structurally different ester and alcohol species. Both reactions are typically catalyzed by lipase and can, therefore, benefit from milder reaction conditions, substrate and/or regiospecificity (Chebil *et al.*, 2006).

Based on current research, some important factors to be considered during enzymatic processing of bioactives include the type of reaction media used, the level and distribution of water in the system, biocatalyst type and loading, reaction temperature, agitation and the type and ratio of substrates selected for use in a particular system. Overall, proper adjustment of the aforementioned parameters can help to push the reaction equilibrium towards the production of modified products and may also result in a more efficient and cost effective reaction set-up (Devi *et al.*, 2008).

With regards to enzymatic processing of bioactive compounds, bringing strongly hydrophilic and hydrophobic substrates together to create an efficient reaction system still poses serious challenges (Villeneuve, 2007; Xanthakis *et al.*, 2010). Most often, some compromise is necessary, such as utilizing intermediate polarity solvents or else co-solvent systems to achieve good contact between the vastly differing substrates and the biocatalyst. Some success has also been achieved employing ionic liquids as novel media, since many ionic liquids can dissolve vast quantities of hydrophilic and hydrophobic substrates (Katsoura *et al.*, 2006; Hu *et al.*, 2009). Additional advantages of ionic liquids includes their ability to activate and protect enzymes in certain environments; however, it is also clear that destabilization of enzymes such as lipase by particular ionic liquids can also have a strong impact on reaction systems (Lau *et al.*, 2004; van Rantwijk and Sheldon, 2007; Zhao *et al.*, 2009). Judicious selection of ionic liquids is therefore of the utmost importance.

Microemulsion systems, also, offer the opportunity for improved miscibility of substrates. For instance, esterification of ferulic acid with pentanol at 40 °C in a water-in-oil microemulsion resulted in a yield of ferulic ester of up to 60% within the short time frame of eight hours using a non-immobilized feruloyl esterase from *Aspergillus niger* (Giuliani *et al.*, 2001). While it is advantageous in maximizing bioconversion yields, it should be noted that microemulsions can pose additional challenges during downstream separation processes following enzymatic conversions.

The presence and distribution of water in the system also has a strong influence on the reaction. Lipases require a small amount of water to maintain their active conformation. Too little or too much water generally results in suboptimal activity. As a by-product of the esterification reaction, increasing quantities of water will shift the reaction equilibrium toward the substrates. Thus, reaction systems containing just sufficient amount of water to maintain the active lipase conformation without contributing significantly to the competing hydrolytic reaction are most efficient. Excess free water is often removed from molecules the reaction system through the use of activated molecular sieves (Duan *et al.*, 2006).

Other parameters of interest in the enzymatic processing of bioactive compounds include reaction temperature, agitation level and the selection of substrate types and ratio. Firstly, temperature can influence not only the activation and denaturation of the enzyme, but also substrate and product solubility and the viscosity of the reaction media (Chebil *et al.*, 2006; Villeneuve, 2007). Particularly when viscous substrates or ionic liquids are involved, a well mixed system is very important to ensure proper contact between the biocatalyst and substrates. From the preceding brief discussion, it is clear that the efficiency of enzymatic processing of bioactive compounds is affected by a number of parameters. Excellent in-depth discussions of some key parameters can be found in various reviews (Lau *et al.*, 2004; Chebil *et al.*, 2006; Villeneuve, 2007).

Table 14.3 provides a glimpse of some of the research which has been carried out to date. One major area of research relates to the enzymatic processing of phenolic acids, whereby acylation of phenolics, such as dihydrocaffeic, *p*-hydroxyphenylpropionic, chlorogenic and ferulic acids, with a range of fatty alcohols have yielded some promising results (Giuliani *et al.*, 2001; Stamatis *et al.*, 2001; Figueroa-Espinoza and Villeneuve, 2005; Duan *et al.*, 2006). More recently, further studies on ferulic acid have been carried out using a chemi-enzymatic approach, whereby chemically produced vinyl ferulate was transesterified with octanol or other compounds, including sterols, for production of feruloyl derivatives with enhanced antioxidative properties (Chigorimbo-Murefu *et al.*, 2009). The biosynthesis with octanol was carried out at 45 °C in *tert*-butyl-methyl ether using *C. antarctica* lipase, while *C. rugosa* lipase was more appropriate for sterol ferulate biosynthesis due to steric hindrance resulting from both the phenylpropenoic derivatives and tetracyclic steroid skeleton.

Literature also details acylation reactions of other polyphenolic compounds including flavonoids. For instance, Stevenson and co-workers (2006) reported biosynthesis reactions of naringin and isoquercetin with a range of acyl donors, including palmitic, cinnamic and phenylpropionic acids. In this case, both purified phenolic glycosides and fruit extracts were investigated in *t*-butanol at 60 °C for up to seven days, resulting in yields from approximately 25% to 95%. High purity lauroyl and palmitoyl rutin esters were also prepared recently by Lue and co-workers (2009) following biosynthesis reactions in acetone at 50 °C for up to 96 h. Generally, use of increasingly bulky or longer chain length acyl substrates has often resulted in the need for longer reaction times (Devi *et al.*, 2008).

Enzymatic glucosylation (the addition of a sugar moiety) and amidation (the addition of an amine moiety) are two additional approaches that can be employed to alter the properties and functionalities of bioactive compounds (Villeneuve, 2007; Khare *et al.*, 2009). Finally, many reports also describe approaches involving the enzymatic hydrolysis of various proteins to yield bioactive peptides with novel antioxidative, antimicrobial and health-promoting properties (Wei and Chiang, 2009; Zhang *et al.*, 2009).

Generally, enzymatic processing of bioactive compounds can improve the bioactive and physicochemical properties of a very diverse group of compounds (López-Giraldo *et al.*, 2009; Lue *et al.*, 2009). However, to do so successfully, it is important to carefully consider the reaction environment, enzyme stability and the sensitivity of the modified and unmodified bioactive substrates during optimization of these enzymatic processes. Although an array of patents exists in the area (Wang *et al.*, 2008, Singh and Jagaveerabhadra, 2010, Tashiro *et al.*, 2010), to date there is little evidence that enzymatic processing of bioactives is being carried out on a large scale. In reality, cost as well as technical challenges relating to product separation, enzyme deactivation and the sensitivity of bioactive substrates and products have limited commercialization of these processes. Nonetheless, demand for

Table 14.3 Literature compilation on enzymatic processing of bioactive compounds.

Bioactive Compound	Acyl Donor	Parameters of Interest	References	Maximum Conversion (%)
Arbutin	lauric acid	reaction media	Watanabe *et al.* (2009)	50
Phloridzin	lauric acid	reaction media	Watanabe *et al.* (2009)	37
Coumaric acid	octanol	lipase type	Katsoura *et al.* (2006)	33
Ferulic acid	fatty alcohols (C1–C10)	lipase, substrate ratio, media, product solubility, reuse of lipase	Katsoura *et al.* (2006)	60
Caffeic acid	octanol	lipase type	Katsoura *et al.* (2006)	12
Sinapic acid	octanol	lipase type	Katsoura *et al.* (2006)	31
Naringin	lauric acid	reaction media	Watanabe *et al.* (2009)	35
Esculin	palmitic acid	reaction media	Hu *et al.* (2009)	78

diverse bioactive products continues to grow and new approaches are continuously being developed towards meeting this demand.

14.4.3 Enzymatic processing of phospholipids

Phospholipid (PL) is an amphiphilic molecule with one or more phosphate groups in its structure. It can be divided into four main classes, namely glycerophospholipids, sphingolipids, ether phospholipids and phonophospholipids, depending on their backbones and bonding types (Guo *et al.*, 2005). Glycerophospholipids such as phosphatidylcholine, phosphatidylethanolamine, phosphatidylserine, phosphatidylinositol, phosphatidylglycerol and cardiolipin contain a glycerol backbone and are most abundant in nature (Hawthorne and Ansell, 1982).

As one of the main structural constituents of biological membranes, PLs play a crucial role in the biochemistry and physiology of cells. PLs have also found applications in numerous industries. For example, PLs are often used as emulsifier, stabilizer and wetting and dispersing agents in the food industry. Pharmaceutical and cosmetic industries utilize PLs in liposomal encapsulation of active ingredients for enhanced delivery and bioavailability (Gabizon *et al.*, 1997).

Research on structural modification of PLs to achieve the aforementioned nutritional and functional properties has increased prominently. The structural modification of PLs can be done enzymatically using phospholipases and lipases. Phospholipase A_1 (PLA_1) cleaves the fatty acid at the *sn*-1 position, while Phospholipase A_2 (PLA_2) cleaves the fatty acid at the *sn*-2 position. Phospholipase C (PLC) hydrolyses the phosphodiester bond between the glycerol backbone and phosphate. On the other hand, phospholipase D (PLD) hydrolyses the phosphodiester bond between the phosphate and the head group. Lipases can also be used to modify the fatty acid composition at sn-1 and sn-2 positions of phospholipids. Guo and co-workers (2005) and Nieuwenhuyzen and Thomas (2008) have provided a detailed review on enzymatic modification of phospholipids for functional applications and human nutrition.

Hydrolysis, alcoholysis, esterification, transesterification and transphosphatidylation are some of the reactions used for enzymatic modification of PLs. Enzymatic hydrolysis of PLs is usually performed to obtain lysophospholipids (LPL). LPL possess superior emulsification properties and are often used in the food, pharmaceutical, agrochemicals and personal care industries (Fujita and Suzuki, 1990). Enzymatic hydrolysis of PL to form LPL can be conducted using phospholipase A_1 (Kim *et al.*, 1997), phospholipase A_2 (Yesair, 1998) or lipase (Kim and Kim, 2000; Virto and Adlercreutz, 2000). In fact, enzymatic hydrolysis of phospholipids has been carried out industrially (Sas and Peys, 2000; Schmitt *et al.*, 2005; Schmitt and Heirman, 2007). Hydrolytic reaction requires the presence of water in the reaction mixture. As PLs have limited solubility in water, the reaction is inefficient. Haas and co-workers (1993) conducted enzymatic hydrolysis of phosphatidylcholine (PC) in the presence of organic solvents. Hydrophobic solvents were found to increase solubility of PLs significantly, which subsequently enhanced the reaction efficiency and yield of LPL. MAG and fatty acids can be added to the reaction mixtures to form a microemulsion system, which produces a higher yield of LPL and minimizes formation of by-product.

Enzymatic alcoholysis of PLs can be used to produce LPLs and fatty acid esters simultaneously. The main factors affecting enzymatic alcoholysis of PLs are the type of lipase and chain length of the alcohol. For example, immobilized RML exhibited significant preference for short chain alcohols like ethanol, whereas lipase TLL preferred long chain alcohol (Sarney *et al.*, 1994). Sarney and co-workers (1994) reported that addition of a small amount of water to the reaction mixture increased the LPLs yield significantly.

Enzymatic transesterification of PLs is used to synthesize structured phospholipids. For example, microbial lipases from *Aspergillus*, *Candida*, RML and TLL have been used for transesterification of PLs with fatty acids of different chain lengths (Peng *et al.*, 2002; Hara *et al.*, 2002; Doig and Diks, 2003). The type of acyl donor affects enzymatic transesterification of PLs significantly. Generally, FFA is a more efficient acyl donor in comparison to esters. Although the reaction kinetics is slightly slower when using FFA as acyl donor, the yield of structured PLs is higher because of the inhibition of hydrolytic reaction. Another important factor is the use of solvent, especially hydrophobic solvent, to improve miscibility of substrates, which subsequently enhances mass transfer and overall reaction rate.

Enzymatic transphosphatidylation of PLs can be conducted using PLC and PLD. PLC specifically hydrolyzes the C–O–P bond in phospholipids yielding sn-1,2 DAG and the phosphate residue bearing the corresponding head group (Durban and Bornscheuer, 2007). DAG produced from enzymatic transphosphatidylation possesses the enantiomeric structure of the natural product, which can be used directly for synthesis of stereospecific compounds. This is not possible when lipases are used for DAG production (Anthonsen *et al.*, 1999). For example, PLC from *Bacillus cereus* has been used to produce biologically active sphingosines, ceramides, sphingosine 1-phosphate (Morigaki *et al.*, 1998), 1-lauroyl-rac-glycerophosphate (Virto and Adlercreutz, 2000) and cyclic five-membered phosphodiesters (Bruzik *et al.*, 1996). PLD is an expensive biocatalyst and is used primarily for production of high value pharmaceutical products. Parameters considered important in enzymatic transphosphatidylation include the nature of acceptor alcohol and reaction media. Primary alcohols can be easily incorporated to PLs with PLD, especially PLD from *streptomyces* species (Hosokawa *et al.*, 2000; Yaqoob *et al.*, 2001; Iwasaki *et al.*, 2003). Small water-soluble alcohols, such as ethanol and glycerol, are also easier to incorporate into PLs. Selection of a suitable reaction medium capable of dissolving substrates that differ in terms of polarity is crucial for the efficiency of enzymatic transphosphatidylation. A biphasic solvent system consisting of ethyl acetate and diethyl ether is commonly used for transphosphatidylation reactions. In some cases, benzene or a mixture of hexane and octanol has been used as well.

The main constraint in enzymatic processing of PLs is the high cost of enzymes. For instance, PLD, which is relatively high cost and available only in free form, is primarily used for development of pharmaceutical products with high added values. Development of low cost and robust enzymes through protein engineering and directed evaluation may offer a solution to the problem (Song *et al.*, 2005). Besides that, separation and purification of pure PLs from a complex natural PL mixture is relatively tedious. To date, a simplified isolation procedure is not yet available. Thus, research for a simplified purification method of PLs is very much in need.

14.4.4 Enzymatic processing of fatty acid alkyl esters

Fatty acid alkyl esters (FAAE) can be prepared from transesterification of vegetable oils or animal fats with aliphatic alcohols (Knothe *et al.*, 2005). The transesterification reaction, also known as alcoholysis, is the exchange of alkoxy group of an ester compound (TAG) with an aliphatic alcohol (the acyl acceptor) in the presence of a catalyst. The overall reaction is a sequence of three consecutive and reversible reactions in which DAG and MAG are formed as intermediate products (Ma and Hanna, 1999). The general reaction scheme is given in Figure 14.5.

At present, industrial production of FAAE is performed using basic chemical catalysis with, for example, sodium or potassium hydroxide. This process is energy consuming and not environmentally friendly (Fjerbæk *et al.*, 2009). In the early 1990s, lipases were firstly used in place of chemical catalysts for transesterification reaction (Mittelbach, 1990). Since then, studies on enzymatic transesterification for FAAE production have increased by leaps and bounds. Some of the benefits of enzymatic FAAE production worth mentioning include easier handling operations, milder reaction conditions and the possibility of using alternative oil sources, such as high FFA containing oils, without problems of soap by-product formation (Kulkarni and Dalai, 2006).

There are many factors affecting enzymatic transesterification (Fjerbæk *et al.*, 2009; Robles-Medina *et al.*, 2009). Among the key factors are type, stability and reusability of lipase, type of the acceptor alcohol, substrate ratios, quality of feedstocks, reaction temperature, water activity and/or water content.

Lipases used for FAAE production are normally of microbial origin, such as CALB and TLL lipase. They are often used in immobilized forms, which are more stable and versatile than their free forms (Shimada *et al.*, 2002; Kojima *et al.*, 2004; Nielsen *et al.*, 2008). The immobilized form is also more industrially feasible, as they can be easily packed and reused in industrial reactors. Nevertheless, they are also more costly in terms of enzyme price (per kg of immobilized enzyme). However, the stable and reasonably high productivity of the enzyme (kg of biodiesel/kg of immobilized enzyme) during a relatively long lifetime is more important than the sole price comparison. Whole-cell biocatalysts which are cheaper and more robust may be appropriate for industrial FAAE production (Antczak *et al.*, 2009). The activity of whole-cell biocatalysts depends on the fatty acid composition of the cell wall membrane. As there may be different lipases bound to the cell wall or membrane, the FAAE yield may vary (Adamczak *et al.*, 2009).

A combination of different enzymes has been shown to increase yield in the enzymatic production of FAAE. For example, Türkan and Kalay (2006) used three different immobilized lipases from RML, *TLL* and *Candida*. They found lipases from RML and *TLL* catalyzed the first step (TAG to DAG) of transesterification faster while lipase from *Candida antarctica* catalyzed the second (DAG to MAG) and third (MAG to FAAE and glycerol) steps faster.

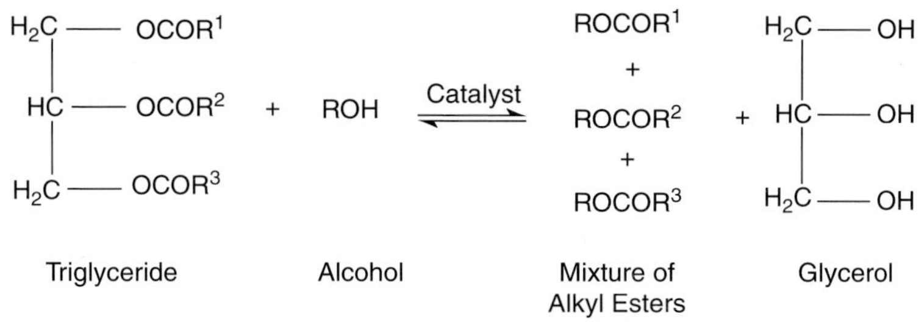

Triacylglycerides + *Alcohol* ⟷ *Diacylglycerides* + *FAAE*

Diacylglycerides + *Alcohol* ⟷ *Monoacylglycerides* + *FAAE*

Monoacylglycerides + *Alcohol* ⟷ *Glycerol* + *FAAE*

Figure 14.5 Preparation of fatty acid alkyl esters by transesterification.

Thus, a combination of different enzymes is found to be more efficient in catalyzing FAAE production in comparison to the single lipase system.

Table 14.4 shows a list of CALB lipase-catalyzed transesterification reactions using methanol as the acyl acceptor. Although methanol is the cheapest source of acyl acceptor, it has a harmful effect on biocatalysts. In countries where ethanol production from renewable feedstocks is economically feasible, ethanol is the more preferable acyl acceptor (Kojima *et al.*, 2004).

Stoichiometrically, a 3:1 molar ratio of alcohol to oil is needed for complete transesterification reaction. Nevertheless, the solubility of short chain alcohols, namely methanol and ethanol, in vegetable oil is only 1/2 and 2/3 of the stoichiometric amount, respectively. The insoluble alcohols form either emulsion droplets under continuous stirring or a heterogeneous alcohol phase in equilibrium conditions, with the oil phase leading to inactivation of enzymes and, subsequently, lower conversion rate (Shimada *et al.*, 1999). To minimize the effect of lipase inactivation, Shimada and co-workers (1999) performed a stepwise addition of methanol. This procedure was found to be effective with close to complete conversion of FAAE (98.4%). The use of organic solvents to improve substrate miscibility is deemed undesirable considering the cost of recovery and also increasing environmental concerns. The cost concern also applies to the use of ionic liquids despite them being green solvents (Ha *et al.*, 2007). On the other hand, recycling or addition of FAAE as the co-solvent seems a promising possibility for keeping the reaction media homogeneous and, thus, increasing mass transfer and reaction rates (Fjerbæk *et al.*, 2009).

The quality of feedstock plays an important role in enzymatic processing. Alternative oil sources, such as waste frying oils, inedible vegetable oils and waste oils from processing industries, which are cheaper than edible vegetable oils, are potential feedstocks for FAAE production. Biocatalysts are preferable for such type of feedstocks. Lipases can simultaneously esterify and transesterify FFA and oils. Thus, it can convert low quality oils, such as high FFA oils and restaurant greases, and inedible oils, such as Jatropha and castor oil, into FAAE at reasonably good yields. Basic catalysts have the tendency of forming soap with such feedstocks. Although acid catalysts, for example hydrochloric or sulfuric acid, do not have such drawbacks, the reaction rates are much slower (Robles-Medina *et al.*, 2009).

Enzymatic FAAE production does have its constraints. The major constraint to enzymatic production of FAAE is substrate immiscibility. Both the substrates, oil and alcohol, have

Table 14.4 Literature compilation on enzymatic transesterification of various oils with methanol for FAAE production.

Oil/Fat Source	Acyl Acceptor	Conversion (c) or Yield (y)(mol- or wt-%)	Conditions and Observations	Reference
Soybean (crude and refined)	Methanol	94 (c)	Stepwise addition; pretreatment of biocatalyst with crude oil for 120 h.	Du *et al.* (2004)
Rice bran	Methanol	98 (c)	T: 50 °C; t: 6 h; 5 wt-% enzyme based on oil weight; batch.	Lai *et al.* (2005)
Tung	Methanol	67.5 (c)	T: 43 °C; t : 18 h; R: 2.2/1.	Gui-Zhuan *et al.* (2006)
Soybean	Methanol	80 (c)	T: 50 °C; t: 12 h; R: 4/1; *C.antarctica* lipase	Ha *et al.* (2007)
Acid oil (rapeseed oil refining by-product)	Methanol	>98 (y)	T: 30 °C; t: 24 h; 77.9,% free fatty acids; 2-step reaction system; glycerol addt. in the 2nd step.	Watanabe *et al.* (2007)
Palm	Methanol	97 (y)	T: 60 °C; t: 35 h; R: 3.2/1; 4 wt-% enzyme based on oil weight; stepwise addition (3-step).	Talukder *et al.* (2008)
Jatropha	Methanol	76 (y)	T: 30 °C; t: 90 h; R: 1/3; stepwise addition (3-step); 4 wt-% enzyme based on oil weight.	Tamalampudi *et al.* (2008)
Vegetable oils (sunflower, borage, olive and soybean)	Methanol	~100 (y)	Refer to the original article.	Hernandez-Martin and Otero (2008)

T: optimum or reaction temperature; t: reaction time; R: molar ratio of alcohol to oil.

limited miscibility, which lead to lower mass transfer, reaction rate and product yield. In addition, the presence of insoluble alcohol also leads to lipase inactivation. As mentioned earlier, the problem of substrate immiscibility can be overcome by using organic solvent, emulsifier, or FAAE as co-solvent.

Another important bottleneck in enzymatic processing of FAAE is adsorption of glycerol onto the biocatalysts beads. Glycerol adsorption inactivates enzymes, particularly in continuous and repeated batch processes which employed same batch of enzyme in the repeated production cycles (Antczak *et al.*, 2009; Fjerbæk *et al.*, 2009). The problem of glycerol adsorption can be solved by using membrane reactors, filters or separators (Belafi-Bako *et al.*, 2002; Fjerbæk *et al.*, 2009), suitable solvents as washing or flushing systems for glycerol removal and addition of porous hydrophilic particles, such as micro-porous sieves and silica gel, to the enzyme bed for glycerol adsorption (Dossat *et al.*, 1999).

In conclusion, robust and economical biocatalysts with improved activities, the use of inedible oils and waste oil as potential feedstocks and optimized reaction conditions for different substrates and biocatalyst pairs will determine the possibility of industrially sustainable enzymatic FAAE production.

14.5 CONCLUSIONS AND FUTURE TRENDS

Production cost and efficiency are the key concerns in industrial applications of biocatalysts. The need for production cost reduction is far more critical than ever. Therefore, concept development is no longer the key issue for enzymatic processing. Rather, optimization of the enzymes for efficient conversions and processes is the goal for broader use of enzyme technology in industrial applications. Although enzymes still comprise a substantial part of the operating costs, their share is relatively small considering the large total production cost of some of the biocatalysis products. Therefore, improvements in process and system design and operation are the key to cost reduction.

Figure 14.6 shows a typical flowchart of engineering enzymes as catalysts for efficient bioprocessing. Major challenges in industrial applications of biocatalysis are the identification and selection of suitable enzymes for the formation of desirable reaction products; this can be done using high-throughput screening technologies, which involve collecting genes from the environment and expressing them in recombinant organisms (Arnold, 2001; Van Beilen and Li 2002; Arnold, 2001). These methods allow faster access to useful catalysts from organisms that cannot be cultured. Once the enzymes of interest are identified or selected, they have to be separated and purified, as most enzymes exist with many isomers or other proteins.

In nature, enzymes play an important role in the survival and reproduction of their source organism. Therefore, many enzymes in their natural form are not suitable for application directly as biocatalysts in bioprocessing. For example, most enzymes are active at relatively mild conditions, thus may not be viable under the harsher conditions encountered in most industrial production systems. Various methods have been used to improve the properties of enzymes, including selectivity, activity and thermostability, in order to enable them to function as efficient industrial biocatalysts. Enzyme engineering, which encompasses rational design and directed evolution, is an efficient method to improve enzyme properties. Rational design seeks for beneficial mutations or protein sequences by applying empirically derived rules or theoretical models. Meanwhile, directed evolution uses a combinatorial approach to create libraries of enzymes from which enhanced variants can be identified

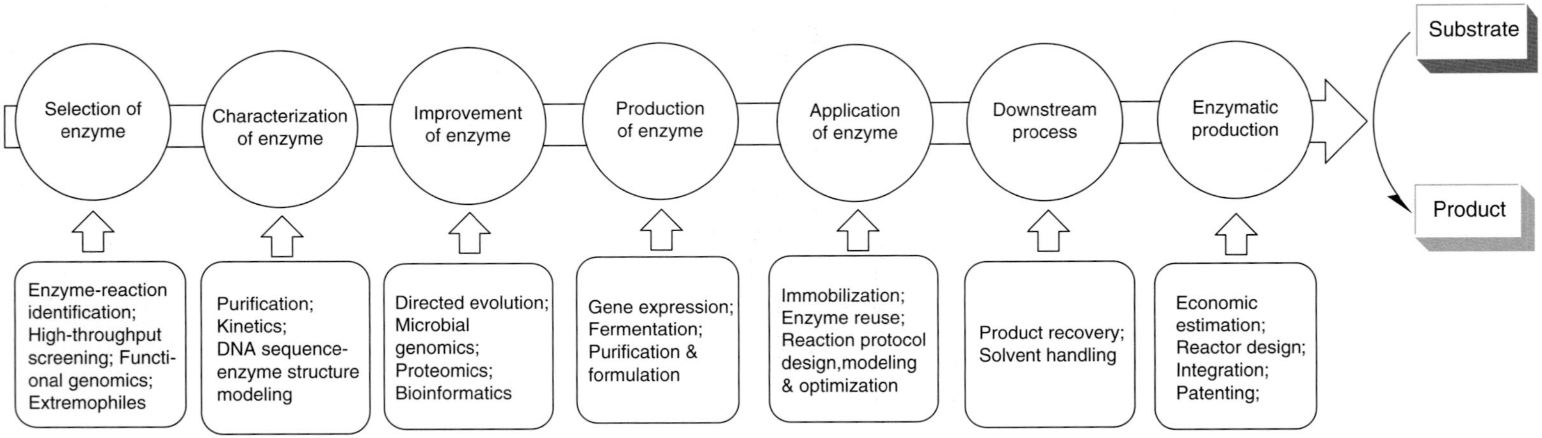

Figure 14.6 Typical flowchart of engineering enzyme as catalyst for efficient bioprocessing.

experimentally (Arnold, 2001; Bornscheur and Pohl, 2001; Dalby, 2007). In both approaches, gene encoding of the enzymes of interest, a suitable expression system and a sensitive detection system are prerequisites. More and more successful demonstrations indicate that laboratory evolution methods are efficient to improve properties of enzymes within a reasonable timescale (Arnold, 2001).

Another challenge for industrial application of enzymes is cost efficiency. New fermentation technologies, such as solid-substrate fermentation, genetically modified microorganism, and improved protein refolding techniques, are being developed to improve the cost efficiency of enzymes (Van Beilen and Li, 2002). Enzymes are usually applied in immobilized form to facilitate reusability and recovery. A successful immobilization method depends largely on the property of enzyme and typical operation conditions to find or design a suitable carrier (Sharma *et al.*, 2001). In addition, development of the reaction system, including formulation of suitable reaction media and optimization of process parameters, plays a significant role in increasing reaction rate and shifting the reaction equilibrium towards desired products and also helps in ensuring cost efficiency of enzymes.

REFERENCES

Adamczak M, Bornscheuer UT, Wlodzimierz B. 2009. The application of biotechnological methods for the synthesis of biodiesel. Eur J Lipid Sci Technol 111:808–813.

Angkawidjaja C, Kanaya S. 2006. Family I.3 lipase: bacteria lipases secreted by the type I secretion system. Cell Mol Life Sci 63:2804–2817.

Antczak MS, Kubiak A, Antczak T, Bielicki S. 2009. Enzymatic biodiesel synthesis – key factors affecting efficiency of the process. Renew Energ 34:1185–1194.

Anthonsen T, D'Arrigo P, Pedrocchi-Fantoni G, Secundo F, Servi S, Sundby E. 1999. Phospholipids hydrolysis in organic solvents catalyzed by immobilized phospholipase C. J Mol Catal B: Enzym 6:125–132.

Arnold FH. 2001. Combinatorial and computational challenges for biocatalyst design. Nature 409:253–257.

Babicz I, Leite SGF, de Souza ROMA, Antunes OAC. 2010. Lipase-catalyzed diacylglycerol production under sonochemical irradiation. Ultrason Sonochem 17:4–6.

Balsano C, Alisi A. 2009. Antioxidant effects of natural bioactive compounds. Curr Pharm Des 15:3063–3073.

Belafi-Bako K, Kovacs F, Gubicza L, Hancsok J. 2002. Enzymatic biodiesel production from sunflower oil by *Candida antarctica* lipase in a solvent-free medium. J Biocat Biotransf 20:437–439. 16

Bornscheur UT, Pohl M. 2001. Improved biocatalysts by directed evolution and rational protein design. Curr Opin Biotechnol 5:137–143.

Bruzik KS, Guan ZW, Riddle S, Tsai MD. 1996. Synthesis of inositol phosphodiesters by phospholipase C-catalyzed transesterification. J Am Chem Soc 118:7679–7688.

Brzozowski AM, Savage H, Verma CS, *et al.* 2000. Structural origins of the interfacial activation in Thermomyces (Humicola) lanuginosa lipase. Biochemistry 39:15071–15082.

Chebil L, Humeau C, Falcimaigne A, Engasser JM, Ghoul M. 2006. Enzymatic acylation of flavonoids. Process Biochem 41:2237–2251.

Cheirsilp B, H-Kittikun A, Limkatanyu S. 2008. Impact of transesterification mechanisms on the kinetic modeling of biodiesel production by immobilized lipase. Biochem Eng J 42:261–269.

Cheong LZ, Lai OM. 2009. Diacylglycerol oil: Healthful or hype? International news on fats, oils and related materials: Inform 20:391–393.

Cheong LZ, Ariffin N, Tan CP, *et al.* 2007. Production of a diacylglycerol enriched palm olein using lipase-catalyzed partial hydrolysis: optimization using response surface methodology. Food Chem 105: 1614–1622.

Chigorimbo-Murefu NTL, Riva S, Burton SG. 2009. Lipase-catalysed synthesis of esters of ferulic acid with natural compounds and evaluation of their antioxidant properties. J Mol Catal B: Enzym 56:277–282.

Cornish-Bowden A. 2004. Fundamentals of enzyme kinetics. 3 rd edn. London: Portland Press.

Dalby PA. 2007. Engineering enzymes for biocatalysis. Recent Pat Biotechnol 1:1–9.

Damstrup ML, Jensen T, Sparsø FV, Kiil SZ, Jensen AD, Xu X. 2005. Solvent optimization for efficient enzymatic monoacylglycerol production based on a glycerolysis reaction. J Am Oil Chem Soc 82: 559–564.

Dastoli FR, Price S. 1967. Further studies on xanthine oxidase in non polar media. Arch Biochem Biophys 122:289–91.

Dastoli FR, Musto NA, Price S. 1966. Reactivity of active sites of chymotripsin suspended in an organic media. Arch Biochem Biophys 115:44–47.

Devi BLAP, Zhang H, Damstrup ML, *et al.* 2008. Enzymatic synthesis of designer lipids. OCL 15:189–195.

Divakar S, Manchar B. 2007. Use of lipases in the industrial production of esters. In: Polaina J, MacCabe A (editors). Industrial enzymes: structure, function and application. Dordrecht, Germany: Springer. pp. 283–300.

Doig SD, Diks RMM. 2003. Toolbox for modification of the lecithin group. Eur J Lipid Sci Technol 105:368–376.

Dossat V, Combes D, Marty A. 1999. Continuous enzymatic transesterification of high oleic sunflower oil in a packed bed reactor: influence of the glycerol production. Enzyme Microb Technol 25:194–200.

Du W, Xu YY, Zeng J, Liu DH. 2004. Novozym 435-catalyzed transesterification of crude soya bean oils for biodiesel production in a solvent free medium. Biotechnol Appl Biochem 40:187–190.

Duan Y, Du Z, Yao Y, Li R, Wu D. 2006. Effect of molecular sieves on lipase-catalyzed esterification of rutin with stearic acid. J Agric Food Chem 54:6219–6225.

Durban MA, Bornscheuer UT. 2007. An improved assay for the determination of phospholipase C activity. Eur J Lipid Sci Technol 109:469–473.

Ericsson DJ, Kasrayan A, Johansson P, *et al.* 2007. X-ray structure of Candida antarctica lipase A shows a novel lid structure and a likely mode of interfacial activation. J Mol Biol 376:109–119.

Esteban L, Munio MM, Robles A, *et al.* 2005. Synthesis of 2-monoacylglycerols (2-MAG) by enzymatic alcoholysis of fish oils using different reactor types. Biochem Eng J 44:271–279.

Fernandez-lafunete R. 2010. Lipase from Thermomyces lanuginosus: uses and prospects as an industrial biocatalyst. J Mol Cat B: Enzym 62:197–212.

Fiametti KG, Rovani S, de Oliveira D, Corazzo ML, Treichel H, Oliveira JV. 2008. Assessment of variable effects on solvent-free monoacylglycerol enzymatic production in AOT surfactant. Eur J Lipid Sci Technol 110:510–515.

Figueroa-Espinoza MC, Villeneuve P. 2005. Phenolic acids enzymatic lipophilization. J Agric Food Chem 53:2779–2787.

Fjerbæk L, Christensen KV, Norddahl B. 2009. A review of the current state of biodiesel production using enzymatic transesterification. Biotechnol Bioeng 102:1298–1315.

Fujita S, Suzuki K. 1990. Surface Activity of the Lipid Products Hydrolyzed with Lipase and Phospholipase A-2. J Am Oil Chem Soc 67:1008–1014.

Fusee MC. 1987. Industrial Production of L-aspartic Acid Using Polyurethane-immobilized Cell Containing Aspartase. Meth Enzymol 136:463–471.

Gabizon A, Goren D, Horowitz AT, Tzemach D, Lasos A, Siegal T. 1997. Long–circulating liposomes for drug delivery in cancer therapy: A review of biodistribution in tumor-bearing animals. Adv Drug Deliv Rev 24:337–344.

Gavrilescu M, Chisti Y. 2005. Biotechnology – a sustainable alternative for chemical industry. Biotechnol Adv 23:471–499.

Giuliani S, Piana C, Setti L, *et al.* 2001. Synthesis of pentylferulate by a feruloyl esterase from *Aspergillus niger* using water-in-oil microemulsions. Biotechnol Lett 23:325–330.

Gui-Zhuan X, Bai-Liang Z, Sheng-Yong L, Jian-Zhi Y. 2006. Study on immobilized lipase-catalyzed transesterification of tung oil. Agric Sci China 5:859–864.

Guo Z, Xu X. 2005. New opportunity for enzymatic modification of fats and oils with industrial potentials. Org Biomol Chem 3:2615–2619.

Guo Z, Vikbjerg AF, Xu X. 2005. Enzymatic modification of phospholipids for functional applications and human nutrition. Biotech Adv 23:203–259.

Guo Z, Kahveci D, Ozcelik B, Xu X. 2009. Improving enzymatic production of diglycerides by engineering binary ionic liquid medium system. New Botechnol 26:37–43.

Ha SH, Lan MN, Lee SH, Hwang SM, Koo YM. 2007. Lipase Catalyzed biodiesel production from soybean oil in ionic liquids. Enzyme Microb Technol 41(4):480–483.

Haas MJ, Cichowicz DJ, Phillips J, Moreau R. 1993. The hydrolysis of phosphatidylcholine by immobilized lipase: optimization of hydrolysis in organic solvents. J Am Oil Chem Soc 70:111–117.

Haeffner F, Norin T. 1999. Molecular modeling of lipase-catalyzed reactions. Prediction of enantioselectivity. Chem Pharm Bull 47:591–600.

Hara S, Hasuo H, Nakasato M, Higaki Y, Totani Y. 2002. Modification of soybean phospholipids by enzymatic transacylation. J Oleo Sci 51:417–421.

Hawthorne JN, Ansell GB. 1982. Phospholipids, New Comprehensive Biochemistry. Amsterdam, The Netherlands: Elsevier Biomedical Press.

Hernandez-Martin E, Otero EC. 2008. Different enzyme requirements for the synthesis of biodiesel: Novozyme 435 and Lipozyme TLIM. Bioresourc Technol 99:277–286.

Hosokawa M, Shimatani T, Kanada T, Inoue Y, Takahashi K. 2000. Conversion to docosahexaenoic acid-containing phosphatidylserine from squid skin lecithin by phospholipase D-mediated transphosphatidylation. J Agri Food Chem 48(10):4550–4554.

Hou CT. 2005. Handbook of industrial biocatalysis. Boca Raton, FL: CRC Press.

Hu Y, Guo Z, Lue BM, Xu X. 2009. Enzymatic synthesis of esculin ester in ionic liquids buffered with organic solvents. J Agric Food Chem 57:3845–3852.

Iwasaki Y, Mizumoto Y, Okada T, Yamamoto T, Tsutsumi K, Yamane T. 2003. An aqueous suspension system for phospholipase-D mediated synthesis of PS without toxic organic solvent. J Am Oil Chem Soc 80:653–657.

Kahveci D, Guo Z, Ozcelik B, Xu X. 2009. Optimization of enzymatic synthesis of diacylglycerols in binary medium systems containing ionic liquids. Food Chem 119:880–885.

Katsoura MH, Polydera AC, Tsironis L, Tselepis AD, Stamatis H. 2006. Use of ionic liquids as media for the biocatalytic preparation of flavonoid derivatives with antioxidant potency. J Biotechnol 123:491–503.

Khare SK, Kumar A, Kuo TM. 2009. Lipase-catalyzed production of a bioactive fatty amide derivative of 7,10-dihydroxy-8(*E*)-octadecenoic acid. Bioresour Technol 100:1482–1485.

Kim JK, Kim MK, Chung GH, Choi CS, Rhee JS. 1997. Production of Lyso phospholipid Using Extracellular Phospholipase A_1 from Serratia sp. MKI. J Microbiol Biotechnol 7:258–261.

Kim J, Kim BG. 2000. Lipase-catalyzed synthesis of lysophosphatidylcholine using organic cosolvent for in situ water activity control. J Am Oil Chem Soc 77:791–797.

Kittikun A, Kaewthong W, Cheirslip B. 2007. Continuous production of monoacylglycerols from palm olein in packed-bed reactor with immobilized lipase PS. Biochem Eng J 40:116–20.

Knothe G, van Gerpen J, Krahl J. 2005. The Biodiesel Handbook. Illinois: AOCS Press.

Kojima S, Du D, Sato M, Park EY. 2004. Efficient production of fatty acid methyl ester from waste activated bleaching earth using diesel oil as an organic solvent. J Biosci Bioeng 98:420–424.

Kristensen JB, Xu X, Mu HL. 2005. Process optimization using responses surface design and pilot plant production of dietary diacylglycerols by lipase-catalyzed glycerolysis. J Agric Food Chem 53:7059–7066.

Kulkarni MG, Dalai AK. 2006. Waste cooking oils – an economical source for biodiesel: a review. Ind Eng Chem Res 45:2901–2913.

Lau RM, Sorgedrager MJ, Carrea G, van Rantwijk F, Secundo F, Sheldon RA. 2004. Dissolution of *Candida antarctica* lipase B in ionic liquids: effects on structure and activity. Green Chem 6:483–487.

Lai OM, Lo SK, Akoh CC. 2004. Patent review on lipid technology. In: Lai OM, Akoh CC (editors). Healthful lipid. Illinois: AOCS Press. pp. 433–507.

Lai OM, Yusoff MSA, Lo SK, *et al.* 2007. Process for the production of diacylglycerol. 5 July 2007. WO 2007/075079 A1.

Lo SK, Cheong LZ, Ariffin N, *et al.* 2007. Diacylglycerol and triacylglycerol as responses in a dual response surface-optimized process for diacylglycerol production by lipase-catalyzed esterification in packed-bed reactor. J Agric Food Chem 55:5595–5603.

López-Giraldo LJ, Laguerre M, Lecomte J, *et al.* 2009. Kinetic and stoichiometry of the reaction of chlorogenic acid and its alkyl esters against the DPPH radical. J Agric Food Chem 57:863–870.

Lowe ME. 2002. The triglyceride lipases of the pancreas. J Lipid Res 43:2007–2016.

Lue BM, Guo Z, Glasius M, Xu X. 2009. Scalable preparation of high purity fatty acid esters. J Am Oil Chem Soc 87:55–61.

Ma F, Hanna MA. 1999. Biodiesel production: a review. Bioresour Technol 70:1–15

Madhuri S, Pandey G. 2009. Some anticancer medicinal plants of foreign origin. Curr Sci 96:779–783.

Marangoni AG. 2003. Enzyme kinetics: A modern approach. New Jersey: John Wiley & Sons, Inc.

Martin JC, Bello JF, Burguillo FJ, Roig MG. 1994. Kinetic behaviour and reaction mechanism of the hydrolysis of p-nitrophenyl palmitate in mixed micelles with Triton X-100 catalyzed by lipase from Candida rugosa. J Mol Catal 93:37–52.

Minoru K, Keiji S, Toshiteru K. 2006. Process for production fat or oil. 2 November 2006. EP Patent 1717318B1.

Minoru K, Masao S, Junya M. 2010. Process for producing oil and fat rich in diacylglycerol. 14 April 2010. EP Patent 2175032A2.
Mitchell DA, Carrière F, Krieger N. 2008. An analytical method for determining relative specificities for sequential reactions catalyzed by the same enzyme: general formulation. Biochim Biophys Act 1784: 705–715.
Mittelbach M. 1990. Lipase catalyzed alcoholysis of sunflower oil. J Am Oil Chem Soc 67:168–170.
Morigaki E, Miura Y, Takahata K, Tada M, Nakajima S, Baba N. 1998. Enzymatic preparation of sphingosine 1-phosphate. J Chem Res 12:774–775.
Naik S, Basu A, Saikia R, *et al.* 2010. Lipases for use in industrial biocatalysis: specificity of selected structural groups of lipases. J Mol Cat B: Enzym 65:18–23.
Nielsen PM, Brask J, Fjerbæk L. 2008. Enzymatic biodiesel production: technical and economical considerations. Eur J Lipid Sci Technol 110:692–700.
Nieuwenhuyzen WV, Thomas MC. 2008. Update on vegetable lecithin and phospholipid technologies. Eur J Lipid Sci Technol 110:472–486.
Pan YH, Bahnson BJ. 2007. Structural basis for bile salt inhibition of pancreatic phospholipase A2. J Mol Biol 369:439–450.
Patil BS, Jayaprakasha GK, Murthy KNC, Vikram A. 2009. Bioactive compounds: historical perspectives, opportunities, and challenges. J Agric Food Chem. 57:8142–8160.
Pazczola DE. 2001. From soybean to spaghetti: the broadening use of enzymes. Food Technol 55:54–62.
Peng L, Xu X, Mu H, Høy CE, Adler-Nissen. 2002. Production of structured phospholipids by lipase catalyzed acidolysis: optimization using response surface methodology. J Enzym Microb Technol 31:523–532.
Plou FJ, Barandiaran M, Calvo MV, Ballesteros A, Pastor E. 1996. High yield production of mono- and di-oleylglycerol by lipase-catalyzed hydrolysis of triolein. Enzym Microb Tech 18:66–71.
Reis P, Holmberg K, Watzke H, Leser ME, Miller R. 2009. Lipases at interfaces: A review. Adv Colloid Interface Sci 147–8:237–250.
Rice-Evans CA, Miller NJ, Paganga G. 1996. Structure-antioxidant activity relationships of flavonoids and phenolic acids. Free Radical Biol Med 20:933–956.
Robles-Medina A, Gonzales-Moreno PA, Esteban-Cerdan L, Molina-Grima E. 2009. Biocatalysis: towards ever greener biodiesel production. Biotechnol Adv 27(4):398–408.
Rosu R, Uozaki Y, Iwasaki Y, Yamane T. 1997. Repeated use of immobilized lipase for monoacylglycerol production by solid-phase glycerolysis of olive oil. J Am Oil Chem Soc 74:445–450.
Rozzell JD. 1999. Commercial scale biocatalysis: Myths and reality. Bioorg Med Chem 7:2253–61.
Salis A, Svensson I, Monduzzi M, Solinas V, Adlercreutz P. 2003. The atypical lipase B from Candida antarctica is better adapted for organic media than the typical lipase from Thermomyces lanuginosa. Biochim Biophys Acta 1646:145–151.
Sarney DB, Fregapane G, Vulfson EN. 1994. Lipase catalyzed synthesis of lyso-PLs in a continuous bioreactor. J Am Oil Chem Soc 71:93–96.
Sas B, Peys E. 2000. Method for conversion of lecithin into lysolecithin. 30 May 2000. US Patent 6068997.
Sasaki K, Alamed J, Weiss J, *et al.* 2010. Relationship between the physical properties of chlorogenic acid esters and their ability to inhibit lipid oxidation in oil-in-water emulsions. Food Chem 118:830–835.
Saura-Calixto F, Gon I. 2009. Definition of the Mediterranean diet based on bioactive compounds. Crit Rev Food Sci Nutr 49:145–152.
Schmid A, Dordick JS, Hauer B, Kiener A, Wubbolts M, Witholt B. 2001. Industrial Biocatalysis today and tomorrow. Nature 109:258–266.
Schmitt H, Heirman M, Bruse F, Schneider M, Vander S. 2005. Enzymatic production of hydrolyzed lecithin products. 27 October 2005. US Patent 2005011784.
Schmitt H, Heirman M. 2007. Enzymatic modification of lecithin. 13 March 2007. US Patent 7189544.
Sharma R, Chisti Y, Banerjee UC. 2001. Production, purification, characterization, and applications of lipases. Biotechnol Adv 19:627–662.
Shimada Y, Watanabe Y, Samukawa T, *et al.* 1999. Conversion of vegetable oil to biodiesel using immobilized *Candida antarctica* lipase. J Am Oil Chem Soc 76:789–793.
Shimada Y, Watanabe Y, Sugihara A, Tominaga Y. 2002. Enzymatic alcoholysis for biodiesel fuel production and application of the reaction to oil processing. J Mol Catal B: Enzym 17:133–142.
Singh CU, Jagaveerabhadra RN. 2010. Esters of capsaicinoids as dietary supplements. 13 May 2010. US Patent 20100120912.
Song JK, Han JJ, Rhee JS. 2005. Phospholipases: Occurrence and production in microorganisms, assay for high throughput screening, and gene discovery from natural and manmade diversity. J Am Oil Chem Soc 82:691–705.

Stamatis H, Sereti V, Kolisis FN. 2001. Enzymatic synthesis of hydrophilic and hydrophobic derivatives of natural phenolic acids in organic media. J Mol Catal B: Enzym 11:323–328.

Stevenson DE, Wibisono R, Jensen DJ, Stanley RA, Cooney JM. 2006. Direct acylation of flavonoid glycosides with phenolic acids catalysed by *Candida antarctica* lipase B (Novozym 435®). Enzyme Microb Technol 39:1236–1241.

Takase H. 2007. Metabolism of diacylglycerol in humans. Asia Pac J Clin Nutr 16(Supp.1):398–403.

Talukder MMR, Min Beatrice KL, Puay Song O, *et al.* 2008. Improved method for efficient production of biodiesel from palm oil. Energy Fuels 22:141–144.

Tamalampudi S, Talukder MR, Hama S, Numata T, Kondo A, Fukada H. 2008. Enzymatic production of biodiesel from Jatropha oil: A comparative study of immobilized-whole cell and commercial lipases as a biocatalyst. Biochem Eng J 39:185–189.

Tashiro Y, Suzuki K, Mishima S. 2010. Manufacture of caffeic acid phenethyl ester using chlorogenate esterase. 22 July 2010. JP Patent 2010158223.

Thomas A, Allouche M, Basyn F, Brasseur R, Kerfelec B. 2005. Role of the lid hydrophobicity pattern in pancreatic lipase activity. J Biol Chem 280:40074–40083.

Tripathi V, Trivedi R, Singh RP. 2006. Lipase-catalyzed synthesis of diacylglycerol and monoacylglycerol from unsaturated fatty acid in organic solvent system. J Oleo Sci 55:65–69.

Türkan A. Kalay Ş. 2006. Monitoring lipase-catalyzed methanolysis of sunflower oil by reversed-phase high-performance liquid chromatography: elucidation of the mechanism of lipases. J Chromatogr A 1127: 34–44.

Uppenberg J, Hansen MT, Patkar S, Jones TA. 1994. The sequence, crystal structure determination and refinement of two crystal forms of lipase B from Candida antarctica. Structure 2:293–308.

Valerio A, Fiametti KG, Rovani S, *et al.* 2009. Enzymatic production of mono and diglycerides in compressed n-butane and AOT surfactant. J Supercritical Fluids 49:216–220.

Van Beilen JB, Li Z. 2002. Enzyme technology: an overview. Curr Opin Biotechnol 13:338–344.

van Pouderoyen G, Eggert T, Jaeger KE, Dijkstra BW. 2001. The crystal structure of Bacillus subtilis lipase: a minimal alpha/beta hydrolase fold enzyme. J Mol Biol 309:215–226.

van Rantwijk F, Sheldon RA. 2007. Biocatalysis in Ionic Liquids. Chem Rev 107:2757–2785.

Villeneuve P. 2007. Lipases in lipophilization reactions. Biotechnol Adv 25:515–536.

Virto C, Adlercreutz P. 2000. Lyso phosphatidylcholine synthesis with *Candida antarctica* lipase B (Novozyme 435). Enzyme Microb Technol 26:630–635.

Wang F, Gong X, Tan T, Deng L, Nie K. 2008. Method for synthesizing ester of natural tocopherol under catalysis of lipase. 13 Auguct 2008. CN Patent 101240303.

Watanabe Y, Nagao T, Nishida Y, Takagi Y, Shimada Y. 2007. Enzymatic production of fatty acid methyl esters by hydrolysis of acid oil followed by esterification. J Am Oil Chem Soc 84:1015–1021.

Watanabe Y, Nagai M, Yamanaka K, Jose K, Nomura M. 2009. Synthesis of lauroyl phenolic glycosides by immobilized lipase in organic solvent and its antioxidative activity. Biochem Eng J 43:261–265.

Wei JT, Chiang BH. 2009. Bioactive peptide production by hydrolysis of porcine blood proteins in a continuous enzymatic membrane reactor. J Sci Food Agric 89:372–378.

Winkler FK, D'Arcy A, Hunziker W. 1990. Structure of human pancreatic lipase. Nature 343:771–774.

Xanthakis E, Theodosiou E, Magkouta S, *et al.* 2010. Enzymatic transformation of flavonoids and terpenoids: Structural and functional diversity of the novel derivatives. Pure Appl Chem 82:1–16.

Xiong J, Wu J, Xu G, Yang L. 2008. Kinetic study of lipase catalyzed asymmetric transesterification of mandelonitrile in solvent-free system. Chem Eng J 138: 258–263.

Xu X. 2004. Biocatalysis for lipid modification. In: Dunford NT, Dunford HB (editors). Nutritionally enhanced edible oil processing. Champaign, IL: AOCS Press. pp. 162–196.

Xu X. 2004. Short path distillation for lipid processing. In: Lai OM, Akoh CC (editors). Healthful lipid. Champaign, IL: AOCS Press. pp. 127–144.

Xu X, Guo Z, Zhang H, Vikbjerg AF, Damstrup ML. 2006. Chemical and enzymatic interesterification in lipid modification. In: Gunstone F (editor). Modifying lipids for use in food. Cambridge, UK: Woodhead Press. pp. 234–272.

Xu X, Kittikun AH, Zhang H. 2007a. Enzymatic processing of omega-3 speciality oils. In: Breivik H (editor). Long-chain omega-3 specialty oils. Bridgewater, UK: The Oily Press. pp. 141–164.

Xu X, Kristensen JB, Zhang H. 2007b. Production of novel lipids with functional and health benefits. In: Rastall B (editor). Novel enzyme technology for food applications. Cambridge, UK: Woodhead Publishing. pp. 270–284.

Xu X, Vilbjerg AF, Guo Z, Zhang L, Acharya AK. 2008. Enzymatic modification of phospholipids. In: Gunstone F (editor). Phospholipid Science and Technology. Bridgewater, UK: Oily Press. pp. 41–82.

Yang T, Rebsdorf M, Engelrud U, Xu X. 2005. Monoacylglycerol synthesis via enzymatic glycerolysis using a simple and efficient reaction system. J Food Lipids 12:299–312.
Yaqoob M, Nabi A, Masoom-Yasinzai M. 2001. Bioconversion of phosphatidylcholine to phosphatidylserine using immobilized enzyme mini-columns. Process Biochem 36:1181–5.
Yasukawa T, Katsuragi Y. 2004. Diacylglycerols. In: Katsuragi Y, Yasukawa T, Matsuo N, Flickinger BD, Tokimitsu I, Matlock MG (editors). Diacylglycerol oil. Champaign, IL: AOCS Press. p. 3.
Yesair DW. 1998. Methods for making lyso phosphatidyl choline. 10 February 1998. US Patent 5,716,814.
Zaks A, Klibanov AM. 1985. Enzyme-catalyzed processes in organic solvents. Proc Natl Acad Sci USA 82: 3192–6.
Zhang SB, Wang Z, Xu SY, Gao XF. 2009. Purification and characterization of a radical scavenging peptide from rapeseed protein hydrolysates. J Am Oil Chem Soc 86:959–66.
Zhao H, Jones CL, Cowins JV. 2009. Lipase dissolution and stabilization in ether-functionalized ionic liquids. Green Chem 11:1128–38.

Index

Food and Industrial Bioproducts and Bioprocessing, First Edition. Edited by Nurhan Turgut Dunford.